THE UNIVERSITY OF
WINCHESTER

Environmental management is a wide, expanding, and rapidly evolving field, which concerns all humans, and plays a crucial role in the quest for sustainable development. Environmental management affects everybody from individual citizens, farmers, administrators and lawyers, to businesses, governments, international agencies and non-governmental organisations.

This updated second edition explores the nature and role of environmental management, covering key principles and practice, and offers a comprehensive and understandable introduction, which points readers to further in-depth coverage. This new edition reflects the rapid expansion and evolution of the field and focuses much more strongly on sustainable development. There has also been extensive rearrangement to make the book more accessible to those unfamiliar with environmental management and lacking a science background and there is greater coverage of topics such as key resources under stress, environmental management tools and urban environmental management. Among the key themes covered are:

- sustainable development
- proactive approaches
- the precautionary principle
- the 'polluter-pays' principle
- the need for humans to be less vulnerable and more adaptable.

With rapid expansion and evolution of the subject it is easy for those starting to study it to get disorientated, but *Environmental Management for Sustainable Development* offers a structured coverage and foundation for further, more-focused interest. The book is a much revised, restructured and updated second edition accessible to all readers. It is illustrated throughout with figures, plates and case studies.

Chris Barrow is Reader in the School of the Environment and Society at the University of Wales Swansea.

Routledge Introductions to Environment

Environmental Management for Sustainable Development

Second Edition

C.J. Barrow

Routledge
Taylor & Francis Group

LONDON AND NEW YORK

First published 1999
by Routledge
2 Park Square, Milton Park, Abingdon, Oxon OX14 4RN

Simultaneously published in the USA and Canada
by Routledge
711 Third Avenue, New York, NY 10017

Second edition 2006

Routledge is an imprint of the Taylor & Francis Group, an informa business

Typeset in Times and Franklin Gothic by
Florence Production Ltd, Stoodleigh, Devon

British Library Cataloguing in Publication Data
A catalogue record for this book is available from the
British Library

Library of Congress Cataloging in Publication Data
Barrow, Christopher J.
  Environmental management for sustainable development /
  C.J. Barrow. – 2nd ed.
      p. cm.
    Rev. ed. of: Environmental management and development.
  1st ed. 1999.
  Includes bibliographical references and index.
  1. Environmental management.   2. Environmental policy.
  3. Sustainable development.   I. Barrow, Christopher J.
  Environmental management and development.   II. Title.
  GE300.B375 2006
  363.7'05—dc22   2005037580

ISBN10: 0–415–36534–1 (hbk)
ISBN10: 0–415–36535–X (pbk)
ISBN10: 0–203–01667–X (ebk)

ISBN13: 978–0–415–36534–5 (hbk)
ISBN13: 978–0–415–36535–2 (pbk)
ISBN13: 978–0–203–01667–1 (ebk)

Contents

Illustrations

Figures

Tables

Boxes

Preface to the second edition

This book explores the nature, scope and role of environmental management, with a strong focus on sustainable development. It offers a foundation for a series of texts which deal with the application of environmental management, including:

Environmental Risk Management
Managing Environmental Pollution
Coastal and Estuarine Management
Countryside Management
Environmental Assessment in Practice

Environmental management is a broad and rapidly evolving discipline. This book explores the subject's core themes and principles, which include:

- a goal of sustainable development;
- a multidisciplinary, interdisciplinary or holistic approach;
- support for the 'polluter-pays principle';
- concern for limits, hazards and potential;
- an attempt to act beyond the local or project level;
- support for long-term not just short-term planning;
- adherence to the 'precautionary principle';
- translation of theory to effective practice;
- the integration of environmental science, planning and management, policy making and public involvement;
- an awareness of the need to change the ethics of peoples, businesses and governments.

The decision was made to prepare a second edition in 2004, five years after the publication of the original. This was prompted by considerable development of the field marked by: the appearance of many new taught courses; the expansion of media coverage; increasing government, agency and citizen interest. This new edition, as well as being updated, seeks to better address sustainable development, key resource issues, urban environments, environmental change and tourism. The evolution of environmental management tools and approaches and the expansion of Internet sources also necessitated some updating. Since the late 1990s there has been increased involvement of social scientists, lawyers, business, politicians and economists in environmental management, and its use has spread beyond developed countries.

CJB January 2006

THEORY, PRINCIPLES AND KEY CONCEPTS

1 Introduction

- Aims and coverage
- Key terms and concepts
- The definition and scope of environmental management
- The evolution of environmental management
- Problems and opportunities
- Summary
- Further reading

Aims and coverage

This book seeks to offer a comprehensive and understandable introduction, which points readers to further, more in-depth sources. *Environmental Management for Sustainable Development* is divided into three parts: Part I deals with theory, principles and key concepts; following this introductory chapter (1), the following five chapters examine: fundamentals (key concepts) and goals (2); the scientific underpinnings (3); social aspects (4); business and law issues (5); and participants (stakeholders) (6). Part II focuses on practice, and includes chapters on: environmental management approaches (7); methods and tools (8 and 9); key resources which have to be 'managed' (10); global challenges (11); pollution and waste management (12); environmental management in sensitive, vulnerable and difficult situations (13); tourism and environmental management (14); urban environmental management (15). Part III looks to the future, and seeks to assess the way ahead (16). A glossary is provided to aid those new to the field.

Environmental management is evolving rapidly; it is important for more and more sectors of human activity and plays a crucial role in establishing sustainable development. As government, business, agencies and citizens become more involved with environmental issues, and with the media giving them more coverage, things can become veiled and distorted by polarised perceptions and the acceptance of inaccurate received wisdom. Environmental managers have to acquire and sift available evidence, and distinguish between accurate and inaccurate data and avoid mistaking symptoms for actual causes. Once a clear understanding is acquired it is usually necessary to advise, lobby and educate stakeholders to win their support for seeking the 'best' environmental management option. There is often a dilemma for environmental management – to reconcile the conflict between a desire to adequately research, and the real-world demands for rapid, economical and clear-cut decisions. Delay may result in costly, even irreparable problems, but mistaken advocacy can prove disastrous. Environmental management demands co-ordination skills, ability to devise trade-offs, negotiation and diplomacy skills, and foresight. To catch problems soon enough to have a chance of satisfactory resolution demands a level of forward vision and monitoring beyond that of many disciplines. Clearly, unpredictable natural disasters and human fickleness mean that even the best prediction and most careful observations will sometimes give little or no warning of problems; environmental management must therefore address such issues as human vulnerability and seek adaptable and flexible strategies.

Environmental management generally demands a multidisciplinary approach, and achieving this in a satisfactory manner can be a challenge because suitable supportive systematic frameworks are still being developed (Hunt and Johnson, 1995). However, there has been progress, and environmental management is acting increasingly as an integrative force, capable of bringing together diverse stakeholders, specialists, levels of administration, different sectors, and even groups of nations, that might otherwise have little inclination to co-operate (O'Callaghan, 1996). It should be noted that a multi-disciplinary approach draws upon various disciplines for information, analytical skills and insight, but does not seek an integrated understanding. An interdisciplinary approach draws upon common themes and goes beyond close collaboration between different specialists to attempt integration, and is very difficult because it involves blending differ-ently derived concepts (O'Riordan, 1995: 2–4). Environmental management demands awareness that issues may be part of complex transnational, even global environmental, economic and social interaction, which is likely to be affected by politics, perception and ethics. In practice those involved in environmental management have some degree of specialisation, and focus on an issue, sector, country, region, environment or busi-ness. Sometimes environmental managers conduct their own research or they apply knowledge generated by others. Some environmental managers work for a firm, body or institution but generally profess a greater degree of responsibility to a wider range of stakeholders ranging up to the global environment. To some extent all people are environmental managers, making choices which affect the quality of their surroundings and sustainability of their lifestyles. However, most have insufficient training, infor-mation and powers to achieve much.

Key terms and concepts

Key concepts and goals of environmental management are explained in more depth in Chapter 2. It is difficult to separate environmental management from the process of development; put crudely, the environmental manager is expected to advise on wise resource use, potential environmental opportunities and threats (linkages between environmental management, the development process, and developing countries are explored by the author in *Environmental Management and Development* – Barrow, 1999). Development is seen increasingly to require reduction of inter-group disparity, or a 'social transformation' (alteration of society and culture), through the use of capital, technology and knowledge. It has often been argued that richer countries, international agencies and non-governmental organisations (NGOs) should 'assist' others to develop. However, some feel that people must do this for themselves, and there are countries which have tried 'decoupling' their development from the rest of the world (Adams, 1990: 72, 83).

Throughout much development activity runs a Western, liberal democratic bias (something also true of environmental concern and environmental management). This currently dominant Western outlook is also anthropocentric, placing human needs (and often profit) before protection of the environment. So, there are increasingly calls to open up to non-Western outlooks, for the development of a less profit-motivated worldview, and from some quarters for less anthropocentrism. Many involved in environmentalism and environmental politics (see Chapters 4 and 6) are calling for radically altered development ethics; most environmental managers operate on the assumption that such changes will be limited – a 'business-as-usual scenario' – with human attitudes and economic forces little altered.

Currently, the predominant view among the environmentally aware is that humankind has a limited time (a few decades) to set in motion development that will sustain indefinitely as many people as the Earth can support, giving them a satisfactory 'quality of life', and causing as little environmental damage as possible (Caldwell, 1977: 98; Berger, 1987: 116; Ghai and Vivian, 1992). *En route* to that goal it will probably be necessary to support too large a global population and to cope with excessive environmental demands, damage and conflicts, perhaps for several decades.

Recent human development has taken place during several thousand years of relatively stable and benign environmental conditions; this is unlikely to last and deterioration may be swift. There is a rapidly increasing human population placing more and more stress on the environment, so even if there are not challenges caused by nature there are some caused by development. Environmental management must assess threats, and if any seem significant and likely, seek avoidance, mitigation or adaptation. Assessing threats is not easy and is imprecise, there may be conflicting advice from experts, and vested interests are likely to lobby for a particular response. There are also biases caused by researchers' personal, political and funding backgrounds. For example, it is often more acceptable to blame land degradation on the local peasantry, rather than accept that it lies with policies promoted by the ruling elite. Misleading data are all too easy to acquire, particularly when researchers and administrators hold particular worldviews which lead to 'polarised perception' (e.g. 'Western, urban, colonialist, commercial', economist, anthropologist, scientist). Apparent causes of a problem may in reality be symptoms, and faulty diagnosis can lead to costly mis-spending on 'solutions' (see Fairhead and Leach, 1996), and for more controversial questioning of received wisdom (Lomborg, 2001, 2004). Lomborg makes a valuable point: that too many people make selective and mistaken or misleading use of environmental and developmental evidence. Discussion, negotiations and policy making must not be based on misconceptions and poor statistics ('myths'). Data and concepts must always be questioned, and whenever possible multiple lines of evidence sought.

After this brief outline of the evolution, characteristics and problems of environmental management, it is useful to present a picture of its scope, definitions and principles, and rules.

The definition and scope of environmental management

Environmental management seeks to steer the development process to take advantage of opportunities, try to avoid hazards, mitigate problems, and prepare people for unavoidable difficulties by improving adaptability and resilience (Erickson and King, 1999; International Network for Environmental Management website http://www.inem.org – accessed January 2005). Environmental management is a process concerned with human–environment interactions, and seeks to identify: what is environmentally desirable; what are the physical, economic, social and technological constraints to achieving that; and what are the most feasible options (El-Kholy, 2001: 15). Environmental issues are so intertwined with socio-economic issues that it has to be sensitive to them, especially in poor developing countries – in the South, environmental management is 'of a single piece with survival and justice' (Athanasiou, 1997: 15).

There can be no concise universal definition of environmental management, given its very broad scope and the diversity of specialisms involved. Definitions of environmental management which I have culled from recent literature are presented in Box 1.1.

Environmental management displays the following characteristics:

- it supports sustainable development;
- it is often used as a generic term;
- it deals with a world affected by humans (there are few, if any, wholly natural environments today – an eminent environmental scientist recently suggested that the current geological unit, the Holocene, should be declared 'ended' and succeeded by the Anthropocene or 'human-altered' period);

Box 1.1

Some definitions of environmental management

- An approach which goes beyond natural resources management to encompass the political and social as well as the natural environment . . . it is concerned with questions of value and distribution, with the nature of regulatory mechanisms and with interpersonal, geographic and intergenerational equity (R. Clarke, Birkbeck College, University of London: personal communication).
- Formulation of environmentally sound development strategies.
- An interface between scientific endeavour and policy development and implementation (S. Macgill, Leeds University, UK: personal communication).
- The process of allocating natural and artificial resources so as to make optimum use of the environment in satisfying basic human needs at the minimum, and more if possible, on a sustainable basis (Jolly, 1978).
- Seeking the best possible environmental option to promote sustainable development (paraphrased from several 1990s sustainable development sources).
- Seeking the best possible environmental option (BPEO), generally using the best available techniques not entailing excessive cost (BATNEEC) (based on two widely used environmental management acronyms).
- The control of all human activities which have a significant impact upon the environment.
- Management of the environmental performance of organisations, bodies and companies (Sharratt, 1995).
- A decision-making process which regulates the impact of human activities on the environment in such a manner that the capacity of the environment to sustain human development will not be impaired (paraphrase from various 1990s 'green development' sources).
- Environmental management cannot hope to master all of the issues and environmental components it has to deal with. Rather, the environmental manager's job is to study and try to control processes in order to reach particular objectives (Royston, 1978).
- Environmental management – a generic description of a process undertaken by systems-oriented professionals with a natural science, social science, or, less commonly, an engineering, law or design background, tackling problems of the human-altered environment on an interdisciplinary basis from a quantitative and/or futuristic viewpoint (Dorney, 1989: 15).

- it demands a multidisciplinary, interdisciplinary or even 'holistic' approach;
- it has to integrate and reconcile different development viewpoints;
- it seeks to co-ordinate science, social science, policy making and planning;
- it is a proactive process;
- it generally embraces the precautionary principle;
- it recognises the desirability of meeting, and if possible exceeding, basic human needs;
- the timescale involved extends well beyond the short term, and concern ranges from local to global;
- it should identify opportunities as well as address threats and problems;
- it stresses stewardship, rather than exploitation.

Most environmental managers aim for an optimum balance of natural resource uses and must decide where that lies, using planning and administrative skills to reach it. This conceptualisation, usually adopted by mainstream environmental management, is clearly biased towards the anthropocentric, i.e. the view that environmental issues are considered after human development objectives have been set (Redclift, 1985). However, there are many who would object to this and advocate other (non-mainstream) approaches, for in environmental management there is a wide diversity of beliefs ranging from anthropocentric to ecocentric. In general, there has been a reshaping of environmental management since the mid 1980s towards greater emphasis on social aspects and links with human geography, environmental economics, environmental law, environmental politics and business management, and there is growing support for sustainable development (Bryant and Wilson, 1998).

Environmental management must do three things: (1) identify goals; (2) establish whether these can be met; (3) develop and implement the means to do what it deems possible. The first (1) is seldom easy: a society may have no clear idea of what it needs. Indeed, some people may want things that are damaging to themselves, to others and the environment, and needs and fashions change over time. Sustainable development demands trade-offs between current enjoyment and investment in ensuring future function; many people find it difficult to be altruistic and forgo something in order to benefit future generations and non-relatives. Environmental managers have to identify goals, and then win over the public and special-interest groups. To pursue (2) and (3) requires the environmental manager to interface with ecology, economics, law, politics, people and so on to seek sustainable development. To co-ordinate such a diversity of factors is difficult because most humans operate on a piecemeal, short-term basis. Much of what is done at a given point in time and space has wider and longer term impacts, so it is desirable for development to be managed at all levels: regional, national and international – the environmental manager must somehow, as Henderson (1981a) advised, 'think globally, act locally' – and encourage a long-term outlook. Figure 1.1 suggests how environmental management is typically conducted.

Environmental management, whatever its approach, is related to, overlaps and has to work with environmental planning. The focus of environmental management is on implementation, monitoring and auditing; on practice and coping with real-world issues (e.g. modifying human habits that damage nature), rather than theoretical planning (Hillary, 1995). While a close integration with environmental planning is desirable, environmental management is dedicated to understanding human–environment interactions and the application of science and common sense to solving problems. General acceptance that economic development and environmental issues should not be approached separately gained widespread acceptance somewhere between 1972 (the UN Conference on the Human Environment, Stockholm) and 1992 (the UN Conference on

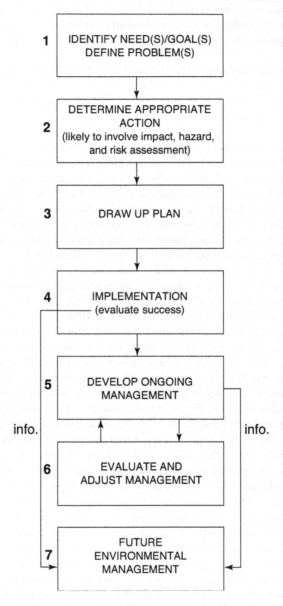

Figure 1.1 A typical scheme of practice adopted for environmental management

Note: Increasingly, stages 1, 2 and 3 are influenced by broad strategic policies, and are accountable to public scrutiny (as is stage 5). Ideally, lessons learned at every stage should be passed on to improve future environmental management – the evaluation of stages 4 and 5 is especially helpful in future management. At stage 1 the public or a developer may not have a clear idea of needs or goals, so the environmental manager may need to establish these.

Environment and Development, Rio de Janeiro – the 'Earth Summit'). By the early 1990s natural resources management had given way, as Wisner (1990) observed, perhaps unfairly, to 'a murky philosophical plunge' towards environmental management.

The evolution of environmental management

Since prehistory, humankind has accumulated environmental know-how and developed strategies for exploiting nature. To help regulate and sustain resource use people often evolved taboos, superstitions and common rights, formulated laws to improve steward-ship, and for centuries some have undertaken resource inventories. Many societies have managed to sustain reasonable lifestyles for long periods. Others have been less careful and suffered hardships or full environmental and social collapse. The idea that pre-modern people were 'close to nature' and caused little environmental damage is often a myth. Indeed, with populations a fraction of today's, some prehistoric peoples, using fire and weapons of flint, bone, wood and leather, managed to alter the vegetation of whole continents and probably wiped out many animal species, including some large and numerous animals (Tudge, 1995). However, in the past, small and scattered popu-lations of mainly non-sedentary and resilient people could move and adapt. Modern populations are huge, much less mobile and adaptable, and are probably more vulnerable – sustainable development strategies have to plan for that.

Developments apparent in the late twentieth century make it critical that environ-mental management is got right; these include human population growth; global pollution; loss of biodiversity; soil degradation; and urban growth. *Laissez-faire*, poorly controlled development is no longer wise in a crowded and vulnerable world. The challenges are great, but there have been advances in understanding the structure and function of the environment, in monitoring impacts, data handling and analysis, modelling, assessment, and planning (see Chapter 3). Environmental management must co-ordinate and focus developments, to improve human well-being, and try to mitigate or prevent further damage to the Earth and its organisms.

In Western societies from the 1750s (AD) the belief gained hold that human welfare could be improved through hard work and the appliance of technology and moral devel-opment ('civilisation'). Natural resources were to be exploited to these ends, and some even believed that humans would conquer nature and control it. Technological opti-mism, apparent in the West from about the 1830s, began to falter by the 1960s as awareness of environmental problems grew and the lessons that people must manage the environment were learned (Mitchell, 1997; citizen and special-interest group aware-ness of environmental issues – 'environmentalism' – is discussed in Chapter 4). Between the mid 1940s and the late 1980s much development effort was 'flavoured' or sidelined by concern and spending on Cold War issues. In that period development was seen to be concerned primarily with the reduction of poverty; environmental concern was often deemed irrelevant, or a 'luxury' poor countries could not afford, or it was even seen to be part of a conspiracy by the rich to hold back the less-developed nations. It was not until after 1987 that it was widely accepted that development needed effective environ-mental management. The shift to serious environmental concern has probably been prompted by a complex of causes which included: increasingly apparent pollution; loss of biodiversity; declining fish stocks; soil degradation; deforestation; a realisation, thanks in part to space exploration, that the world was finite, closed, and easily irreparably damaged; concern at the rate of human population growth; and worries about the threat of nuclear warfare and inadvertent technological disasters (see McNeill (2000) for a readable environmental history).

Before the 1970s some limited efforts were made to integrate natural resources exploitation with social and economic development (e.g. integrated river basin planning and management (Barrow, 1997)). Modern urban and regional planning have some roots in holistic, ecosystem approaches (see Chapter 3's discussion of the ecosystems approach) (Slocombe, 1993: 290). Natural resources management evolved before environmental management (by the 1960s) and deals more with specific components of the Earth – resources – which have utility and can be exploited, mainly for short-term gain and the benefit of special-interest groups, companies or governments (environmental management stresses stewardship rather than exploitation – although natural resources management has moved that way also). Natural resources management responses to problems tend to be reactive, and often seek a quick-fix technological solution and adopt a project-by-project approach. Natural resources managers have generally been drawn from a limited range of disciplines, before the 1980s with little sociological and limited environmental expertise. Their approach has often been authoritarian and has failed to involve the public; they have also tended to miss off-site and delayed impacts. Natural resources management has lost ground to environmental management in the past forty years, but recently both have developed more participatory and socially aware approaches.

There are a wide range of bodies and professionals involved in environmental management: government agencies, international bodies and aid organisations (e.g. the UNEP, FAO, World Bank, USAID), research institutes (e.g. the Worldwatch Institute, IIED), NGOs (e.g. WWF, IUCN, Friends of the Earth; the public). Identifying a single environmental manager in a given situation may be like trying to identify which individual built a Boeing 747 aircraft. What motivates environmental management? One or more of the following may lead to its adoption:

- *Pragmatic reasons* – fear or common sense makes people or administrators seek to avoid a problem.
- *Desire to save costs* – it may be better to avoid problems or counter them than suffer the consequences: pollution, species extinction, human deaths, costly litigation. There may also be advantages in waste recovery, energy conservation and maintaining environmental quality.
- *Compliance* – individuals, local government, companies, states and so on may be required by laws, national or international agreement to care for the environment.
- *Shift in ethics* – research, the media, individuals or groups of activists may trigger new attitudes, agreements or laws.
- *Macro-economics* – promotion of environmental management may lead to economic expansion: a market for pollution control equipment, use of recovered waste, more secure and efficient energy and raw materials supply; or there may be advantages in 'internalising externalities'.

Those involved with environmental issues have generally shifted their emphasis since the 1970s, from listing problems, issuing warnings and voicing advocacy, towards environmental management: problem solving, creating practical tools, developing governance, and policy formulation. Environmental management has, or is developing, a more flexible and sensitive style: assessment of a situation leading to an appropriate approach, emphasising stewardship rather than exploitation; managing a situation with the goal of long-term sustainable use; multidisciplinary, interdisciplinary, or even holistic (see Chapter 7). It is also usual to adopt a precautionary and participatory stance (Dorney, 1989).

Most environmental managers would accept that sustainable development is one of their key goals, but providing a universally acceptable definition of sustainable development is not easy. Most would accept that sustainable development demands the maintenance of environmental quality and ensuring resource-use benefits are shared equitably between all groups of humans at present, and that current activities do not damage the range of livelihood options or degrade the environment for future generations. Sustainable development is about improving the lot of people and avoiding environmental degradation. In a quest for sustainable development the goal of environmental management may be said to be: *to stretch what nature provides to the optimum and maintain that expansion indefinitely without environmental breakdown, in order to maximise human well-being, security and adaptability.* This demands high-quality management of the environment and human institutions, and the ability to recognise and avoid, mitigate or adapt to socio-economic and physical threats. Many are now 'dancing to the same sustainable development tune', reflecting the greening of politics in Western countries since the 1980s (Adams, 2001: 1–3). One problem faced by environmental managers is that the goal of sustainable development is not fully formed and its fundamental meaning is still debated (see Box 1.2 – sustainable development is discussed further in Chapter 2). The concept appeared in the 1970s, and was disseminated in the early 1980s in the *World Conservation Strategy* (IUCN, UNEP and WWF, 1980), which called for the maintenance of essential ecological processes; the preservation of biodiversity; and sustainable use of species and ecosystems. The Brundtland Report, *Our Common Future* (World Commission on Environment and Development, 1987), placed sustainable development on the world's political agenda and helped rekindle public interest in the environment. It also spread the messages that global environmental management was needed; and that without a reduction of poverty ecosystem damage would be difficult to counter. Twenty-six years after the *World Conservation Strategy* the same three bodies published *Caring for the Earth* (IUCN, UNEP and WWF, 1991), which proposed principles intended to help move from theory to practice. Interest in sustainable development is now well established and it is a professed goal of many governments, agencies and companies.

Sustainable development was in part generated by fears that the materially comfortable way of life enjoyed in some countries probably cannot be maintained on anything like a global scale with likely population growth (Pirages, 1994). Caution is needed; *sustainability* and *sustainable development* are not the same, but are often used without caution as if they were. The former is the ongoing function of an ecosystem or use of a resource, and implies steady demands; the latter implies increasing demands for improving well-being and lifestyles and probably, in the foreseeable future, for a growing population. As a concept, sustainable development draws upon two, often opposed, intellectual traditions: one concerned with the limits nature presents to humans, the other with the potential for ever-increasing human material development (Redclift, 1987: 199; Barrow, 1995b). Interpretation varies considerably:

- Some see it as a quest for harmony between humans and their environment.
- Some fail to accept that in a finite world there cannot be unlimited demand on resources.
- Some feel there can be a shift to less environmentally damaging improvements in the quality of human life.
- Some hope technology will allow limits to be stretched in a sustained manner.

There are many situations where naive, ill-thought-out appeals for sustainable development are made. This harms the concept, risking its dismissal by the public and decision

Box 1.2

Some definitions of sustainable development

- Environmental care 'married' to development.
- Improving the quality of human life while living within the carrying capacity of supporting ecosystems.
- Development based on the principle of inter-generational (i.e. bequeathing the same or improved resource endowment to the future that has been inherited), inter-species and inter-group equity.
- Development that meets the needs of the present without compromising the ability of future generations to meet their own needs.
- An environmental 'handrail' to guide development.
- A change in consumption patterns towards more benign products, and a shift in investment patterns towards augmenting environmental capital.
- A process that seeks to make manifest a higher standard of living (however interpreted) for human beings . . . that recognises this cannot be achieved at the expense of environmental integrity.

Source: Barrow (1995b: 372)

makers as shallow, unworkable and so on. Worse, there are cases where sustainable development is being used as rhetoric or cunning deceit to mislead people (see later discussion of greenwash). There have been complaints that calls for sustainable development are often unworkable and cause the side-stepping of necessary radical socio-economic reform. Environmental management must police the use of the concept and try to develop workable strategies without too draconian controls.

Currently, 'mainstream' sustainable development typically urges:

- the maintenance of ecological integrity;
- the integration of environmental care and development;
- the adoption of an internationalist (North–South interdependence) stance;
- the satisfaction of, at least basic, human needs for all;
- 'utilitarian conservation';
- concern for inter-generational, inter-group and inter-species equity;
- the application of science, technology and environmental knowledge to world development;
- the acceptance of some economic growth (somehow without exceeding environmental limits);
- the adoption of a long-term view.

The question is whether sustainable development is going to act just as a guiding principle (which in itself is valuable) or whether it can generate practical workable strategies that improve human well-being and prevent environmental degradation. As a principle and way of integrating diverse interests it is already established, but practical strategies need more development, and there is much misuse of the concept, making it something of a shibboleth.

A number of developments have helped to establish environmental management:

1   In an increasing number of countries the public have become environmentally aware and unwilling to trust government and corporations to protect the environment. This has largely grown out of their witnessing accidents, misuse of resources, and from concern about ecological threats.
2   NGOs, international agencies, businesses and governments have started to pursue environmental management.
3   The media monitor and report on environmental issues.
4   International conferences, agreements and declarations have publicised issues and supported environmental management.
5   The establishment in 1973 of the UN Environment Programme (UNEP) and other environmental agencies.
6   The 1969 US National Environmental Policy Act (passed 1970) and the creation of the US Environmental Protection Agency (EPA) in 1970.
7   Publications in North America and Europe which raised environmental concern after the mid 1960s.
8   The development of environmentalism and green politics since the 1970s.
9   Aid and funding agencies in the late 1970s began to require environmental assessments and environmental management before supporting development.
10 The Brundtland Report (World Commission on Environment and Development, 1987) increased awareness of the need for environmental care.

At the time of the UN Conference on the Human Environment, Stockholm (1972), few countries had environmental ministries, few newspapers had environmental editors, or broadcasting companies environmental producers. By the 1992 UN Conference on Environment and Development, Rio (the Earth Summit), most countries had environmental ministries and media interest had vastly increased. The release of *Agenda 21* (UN, 1992; Keating, 1993; Local Government Management Board, 1994) encouraged governments and other bodies to seek sustainable development and progress environmental management. For example, *Agenda 21* has been adapted to local needs in a number of countries (Evans, 1995; Patterson and Theobald, 1995). Since the early 1990s the European Union (EU) and the UK have published policy documents on sustainable development (Commission of the European Community, 1992; Department of Environment, 1994), Europe has established an Eco-Management and Audit System (EMAS), international environmental standards have been developed, and most countries now require impact assessments before significant developments proceed.

Broadly, the main principles of environmental management are prudence and stewardship. These are pursued via:

● forward-looking, broad-view policy making and planning (mainly left to various planners to undertake);
● establishing standards and rules, monitoring and auditing;
● co-ordination (the environmental manager adopting a multidisciplinary, interdisciplinary or holistic approach);
● operationalisation/implementation.

Sustainable development, a key component of environmental management, is linked to prudence and stewardship as a goal; another is human welfare, though there may be situations where long-term human survival or conservation aims overrule this.

Since the mid 1980s new branches have appeared on the evolutionary tree of environmental management, including:

- environmental law (see Chapter 5);
- green business (see Chapter 5);
- impact, risk and hazard assessment (see Chapters 8 and 9);
- total quality management (TQM), which has led to total environmental quality management (see Chapters 5, 7 and 8);
- environmental standards (see Chapter 8);
- eco-auditing (see Chapter 9);
- environmental management systems (see Chapters 5 and 8).

Problems and opportunities

Some dismiss much of present-day environmental management as 'environmental managerialism' which pays insufficient attention to human–environment interaction, has become institutionalised, and is essentially a state-centred process concerned with formulating and implementing laws, policies and regulations which relate to the environment (Bryant and Wilson, 1998). Whatever one might wish for environmental management as a theoretical subject, it is being used to address real-world problems, and consequently managerialism and other shortcomings may creep in. It should be stressed that environmental management is currently evolving and is far from being fixed in form.

Some people are sufficiently aware of pollution, soil erosion, over-fishing, loss of forests and other changes in their physical surroundings, and are prepared to voice concern. Environmental management activities are often prompted by such people, by those monitoring developments, and also by historians, palaeoecologists, archaeologists, geologists and others interested in human–environment interactions and environmental change. Recently, the focus has been more on how humans affect the environment rather than on how environment affects humans, which is unwise.

There is currently widespread complacency, and many assume that current living standards, patterns of governance and technological progress will continue and even improve without much upheaval. This is unwise, given that few nations have had more than 150 years without serious famine, less without large fatalities to epidemic diseases or warfare, and that the past 200 years have been one of the most climatically favourable periods during the past two million years of marked changes and often inclement environment. There has been no global catastrophe during recorded history to provoke caution, yet over the last 500,000 years there is evidence of mega-eruptions and other hugely damaging environmental disasters. Humans are more numerous than ever before, they are upsetting their environment and adding anthropogenic global changes to natural threats. Although it appears that there has been huge progress there is only a thin veneer of technology and governance protecting today's humans from disaster.

Environmental managers should be aware of these threats and seek to reduce human vulnerability and enhance adaptability – some worthwhile strategies should be relatively cheap and easy. Awareness of the past helps in scoping and planning future scenarios, and it can also interest the public in environmental forecasting (Pest and Grabber, 2001).

Environmental stress may be caused by human activities (e.g. resource exploitation, urban growth, warfare, globalisation, capital penetration and technological change), and since the 1980s structural adjustment programmes, rising oil prices and debt have reduced the funds available to deal with pollution, conservation and other challenges.

Socio-economic factors can degrade social capital, causing environmental and human welfare problems. Those warning of crisis have frequently been branded 'Cassandras', while some are the opposite, being over-optimistic or 'cornucopians'; however, the majority of people in rich and poor countries do not think much about threats ('apathetic'). One of the tasks of environmental management is to offer carefully weighed warnings in a persuasive manner. This demands sound judgemental, negotiating and diplomatic skills and an ability to take risks and survive. Should a problem flagged by an environmental manager not materialise (or if it develops in an unexpected way) there will be accusations of 'crying wolf', and there will be a wider impact when future warnings are issued. There is no way for an environmental manager to avoid risk-taking; but reliance on sound data from more than one source, careful checking and seeking win–win solutions helps. Win–win solutions are situations where a beneficial outcome results, even if the problem addressed fails to develop as expected.

Often considerable effort and much money have been expended treating symptoms of a problem but not the causes, which may be difficult to identify because they are complex, inadequately understood, or are located at some distance (in space and/or time) along a chain of causation. The risk of making this sort of mistake should be reduced by the adoption of a careful approach. Unfortunately, decisions may sometimes have to be based on 'snapshot' information; but it is important whenever possible to use broad-view, long-term and, if possible, gap-free monitoring and auditing (Born and Sonzogni, 1995).

Environmental management may need to modify the activities and ethics of individuals, groups and societies to achieve its goals. There are three main approaches which can be adopted to try to do that:

1 *Advisory*

- through education;
- through demonstration (e.g. model farms or factories);
- through the media (advertisements or covert approaches – the latter includes subtle 'messages' incorporated in entertainment);
- through advice (e.g. leaflets, drop-in shops, helplines).

2 *Economic or fiscal*

- through taxation ('green' taxes);
- through grants, loans, aid;
- through subsidies;
- through quotas or trade agreements.

3 *Regulatory*

- through standards and laws;
- through restrictions and monitoring;
- through licensing;
- through zoning (restricting activities to a given area).

Environmental problems often do not have a single simple workable solution. Attempts to address a problem may present alternatives and challenges. Bennett (1992: 5–9) explored such environmental management difficult choices, recognising: (1) Ethical dilemmas – e.g. what to conserve: Inuit hunters or whales? (2) Efficiency dilemmas – e.g. how much environmental damage is acceptable? (3) Equity dilemmas – e.g. who benefits from environmental management decisions, and who pays? (4) Liberty dilemmas

– e.g. to what degree must people be restricted to protect the environment? (5) Uncertainty dilemmas – e.g. how to choose a course of action without adequate knowledge or data. (6) Evaluation dilemmas – e.g. how to compare different effects of various options or actions.

Environmental managers may be forced into crisis management situations, which in turn force hasty ad hoc responses. Human beings often respond to perceived crises, rather than carefully assessing the situation and acting to prevent problems. With sustainable development as a goal, crisis management is a dangerous practice, for, once manifest, problems may not be easily solved. The solution is to adopt the precautionary principle and spot problems early (see Chapter 2 for further discussion) (Bodansky, 1991; Costanza and Cornwell, 1992; O'Riordan and Cameron, 1995; Francis, 1996). The precautionary principle shifts the burden of proof that a proposal is safe from the potential 'victim' to the 'developer' (O'Riordan, 1995: 8–10). It also makes sense because environmental management often deals with inadequate data, may have to rely on modelling that is deficient, and frequently has to cope with issues that are complex and not fully understood. Politicians, some NGOs, movements, lobby groups and individual 'gurus' may get away with advocacy, but environmental managers have to 'produce the goods', and perfect and carry through policies, programmes and projects which work.

The problem of 'polarised perceptions' (ideas based more on stakeholders' prejudice, misconception or greed than objectivity) is something environmental management often has to address (Baarschers, 1996; Pratt, 1999). Even if the environmental manager is objective, powerful special-interest groups such as the rich; government ministers; lobby groups; non-governmental organisations (NGOs), industry, the military and so on may not be. Where environmental managers have only advisory powers, powerful special-interest groups or even individuals are likely to override or side-step them. Governments and multinational companies can be very powerful opponents or allies. Sovereignty, political, cultural or strategic need arguments can threaten common-sense decisions and make transboundary issues difficult to resolve. Environmental managers must recognise, and whenever possible manipulate, these forces. Little remains fixed: demands from various stakeholders alter, the environment changes, public attitudes shift, human capabilities vary – so environmental management must be flexible, adaptive and perceptive (Holling, 1978).

Successful co-ordination of environment and development requires awareness of environmental and human limits and potential threats. For most of human history worries have mainly been caused by the acquisition of inputs – food, water, fuel and so on. But additional problems have appeared since the 1750s: outputs (pollution and waste), population expansion and technological impacts. Environmental problems are caused by human behaviour, notably consumerism, and poverty as well as natural processes and events.

To summarise, environmental management is faced with 'real-world' challenges, which include:

- greed, corruption and foolishness;
- knowledge and technical skills which are still too limited;
- increasing numbers of people who demand more and more material benefits;
- the time available to make real progress in resolving key environmental degradation is probably limited (quite possibly less than fifty years).

Environment and development problems are increasingly transnational (they cross borders) and often have to be dealt with on a global scale. Law, governance, the sciences and management are still trying to adapt to meet those demands. In the past scientists have been able to research problems thoroughly and then suggest solutions, but increasingly advice has to be offered before there is adequate data or knowledge, otherwise the challenge could become an uncontrollable or costly problem. Environmental management may face unexpected and rapid changes, and also situations which develop so slowly that novel inter-generational approaches are required to identify and address them.

Environmental management has to research, model and monitor to gain sufficient knowledge to try and give early warning. Some threats are random and difficult to recognise in advance; others develop in an insidious way and can be easily overlooked. Worse, a problem may have indirect and cumulative causes – a number of unrelated factors suddenly conspire to cause trouble – or a process develops positive or negative feedback which (respectively) quickly accelerates or slows down developments.

Environmentalism, environmental management practices, environmental ethics, environmental legislation, and techniques for monitoring and forecasting have in large part originated from the Western 'liberal democracies'. Consequently, things often need to be adapted to suit other countries' laws, attitudes, business, trade and so on (Lafferty and Meadowcroft, 1996; Gupta and Asher, 1998). Given that the spread of environmental management has taken place only in the past 30 years or so, there has been much progress. However, tools and methodology are still evolving, and the database of environmental and social knowledge for many countries is still woefully inadequate.

Environmental managers frequently find that they face:

- a poorly researched threat;
- transboundary or global challenges;
- problems demanding rapid decisions;
- an increasing exchange of information with NGOs via the Internet and various other networks (this means that environmental managers must keep abreast of the activities of many bodies, but it also offers possibilities for alliances and data gathering from different sources).

Modern science has traditionally adopted a reductionist approach, with disciplinary specialists studying components of a problem and avoiding giving any judgement or advice to managers or planners before there is adequate proof. Environmental managers have to deal with uncertainty and complex problems, and, as discussed above, often cannot afford to wait for proof (Funtowicz and Ravetz, 1991). Something may have the potential to cause serious, possibly irreversible problems unless appropriate and prompt action is taken but it has not actually been proved to be a threat (the classic case was global warming). Environmental management must often rely on modelling, simulation and forecasts rather than factual predictions – it may be necessary to resort to advocacy without proof, and to identify the agency and the mechanism for a solution from such an insecure basis (Redclift, 1984: 44).

The past few decades have seen the manifestation or recognition of more and more transboundary or global threats. Before the 1970s, environmental problem solving seldom involved international negotiation. However, there have been helpful developments: environmental management can now draw upon improved knowledge of the structure and function of the environment, and of human institution building, group interaction and perceptions. There are also powerful new tools available that improve monitoring, data gathering, impact assessment, information processing, decision making and communication. Although environmental managers face growing problems, they have more powerful aids to draw upon and growing public and institutional support.

These developments mean that it is sometimes possible for environmental management to move away from corrective to anticipatory action.

With something as broad and ambitious as environmental management, criticism is inevitable (Trudgill, 1990). A frequently voiced worry is that it is prescriptive and insufficiently analytical. It also attracts the complaint that it involves subjective judgement, and so is not reliable scientific enquiry. Sometimes it is the approach to environmental management that causes offence – over-zealous efforts have been seen to be tantamount to 'eco-fascism' (Pepper, 1984: 204). Redclift (1985) warned of 'environmental managerialism', symptoms of which include: the consideration of the environment after development objectives have been set; the tendency to plunge into techniques regardless of whether they are needed; and failing to see the wood for the trees. Too often environmental management is pursued as a reactive, piecemeal approach, working on projects that have components designed to mitigate, rather than avoid, environmental impacts (Schramm and Warford, 1989: 8). Environmental management must go beyond monitoring and reacting and adopt a longer term and proactive view – most planners and politicians do not. A longer term view improves the chances of avoiding problems and allows time to develop contingency plans, acquire technology and so on.

Environmental management has so far developed mainly where there is relative freedom of access to information (e.g. USA Freedom of Information Act; the European Directive 82/501/EEC – Article 8 of which requires that local communities have information about any hazardous installation; and recent UK access to information legislation) (Haefele, 1973). Environmental management needs to be adapted to suit different social, cultural, economic and political conditions (Russo, 1999).

Summary

- Environmental management is evolving and spreading. It has still to be adequately adapted to suit all conditions, and will continue to have to be improved.
- Environmental management demands a proactive approach to development and must integrate closely with other disciplines.
- Without proactive environmental management, development is unlikely to be sustainable and people will be more vulnerable to disasters.

Further reading

Lomborg, B. (2001) *The Sceptical Environmentalist: measuring the real state of the world.* Cambridge University Press, Cambridge (published in Danish 1998).
Controversial interpretation of 'polarised perception' and received wisdom.

Lomborg, B. (ed.) (2004) *Global Crisis: global solutions.* Cambridge University Press, Cambridge.
Challenges perceptions on environmental management issues.

McNeill, J.R. (2000) *Something New Under the Sun: an environmental history of the twentieth century.* W.W. Norton & Co, New York. Penguin edition available.
Readable, thought-provoking historical introduction to environmental issues.

O'Riordan, T. (ed.) (1995) *Environmental Science for Environmental Management.* Addison Wesley Longman, Harlow.
Good interdisciplinary introduction – covers interactions between the Earth, life and human socio-economic activity.

O'Riordan, T. and Turner, R.K. (eds) (1983) *An Annotated Reader in Environmental Planning and Management.* Pergamon, Oxford.
A good but dated introduction.

Owen, L. and Unwin, T. (1997) *Environmental Management: readings and case studies.* Blackwell, Oxford.
A good introduction.

Schumann, R.W. III (ed.) (1994) *Eco-data: using your pc to obtain free environmental information.* Government Institutes Inc., Rockville, MD (MD 20850).

Theodore, L., Dupont, R.R. and Baxter, T.E. (1998) *Environmental Management: problems and solutions.* CRC Press, Boulder, CO, and Springer, New York.
Part 1 is especially useful.

UNDP (1992) *Handbook and Guidelines for Environmental Management and Sustainable Development.* United Nations Development Programme, New York.
Points to further sources.

Wilson, G.A. and Bryant, R.L. (1997) *Environmental Management: new directions for the twenty-first century.* University College London Press, London.
Environmental management presented as a multi-layered process.

WWW sources

This is a small selection, and it should be noted that these sources change, may disappear, and are often of unknown provenance (compared with refereed journals).

Journal of Environmental Management http://www.elsevier.com/wps/find/journaldescription. cws_home/622871/deser (accessed November 2005).

The following were accessed June 2005:

http://www.eea.dk/frames/main.html – European Environmental Agency (EU).
http://www.epa.gov/global warming – global warming (USA).
http://www.gn.apc.org – Green Net home page.
http://www.iied.org – International Institute for Environment and Development (UK, tel. 0171 388 2117).
http://www.iucn.org – IUCN home page.
http://www.sosig.ac.uk – Social Science Information Gateway (social science and environmental management sources UK).
http://www.Panda.org/home.htm – WWF International.
http://www.unep.ch – UNEP site, conference information.
http://www.wyw.ac.uk – University of London, Wye College postgraduate environmental management courses by distance learning.

Professional bodies

Chartered Institution of Water and Environmental Management (UK)
EIA Centre, Manchester University (UK)
Environmental Auditors Registration Association (UK)
Institute of Ecology and Environmental Management (IEEM) (UK) http://www.ieem.co.uk (accessed July 2005)
Institute of Environmental Assessment (UK)
Institution of Environmental Sciences (UK)
Institute of Environmental Management and Assessment (IEMA)(UK) http://www.iema.net (accessed June 2005)
International Association of Impact Assessment (USA)
World Federation of National Associations for Environmental Management: this offers tools, links, case studies and so on. International Network for Environmental Management (INEM) http://www.inem.org (accessed October 2005).

2 Environmental management fundamentals and goals

- The nature of environmental management
- Key terms and concepts
- Environmental management challenges
- Summary
- Further reading

The nature of environmental management

Environmental management appeared by the 1970s as a problem-solving field, providing practical assistance mainly to state officials. Before the 1990s it paid limited attention to social issues. Effectively, it was state stewardship of the environment undertaken on behalf of citizens largely by experts trained in the sciences (Bryant and Wilson, 1998: 321–322). It was applied in a largely 'top-down' manner, implementing and enforcing environmental policies in the main by coercion (through laws, fines and closure for breaches of regulation). 'Management' is difficult to define precisely – it is a dynamic process which can include many aspects: reduction of uncertainty, leadership and motivation. The past twenty years have seen environmental management, along with many other businesses and government departments, shift from a command ('top-down') and technocratic ('trust me, do not question, I am a professional') approach to one where the public demand accountability and consultation, and social and economic issues are considered (Martin, 2002). In addition, ethics, management skills, quality standards, codes of conduct and transparency are increasingly important.

Since the 1970s environmental management has become more multidisciplinary or interdisciplinary, even holistic, with less disciplinary compartmentalisation, often encouragement and support rather than enforcement, and sometimes citizen involvement ('bottom-up' approach). Environmental managers once consulted mainly with natural science advisers, planners and administrators. Nowadays, the input of the social sciences has markedly increased, and environmental managers now commonly deal with historical data, policy formulation, social capital and institutional issues, qualitative socio-economic information, social development, social impact assessment, political ecologists, economists, lawyers, business personnel, anthropologists and others. A growing number of businesses and institutions employ environmental managers and promote the field. On the whole, environmental management has become more co-ordinatorial and participatory and much more integrative; and it has also spread widely beyond the Western 'liberal' democracies where it originated. The ongoing dissemination from Western developed countries means that it often needs to evolve to suit new situations.

Some people are sufficiently aware of pollution, soil erosion, natural disasters, over-fishing, over-hunting, loss of forests, and other changes in their physical surroundings to voice concern. However, more often environmental managers have to prompt awareness by using evidence from environmental historians, palaeoecologists, archaeologists, geologists, those modelling and forecasting future social, economic and social changes, and others interested in human–environment interactions. For the past seventy-five years or so, the focus has been more on how humans affect the environment, rather than on how environment affects humans; the 2004 late December tsunamis around the Indian Ocean may have helped shake up that complacency. Awareness of the past helps scoping and planning for future scenarios, and hindsight can also interest the public in environmental forecasting and encourage them to support expenditure on disaster warning (Pest and Grabber, 2001; Barrow, 2003).

Definitions reflect the current values of those making them; however, most of the world's population today probably see development as the goal they aspire to – a drive for the material lifestyles and consumption patterns apparent in richer nations. Some may look forward to non-material 'development', an increase in contentment, sense of security, religious or cultural enrichment, or whatever. Given that the former, material, outlook is dominant and probably increasing, the questions arise: Will the Earth's environment support these people's hopes? What can be done to improve the chances of a better lifestyle for those seeking it, given the structure and function of the environment? Some countries have achieved what they and others see as development through agricultural and industrial development, others may follow a similar pattern, but there may be societies that take different routes. And as world population grows, some may struggle to sustain current lifestyles, let alone develop further. Development is thus a goal and an ongoing process, but there is uncertainty over its exact meaning, the strategy that is best adopted to pursue it, or how it functions. Providing a universally acceptable and precise definition of development is impossible; most would accept that it is a process of change (which can progress, regress or stagnate at varying speeds). Planners, managers and individuals may try to drive it forward in a wide range of ways, such as development planning, key speeches, books, fashions, inspirational acts (including terrorism), and by many other actions. Efforts to improve human material well-being and security have rarely been well planned, intended to benefit a broad swathe of society and avoid environmental damage. Civilisations have seldom lasted many centuries before human or environmental problems or both have confounded them. Hopefully, environmental management will change things.

Before the modern era (c. 1700 to the present), social patterns and lifestyles were seldom questioned – Francis Bacon in The New Utopia (1627) was one of the first to suggest that science would allow humans to dominate nature and achieve better conditions. The acceptance that fortunes could be improved by humans themselves, through material rather than religious works, owes much to the appearance of scientific enquiry and Western rationalism in Europe mainly after the start of the eighteenth century (Uglow, 2002). As the twenty-first century unfolds it is by no means certain that democracy, rationalism and science will remain strong. Generally, the group in power decides fashions and desirable goals, the latter not always material and worldly things (Barrow, 1999). For the past five centuries or so the West has been dominant, so there runs a Eurocentric, democratic bias throughout much of the world's development activity. This is also true of environmental concern that has also largely evolved in Western democracies since the mid 1960s. The predominant outlook is anthropocentric, and places human needs (and often profit) before protection of nature. In many non-Western countries the established legal system, civil engineering regulations and methods of governance are influenced by the West, and some of their environmental problems are a consequence of this. For example, water laws transferred from wet temperate Europe are often unsuitable for seasonally dry tropical states. The benchmarks used to judge the progress and success of development have commonly been inappropriate, paying attention to economic or engineering criteria, and giving too little attention to environmental, social and local issues. Developers may have inadequate local knowledge because they are frequently expatriates or overseas-trained city folk, and may be insensitive to poverty, social issues, biota and environment.

Development is widely conducted against the clock: in order to achieve goals before a government runs out of its term of office, or to cut costs, or because there is a genuine sense of haste to achieve development. Hindsight experience is often not adequately shared because it is restricted to limited-circulation consultancy reports or academic journals which poor countries cannot access; also, post-development appraisals are seldom satisfactory because there is scarce funding, and those involved do not want to highlight 'shortcomings'. Consequently mistakes are repeated.

Development management has evolved independently of environmental management, but commonly overlaps. Development management is essentially the manipulation of interventions aimed at promoting development. Adopting a theatrical analogy – it has largely been as if only the actors were involved, and the theatre, lighting and stage attracted little concern. For much of the history of the Western nations' struggle to develop there was strong support for *laissez-faire*, rather than development management interventions. *Laissez-faire* strategies are no longer wise in a crowded and vulnerable world. Few now question the importance of caring for the environment; but in reality the world's governments often refuse to spend. People frequently resist changing environmentally damaging lifestyles, or paying more for necessities or even luxury items, or through poverty are unable to do so. Many governments and businesses have genuinely embraced environmental concern; however, some are ineffective, some hijack environmental concern for their own ends, and others ignore environmental issues for 'strategic' reasons. Societies, governance and law have to evolve to support environmental management. Some optimistic forecasts assume progress towards less damaging habits will be adequate; others are more pessimistic and, perhaps realistically, reckon it is likely there will be 'business-as-usual' scenarios. Those seeking to manage the environment thus have many challenges.

It has been argued that a crisis or turning point has been reached, and that there is limited time available for humans to get environmental management right and avert disaster. Various estimates suggest there is no more than a generation or two available – the 'Brundtland Report' (World Commission on Environment and Development, 1987: 8) observed that: 'Most of today's decision-makers will be dead before the planet feels the heavier effects of acid precipitation, global warming, or ozone depletion. . . . Most of today's young voters will still be alive.' Humankind must set in motion development that will sustain indefinitely as many people as the Earth can support with a satisfactory 'quality of life' (Caldwell, 1977: 98; Berger, 1987: 116; Ghai and Vivian, 1992). *En route* to that goal it will probably be necessary to 'overshoot' and support too large a population and cope with excessive environmental damage and conflicts, perhaps for several decades.

There are various reactions to the idea a crisis is approaching or has been reached: (1) ignore the threat; (2) promote abandonment of technology and a return to simple ways; (3) use all 'tools' available, including technology, to achieve sustainable development. The first is foolhardy, the second would mean disaster for most of the current world population, and cultural and intellectual regression for survivors. In addition, humans have caused great damage and rehabilitation will demand their efforts, rather than reliance on nature. There seems to me to be no choice other than to adopt the third way (no. 3 above).

The idea that the world faces an environmental crisis may provoke needed change, but it may also encourage emotive, journalistic debate and 'fire-fighting' solutions – ill-considered short-term focus approaches and activities that divert attention from other important tasks. A crisis attitude may actually prompt things to get worse, and, if causes and treatments are not carefully researched, little will be achieved. Preoccupation with global carbon emissions controls may blinker the world to other challenges.

Key terms and concepts

This is written by a white, relatively affluent, Westerner with light-green sympathies (cautious use of technology is welcomed).

The process and goals of environmental management

Environmental managers make deliberate efforts to steer the development process to: take advantage of opportunities, try to avoid hazards, mitigate problems, and prepare people for unavoidable difficulties by improving adaptability and resilience (Erickson and King, 1999; International Network for Environmental Management website http://www.inem.org – accessed February 2005). In 1975 Sewell (1975: ix) felt that the environmental manager should 'be able to manipulate both social institutions and appropriate technologies but must do this with the sensitivity of an artist, the insights of a poet, and, perhaps, the moral purity and determination of a religious zealot'. Advice that is still relevant.

Environmental management is still a relatively young discipline, so judging how successful it has been and in what ways it should be 'tuned' to better serve the quest for development is difficult. Environmental management has to cope with natural threats and problems caused by human activity; it has to do this in a world where nature is being degraded, and it has to support livelihoods and steer these to ensure sustainable development. Although it appears to the rich that there has been huge progress, there is only a thin veneer of technology and governance protecting them. In the past human survival was largely aided by intelligence and adaptability but many people today have lost these qualities. Humankind has also increased in numbers far beyond anything in the past, which with other developments probably makes us more vulnerable than our ancestors. One key task of environmental management is to reduce human vulnerability and improve adaptability.

Environmental management seeks to improve environmental stewardship by integrating ecology, policy making, planning and social development, and whatever else is needed. Its goals include:

- sustaining and, if possible, improving existing resources;
- the prevention and resolution of environmental problems;
- establishing limits;
- founding and nurturing institutions that effectively support environmental research, monitoring and management;
- warning of threats and identifying opportunities;
- where possible improving 'quality of life';
- identifying new technology or policies that are useful.

To adequately pursue such goals demands a focus, which stretches from local and short term to global and long term (Dorney, 1989: 5). Without overall vision it is difficult to avoid fragmented decision making, or to prioritise and identify urgent tasks. Effective environmental management also demands 'scoping' (deciding goals and setting limits on efforts) before starting to act; however, this is often neglected. Some environmental managers express their overall vision and goals by publishing environmental policy statements – to show intent, identify priorities and principles, and to give a sense of purpose. While this informs the public, it does not guarantee sound environmental management.

Environmental managers must ensure there is an optimum balance between environmental protection and allowing human liberty. Establishing where that balance lies depends largely on accepted ethics. Clark (1989) argued that at its core environmental management asks two questions: (1) What kind of planet do we want? (2) What kind of planet can we get? Even if agreement on an optimum balance can be reached, the approach to environmental management goals may take different paths (see Chapter 7). For example, environmental management may adopt a human ecology approach, or a systems analysis or a political ecology approach, or a bioregional approach, an ecosystem approach, or others. McHarg (1969) used river basins (a bioregional approach), and Doxiadis (1977) tried to develop a science of planning settlement in balance with nature – ekistics. Rapoport (1993: 175) recognised two main groupings: those who adopt a horticultural metaphor – Garden Earth – and those who prefer one that is more technological – Spaceship Earth. The diversity of challenges, and the fact that the public, commercial interests, professions, local and national government, special-interest groups, the voluntary sector, and other stakeholders are involved, means that in practice environmental managers often focus on a region, ecosystem, sector of activity or resource (Box 2.1).

Environmental managers may not achieve their objectives, might be criticised (even sued), fall into disrepute with those who employ them, and lose public trust. So, like many other professionals, environmental managers tend to follow risk-aversion strategies, including:

● working to safe minimum standards;
● adopting sustainability constraints;
● following a 'win–win' or 'least regrets' approach (i.e. actions which seek benefits *whatever the outcome* or seek to reduce unwanted impacts, respectively).

The argument may be made that 'what cannot be measured cannot be managed' – the development of reliable indicators and effective monitoring and forecasting techniques is vital (Jeffrey and Madden, 1991). Environmental management also demands skill in reading the public mood, so as to win support. Partly related to the former point, discrete problems are more likely to attract public support than slow-onset, often insidious ones (even if these are seriously threatening). It also helps if environmental management can point to clear benefits from its actions and not just flag threats.

Environmental management may be subdivided into a number of fields, including (not in any particular order):

● sustainable development issues;
● environmental assessment, modelling, forecasting and 'hindcasting' (using history or palaeoecology for future scenario prediction), and impact studies;
● corporate environmental management activities;
● pollution recognition and control;

Box 2.1

Approaches to environmental management

There may be some overlap between groupings and within categories. Environmental managers may be more or less anthropocentric or ecocentric, more or less 'green', more or less supportive of technology. There is also a wide spectrum of political and philosophical stances, all of which colour the approach adopted.

1 *Ad hoc approach:* approach developed in reaction to a specific situation.

2 *Problem-solving approach:* follows a series of logical steps to identify problems and needs and to implement solutions (see Figure 1.1).

3 *Systems approach:* for example,
   ● ecosystem (mountain; high latitude; savanna; desert; island; lake and so on) (Dasmann *et al.*, 1973; Ruddle and Manshard, 1981)[†]
   ● agro-ecosystem (Conway, 1985a and 1985b).

4 *Regional approach:* mainly ecological zones or biogeophysical units, which may sometimes be international (i.e. involve different states, e.g. an internationally shared river basin). For example,
   ● watershed (Easter *et al.*, 1986)[†]
   ● river basin (Friedman and Weaver, 1979; Barrow, 1998)[†]
   ● coastal zone[†]
   ● island
   ● command area development authority (irrigation-related)
   ● administrative region
   ● sea (e.g. Mediterranean; North Sea; Baltic; Aral Sea, etc.).[†]

5 *Specialist discipline approach:* often adopted by professionals. For example,
   ● air quality management
   ● water quality management
   ● land management
   ● environmental health
   ● urban management
   ● ocean management
   ● human ecology approach
   ● tourism management/ecotourism
   ● conservation area management.

6 *Strategic environmental management approach:* see Chapter 7.

7 *Voluntary sector approach:* environmental management by, or encouraged and supported by, NGOs. For example,
   ● debt-for-nature swaps
   ● private reserves
   ● 'ginger groups' which try to prompt environmental management
   ● private funding for research or environmental management.

8 *Commercial approach:* environmental management for business/public bodies.

9 *Political economy or political ecology approach:* see Chapter 7 (Blaikie, 1985).

10 *Human ecology approach:* see Chapter 7.

Note: [†] = biogeophysical systems

- environmental economics;
- environmental enforcement and legislation;
- environment and development institutions (including NGOs) and ethics;
- environmental management systems and quality issues;
- environmental planning and management;
- assessment of stakeholders involved in environmental management;
- environmental perceptions and education;
- community participation for environmental management/sustainable development;
- institution building for environmental management/sustainable development;
- biodiversity conservation;
- natural resources management;
- environmental rehabilitation/restoration;
- environmental politics;
- environmental aid and institution building.

The concept of 'limits' to development

In the past, various societies have sought to control their population to reduce environmental damage (e.g. by enforcing late marriage). For much of the past 400 years or so Westerners have tended to see themselves as being at war with nature, rather than seeking to understand it and then trying to exist within its constraints. The environment was to be 'tamed' and unspoilt lands were 'wastelands'. A few romantics, proto-environmentalists and anarchists bemoaned the 'rape of nature' by industrial development, deforestation and hunting – the English novelist Mary Shelley even went so far as to warn that humans could become extinct through science and greed (*The Last Man*, 1826). During the 1930s the last land frontiers were obviously closing, some ocean fisheries were stressed, unsettled areas capable of giving a good livelihood were becoming difficult to find, and Midwest USA was suffering severe soil degradation. By the mid 1960s the limitless world was seen to have shrunk; Spaceship Earth was increasingly seen to be a finite and delicate system which needed to be taken care of if it was to support humanity. The Gaia viewpoint (see Chapter 3) emerged in the late 1960s (similar views had been expressed in the eighteenth century, but not widely), regarding the Earth as a complex system which, if upset by careless development, might adjust in such a way as to make current lifestyles impossible or even eliminate humans.

In late eighteenth century England Thomas Malthus offered the thesis that human population growth puts pressure on the means of subsistence, throwing it out of balance with the environment so that there is population collapse. Interest in the limits to human population was rekindled in the 1970s by a group of ecologists, systems analysts, demographers and 'environmentalists' – neo-Malthusians (e.g. Ehrlich *et al.*, 1970). Neo-Malthusians argued that, for a given species and situation, population tends to grow until it encounters a critical resource limit or controlling factor, whereupon there is a gradual or sudden, limited or catastrophic decline in numbers, or a shift to a cyclic boom-and-bust pattern. Neo-Malthusians saw population growth as the primary cause for concern, although a few also focused on the growing threat from 'careless technology' (Farvar and Milton, 1972). One neo-Malthusian, Hardin (1968), argued in his 'tragedy of the commons' essay (and related works) that commonly owned natural resources under conditions of population growth would be damaged because each user would seek to maximise their short-term interests. This thesis that population increase invariably causes environmental degradation and poverty is now largely dismissed as simplistic, together with much other neo-Malthusian theory, because it failed to examine the social and historical context of demographic growth.

When neo-Malthusians were drawing attention to limits, the Club of Rome (an informal international group concerned about the predicament of humanity) reported on a systems dynamics computer world model (Meadows *et al.*, 1972 – *The Limits to Growth*). This model tried to determine future scenarios, using global forecasts of accelerating industrialisation; population growth; rates of malnutrition; depletion of non-renewable resources; and a deteriorating environment. The report was designed to promote public interest, and concluded that 'If present growth trends ... continue unchanged, the limits to growth on this planet will be reached within the next hundred years' (by 2072). Meadows and her colleagues concluded that effective environmental management could sustain a condition of adequate 'ecological and economic stability'.

Concern for limits and demand for material growth clearly conflicted and some began calling for reduced or even 'zero growth'. In the early 1970s a much more palatable alternative was proposed – sustainable development. This seemed to offer a way for continued growth to avoid conflict with environmental limits (Barrow, 1995b). The goals of sustainable development and the Club of Rome are broadly the same – adequate sustained quality of life for all without exceeding environmental limits. *The Limits to Growth* message was that it is possible to stretch some limits, using technology, and/or alter people's demands, and/or find resource substitutes. Even unsustainable 'overshoot' could be survived for a while until sustainable development is achieved – but it will have to be well managed (no more resort to *laissez-faire*) on a global scale.

In a sequel to *The Limits to Growth* two decades later the same principal authors refined their original systems dynamics model and fed in much-improved data. *Beyond the Limits* (Meadows *et al.*, 1992) argued that the 1972 warnings were broadly correct, that some of the limits have already been exceeded, and that, if current trends continue, there is virtually certain to be global collapse within the lifetime of children alive today (see Figure 2.1). They argue that it is possible to have 'overshoot but not collapse', and to achieve the goal of sustainable development in spite of excessive population growth in the short term, provided demands are cut and there is an increase in efficiency of materials and energy use soon. *Beyond the Limits* threw down an urgent challenge to environmental management and indicated an approximate timescale for action.

While 1970s environmentalist arguments were largely dogmatic warnings or pleas for change, weak on proof and workable strategies, they did trigger an awareness that in a finite world there were limits, complex environment–population linkages, and the risk of unexpected feedback. The speed of population growth related to the ability to upgrade technology is going to be crucial. It also became clear that damage to the environment is a function of:

● levels of consumption of the population (i.e. lifestyle);
● the type of technology used to satisfy consumption and dispose of waste (Harrison, 1990);
● environmental conditions and/or environmental change.

Boserüp (1965, 1981, 1990) explored how, provided it does not overwhelm the adaptive ability of people, population increase may prompt social and technological changes leading to improved quality of life (see also Turner and Ali, 1996). Tiffen (1993, 1995; Tiffen *et al.*, 1994) documents situations where not only has population growth led to innovation that improved quality of life, it has also reduced environmental degradation. While there are grounds for tempering Malthusian and neo-Malthusian pessimism, the past four or five decades have witnessed a worldwide breakdown of established livelihood strategies, often triggering environmental degradation. Some of these situations involve a relatively low human population (e.g. parts of

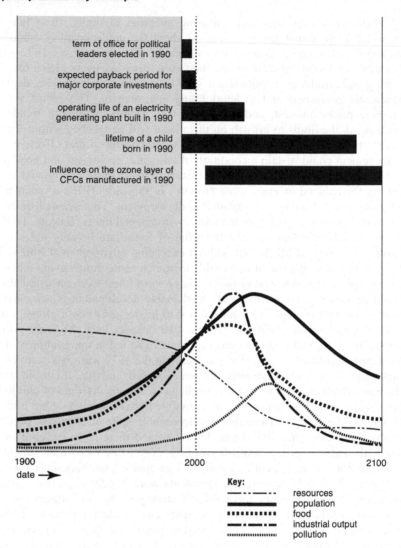

term of office for political
leaders elected in 1990

expected payback period for
major corporate investments

operating life of an electricity
generating plant built in 1990

lifetime of a child
born in 1990

influence on the ozone layer of
CFCs manufactured in 1990

1900   2000   2100
date ➤

Key:

— · — · — resources
▬▬▬▬▬ population
▪▪▪▪▪▪▪▪▪ food
▬ · ▬ · ▬ industrial output
▪▪▪▪▪▪▪▪▪▪ pollution

Figure 2.1 Time horizon of the World3 model

Source: Meadows *et al*. (1992: 235, Fig. 8.1)

Australia). Transboundary pollution of otherwise virtually undisturbed areas is also becoming a threat.

There are too many humans, so environmental managers will have to establish what an ultimate optimum sustainable population is, and how it can best be reached. Some demographers argue that only if effective action is taken within the next decade or so will it be possible to achieve a sustainable population with humane approaches (Hartshorn, 1991: 401).

Sustainable development

Before the 1950s the prevailing viewpoint in the West was that humans could and should modify nature to their advantage, and that the environment was virtually limitless and resilient. Nature was to be studied, catalogued, tamed and exploited. The frontier was still 'open', with land to settle and relatively few signs of environmental stress, other than localised pollution and some loss of biodiversity. The pre-1960s outlook, still not fully extinguished, was essentially mechanistic – that nature was relatively easy to understand, model and control – like 'clockwork', and there was little awareness of the complexity, vulnerability and limitations of the Earth's ecosystems. In 1965 US Ambassador to the UN, Adlai Stevenson, popularised the catch phrase 'Spaceship Earth', which became an icon for many. It represented the world as a fragile, unique, closed environment in which first-class passengers (the developed countries and other rich people) were greedy and profligate, and the more numerous lower-class passengers (the poor, mainly non-Western, nations – who each consumed far less, but *en masse* caused growing stress), were multiplying beyond the life-support capacity. Neither first- nor lower-class passengers were in control and both were vulnerable to each other and increasingly to natural hazards.

Although the concept was first voiced in the early 1970s (e.g. by the World Council of Churches in 1974), serious interest in sustainable development was limited before publications such as *The Limits to Growth* in 1972 and Schumacher's *Small is Beautiful* in 1973. After that it seemed to offer a way to heed limits *and* develop (have economic growth) – preferable to 'zero growth' (Meadows *et al.*, 1972). Sustainable development has three component goals: economic development (especially poverty reduction); social development; environmental protection. The Brundtland Report greatly boosted interest in sustainable development (World Commission on Environment and Development, 1987), and the concept is now so well established that it is unlikely to pass out of fashion. Sustainable development is now a key goal for environmental management. A huge diversity of agencies and groups are now 'dancing to the same sustainable development tune' (with varying degrees of sincerity and effort), following the greening of politics in Western countries in the 1980s (Adams, 2001: 1–3).

There are two overlapping areas of debate about sustainable development: one focuses on its meaning, the second on practical aspects (implementation). A goal of sustainable development may be used to help integrate diverse interests that would probably not otherwise co-operate. Even if it is achieved in only a limited way, it may nevertheless prove to be a valuable 'guiderail for development'. There are parallels with judges seeking justice, citizens wanting liberty, and philosophers and scientists pursuing truth – the goal may be elusive but efforts to reach it have to be maintained. A more pessimistic view is that of Dresner (2002: 4), who suggested that sustainable development marks the end of the West's faith in progress – a sort of post-industrial loss of confidence. There are a huge number of definitions of sustainable development. It is many things: a goal, a paradigm shift, above all difficult to achieve and often complex. A definition which has become well known is: 'to meet the needs of the present without compromising the ability of future generations to meet their own needs' (World Commission on Environment and Development, 1987) (some broad definitions of sustainable development were offered in Chapter 1 – see Box 1.2). Most definitions stress inter-generational equity (passing to future generations as much as the present enjoys) and intra-generational equity (sharing what there is between all groups).

Caution is needed: *sustainability* and *sustainable development* are not the same, but are often used as if they were. The former is the ongoing function of an ecosystem or use of a resource (i.e. maintenance of environmental quality). So, sustainability is the quantification of status and progress (environmental or social) and the goal of the sustainable development process (Becker and John, 1999: 22). Ecologists, fisheries managers, foresters and biologists developed measures such as carrying capacity and maximum sustainable yield by the 1940s – the idea being that an ecosystem can sustain a given level of demand. It should be noted that, even if demand is sustainable, un-expected environmental changes may upset things. Those seeking sustainability often assume steady demands, which may not be the case. Sustainable development to many implies increasing demands in order to improve well-being and lifestyles, and probably in the foreseeable future to cope with a growing population. Perhaps there will be some leeway if new technology, altered tastes and substitution of resources enable increasing demand to be met without greater environmental impact – many hope so.

Environmental economists often split sustainable development into two (unsatisfactory) extremes: *strong* and *weak*:

- *Strong* – belief that the existing stock of natural capital should be maintained or improved. Rejection of strategies such as substitution (e.g. not burning oil, which is non-renewable, and then invest some of the profit in sustainable energy sources such as wind generators). The same amount of natural capital is passed on to future generations. Human misery is acceptable as a cost of reaching sustainable development (Pearce and Barbier, 2000: 24; Schaltegger *et al.*, 2003: 23). This means that development must be based on natural capital that can be regenerated.
- *Weak* – the costs of attaining sustainable development are carefully weighed in human terms – unpleasant impacts are resisted, even if sustainable development is delayed or endangered. Substitution is possible – i.e. if need be it is permissible to trade natural capital through substitution (future generations receive about the same total capital, but it may have been changed). What cannot yet be substituted is protected. Broadly, this viewpoint concedes that existing economics and development strategies may be used (Neumayer, 2004).

Variants of the latter (weak) interpretation are currently dominant (mainstream), and in reality few hold to strong sustainable development.

There is also a split in opinion as to whether sustainable development can be achieved gradually, or demands rapid and radical change (e.g. of ethics, habits, economics). Sustainable development may be pursued at local, regional, national and supranational levels, the approach being 'top-down' or participatory. It is possible for a region, city, country or company to win a false sustainable development at the expense of somewhere else. For example, a town improves its sewage pollution by dumping it far away; or a poor nation may bear the environmental impacts of resource processing, while rich consumers do not – this is a sort of export of unsustainable development.

Sustainable development is widely held to have three goals: economic growth, environmental protection, and the health and happiness of people. Plenty of academics have noted the conflict within the concept of sustainable development – between wishing to remain within environmental limits and seeking growth or development. Supporters of sustainable development do not pursue environmental quality in isolation from addressing social disintegration and poverty. Tough environmental standards are not acceptable if they cause poverty or for richer people a resented decline in well-being. The question is: Can the ambitious goals of sustainable development be achieved in real-world situations and within environmental limits? It might be argued that it is better to set sights a little lower and pursue *survivability* rather than sustainability (i.e. a development approach which does not risk human survival).

Sustainable development is a prime objective of environmental management, but it is a challenge to find effective and workable strategies (Figure 2.2). Such strategies will frequently overlap and interact, so it is vital to ensure that they do not interfere with each other and, if possible, are mutually supportive – which necessitates both a local knowledge and strategic co-ordination, ultimately at the global scale. There have to be

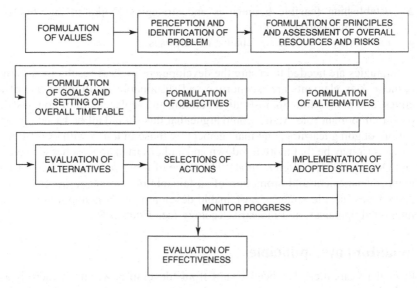

Figure 2.2 Major decision-making steps in a typical environmental management process
Source: Part-based on Matthews *et al.* (1976: 10, Fig. 1)

supportive human institutions, which must be resilient and adaptable to meet unforeseen challenges; there must be adequate information about the past, the present and the future (environmental, social, technical and cultural); and a drive and willingness to make sustainable development work. It is highly unlikely that all constraints and challenges will ever be fully assessed in advance, so resilience and adaptability are crucial to any strategy.

There has been no shortage of international meetings, agencies and NGOs promoting sustainable development since 1990 (Welford and Gouldson, 1993). The UN Commission on Sustainable Development was established in 1993 by the UN Economic and Social Council to follow up proposals made at the 1992 UN Conference on Environment and Development (the Rio 'Earth Summit' – widely seen to have had limited success); it is now promoting sustainable development but lacks 'teeth'. Other bodies include the World Business Council for Sustainable Development (http://www.wbcsd.org – accessed March 2005), and the International Institute for Sustainable Development (http://www.iisd.ca – accessed March 2005). The former is also widely considered to be ineffectual. *Agenda 21* (a 500-page-plus publication released in 1992 at the Earth Summit to act as a framework for achieving sustainable development), and derivative initiatives have helped promote sustainable development. Support is also provided by the Commission on Sustainable Development (Dresner, 2002: 46). In 2002 a World Summit on Sustainable Development was held at Johannesburg (WSSD, 2002).

Poverty is getting worse in some countries and the problems it generates undermine chances to achieve sustainable development. The factors leading to impoverishment are complex and the way forward is unclear – certainly part of the struggle will be to spread ethics which value sustainable development, foster productive social interaction and make better use of knowledge. Environmental management can support sustainable development by:

● identifying key issues;
● clarifying threats, opportunities and limits;

- establishing feasible boundaries and strategies (Nijkamp and Soeteman, 1988; Pearce and Barbier, 2002);
- monitoring to reduce the chance of surprises.

Strategies are needed to ensure the development process results in sustainable development. Key elements are: supportive and sustainable institutions; use of appropriate ethics; and motivation. So, institution building and replacement of non-sustainable ways play an important role. Many would argue that this should all be based on careful observation, ongoing learning, and management by those in touch with or even working with citizens (it may be that more top-down and authoritarian approaches also work) (Connor and Dovers, 2004; Lafferty, 2004). The range of tools and approaches for the measurement, management and promotion of sustainable development are growing. Promising approaches include strategic sustainable development (Nijkamp and Soeteman, 1988); integrated appraisal; and industrial ecology (see Chapter 9).

The 'polluter-pays' principle

Over the years there has been a shift from 'develop now, and if there is a problem – seek abatement and clean up', to 'avoid problems if possible' (see below on the precautionary principle and proactive approach). There has also been a shift from the burden of problems being borne by those affected, to its being handled by the public in general, or better still, to making the 'polluter-pay'. If forced to pay for errors the polluter is, hopefully, less likely to cause problems. It is also more just that bystanders, consumers or workers should not pay for developers' mistakes. In the past penalties for pollution were often hard to enforce and were relatively light; consequently, organisations motivated by profit would be tempted to 'push the envelope' and try to get away with sometimes getting caught and paying limited damages. There is also a risk that licensing and penalties for infringements will have similar outcomes. In an ideal world environmental managers educate and motivate potential polluters to seek genuinely to avoid polluting.

Sometimes environmental damage has already been done, and it becomes evident only years after; meantime, the body responsible has closed down or it is too late to use current law to claim damages. The 'polluter-pays' approach seeks to make it difficult for the responsible parties to escape damages and ensure the penalty is enough to deter. If this is done through licensing, the potential offender has to convince the authorities that difficulties will not arise (with a safe margin of error); thus risk assessment, hazard assessment and impact assessment become important.

The 'polluter-pays' principle is widely seen as a distinctly separate 'twin' of the precautionary principle (see next section). Impetus to adopt the 'polluter-pays' principle has been given by disasters such as Bhopal in 1984. In addition, the development of eco-efficiency ideas has further prompted adoption of the 'polluter-pays' principle – which enables waste and the costs of its control to be shifted to become useful by-products and recovered profits.

The precautionary principle

The art of precautionary (proactive) planning has evolved quite recently and is still being adapted to real-world conditions; consequently, it is easily side-stepped, or is applied too late to select the best development option, is misused or neglected, or simply lacks the power to identify impacts well enough. The precautionary principle has no precise definition (EU Commission document COM 2000 offers a twenty-nine-page definition); it has been described as 'institutionalised caution', and is constructed around the goal of preventing, rather than reacting to, environmental harm (Applegate, 2000). Cynics see it as a manifestation of bureaucrats covering their backs. Supporters see it as crucial for policies leading to sustainable development. Over the past twenty years it has been part of a number of international disputes and debates (Gee *et al.*, 2002). The precautionary principle has (according to Kriebel *et al.*, 2001) four central components:

1   taking preventive action in the face of uncertainty;
2   shifting the burden of proof to the proponents of a development;
3   exploring a wide range of alternatives to try and avoid unwanted impacts;
4   increasing public participation in decision making.

Acceptance of the precautionary principle in environment, healthcare, economic and other policy implementation means that regulatory action can precede full scientific certainty about an issue – lack of evidence is no reason for inaction (Harremoës *et al.*, 2002). Consequently it risks costs which may not be justified, inappropriate responses, and accusations of 'crying wolf' or delaying development. The precautionary principle is widely accepted in European and international law, but is less well established in the USA, although elements of it are familiar there. Because there is no universal agreement on a definition its status is one of a broad approach, rather than a firm and precise principle of law. The 1992 Rio Declaration (Principle 15) of the UN Conference on Environment and Development (the Rio 'Earth Summit') urged widespread use of the precautionary principle (O'Riordan and Cameron, 1994). Following the precautionary principle is not costless. For example, in many situations some things have to be forgone to keep open escape options (Earll, 1992; Pearce, 1994: 1337).

In the last few years there have been a lot of appraisals of the use of the precautionary principle; a few suggest it be discarded for all but general policies (Keeney and von Winterfeldt, 2001), although most accept that it is valuable when serious, possibly irreversible, impacts are likely. A precautionary principle approach is useful for social development as well as environmental issues, especially in developing countries, where poor people have little in the way of security to fall back upon if measures such as land reform or agricultural innovation fail – efforts have to be right first time or some preparations should be made to provide prompt aid if there are problems. There is also the question of how much a society can afford to pay to support a precautionary principle approach. In practice (for example, in trade–environment disputes), its application is usually triggered by a risk assessment.

Although the USA has lagged behind some other countries and international law in accepting the precautionary principle, it has nevertheless played a pivotal role in promoting some key tools which support the precautionary principle; for example: environmental impact assessment (EIA) and social impact assessment (SIA). Impact assessment effectively forces developers to 'look before they leap' and, if problems are anticipated, to delay acting until there can be effective avoidance or mitigation (Applegate, 2000: 421). The USA has also promoted legislation, which seeks to ensure that there are 'margins of safety' for the environment and human well-being. So, one aspect of the precautionary principle approach is to ensure margins of safety when designing technology, certifying drugs or pesticides, developing areas at risk from natural hazards, setting standards for pollution measurement, and much more. Clearly, in many fields, margins of safety are far from reliably established, so precautionary planning is not a precise art.

The precautionary principle approach demands 'upstream thinking' – looking for underlying causes of problems, rather than fixing on symptoms and 'backcasting' – visualising a desirable future scenario, identifying likely constraints, and then deciding what must be done now to move towards such a situation. Backcasting must not be confused with hindcasting – using historical and palaeoecological evidence to understand past events and human responses to help forecasting and assess reactions to problems (note that forecasting starts from today's situation and projects current solutions and responses into the future) (Robért, 2000). Lomborg (2001: 348) warned that the precautionary principle may be undesirable, because it encourages pessimism, which can cause planners to abandon a proposal rather than go ahead and build in what they think is 'a margin of safety'. He also warned that funds spent because of unsound use of the precautionary principle could mean less to spend on other things. There is also a possibility that environment and development assumptions are incorrect and could prompt inappropriate policies; Lomborg (2001: 31) noted: 'It cannot be in the interests of our society for debate about such a vital issue as the environment to be based more on myth than on truth.' Another problem with the precautionary principle is that it can be anti-democratic, because it demands action before a law or regulation has been broken or damage done without the state or anyone else necessarily proving there is a problem.

All of this presents the environmental manager with a dilemma: there is often a need to act before knowledge and proof is available, which is something scientists have shunned in the past. It means that risky decisions have to be made. Politicians, financiers and most professionals are reluctant to take such risks. So, whenever possible, efforts are made to find 'win–win' paths (i.e. which pay off beneficially even if predictions prove to be wrong). Sometimes the pay-off is direct and sometimes a useful opportunity is created. For example, a carbon-sink forest will have amenity, conservation and timber value, even if global warming proves to be a false alarm (Karagozoglu and Lindell, 2000).

Phillimore and Davidson (2002) provided a case study of the application of the precautionary principle – the 'millennium bug' (Y2K) experience. In this, they ask whether, given the minimal disruption which actually occurred, precautionary expenditure of huge sums of money to address fears was worth while. Little research has been done on the misjudgement of the Y2K threat, yet it probably cost over US$580 billion (Ravetz, 2000a). Another interesting aspect of Y2K was that bodies, which often reject a precautionary principle approach, spent huge sums using it to counter a threat. Organisations believed that there were effective technical solutions and that it was a discrete issue – they were clearly told what might happen and the computer industry is powerful and persuasive. Advocacy for many environmental management problems is likely to be less persuasive and the issue is unlikely to be so well defined.

While governments may prepare against perceived threats posed by non-sudden events, studies suggest that, even if they are known to be likely, sudden events (e.g. floods, earthquakes) tend to be dealt with after they have happened (Dery, 1997). (Reviews of precautionary principle applications may be found in the *Journal of Risk Research* or *Journal of Environmental Law*.)

Environmental management challenges

Environmental management involves making decisions (Figure 2.2). How these are made depends on whether a technocratic or a consultative (bottom-up) model is adopted. The latter has become the usual pattern in the USA and Canada, and is increasingly being chosen in Japan and Europe, reflecting the trend towards democracy and free-dom of information. Whatever the overall approach, environmental management is, as Matthews *et al.* (1976: 5) noted, a 'myriad of individual and collective decisions by persons, groups, and organisations', and 'together these decisions and interactions constitute a process – a process that in effect results in management (unfortunately, sometimes mismanagement) of the environmental resources of a society'.

Of the many problems that beset environmental management (Chapter 10 explores key resources, and Chapter 11 global challenges), inadequate data is a common hindrance: there are still huge gaps in knowledge of the structure and function of the environment, the workings of global, regional and local economics, and of how soci-eties and individual humans behave. The ideal is adequate data that may be presented in real time, so that the scenario can be observed as it changes. With improved computers, software and the development of tools such as geographical information systems (GIS), this may one day be possible, but often all that is available today is an occasional, incomplete snapshot view (i.e. limited in time and space, which can be misleading). Decision making is often made difficult by politics; lobbying; media, public and NGOs' attention; lack of funding and expertise. Environmental managers are faced with two temporal challenges: (1) problems may suddenly demand attention and allow little time for solution; (2) the desirability that planning horizons stretch further into the future than has been usual practice. Decisions are easier to make and policies more easily adjusted if there is time available – for example, a 3°C climate change over a hundred years may not be too much of a challenge, but if it happens over twenty years it would be (Chiapponi, 1992). Predictions are difficult enough with stable environ-ments, but many are unstable and some are becoming uncertain; once the stability has been upset there may be unexpected and sudden feedbacks or shifts to different states, all of which are difficult to forecast. The behaviour of economic systems is even more challenging to predict, and human behaviour is especially fickle, with tastes and atti-tudes often suddenly altering. The unpredictability and rapidity of challenges prompted Holling (1978) to argue for adaptive assessment and management.

The need to be adaptable and to seek to reduce human vulnerability

Human development over the past several thousand years has enjoyed relatively stable and benign environmental conditions. Modern societies have not even had to face rela-tively gradual and limited challenges, like that of the Little Ice Age (*c.* AD 1500 to 1750), but this is unlikely to last. The rapidly increasing human population is placing more and more stress on the environment, which could trigger sudden changes. Estimates place world human population between twenty and twenty-six million about 1,000 years ago; by AD 1500 it had probably risen to between 400 and 500 million; now it has exceeded 6,500 million (McNeill, 2000: 7). That is a lot of mouths to feed if there are poor harvests. Globalisation of trade and complex technology means that a disaster in one nation can have worldwide impacts; for example, disruption of computer chip production in the Far East soon affects Europe. World food supplies are increasingly obtained from a few key regions. The challenges faced by development management and environmental management are growing fast.

Environmental management is not just about coping with challenges; it has to model and monitor to gain sufficient knowledge and give early-warning signs to have any chance of coping. Some threats are random and difficult, if not impossible, to recog-nise in advance; others develop in an insidious way and may be easily overlooked.

Worse, a problem may have indirect and cumulative causes – a number of unrelated factors suddenly conspire to cause trouble, or a process develops a positive or negative feedback which (respectively) quickly accelerates or slows down developments.

The need to be multidisciplinary and integrative

Environmental management has to deal with humans and natural processes, and it has to cope with changes of fashion, economic variations, changing technological capabilities, alterations of attitudes, social capital, social values, skills, confidence, and many other variables. Monitoring and responses have to be multidisciplinary to recognise challenges and determine how environment, biota and people will be affected and react, and to weigh up the best way to cope. In the past environmental management was mainly practised by those with a science background (e.g. environmental scientists, ecologists, pollution specialists, technologists), and those concerned with monitoring and enforcement. There has been a very marked broadening out during the past decade or so, to the extent that environmental management is now more than half staffed by social studies specialists. In academic institutions sociologists, anthropologists, economists, geographers, physical scientists, planners and engineers come together increasingly in departments of environmental management and work as teams.

For effective environmental management there must be the means of resolving controversies regarding proper conduct (Cairns and Crawford, 1991: 23). Ethics can guide this. Ethics may be defined as a system of cultural values motivating people's behaviour (Rapoport, 1993). It draws upon human reasoning, morals, knowledge of nature and goals to act as a sort of plumb line for development and help shape a worldview. Ethics operates at the level of individuals, institutions, societies and internationally. Some have blamed Judaeo-Christian ethics for the tendency over the past 2,000 years for Western peoples to see themselves as having dominion over nature, and to pursue strategies of exploitation rather than of stewardship (White, 1967). In the West since the sixteenth century there gradually evolved utopian ideas, whereby it was no longer considered blasphemy to look for the improvement of human conditions through social development, 'civilisation' and technology. Especially in Britain, France, Holland and the German states from the early eighteenth century conditions were more liberal and democratic, and there was a respect for rational, objective questioning and a gradual rejection of superstition – an enlightenment or age of reason. From the late sixteenth century what has been termed the Protestant ethic spread through the West – broadly this encouraged the individual to be responsible for self-improvement through good acts and hard work (Weber, 1958; Hill, 1964), the stress being on encouragement. Because the predominant attitude remained one of *laissez faire*, states hesitated to dictate economic and other policies too closely. By the 1930s the Soviet Union, fascist regimes (e.g. Germany), and even briefly the USA (with the Tennessee Valley Authority in 1933) explored state manipulation of development. However, few tried to shift *laissez-faire* attitudes towards environmental management before the 1960s. It was really left to Western (at first mainly American) environmental activists in the 1960s and 1970s to prompt the quest for new development and environment ethics (Cheney, 1989; Dower, 1989; Barrow, 1995a: 14–16). New books, lobby groups, journals (e.g. *Environmental Ethics*; *Ethics and Behaviour*; *Ethics, Place and Environment*; *Environmental Values*; *Science, Technology and Human Values*) and NGOs were spawned and have continued to appear. Some environmental ethics literature has come from non-academic sources: businesses, religious thinkers, and others. Carley and Christie (1992: 78) tried to summarise the range of environmental ethics, dividing them into four groups:

1   *Technocratic* environmental ethics = resource-exploitative, growth-oriented;
2   *Managerial* environmental ethics = resource-conservationist, oriented to sustainable growth;
3   *Communalist* environmental ethics = resource-preservationist, oriented to limited or zero growth;
4   *Bioethicist* or *deep ecology* environmental ethics = extreme preservationist, antigrowth.

Group 1 is anthropocentric and places faith in the capacity of technology to overcome problems. Group 4 is unlikely to attract support from enough people to be a viable approach, and offers little guidance to environmental managers. Carley and Christie felt the ethics of groups (2) and (3) were more likely to support sustainable development and provide guidance for environmental management.

Another grouping of environmental ethics is:

1   *Anthropocentric* – human welfare is placed before environment or biota;
2   *Ecocentric* – focused on ecosystem conservation (holistic outlook);
3   *Biocentric* – organisms are seen to have value per se.

Inevitably, the first of these three groups predominates.

How will environmental management achieve its policy goals? Probably through a mix of moral pressure, the spread of appropriate ethics, and by ensuring that governments, citizens, economics, business and law are sufficiently sensitive to the needs of the environment. Environmental management will need to make use of education and the media to alter social attitudes so that there is awareness of environmental issues and an acceptance of a new ethics. It will also have to draw upon other fields to achieve its goals, and must develop effective institutions. Manuals, checklists, conventions and agreements can help guide the identification of goals and preparation of action plans and their implementation. The reductionist approach of splitting problems into component parts for study and solution lies at the core of Western rational, scientific study (which the modern world owes a great deal to). Some feel that a holistic 'overall view' approach should replace 'compartmentalised and inflexible science'; that is a mistake – there is a need for both without weakening rationalism (Risser, 1985; Savory, 1988; Atkinson, 1991a: 154; Rapoport, 1993: 176).

Sustainable development calls for trade-offs. For example, it may be necessary to forgo immediate benefits to secure long-term yields – which may far outweigh the former. Such trade-offs can be a cruel choice for individuals, groups or countries, and a minefield for the environmental manager. For poorer nations, foreign aid could be focused to cushion trade-offs. Institutional problems present difficulties for environmental management as much as technical or scientific challenges (Cairns and Crawford, 1991). Human institutions change and can be difficult to understand and control, and building new ones may be hard. It is vital that the institutions involved in environmental management are effective. Even if there is technology and funding and a will to solve a problem, success will be unlikely without the right type of effective sustainable institutions.

A growing number of social scientists have been focusing on institution building, which gives some grounds for optimism. However, a key international body charged with environmental management, the United Nations Environment Programme (UNEP), was designed in the 1960s (founded in 1973) and needs remodelling to be more effective (Von Moltke, 1992). The UNEP was located away from Paris, Geneva, New York and the Pacific Rim, in Nairobi, which has had mixed results. The location is rather off the beaten track and probably partly explains its poor funding and lack of power, but it does help counter a Western bias and represent developing countries. To be fair to institutions such as the UNEP, they must rely on the quality of their arguments to convince countries and multinational companies (MNCs) or transnational companies (TNCs) to accept a strategy, and have been given little in the way of sanctions to enforce policies.

Summary

- This chapter seeks to clarify the meaning of 'development', 'sustainable development' and 'environmental management', and explores how they interrelate.
- The world and its resources are finite, yet human demands continue to increase. The ultimate goal of environmental management is to address this issue and to seek sustainable development.
- Modern humans are more numerous than at any point in the past and are less adaptable.
- There is no one single approach to environmental management, but there are key concepts.
- Environmental management has many tools to choose from. These are often still evolving and may not be tuned to non-Western country needs and new challenges. Environmental managers have to select suitable strategies and tools best suited for a given situation.
- A precautionary and proactive approach is wise if sustainable development is a serious goal, and because humans appear to be more vulnerable than many admit.

Further reading

Dresner, S. (2002) *The Principles of Sustainable Development.* Earthscan, London.
A good introduction, especially to social and ethical aspects.
Kirkby, J., O'Keefe, P. and Timberlake, L. (eds) (1995) *The Earthscan Reader in Sustainable Development.* Earthscan, London.
Selected readings on sustainable development.
Meadows, D.H., Randers, J. and Meadows, D.L. (2004) *Limits to Growth: the 30-year update.* Earthscan, London.
A pragmatic, at times optimistic, up-to-date summary of the limits to growth debate.

WWW sources

International Institute for Sustainable Development http://www.iisd.org (accessed March 2005).
Sustainable Development Communication Network http://sdgateway.net (accessed February 2005).
UN Division for Sustainable Development http://www.u.org/esa/sustdev (accessed February 2005).

3 Environmental management and science

> Because science carries us toward an understanding of how the world is, rather than how we would wish it to be, its findings may not in all cases be immediately comprehensible or satisfying.
>
> (Sagan, 1997: 31)

It is important to understand the structure and function of the environment to be able to assess the impacts of human activities and the viability of development efforts (Adger *et al.*, 2004: 21–24). Knowledge about the Earth, its organisms and human affairs is incomplete and data collection has often been inadequate, so forecasting and decision making are frequently far from perfect. Nevertheless, compared with the situation before the International Geophysical Year (1957–1958), there is now much more understanding of the environment and humans, and a vastly improved ability to monitor and forecast.

Environment and environmental science

When environmental management makes use of science it can adopt one of two broad approaches: (1) multidisciplinary – which involves communication between various fields but without much of a breakdown of discipline boundaries; (2) interdisciplinary (even holistic) – the various fields are closely linked in an overall, coherent way. The interdisciplinary approach is widely advocated as a cure for the fragmentation of science (what some would see as unwelcome compartmentalisation), but of the two it is much the more difficult to achieve (De Groot, 1992: 32). Environmental science often has to be problem oriented, and this helps promote attempts at multidisciplinary and interdisciplinary study.

Environment may be defined as the sum total of the conditions within which organisms live. It is the result of interaction between non-living (abiotic) – physical and chemical – and living (biotic) components. Interest in the interaction of organisms, including people, with one another and with their surroundings, was stimulated by the publication of *The Origin of Species by Means of Natural Selection* (1859) by Charles Darwin. 'Natural environment' is often used to indicate a situation where there has been little human interference, and 'modified environment' where there has been significant alteration (see Figure 3.1). However, nowadays little of the world is a wholly natural environment. Many organisms alter the environment, and the change they cause may be slow or rapid, localised or global. In the past few thousand years, humans have become such a major force in modifying the Earth's ecosystems that an environmental scientist recently suggested the current geological unit, the Holocene, should be succeeded by the Anthropocene or 'human-altered' period. Much of the alteration is unwitting degradation rather than improvement; however, humans have the potential to recognise and to respond consciously and appropriately to opportunities and threats. Whether we will successfully exploit that potential remains to be seen. It is environmental managers who will play a key part in prompting and supporting a better response. If environmental management is to develop strategies and exploit opportunities effectively it must be much more than applied science; it is also an art which requires understanding of human–environment interactions, considerable management skills, diplomacy and powers of persuasion (Figure 3.2).

There have been attempts to establish ecologically sound planning and management since the 1960s (McHarg, 1969; Dasmann *et al.*, 1973; Aberley, 1994). Before the 1990s social studies and sciences found it undesirable and difficult to communicate effectively because they had different traditions and languages. Today there are stronger links

Figure 3.1 Cape Disappointment, South Georgia. A relatively simple flora and fauna, which, with the exception of larger marine mammals, has been relatively little disturbed by humans, and so offers opportunities for ecosystem studies.

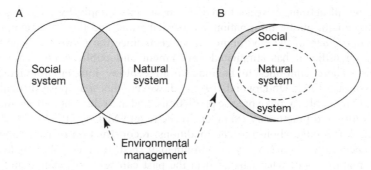

Figure 3.2 Natural system and social system relationship

Note: The social system is likely to affect the management of the natural system – usually the relationship of the two systems is visualised as in (a), with the shaded section representing the main field of activity for environmental management. However, few environments are 'natural': most are to some extent altered by human activity, so the social system and natural system are not largely independent – the pattern shown in (b) is more likely.

Source: Bormann *et al.* (1993: 3, Fig. 1.1)

between environmental science and social studies. Much of this broadening of environmental interest has taken the form of crusading and advocacy; nevertheless, there has also been development of new tools, concepts and practical approaches.

Environmentalism is a generic term for a range of interests directed at achieving better environmental management (the environmental movement and environmentalism are examined in Chapter 4). It must be stressed that while many environmentalists listen to scientific reason, others take little heed or strongly oppose it. Environmental managers may sometimes be confronted by less rational environmentalists who present their interests as 'scientifically sound', and in so doing degrade scientific rigour and truth. Efforts must be made to withstand misapplication, the lobbying of special-interest groups, and demands of policy makers. Environmental management has to be practised in the real world, and it has to sell important issues to people who decide policies. There will be demands for firm answers that may be difficult to come by, and a public which can switch off from crucial issues that fail to catch their eye. For a study of how people respond to threat warnings, see Posner (2005); and for discussion of how individuals view nature and risk, see Maslin (2004: 36–42).

Those involved in environmental management each have their own worldviews, which affect how they proceed. Environmental managers, whatever their worldview, are likely to face: (1) data problems; (2) modelling difficulties; (3) analytical difficulties; (4) insufficient time for adequate research; (5) lobbying from various stakeholders; (6) funding limitations. For example, there may be little baseline data, and what there is may be inaccurate, have gaps, or may be in an unsuitable form. Models may not have been developed or may have deficiencies. Modelling cannot be effectively applied to random processes. The problem under study may also be complex and difficult to understand. Increasingly, environmental scientists are asked to provide advice before they have proof (Funtowicz and Ravetz, 1993). Faced with uncertainty, it makes sense to try to adopt the precautionary principle (see Chapter 2), making recommendations that enhance adaptability to the unforeseen. The precautionary principle generally means that the burden of proving the case for development is shifted to the developer, who must increasingly use science to show that a proposal is safe before proceeding. O'Riordan (1995: 9) argued that 'prevention is simply a regulatory measure aimed at an established threat. Precaution is a wholly different matter. It introduces the duty of care on all actions, it seeks to reduce uncertainty simply by requiring prudence, wise management, public information and participation, and the best technology.'

While much of the environmental activism since the 1960s has been more messianic than scientific, it has stimulated government and public concern for nature (Bailey, 1993). There continues to be input from ecology into environmental management (Troumbis, 1992; Underwood, 1995). However, this may pose difficulties: ecologists are often unable to make precise predictions, and are reluctant to trade thorough research for utility and practicality (Shrader-Frechette and McCoy, 1994: 294–295). In addition to the knowledge about present conditions, reconstruction of past conditions (through palaeoecology, archaeology, environmental history and other fields) is valuable. Information about what happened in the past can warn of change and hazards, establish trends, and suggest possible future scenarios and human reactions.The expression 'backcasting' has been applied to such studies (Mitchell, 1997: 99). It is also possible that the study of other planets may yield knowledge useful for managing the Earth's environment.

There have been a number of developments which aid environmental science and environmental management; these include:

- growing international co-operation;
- standardisation of measurements and definitions;
- remote sensing and computing/data processing advances;
- the diminishment of Cold War rivalry and restrictions;
- the spread of the Internet which facilitates exchange of information and makes it hard for individuals, companies or national authorities to hide environmental problems;
- improved communications between environmental science and social studies.

New approaches to environmental science and environmental management

There has been a marked trend towards supporting holistic approaches in recent years. Smuts (1926) proposed the concept of holism in the 1920s. Eighty years on modern holism is still poorly defined, although it implies acceptance that 'the whole is greater than the sum of the parts' and the idea that modern science has unwisely tended towards excessive reductionism (the standard, modern, scientific view that everything is explainable from the basic principles, and by focused, objective research), empiricism (use of data to prove a case) and compartmentalisation (isolation of fields of study from each other). In short, holistic research seeks to understand the totality of problems rather than their components. In all but the simplest environments problems tend to be so complex that an effective holistic approach is difficult. There are situations where a holistic approach is to be welcomed; unfortunately, there are many situations where it will not work and there are dangers in over-enthusiastic use (in 1998 the University of Plymouth, UK, launched an M.Sc. in holistic science). As already stressed, established 'reductionist' science has yielded a great deal – modern society owes its well-being to it – so it is very unwise to think of wholly abandoning it (Atkinson, 1991a: 154). With pressures for holistic approaches and popular interest in pseudo-science and anti-scientific theories, commonly presented by the media as truth, care is needed to ensure that support for science is not eroded. Popular pressure also tends to polarise support for research – some fields are attractive to citizens and politicians, and others (even though they may be vital) are not. Another pressure is the growing demand, and therefore funding, for applied research rather than studies into what has no obvious practical outcome.

Ironically, many of the practical benefits we enjoy have been generated through pure, not applied, research.

Various environmental researchers argue that it is possible to recognise a postmodern period, beginning in the early 1960s, characterised by a collapse of 'normality'. Increasingly, post-industrial activity, and reduction of confidence in human ability to control nature, have increased environmental awareness and support for a holistic worldview (Kirkpatrick, 1990; Stonehouse et al., 1997). In recent decades some mathematicians, fundamental physicists and other scientists have shifted from approaches based on Cartesian order and systematic, reductionist analysis, to trying to understand chaotic complexity using postmodern holism that embraces chaos theory (Cartwright, 1991), fractals, and other new ideas (Lewin, 1993). Interest in a holistic approach affects a range of subjects, and may prove useful when it is difficult to maintain a separation between science and politics, a point made by Bond (*New Scientist*, 30 May 1998: 54), who noted 'science without the bigger picture is simply bad science'. Holism is valuable as a support for established science but must not be seen as a replacement and must be treated with some caution.

Structure and function of the environment

Since the early 1970s popular texts have occasionally published 'laws of ecology' (often based on those published by Commoner, 1972); three of these are as follows, with environmental management implications in brackets:

1   Any intrusion into nature has numerous effects, many of which are unpredictable (environmental management must cope with the unexpected).
2   'Everything is connected'; therefore, humans and nature are inextricably bound together and what one person does affects others and a wider world (environmental management must consider chains of causation, looking beyond the local and short term).
3   Care needs to be taken that substances produced by humans do not interfere with any of the Earth's biogeochemical processes (environmental management must monitor natural processes and human activities to ensure that no crucial process is upset).

Living organisms, including humans, and non-living elements of the environment interact, frequently in complex ways. Ernst Häckel founded the study of these interactions – ecology – as an academic subject (oecology) in 1866. In 1927 Charles Elton described ecology as 'scientific natural history'. Modern definitions include: the study of the structure and function of nature; the study of interactions between organisms (biotic) and their non-living (abiotic) environment; the science of the relations of organisms to their total environment (Fraser-Darling, 1963; Odum, 1975; Park, 1980). Synecology is the study of individual species–environment linkages – and autecology is the study of community–environment linkages. Ecology is often a guide for environmental management, environmentalism and environmental ethics, suggesting limits and opportunities, and providing many key concepts and techniques (e.g. carrying capacity). Since the early 1970s 'ecology' has also come to mean a viewpoint – typically a concern for the environment – as much as a discipline (O'Riordan, 1976).

Humans either adapt to, or seek to modify, their environment to achieve security and well-being or to satisfy greed and cultural goals. In making modifications people create a 'human environment' (Treshow, 1976). Human ecology developed in the early twentieth century to facilitate the study of people and their environment, expanding in the 1960s and 1970s, and then dying back (Sargeant, 1974; Richerson and McEvoy, 1976; Marten, 2001). A field that currently seems to be expanding, and which can be very useful for environmental management, is political ecology. Political ecologists seek to build foundations for sustainable relations between society and the environment in the real world (Blaikie, 1985; Atkinson, 1991b).

The global complex of living and dead organisms forms a relatively thin layer, the biosphere. The term 'ecosphere' is used to signify the biosphere interacting with the non-living environment, biological activity being capable of affecting physical conditions even at the global scale; for example, through the formation of oxygen, and the sequestration carbonates in the oceans. The global ecosphere can be divided into various climates, the pattern of which has changed in the past and will doubtless do so in the future. Climate may be affected by one or more of many factors, including:

- variation in incoming solar energy due to fluctuations in the Sun's output or possibly dust in space;
- variation in the Earth's orbit or change in its inclination about its axis;
- variation in the composition of the atmosphere: alterations in the quantity of dust, gases or water vapour present (which may be caused by factors such as biological activity, human pollution, volcanicity, and impacts of large comets or asteroids);
- altered distribution of continents, changes in oceanic currents, or fluctuation of sea-level that may expose or submerge continental shelves;
- formation and removal of topographic barriers;
- environmental managers must not assume that climate is fixed and stable (Figure 3.3).

Figure 3.3 A glacier calving into the sea, Cumberland Bay, South Georgia. Evidence shows considerable change in extent of glaciers on this island over the past 10,000 years. Climate is not static.

Climate is by no means the only ecosystem variable affecting human fortunes. Some of the greatest losses of human life have resulted from epidemic diseases; new forms of these may appear through evolution, altered communications and other shifts. Human societies have repeatedly been affected by earthquakes; the migration or decline of fish stocks; volcanic eruptions (the Toba explosive eruption in Sumatra *c.* 74,000 BP may have come close to exterminating *Homo sapiens* – Oppenheimer, 2003: 72–78); and tsunamis. Environmental management should consider the threat of infrequent but severe events, and whenever possible steer development to reduce human vulnerability and conserve biodiversity and cultural riches.

Trophic level and organic productivity

Organisms in an ecosystem may be grouped by function according to their trophic level – the position in the food supply chain or web at which they gain nourishment. Each successive trophic level's organisms depend upon those of the next lowest for their energy requirements (food). The first trophic level, primary producers (or autotrophs), in all but a few cases convert solar radiation (sunlight) into chemical energy. The exceptions which do not depend on sunlight include hydrothermal-vent communities and some micro-organisms deep below ground level. Seldom are there more than four or five trophic levels because organisms expend energy living, moving, and in some cases generating body heat – and transfer of energy from one trophic level to the next is unlikely to be better than 10 per cent efficient. Given these losses in energy transfer, it is possible to feed more people if they eat at a low rather than high trophic level. Put crudely, a diet of grain supports a bigger population than would be possible if it were used to feed animals for meat, eggs or milk (it has been calculated that only about one part in 100,000 of solar energy makes it through to a carnivore).

The sum total of biomass (organism mass expressed as live weight, dry weight, ash-free dry weight or carbon weight) produced at each trophic level at a given point in time is termed the standing crop. This needs to be treated with caution; if taken at the end of an optimum growing period it indicates full potential; if taken during a drought, cool season, period of agricultural neglect or insect damage, it is an underestimate of possible production. Primary productivity may be defined as the rate at which organic matter is created (usually by photosynthesis, although in some situations by other metabolic processes) at the first tropic level. It may be established in several ways. The total energy fixed at the first trophic level is termed gross primary production. Minus the estimated respiration losses, this gives net primary productivity (in $g\ m^{-2}\ d^{-1}$ or $g\ m^{-2}\ y^{-1}$). Net primary productivity gives a measure of the total amount of usable organic material produced per unit time. Most cultivated ecosystems, i.e. efforts to stretch food and commodity production, are well below the net primary production of more productive natural ecosystems. There is thus, in theory, potential for the improvement of existing agriculture.

Ecologists have developed a number of concepts and parameters, some of which have been adopted by those seeking to manage the environment. The most widely used are maximum sustainable yield and carrying capacity (Box 3.1). These should be treated with caution. Maximum sustainable yield may be correctly calculated, but if the environment changes a reasonable resource exploitation strategy could lead to over-exploitation. Maximum sustainable yield calculations can thus give a false sense of security. A given ecosystem may have more than one carrying capacity, depending on factors such as the intensity of use and the technology available. Some organisms, including humans, adjust to their environment through boom and bust, feeding and multiplying during good times, and in bad suffering population decline, migrating or hibernating; calculating carrying

Box 3.1

Ecological concepts and parameters which are useful for environmental management

- *Maximum sustainable yield*
  The fraction of primary production (as organic matter) in excess of what is used for metabolism (net primary production) that it is feasible to remove on an ongoing basis without destroying the primary productivity (i.e. 'safe harvest'). Under US law, maximum sustainable yield would be defined as: maintenance in perpetuity of a high level of annual or regular periodic output of renewable resources.

- *Carrying capacity*
  Definitions vary and can be imprecise. Examples include: the maximum number of individuals that can be supported in a given environment (often expressed in kg live weight per km^2); the amount of biological matter a system can yield, for consumption by organisms, over a given period of time without impairing its ability to continue producing; the maximum population of a given species that can be supported indefinitely in a particular region by a system, allowing for seasonal and random changes, without any degradation of the natural resource base.

- *Assimilative capacity*
  The limiting resource may not be an input such as food or water, it may be inability to deal with outputs (waste products). A given environment has some capacity to purify pollutants up to a point where the pollutant(s) hinder or wholly destroy that capacity – this is termed the assimilative capacity.

capacities for such situations can be difficult. Biogeophysical carrying capacity may differ from the behavioural carrying capacity, such that a population could be fed and otherwise sustained but feel crowded and stressed to a degree that limits their survival.

The more people the Earth supports, the lower the standard of living they are likely to enjoy, and the more conflict and environmental damage are probable. However, there may be situations where human population increase does not exacerbate environmental degradation or lower standards of living (see discussion of Boserüp in Chapter 2). With foreseeable technology, sustaining adequate standards of living and satisfactory environmental quality probably demand that human population on the Earth be less than today's 6,500 million plus.

The carrying capacity of an ecosystem may be stretched by means of trade, human labour and ingenuity, technology and military power (the latter ensures tribute from elsewhere – assuming it is available to be taken). Net primary productivity often increases at the cost of species diversity. The timing of resource use may be crucial: for example, rangeland might feed a certain population of livestock, provided that grazing is restricted during a few critical weeks (at times when plants are setting seed, becoming established or are otherwise temporarily vulnerable). If this is not done, or a disaster like a bushfire strikes, land degradation occurs and far fewer livestock can be supported in the future. Within even the simplest ecosystems there are complex relationships among organisms and between organisms and environment. There are often convoluted food webs; complex pathways along which energy (food) and perhaps pollutants are passed; subtle interdependencies for pollination, seed dispersal and so forth.

Certain pesticides, radioactive isotopes, heavy metals and other pollutants can become concentrated in organisms feeding at higher trophic levels, so that apparently harmless background contamination could, through such biological magnification (bio-accumulation), prove harmful to man and other organisms without assimilative capacity having obviously broken down.

The ecosystem

The biosphere is composed of many interacting ecosystems (ecological systems), the boundaries between which are often indistinct, taking the form of transition zones (ecotones), where organisms from adjoining ecosystems may be present together. It is possible for some organisms to be restricted to an ecotone only. Large land ecosystems or biomes (synonymous with biotic areas) are areas with a prevailing regional climax vegetation and its associated animal life, in effect regional-scale ecosystems. Biomes, such as desert biomes or grassland biomes, often mainly reflect climate, but can also be shaped by the incidence of fire, drainage, soil characteristics, grazing, trampling and so on (Watts, 1971: 186).

Ecosystems have long been recognised as environmental or landscape units (e.g. the *maquis* scrubland of southern France or the *taiga* forests of Siberia). The ecosystem has become the basic functional unit of ecology (Tansley, 1935; Golley, 1991). There are various definitions, which include: 'an energy-driven complex of a community of organisms and its controlling environment' (Billings, 1978); 'a community of organisms and their physical environment interacting as an ecological unit' (Dickinson and Murphy, 1998); 'an integration of all the living and non-living factors of an environment for a defined segment of space and time' (Golley, 1993). According to Miller (1991: 112), ecosystems have six major features: interdependence, diversity, resilience, adaptability, unpredictability and limits. They also have a set of linked components, although the linkages may not be direct – a network or web with organisms as nodes within it (Figure 3.4). Table 3.1 suggests two ways of classifying ecosystems, by function or degree of disturbance. An ecosystem boundary may be defined at organism, population or community level, the crucial point being that biotic processes are sustainable within that boundary. It is also possible to have different physical and functional

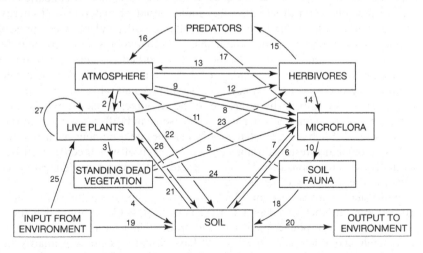

Figure 3.4 The relations between ecosystem components
Source: Van Dyne (1969: 83, Fig. 3)

Table 3.1 *Classifications of environmental systems*

(A) BY FUNCTION

Isolated systems	Boundaries are closed to import and export of material and energy.
Closed systems	Boundaries prevent import and export of material, but not energy. For example, 'Biosphere 2' receives sunlight but is supposed to function with no other exchanges. The Earth is largely a closed system, although it receives dust, meteorites and solar radiation.
Open systems	Boundaries allow free exchange of material and energy. Many of the Earth's ecosystems are of this form and may actually be interdependent.

(B) BY DEGREE OF HUMAN DISTURBANCE

Park (1980: 42) suggested environmental systems could be classified as:

Natural systems	Unaffected by human interference.
Modified systems	Affected to some extent by human interference.
Control systems	Human interference, by accident or design, plays a major role in function (includes most agricultural systems).

Note: Biosphere 2 is an enclosed environment experiment constructed some years ago in the Sonoran Desert, USA.

boundaries to an ecosystem. No two ecosystems are exactly the same, but one may recognise general rules and similarities.

There are two ways of viewing ecosystems: (1) as populations – the community (biotic) approach, in which research may be conducted by individuals; (2) as processes – the functional approach (studying energy flows or materials transfers), best investigated by a multidisciplinary team. Once understood, ecosystems can often be modelled, allowing prediction of future behaviour. It is possible to recognise three broad types of ecosystem: (1) Isolated systems – boundaries recognisable and more or less closed to input and output of materials and energy; (2) Closed systems – boundaries prevent input/output of materials, but not energy; (3) Open systems – boundaries may be difficult to recognise and these allow free input/output of materials and energy. Many of the Earth's ecosystems are type-3 and are often interdependent, which presents environmental management with huge challenges. Alternatively, ecosystems may be classified as (1) Natural – unaffected by humans; (2) Modified – some change due to humans; (3) Controlled – whether by accident or design humans play a dominant role (e.g. agriculture – agroecosystems or urban ecosystems). A naturalist might map the ecosystem of an animal, say a bear, by reference to the resources it uses (i.e. as a function of the organism), so the area may alter with the seasons and differ according to the age or sex of the animal. Such an ecosystem would incorporate a number of distinct components: valley, mountain forest, coastlands and so on, each of which could itself be recognised as an ecosystem (Gonzales, 1996). Alternatively, ecosystem delineation could be by function (i.e. as a sort of landscape unit).

Ecosystems may be recognised across a great range of spatial scales: one may cover 10,000 km^2, another less than 1 km^2; it is even possible to argue that the half-litre of water trapped in a pitcher-plant or a clump of lichen on a tombstone are ecosystems. In a stable ecosystem each species will have found a position, primarily in relation to its functional needs for food, shelter and so on. This position, or niche, is where a given organism can survive most effectively. Some organisms have very specialised demands and so occupy very restricted niches; others can exist in a very wide niche. Niche demands are not always simple: in some situations a species may be using only a portion of its potential niche, and alteration of a single environmental parameter may suddenly open, restrict or deny a niche for an organism. Competition for the niche with other organisms is one such parameter.

Ecosystems can be subdivided, according to local physical conditions, into habitats (places where an organism or group of organisms live), populated by characteristic mixes of plants and animals (e.g. a pond ecosystem may have a gravel bottom habitat and a mud bottom habitat). Within an ecosystem change in one variable may affect one or more, perhaps all other variables.

There are few ecosystems where there are no complex energy flows and exchanges of materials across their boundaries. Even something as well defined as a cave may exchange water and nutrients with regional groundwater or capture debris blown from outside (Bailey, 1986). To simplify study, ecologists have attempted to enclose small natural ecosystems, create artificial laboratory versions (e.g. phytotrons, growth chambers), and study very simple types such as those of the Antarctic 'dry valleys'. A huge hermetically sealed greenhouse complex with a crew of eight, designed to study the function and interaction of several ecosystems, was established in Arizona, USA, in 1991. It was named 'Biosphere 2' to emphasise its separation from the Earth's biosphere, and to reflect one motive of the experiment, which was to test the feasibility of such facilities for life-support on Mars. It was managed to try to maintain a more or less breathable atmosphere and provided almost enough air, water and food for the crew for two years (Allen, 1991). Controlled environment experiments are valuable for those seeking to establish what effects changing global climate and carbon dioxide will have on crops and wild flora and fauna.

In the late 1940s, systems diagrams were constructed to show energy flows between components of ecosystems. Soon similar approaches had been adopted by many social scientists and business managers as frameworks for study and as means of prediction. For example, a systems approach was used in the early 1970s by the Club of Rome to try to model global limits (Smith and Reeves, 1989). Applied systems theory and systems modelling have been steadily improving and are widely used in environmental management (Odum, 1983; Perez-Trejo et al., 1993; Brown and MacLeod, 1996; Dickinson and Murphy, 1998: x). So, while the ecosystem approach may not give precise modelling results, it can often provide a valuable framework for analysis. However, it can be difficult to recognise boundaries; measurement of what goes in and comes out can be problematic; establishing whether an ecosystem is natural, rather than modified, can be challenging, and organisms may drift or migrate in or out. In addition, the assumption that an ecosystem will behave in a linear, predictable manner may be over-optimistic, because some of the processes that are operating work at random. According to systems theory, changes in one component of a system will promote changes in other, possibly all, components. As subsystems may interact in different ways, the ecosystem approach needs to be essentially holistic. Nevertheless, it is often possible to get some idea of an ecosystem's energy and material distribution, and perhaps model its behaviour, although with complex ecosystems this becomes more difficult (Figure 3.5). Ecosystem researchers must ensure that they are looking at realistic assumptions, not over-simple abstractions or misconceptions. One cynical observer noted that 'artists and scientists tend to fall blindly in love with their models'. In practice, adopting an ecosystems approach can be difficult and, when it is possible, results may sometimes be disappointing.

Given time, natural, undisturbed ecosystems theoretically reach a state of dynamic equilibrium or steady state. Regulatory mechanisms (checks and balances) counter

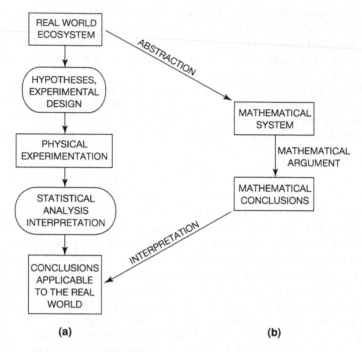

Figure 3.5 Two approaches to studying ecosystems: the conventional approach (a); and a second approach (b), which involves abstraction of the system into a model leading to interpretation of mathematical conclusions

Source: Adapted from Van Dyne (1969: 337, Fig. 7)

changes within and outside the ecosystem to maintain the steady state. However, since each ecosystem has developed under a different set of variables, each has a different capacity to resist stresses and to recover. In addition, humans often upset regulatory mechanisms, so response may be distorted. When ecosystems are exposed to stress, some responses may be immediate and others delayed, perhaps for decades. So, to manage ecosystems effectively it is necessary to know longer-term behaviour as well as short-term response. This means that palaeoecology and historical records have a part to play, as has long-term monitoring.

Ecosystems adjust to perturbation through regulatory mechanisms. When the relationship between input and output to the system is inverse (for example, increased sunlight causes more cloud which reduces the impact of that sunlight on the surface), it is termed a negative feedback. The opposite is a positive feedback, whereby an effect is magnified. There is a risk that a positive feedback may result in a runaway reaction, which is especially dangerous if it damages a crucial biogeochemical or biogeophysical cycle.

Urban ecosystems are of growing importance. Until quite recently the world population was mainly non-urban; now, after rapid urbanisation since the 1800s, over 50 per cent of people live in cities, and the percentage is increasing. Many of the largest, fastest growing cities are in poor countries and pose severe environmental problems. Urban ecosystems have far-reaching 'roots' drawing inputs from a huge catchment; cities also influence decision making that affects rural areas, discharge polluted effluent, contaminate the air flowing past, and generate huge amounts of refuse. Even in developed countries urban environments are a challenge for environmental management (see Chapter 15). In recent years there has been a shift in interest from just coping with city problems to seeking strategies for 'sustainable cities' – however, there is a long way to go before there are practical solutions. Engineering and institutional developments alone will not provide solutions for urban transport, water supply, sanitation, control of crime, or improving social cohesion. For effective environmental management there must be a better understanding of urban and peri-urban environments, societies and economies, and how they interact with rural surroundings.

Environmental and ecosystems modelling, the ecosystem concept, environmental systems and ecosystem management

Once understood and monitored, environmental systems may be modelled using a variety of approaches, including theoretical, physical, analogue or computer models. A large and diverse environmental modelling field has emerged, specialising in anything from sediment transport to hydrology, groundwater, global climate change, carbon sequestration, ocean–atmosphere energy and chemical flux – and many other specialisms (Jakeman et al., 1993). Ecological systems modelling is examined later in this chapter, and includes pastoral systems, agroecosystems, ecotoxicology and biodiversity.

The ecosystem concept became a widely used tool for research after 1945; for example, it has been adopted by the International Biological Program (Myers and Shelton, 1980). The approach focuses on energy flows or nutrient transformations. Biotic activity within an ecosystem can be divided into that of producers, consumers and decomposers, and efforts to study these may focus on population dynamics and productivity, predator–prey relations, parasitism and so on. Study of non-biotic aspects of an ecosystem may focus on estimation of biomass or micrometeorology. In the past three decades there has been a shift from description of the structure of ecosystems to a focus on trying to understand function, processes, mechanisms and systems behaviour.

The ecosystem boundary is often adopted as the spatial and temporal limits to an environmental task, and the ecosystem concept allows the environmental manager to look at portions of complex nature as an integrated system (Van Dyne, 1969: 78; Holling, 1987). The concept may be applied to cities, agriculture (urban ecosystems and agro-ecosystems respectively) and many other situations, although these are not actually true, discrete units in terms of energy flows or function. An ecosystems approach allows a holistic view of how complex components work together, and it can enable the incorporation of human dimensions into assessments of biosphere functioning (Samson and Knopf, 1996; Vogt et al., 1997). This requires a multidisciplinary or, better, inter-disciplinary teamwork that includes consideration of science and social science issues (DiCastri and Hadley, 1985; Roe, 1996; Yaffee, 1996). If the ecosystems approach is pursued with a holistic perspective, that may be interpreted in either a comprehensive or an integrated manner. The integrated approach does not try to research all ecosystem components, only those deemed crucial by planners (Barrett, 1994; Bocking, 1994; Margerum and Born, 1995). A comprehensive approach seeks to research in much greater depth with wider focus, taking time and costing more, so it may be less practical for planning and management. It is important that planners and analysts have a clearly thought-out interpretation of what an ecosystem approach means before using it. Environmental managers may treat an ecosystem rather like a factory: they seek to improve and sustain output and reduce costs but, unlike a factory, there are often several different 'products' such as agricultural produce, tourism, water supply and conservation.

A precise, universally acceptable definition of ecosystem management is impossible, partly because it depends on the stance and outlook of the definer, partly because it is still evolving, and also because it involves a diversity of actors – scientists, policy makers, commerce, citizens and others (Golly, 1993). It is not a science, nor it is a simple extension of traditional resource management; rather, it seeks a synthesis of eco-system science and ecosystem approaches, to provide a framework that links biophysical and socio-economic research and practice in a region or ecosystem through a holistic ecological and participatory methodology (although how it might achieve these goals is usually less than clearly stated) (Grumbine, 1994, 1997). Sustainable ecosystem management seeks to maintain ecosystem integrity and, if possible, produce food and other commodities on a sustained basis. Many of the principles used by ecosystem management are normative (i.e. moral and ethical rather than strictly scientific), which has attracted criticism (Likens, 1992; Haeuber, 1996). Concern has been voiced over the lack of satisfactory established principles for ecosystem management, and that it may lead to a broad and possibly superficial approach in the effort to break down an over-sectoral treatment. To address this it has been suggested that ecosystem manage-ment be integrated with organisational structures that continue along sectoral lines (Mitchell, 1997: 62). Other problems of ecosystems management are that experience gained in one ecosystem may be of limited value for other, even similar, ecosystems; the character of natural ecosystems may be difficult to establish where there has been disturbance, so it is difficult to agree what conservation or land restoration should aim for (Slocombe, 1993: 294; Brunner and Clark, 1997). There are still gaps in knowledge, leading to strong criticisms of lack of scientific rigour and vagueness (Armitage, 1995: 470). For a critique of the ecosystem approach see Pepper (1984: 107–110). Sometimes an environmental systems, rather than ecosystems, approach is adopted.

O'Neil (in Cairns and Crawford, 1991: 39) suggested that the ecosystem approach could be seen as methodology (with models to simulate the ecosystem) and mindset (with a focus on function and properties of ecosystems), the strength of this approach being synthesis of the complexity of problems faced, enabling assessment of con-sequences. In practice there has been an understandable specialisation, for example: ecosystem studies of risk; ecosystem quality management; assessment of ecosystem potential; ecosystem conservation, and so on. It is not only ecologists and environmental managers who have adopted an ecosystems approach: many other disciplines frequently do so, including human ecology (perhaps the first to do so), cultural anthropology (Moran, 1990), planning, management and urban studies. Because the ecosystem approach means different things to various disciplines it is a useful generalisation rather than a precise term.

The decision to adopt an ecosystems approach will usually be based on an assessment of whether its advantages outweigh its disadvantages (Box 3.2). As many institutions are commodity or service orientated, rather than ecosystem orientated, data collection and personnel training may need changing. A commodity or service orientation may be fine if the goal is to maximise production of a single product or service; it is less satis-factory where the ecosystem yields several 'products', and it is important to know hazards, limits and opportunities (Box 3.3).

Two needs often confront the environmental management: the first is a search for ways of integrating environmental and socio-economic planning, and the second is to define and bound areas (e.g. ecosystems) of interest and value to managers and plan-ners. Sometimes ecosystem boundaries coincide with clear physical features (e.g. an island or a forest), but often they are less well delineated. Gonzalez (1996) noted the need to define an ecosystem in 3-D, not just mapping area, but also establishing its 'top' and 'bottom'. The quest is for an eco-socio-economic planning unit, which is stable, clearly defined and likely to support sustainable development. Some comprehensive or integrated regional approaches evolved as early as the 1930s and 1940s and used units

---

Box 3.2

Advantages and disadvantages of the ecosystem approach

Advantages	*Disadvantages*
Comprehensive, holistic approach for understanding whole systems.	May neglect sociocultural issues such as politics, power and equity.
Different view of science that recognises diversity of cause and effect, uncertainty, and probabilistic nature of ecosystems.	Ecological determinism: danger of generalising from biophysical to socio-economic systems.
Draws on theory and methods from different fields to generate models and hypotheses.	Nebulous: a vague, superorganismic theory of poor empirical foundation that relies on analogy and comparison.
Contributes to understanding limits, complexity, stresses and dynamics.	Non-standard definition of 'ecosystem'.
Encourages preventive thinking by placing people within nature.	Reification of analytical systems; in some approaches linked to reductionist and equilibrium views.
Facilitates locally appropriate, self-reliant, sustainable action.	Narrow spatial focus on local ecosystem structures and processes.
Facilitates co-operation, conflict reduction, institutional integration.	Functionalist and/or energy analysis are overemphasised.
Requires recognition of mutual dependence on all parts of a system (e.g. natural/cultural, person/family).	Duplicates and/or overlaps other disciplines without a special contribution of its own.
Results in criteria for management actions.	If ecosystem approaches can apply to everything they may be meaningless.
Facilitates studies that integrate a range of disciplines (holistic).	

Source: Slocombe (1993: 298, Table 3 (with modification))

---

such as river basins. Interest was renewed in the 1970s with attempts to marry ecological concern with regional planning and policy making (McHarg, 1969; Isard, 1972; Nijkamp, 1980). Slocombe (1993) was optimistic that the ecosystem concept might offer a route to integrating environmental management and development planning that would lead to sustainable development; Mitchell (1997: 51) was less enthusiastic, and felt that basic concepts of ecosystem diversity and stability did not adequately describe complex reality: ecosystems were inherently complex, there were unlikely to be simple answers, and environmental managers must accept that they could not just manage ecosystems, but that they were managing human interactions with them.

Box 3.3

How the ecosystem approach can advise the environmental manager – three selected situations

Range management

● what type of stock;
● stocking rate;
● the state of the range;
● how to manage grazing rotation;
● whether to augment with seeding or fertiliser;
● potential threats;
● parallel usage opportunities (e.g. recreation, conservation, forestry).

Forest management

● whether the forest trees are healthy and regenerating;
● whether the mix of species is steady or in decline;
● whether there are threats;
● what harvesting is possible and how;
● parallel usage opportunities (e.g. forest products, conservation, tourism); whether forest can be established/re-established in currently unforested areas.

Conservation management

● whether conservation is viable in the long term;
● what mix and number of species can be carried;
● whether a cull or improvement in breeding is needed;
● whether there are threats;
● what parallel uses are possible (e.g. ecotourism);
● whether there are alternative ecosystems to provide back-up.

Ecosystems analysis, modelling and monitoring

Ecosystems (and environmental systems) may be analysed using systems theory, which enables complex, changing situations to be understood and predictions made. Systems theory assumes that measurable causes produce measurable effects. There have been attempts to combine ecological and economic models in systems analysis. For example, a systems analysis approach to environmental assessment and management was used in the Oetzertal (Valley of the River Oetz, Austria) from 1971, as part of the UNESCO Man and Biosphere Program. This alpine valley ecosystem has experienced great change as a consequence of tourism, especially skiing, and, with the help of the modelling, managers now have a clear idea of what is needed to sustain tourism and maintain environmental quality (Moser and Peterson, 1981). In the early 1990s the USA established a nation-wide Environmental Management and Monitoring Program (EMAP) to aid ecological risk analysis by assessing trends in condition of ecosystems – so far a controversial and expensive exercise. Natural disaster risk assessment is attracting interest, especially since the late 2004 Indian Ocean tsunami (Chen, 2005).

The Millennium Ecosystem Assessment (MEA) is an international programme launched by the UN Secretary General in 2001 and completed in 2005. It is designed to serve decision makers and the public, providing information on ecosystem change and human well-being impacts that this causes. The information is also likely to assist the Convention on Biological Diversity, the Convention to Combat Desertification, the Ramsar Convention on Wetlands, and the Convention on Migratory Species. The UN hopes that if the MEA is successful it will continue similar programmes every five to ten years, plus assessments at national and subnational scale if needed. The MEA seeks to:

- establish priorities for action;
- provide benchmarking;
- offer tools and information;
- give foresight;
- identify response options (especially for sustainable development goals);
- help build institutional capacity;
- guide future researchers.

For further MEA details see http://www.millenniumassessment.org/len/About. Overview.aspx (accessed March 2005).

Environmental system and ecosystem planning and management – biogeophysical units

The first step taken by most planners and managers is to determine the limits of their task so that they can do an effective job, given the time and resources available. A suitable sized and stable unit is needed which reflects the structure and function of nature, but which as far as possible goes beyond being a biogeophysical unit to facilitate consideration and management of social, economic, cultural and other aspects of human–environment interaction. In this section a number of ecosystem-based frameworks are considered. Diamond (2005: 277–308) has examined societies which have sustained themselves and those which have failed, and he notes that social factors tend to dominate natural ones in determining success; also, a number of small regional units with adaptable bottom-up organisation fared well. Some sort of biogeophysical canton or county may be a promising route to sustainable development.

Ecozones

Various researchers have attempted to divide the Earth into ecozones or life zones for study, planning and management (Schultz, 1995). One of the best-known and most widely used systems for land use classification is the Holdridge Life Zone Model. This is based on the relationship of current vegetation biomes to three parameters: annual temperature, annual precipitation, and potential evapotranspiration (Holdridge, 1964, 1971). The Holdridge Model is often used in land use classification and predicts eco-climatic areas but does not directly model actual vegetation or land cover distribution. For an introduction to application of the Holdridge Model see Fennel (1999: 81–82, 124–126) and Dowling (1993).

Zoning should be done with an eye for 'dovetailing' mutually supportive activities and encouraging co-operation between sectors, agencies, NGOs and local people. Areas can also be zoned according to their biodiversity, conservation needs, vulnerability,

Table 3.2 *Hierarchical ecosystem classification used in The Netherlands*

Unit	Area (km²)	
Eco-zone	< 62,500	
Eco-province	2,500–62,500	
Eco-region	**100–2,500**	} best suited
Eco-district	**6.25–100**	} to most needs
Eco-section	0.25–6.25	
Eco-series	0.015–0.25	
Eco-tope	0.0025–0.015	
Eco-element	< 0.0025	

Source: Based on Klijn *et al.* (1995: 799, Table 1)

resilience, susceptibility to hazards, aesthetic value, tourism usage and much more. GIS techniques allow environmental managers to zone with virtually any variable that suits their needs.

Ecoregions and ecodistricts

The Netherlands National Institute of Public Health and Environmental Protection has developed a framework for hierarchical ecosystem classification to try to overcome the confusion resulting from the use of many different geographical regionalisations by various bodies. This is known as 'standardised regionalisation', a hierarchical mapping of nested ecosystems started in 1988 (Table 3.2), and is used for regional environmental policy. It ties in with GIS, is useful for state-of-environment reporting, and has been quite successful. Similar approaches have been tried or adopted in several countries, such as Canada, the USA and Belgium (Omernik, 1987).

Coastal zone planning and management

There has been growing interest in coastal zone management (Carter, 1988; OECD, 1993; Brower *et al.*, 1994; Viles and Spencer, 1995; Clark, 1996; Prestcott, 1996; French, 1998; there are also journals dedicated to the field, e.g. *Coastal Zone Management*). In many parts of the world it is in the coastal zone that most human activity is concentrated and environmental management is required, especially for coastlands subject to flooding or erosion, and regions where mangrove forests are being exploited. Many countries have invested in tourism development in their coastal zones. With the threat of global warming and rising sea-levels, coastal zone management is likely to grow in importance. The late 2004 tsunami disaster is also likely to prompt interest.

Marine ecosystem planning and management

An ecosystems approach has been explored for managing the Baltic Sea (Figure 3.6) (Jansson, 1972), the Mediterranean (and more especially the Aegean), the North Sea and the Japanese Inland Sea. Although not strictly marine, but with similarities, the same may be said for the Great Lakes of North America, the Aral, Caspian and Black Seas, and Lake Baikal. These ecosystems involve several countries, and in order to control pollution management must extend inland to incorporate regions that pollute and ensure control of the hydrology of the whole basin.

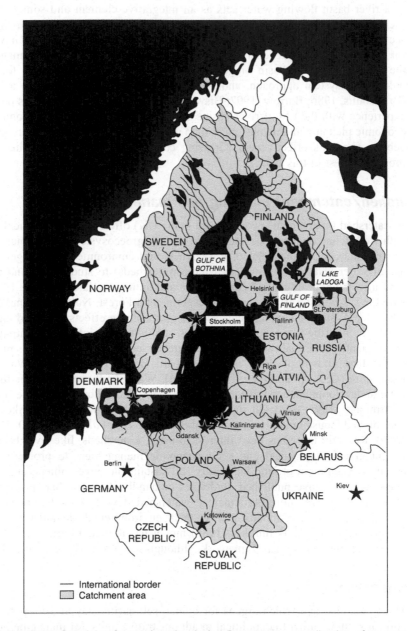

Figure 3.6 Baltic Sea Joint Comprehensive Action Programme – application of an ecosystem approach for management of a sea bordered by several countries

Note: catchment area to ensure jurisdiction over things which affect the ecosystem (or might be argued to be part of it).

Source: Mitchell (1997: 64, Fig. 3.2) – which was based on material from the Helsinki Commission (1993) *The Baltic Sea Joint Comprehensive Action Programme*. Government Printer, Helsinki.

River basin planning and management

In a river basin flowing water acts as an integrative element and something to use to seek development. Watersheds offer a similar management unit (see the following section), but focus more on moisture and soil conservation. River basins have been used for integrated or comprehensive regional development planning and management since the 1930s. The river basin biogeophysical landscape unit is suitable for applying a holistic, ecosystem approach, and is useful when several states share a river system (Briassoulis, 1986; Barrow, 1997; Biswas and Tortajada, 2001). There is probably more experience with the use of river basins as a means for integrated environmental-socio-economic planning and management than with any other ecosystem approach and many debates about its effectiveness. Growing competition for water supplies is likely to prompt interest in this field.

Watershed/catchment planning and management

A watershed ('catchment' is often used in the UK) offers a biogeophysical unit usually with well-defined boundaries and within which agroecosystem use, human activity and water resources are interrelated. Researchers and environmental managers have made use of watersheds or subdivisions (micro-watersheds) to study how land use changes have affected hydrology, soil conservation and human welfare since the 1930s (starting with the US Forest Service Coweeta Experimental Forest, North Carolina) (Vogt et al., 1997: 40). Watershed experiments seek to establish the effects of disturbing vegetation or soil, monitoring inputs to the basin (e.g. sunlight and rainfall) and outputs (by measuring quantity and quality of flows from streams or material removed as produce). One of the best known of these is the Hubbard Brook Experimental Watershed, New Hampshire (USA) (Van Dyne, 1969: 53–76). Watersheds are useful for forestry, agricultural development, erosion control, water supply, pollution and fisheries management.

Armitage (1995: 470) felt that integrated watershed management, like soil erosion control, had focused mainly on technical issues (Easter et al., 1986; FAO, 1988; Pereira, 1989). Recently there has been interest in using watersheds, like river basins, as units for integrated biophysical and socio-economic management to promote better community development or land husbandry and sustainable development, and a number of major agencies have published guidelines or handbooks (Bouchet, 1983; FAO, 1986; Naiman, 1992). In India participatory watershed development has been used to try to improve rural livelihoods and counter environmental degradation (Turton and Farrington, 1998). Hufschmidt's (1986) model has attracted particular attention as an integrative methodological framework, although it is not really ecologically focused.

Bioregionalism

This is an approach which argues for human self-sufficiency at a local scale and support for natural, rather than political or administrative units, for managing human development (Sale, 1985). Definitions are not precise; however, it is usually seen as a 'sense of place' adopting both a life region and a social unit, and places emphasis on local adaptation to environment. Bioregionalism has been described as 'awareness of the ecology, economy and culture of a place and a commitment to making choices to enhance them' (http://www.greatriv.org.bioreg.htm – accessed February 2005). It is often possible to 'nest' a series of bioregions within each other. Bioregionalists embrace the adage 'think globally act locally', generally seek community development in environmentally friendly ways, and strive for self-sufficient sustainable units. It has been called the politics of place, and is generally ecocentric in outlook. Supporters of perma-culture (a form of organic farming) often advocate bioregionalism and a 're-tribalisation' of society (Mollison and Holmgren, 1978).

Bioregionalism appeared in the early 1970s when it was promoted by biogeographers and environmentalists in California such as Peter Berg (Parsons, 1985; Aberley, 1993; Alexander, 1990). The bioregion was seen to be intermediate between biogeographical provinces and ecosystems or groups of ecosystems, and to support 'cultural awareness'. There is a considerable and diverse following of supporters in the USA (with an annual North American Bioregionalism Congress since 1984); however, some advocates are less than rational, and a few are virtually eco-fascist. There is much to be said for adopting manageable-scale stable biogeophysical units within which suitable socio-economic activities may be studied and managed; for example: river basins (or rather sub-basins), watersheds, coastal zones, cantons, parishes, counties, municipalities and so forth.

Agroecosystem analysis and management

Basically this is a form of rapid rural appraisal (Conway, 1985a, 1985b; Conway and Barbier, 1990: 162–193) and a type of ecosystem approach. An agroecosystem is an ecosystem modified by humans in order to obtain food or other agricultural produce. Four agroecosystem properties were recognised by Conway (1985b):

1   *Productivity* – output, yield or net income from a valued product per unit of resource input. This may be measured as yield or income per hectare, total production per household or farm, or at a regional or even national scale. Alternatively, it may be expressed as calories.
2   *Stability* – the constancy of productivity in the face of climatic fluctuations, market demand, and other variables.
3   *Sustainability* – the capacity of an agroecosystem to maintain productivity in the face of environmental challenges and degradation arising from its exploitation.
4   *Equitability* – the evenness of distribution of the productivity benefits among humans.

The agroecosystem may be managed in ways that give optimum levels of each of these properties: maximising productivity is likely to reduce agroecosystem sustain-ability; ensuring sustainability might reduce productivity. Most of the efforts to modernise agriculture so far focus on (1) above. This demands an understanding of ecosystems and of how natural processes are modified by agricultural objectives. To manage agroecosystems effectively requires application of knowledge from a range of disciplines, and this approach supports that (Risser, 1985; Gliessman, 1990). Because the main goal is to improve socio-economic conditions, some feel the agroecosystem approach is more socio-economic than ecological in orientation (Armitage, 1995).

Landscape ecology approach

The landscape ecology approach has its origins in the work *The Theory of Island Biogeography* (MacArthur and Wilson, 1967). It focuses on spatial patterns at the landscape scale (e.g. hedges, fields, rivers), and how their distribution determines the flow of energy and materials and affects organisms (Vink, 1983; Forman and Godron, 1986; Vos and Opdam, 1993; Ze'ev, 1994; http://www.usiale.org – US Inter-national Association for Landscape Ecology – accessed February 2005). The response of an ecosystem to disturbance frequently depends on its neighbouring ecosystems: for example, organisms may escape if there are suitable nearby ecosystems and recolonise after disturbance ceases; also, energy or materials may be transferred between ecosystems. An ecosystem seldom functions in isolation and its ability to withstand stress may depend on how a nearby ecosystem is being managed, or on whether the boundaries are altered – a road or cleared area of forest may prevent animal or plant dispersal to a favourable alternative site. The landscape ecology approach extends ecosystem management to a group of more or less neighbouring or linked ecosystems, and problems are generally dealt with in a holistic way (Jensen *et al.*, 1996; http://www. landscape-ecology.org/; journal: *Landscape Ecology* http://www.kluweronline.com/issn/ 0921–2973 – accessed March 2005).

GIS and quantitative techniques have been applied to the landscape ecology approach (Hassan and Anglestam, 1991; Turner and Gardiner, 1991; Bunce *et al.*, 1993; Haines-Young *et al.*, 1993). Interesting applications of landscape ecology and GIS have been the prediction of the occurrence of Lyme Disease, a growing public health problem in the USA (*New Scientist*, 15 November 1997), and the spread of Chagas Disease in South America. In the UK the Countryside Commission has been exploring the value of landscape character mapping.

Ekistics

Ekistics is described as the 'science of human settlements': it draws upon human ecology and regional planning and treats urban territory as a living organism, adopting an interdisciplinary, problem-solving approach – in some respects similar to an ecosystems approach, especially in its focus on networks (Doxiadis, 1968, 1977). Since 1955 the *Ekistics* journal has worked to establish the field (http://www.ekistics.org/E.Journal.htm – accessed March 2005).

Applying the ecosystem concept to tourism, conservation and heritage management

The application of environmental management to tourism and heritage features has grown since the 1970s (see Chapter 14 for further coverage). It has mainly involved the application of impact assessment, eco-auditing and the exploration of sustainable development strategies (Edington and Edington, 1986; Butler, 1991). Tourism and heritage features management may be divided into: (1) natural history-oriented tourism (environmental features may be the tourist attraction, but there may be little effort to control impacts and limited investment in environmental management); (2) eco-tourism (tourism based on visits to areas of unspoilt natural beauty or rich wildlife), which seeks to minimise impacts and which invests a significant portion of profits into environmental management; (3) tourism actively involved in assisting conservation or heritage management, and/or gathering environmental information (e.g. tourists pay to assist on a wildlife survey or archaeological dig).

Tourism often takes place in a sensitive environment: coastal zones; alpine areas; coral reefs; and where walkers or off-road vehicles cause damage. The value of the ecosystems approach is that it can highlight vulnerable features and threatening human behaviour, which may be easily overlooked. For example, in parts of Australia and South Africa there have been calls to cull sharks. Before doing so it would be wise to study shark behaviour and role in the ecosystem to see whether their value outweighs their threat, and also to check whether they move about so much that local removal is pointless. Similar situations may arise in tropical rainforest environments where apparently minor disturbance of bird or bat roost sites might have serious regional effects, through reduced pollination, seed dispersal or insect predation. Where Alpine farmers turn to tourism and relax their management of summer grazing, the under-grazed grass may fail to anchor winter snow and increase the threat of avalanche.

Tourism may become important as a means of financing and encouraging respect for conservation, and of generating income for local peoples, and a way to help fund transition to sustainable agriculture. Ecosystem management can help ensure that tourism provides optimum support for conservation. Heritage sites can be established to conserve cultural and natural features, including wildlife and old crop varieties in arboreta and the gardens of large estates. In many parts of the world some of the last remaining stands of ancient trees are found as sacred groves, around burial areas and in temple gardens. There is a need to apply ecosystem studies to determine how such refuge areas may be sustained and augmented. Caves are especially vulnerable: visitors can introduce moulds and other organisms which damage delicate structures or fauna, and ecologists can advise to help reduce these problems (Cigna, 1993).

Applying the ecosystem concept to urban and peri-urban management

More than half the world population now live in conurbations, and the effects of urban settlement, in the form of fuelwood demand, air pollution and contamination of watercourses, are increasing and are felt at growing distances into the surrounding regions (White, 1994). There is a growing literature on the urban ecosystem approach which can help to identify strategies to reduce pollution, aid safe disposal of pollutants and production of food, and provide employment. At a regional or national scale it may be possible to understand the linkages that have driven people to settle urban areas, often abandoning once sustainable rural livelihoods (Dorney and McLellan, 1984).

Applying the ecosystem concept to conservation management

Forest management and wildlife conservation make extensive use of the ecosystem approach (Lajeunesse *et al.*, 1995; Bailey, 1996; Samson and Knopf, 1996; Boyce, 1997; Weeks, 1997). Nature reserves are essentially islands in a sea of disturbance, so the study of island ecosystems by biogeographers such as Simberloff, Wilson and MacArthur provides key information on rates of extinction and evolution; minimum size of habitat and linkages between habitats necessary for sustained conservation; whether to conserve selected species or a whole ecosystem; assessment of likely impacts of climate change or acid deposition; clarification of vital pollination and seed dispersal needs; information on predator–prey relationships, and so on (MacArthur and Wilson, 1967; Mueller-Dombois *et al.*, 1981; Di Castri and Robertson, 1982). A conservation area may also fail to sustain biodiversity because disruptive effects penetrate too far towards its core (Soule, 1987). Caution is needed, since some of the island biogeographic theory which conservation managers draw upon is incomplete, imprecise, or has been little tested (Shrader-Frechette and McCoy, 1994).

Studies have been in progress for some years in the Amazonian forests of northern Brazil to improve understanding of the impact of various intensities of disturbance and ecosystem fragmentation on biodiversity survival using different sizes and patterns of forest reserves. This Minimum Critical Size of Ecosystem Study undertaken in Amazonia, and similar ones elsewhere, is vital for establishing what are viable locations, ideal size and pattern of conservation areas (Quammen, 1997).

How stable are environments?

'Stability' can have a number of meanings, including: lack of change in the structure of an ecosystem; resistance to perturbations; or a speedy return to steady state after disturbance (Troumbis, 1992: 252). Environmental managers are likely to want to know whether an ecosystem is stable, and what would happen if it were disturbed. As discussed above, the concept of ecosystem stability has provoked much debate which is not yet fully resolved (Hill, 1987). It is clear that natural ecosystems are rarely static: the best environmental management can expect is a sort of dynamic equilibrium, not a fixed stability (Smith, 1996). Furthermore, human activity is increasingly disrupting ecosystems. Equilibrium is in part a function of sensitivity and resilience to change. Sensitivity may be defined as the degree to which a given ecosystem undergoes change as a consequence of natural or human actions. Resilience refers to the way in which an ecosystem can withstand change. Originally it was proposed as a measure of the ability of an ecosystem to adapt to a continuously changing environment without breakdown. It would be misleading to give the impression that these concepts of stability and resilience are straightforward and fully established.

Ecosystems are subject to natural and anthropogenic changes, some catastrophic and sudden, others gradual and less marked (Stone *et al.*, 1996). It is widely held that, given long enough, a steady state will be reached by an ecosystem because a web of relationships allows it to adjust to serious localised or moderate widespread disturbances. Such an ecosystem is supposed to remain in steady state unless a critical parameter alters sufficiently. If change then occurs, it is termed 'ecological succession' or 'biotic development' (Johnson and Steere, 1974: 8). Some economists and political studies specialists have suggested economics, politics and social development follow predictable evolutionary paths to steady states.

There is debate as to whether an ecosystem: (1) evolves in the long term towards a steady state with equilibrium of its biota through slow and steady evolution of species (phyletic gradualism); or (2) experiences generally steady, slight and slow evolution punctuated by occasional sudden catastrophes and extinctions, after which there may be comparatively rapid and considerable biotic change 'punctuated by equilibrium' (Gould, 1984; Goldsmith, 1990). Whatever the process, the end result is widely held to be a 'climax stage', reached via more or less transient successional stages, at any of which succession might be halted by some limiting factor. The concept of ecological succession, pioneered by Clements (1916), is complex and still debated. According to the concept, organisms occupying an environment may modify it, sometimes assisting others – a birch wood may act as a nursery for a pine forest, which ultimately replaces the birch – thus birch is a successional stage *en route* to a pine stage. These transitional stages leading to a mature climax community are known as seres. Two types of succession are recognised: (1) primary succession and (2) secondary succession. The former is the sequential development of biotic communities from a bare, lifeless area (e.g. the site of a fire, volcanic ash, or newly deglaciated land). The latter is the sequential development of biotic communities from an area where the environment has been altered but has not had all life destroyed (e.g. cut forest, abandoned farmland, land that has suffered a flood or been lightly burnt). Many communities do not reach maturity before being disturbed by natural forces or humans, and so type (2) situations are common. Where succession is taking place from a bare area, the first stage is known as the pioneer stage; although, in practice, the expression may be applied to growth taking place in areas that do have some life – such as regrowth after logging. Natural forests may be assumed to maintain maturity, rather than becoming senile and degenerating, through 'patch-and-gap' dynamics – clearings caused by storms and other disturbances allow regeneration. Pioneer communities usually have a high proportion of plants and animals that are hardy, have catholic niche demands, and disperse well (e.g. weeds with wind-carried seeds, and insects which can fly). Mature, climax communities are supposed to have more species diversity, recycle dead matter better, and be more stable.

Stability (some prefer to use 'constancy') is often invoked by those interested in establishing whether conditions will remain steady or will return via a predictable path to something similar to the initial steady state after disturbance. It is widely held that ecosystem stability is related to biological diversity: the greater the variety of organisms there is, the less likely there is to be change in biomass production, although population fluctuations of various species may still occur (Tilman, 1996). However, it is quite possible that a change in some parameter could have an effect on all organisms. Thus, diversity may help ensure stability, but does not guarantee it. An ecosystem may not have become stabilised when disturbed: it may be close to a starting point, or it could be undergoing cyclic, more or less constant or erratic change. An ecosystem may return to stability after several disturbances but fail to return after a subsequent upset for various reasons (Holling, 1973). Some ecosystems are in constant non-equilibrium or frequent flux, rather than in a stable state at or near carrying capacity. Return to a pre-disturbance state is therefore uncertain.

Resilience is displayed by many things; for example: organisms, ecosystems, communities, regions, individuals, societies, institutions, and nations. Resilience may be defined in many ways:

- The ability to return to maintain a steady-state ecosystem.
- The facilitation of adaptive behaviour.
- The speed of recovery of a disturbed ecosystem.
- The number of times a recovery may occur if disturbance is repeated.

The concept of resilience has been applied by human ecologists: some societies absorb or resist social change and continue with traditional skills and land uses or develop satisfactory new ones; other societies fail, and their resource use and livelihood strategies degenerate. In humans, resilience and vulnerability are not fixed or predetermined; they vary as a consequence of environmental factors, institutions, attitudes, innovation and so on. In particular, poverty can make people more vulnerable and less resilient.

Referring to sensitivity and resilience, Blaikie and Brookfield (1987: 11) suggested a simple classification of land, which may be modified to apply to ecosystems in general:

1 *Ecosystems of low sensitivity and high resilience* – These suffer degradation only under conditions of poor management or natural catastrophe. Generally these are the best ecosystems to stretch to improve production of food or other commodities.
2 *Ecosystems of high sensitivity and high resilience* – These suffer degradation easily but respond well to management and rehabilitation efforts.
3 *Ecosystems of low sensitivity and low resilience* – These initially resist degradation but, once a threshold is passed, it is difficult for any management and restoration efforts to save things.
4 *Ecosystems of high sensitivity and low resilience* – These degrade easily and do not respond readily to management and rehabilitation efforts. It is probably best either to leave such ecosystems alone or to alter them radically. For example, forest

might be converted to rice paddy-field and suffer less ongoing degradation than if it were converted to tree crops. Examples of high sensitivity and low resilience ecosystem type 4 include the loess soil region of China, and much of Australia's interior (due to its weathered infertile soils, harsh climate, widespread salinity, and vulnerable flora and fauna).

Managers or researchers often wish to establish in advance, or sometimes after a disturbance, what the consequences will be:

1 Will the ecosystem re-establish its initial state?
2 Will there be a shift to a new state?
3 If (1) takes place, how rapid will the recovery be and how complete?
4 What path does the recovery take?
5 How often can recovery occur?
6 Will the same recovery path always be followed?
7 Will successive, similar disturbance have the same effect?
8 Would change still occur if there were no disturbances?

It is often argued that ecosystems with greater species diversity are more stable. In practice, many variables are involved in determining ecosystem stability, and in a given situation the path of succession can be unpredictable (Figure 3.7). In a stable ecosystem each species is assumed to have found a position, primarily in relation to its functional needs: food, shelter and so on. This position, or niche, is where a given organism can operate most effectively. Some organisms have very specialised demands and so occupy

Figure 3.7 Abrupt boundary between cleared lowland tropical rainforest and young oil palm plantation, Peninsular Malaysia. A contrast between rich diversity of plant species in the forest, and the oil palm/ground-cover species (planted to try to reduce erosion and weed growth) of the plantation.

very restricted niches (e.g. the water-filled hollow of a particular bromeliad plant, itself with a restricted niche), while others can exist in a wide range of niches. A species may be using only a portion of its potential niche, and alteration of a single parameter affecting competition with other organisms may suddenly open, restrict or deny a niche for an organism.

Threatening environmental events

There are environmental threats which have a predictable pattern of recurrence; others may be impossible to predict with current knowledge, although some of the latter may give warning signs as they manifest, allowing some time for evacuation or mitigation (see also Chapter 11). Catastrophes may be unmistakable and sudden, less obvious and sudden, or they may unfold gradually and obviously or in an insidious way. Sometimes a system is stressed and changes virtually imperceptibly until a threshold is reached, whereupon there may be sudden and possibly drastic effects. Current knowledge only allows some degree of prediction, recognition of threat and appropriate response. Given long enough, chance events probably affect the survival of organisms at least as much as evolution – the process has been described as 'contingency' (Gould, 1984). Events which challenge life but give insufficient time for adaptation would allow some organisms and humans to prevail for quite fortuitous reasons rather than 'survival of the fittest'. Some threats may be predicted, giving early warning, or contingency plans may be made.

Environmental managers must not neglect to assess the threat of rare but highly damaging events. The problem is to convince people it is worth spending money on monitoring to spot threats, which may not have been manifest within living or historical memory. In addition, people must be persuaded that it is worth spending money on vulnerability reduction and improving chances of recovery from unforeseen disasters.

Some nineteenth-century earth scientists invoked catastrophic events to explain erosive landforms, prehistoric extinctions and geological unconformities (Thomas Huxley probably coined the term 'catastrophism' in 1869). In 1755 Europe was horrified when Lisbon was destroyed by earthquake and tsunami; clearly the Earth was not unchanging. With his publication of *Principles of Geology* in 1830, Charles Lyell helped uniformitarianism (the idea of continuing gradual change, involving processes operating in the past that operate today) to prevail over catastrophism. Since the mid nineteenth century there have been various attempts to revive catastrophism (Smith and Dawson, 1990; Ager, 1993). One prompt has been the recognition of a number of mass extinctions, at least fifteen significant events in the past 600 million years; (e.g., *c*. 440 million years BP, *c*. 390 million years BP, *c*. 220 million years BP, and (the K/T boundary event) *c*. 65 million years BP (Raup, 1988, 1993). The cause of mass extinctions is debated, and some question whether there really is adequate evidence, suggesting instead a more gradual loss of species.

Large meteors have clearly struck the Earth – many craters are obvious. In the early 1980s Walter Alvarez noted the widespread occurrence across the Old and New World of iridium (a rare metal), glass spherules and 'shocked quartz' grains in a thin clay layer of K/T boundary age (Alvarez and Asaro, 1990). This, together with tsunami beds around the Gulf of Mexico has been interpreted as evidence of an asteroid of approximately 10 km in diameter impacting the Earth (Kerr, 1972). These strata seem to coincide with the K/T mass extinction, and suggest the strike was into limestones and rocks that led to more acidic fallout and climatic cooling than other strata would have caused. Others suggest that a very large sheet-lava eruption, such as the outpouring of the Deccan Plateau Basalts of India, triggered the extinctions. These are by no means the only causes suggested by those recognising a K/T mass extinction. Others include climate change, sea-level falls, reduction of atmospheric oxygen levels, and disease. Whether or not an asteroid strike caused the extinction of the dinosaurs, or earlier and subsequent disruptions, there is enough evidence of impacts to indicate a threat that environmental managers should seriously consider. Over a hundred ancient craters, a few of more than 100 km in diameter, are known on the Earth, while smaller craters are as recent as 1500 BP (Huggett, 1990). A body estimated to have been about 100 metres in diameter probably exploded at an altitude of about 8 km at Tunguska, Siberia, in AD 1908, flattening 1200 to 2200 km^2 of *taiga* forest. A similar strike probably occurred in South Island, New Zealand, *c.* 800 BP (Hecht, 1991), and a blast (of about 100 kilotons yield) in the South Atlantic in 1978 may have been caused by a comet or asteroid (Lewin, 1992). Strikes are thus a real threat – if such a body were to strike a major food-producing area, settled area, a nuclear power station or waste repository, or caused a major tsunami, there would be a disaster. The impact of anything larger than 1 km in diameter would probably endanger civilisation wherever it hit and whatever angle of approach. Impact of a body over 15 km in diameter would probably destroy all higher life on Earth. We have technology capable of warning us of at least some approaching bodies, and could probably easily develop means of diverting or destroying those presenting a threat – the question is whether there is a willingness to invest in it in advance of a disaster.

Historical volcanic eruptions have been devastating on a regional scale but their wider impacts on climate have not been widely accepted until recently (e.g. Pompeii and Herculaneum – AD 79; Hekla, Iceland – AD 1636; Krakatoa, Indonesia – AD 1883). One historical eruption – Tambora (Indonesia, AD 1815) – caused crop failures for a few years as far away as Europe and North America, but few recognised this. Some recent blasts have had slight global impacts; for example, El Chichon (Mexico – 1982) and Mt Pinatubo (Philippines – 1991) caused a temporary lowering of global temperatures. No volcanic event in historical times has been sufficiently serious to scare governments enough to prepare for global impacts (loss of one or more harvests worldwide). Palaeoecologists and archaeologists have linked many past moderate-sized eruptions with acid deposition in Greenland ice and alteration of climate, which seem to have affected human fortunes. A fairly large eruption – Toba (Sumatra – *c.* 74,000 BP) probably seriously threatened human survival. The geological record shows that there have been plenty of much larger eruptions and catastrophic mega-tsunami (much bigger than the 2004 Indian Ocean waves).

The recurrence of catastrophic events may not be random. Asteroid strikes, variation in the Earth's solar radiation receipts, gamma-ray bursts, increased cosmic radiation, geomagnetic weakening and reversals, and perhaps vulcanicity and seismic activity may be more likely at certain alignments in the orbits of the Earth and other planets. There has been speculation that some ancient extinction events may have occurred when the solar system passed through dust and gas clouds at the galactic plane every twenty-six to thirty-three million years (possibly causing 'ice-house' conditions – the world frozen totally or ice-clad almost to the Equator). Mass extinction could also result from oceanic salinity changes or biological activity causing altered atmospheric gas mix, or changing sea-water circulation.

There have clearly been ancient extraterrestrial and geological or gaian (involving biota) disasters. Climate change is now accepted to be a threat by policy makers and some of the world's public. Most palaeoclimatologists now accept that there have been hot global 'greenhouse', and cold global 'ice-house' conditions (causes are debated). The last ice age and a number of earlier ones have cooled the world, but not sent ice sheets all the way down to the Equator in the way ice-house events may have done.

Nor have Quaternary interglacials been as warm as greenhouse conditions of, say, the Jurassic. Ice ages (cold glacial phases or glacials) alternate with warmer interglacial or less cold interstadial phases, and have happened at several points during the Earth's history. During glacials ice expanded from the poles and worldwide to lower altitudes on high ground – reflecting global cooling of a few degrees C. The most recent cooling began approximately 40 million BP, became more pronounced from about 15 million and reached glacial maximum in the last 1.8 to 2.4 million years (the Quaternary Era). The Quaternary 'ice age' has so far comprised over twenty major glacial–interglacial oscillations. The major interglacials each lasted between 10,000 and 20,000 years and the glacials spanned about 120,000 years. The peak of the last interglacial was about 132,000 to 120,000 BP and the last glacial maximum was about 18,000 BP. The post-glacial seems to have begun quite rapidly, around 13,000 BP in Europe, and ice had retreated to broadly its present limits worldwide by around 10,000 BP (between 7,000 and 3,000 BP average conditions may have been as much as 2°C warmer than today). There are well-established links between glacial conditions and low levels of carbon dioxide in the atmosphere (approximately 25 per cent reduction compared with the present), low levels of methane in the atmosphere, and low sea-levels (which may drop to perhaps 140 metres below those of today). During warm interglacials, carbon dioxide and methane in the atmosphere were higher than currently, and sea-levels perhaps 40 metres above today's.

Humans have enjoyed relatively stable post-glacial conditions over the past few thousand years, but palaeoecology warns this cannot be expected to last, and that sudden serious changes are possible (moderated or exaggerated by our pollution). Pollution is distorting natural climate change, so predictions are being made more difficult. Even during the past few thousand years, drought and the patterns of monsoon rainfall have frequently fluctuated enough to affect humans. Many of these shifts have been linked to ocean–atmospheric processes, which show periodicity or quasi-periodicity (e.g. the El Niño–Southern Oscillation (ENSO) and related El Niña events). ENSO is believed to function in the following manner: a low-pressure, high-temperature weather system lies over Indonesia; thousands of miles away over the southwestern Pacific is a related high-pressure, low-temperature system. It has been established that if pressure in one increases, it falls in the other. These pressure differences cause the southeast trade winds to blow steadily and move water away from the western coast of South America. This causes upwelling of nutrient-rich cold seawater marking the start of an ENSO event. Every year in spring and autumn there is a weakening, even cessation of the trade winds, peaking in the middle of the austral summer and, if it is fully manifest, the eastern tropical Pacific will warm up markedly (Diaz and Markgraf, 1992). ENSO events cause increased rain along the Pacific coast of South America and, months later, drought in Brazil, Australia and Australasia, reduced austral summer rainfall and cloud cover in southern Africa, and other changes around the globe (Diaz and Markgraf, 1992; Hamlyn, 1992). Study of the phenomenon enabled prediction of recent weather shifts in some regions nine months or more in advance. There is evidence that global warming at various points during the past 12,000 years sent surges of fresh meltwater into the Atlantic and altered salinity enough to affect oceanic circulation – effectively the Gulf Stream was turned off, suddenly dropping average annual temperatures in western Europe and eastern USA by several degrees and altering the climate in other ways, perhaps for decades. There are currently fears that global warming could cause this again with serious impacts on civilisation.

The environment poses serious threats, yet planners, administrators and citizens have too little awareness. There is a widespread feeling that technology has reduced vulnerability; however, modern communities are arguably more threatened than ever before.

This is a consequence of population increase, and because most people depend on complex and far-ranging linkages for food, water, livelihood and governance, and so are less adaptable than their forebears (Barrow, 2003).

Biodiversity

Ecological diversity refers to the range of biological communities that interact with each other in a given environment. Biodiversity (biological diversity) refers to species diversity plus genetic diversity within those species. Loss of biodiversity occurs when species extinction exceeds the rate of species creation. Extinction is a natural process, sometimes sudden, perhaps catastrophic, otherwise an ongoing, gradual process. However, humans have greatly accelerated the rate of extinction. Loss of biodiversity is one of the most serious problems facing environmental managers. The consequences, in addition to the immorality of causing loss, are reduction of potential for new crops and pharmaceuticals; and possibly a less stable and resilient environment. Biodiversity issues are examined in Chapter 10.

Biosphere cyclic processes

Within the biosphere, numerous cyclic processes move and renew supplies of energy, water, chemical elements and atmospheric gases. These cycles affect the physical environment and organisms, and some are affected by lifeforms. Although upset by occasional catastrophic events such as volcanic eruptions or asteroid strikes, biogeochemical and biogeophysical cycles are assumed to reach a state of dynamic stability. Nevertheless, environmental managers must not assume an unchanging natural environment; also, human activity is affecting some global cycles, and may trigger serious runaway problems which could be difficult to solve.

Organisms play a key part in some of the cycles, of which the most critical include the maintenance of atmospheric gas mix and ensuring global temperature remains within acceptable limits. There are over thirty known biogeochemical cycles; some have a turnover of as little as a few days and others are so slow, with turnovers of perhaps millions of years, that the material is non-renewable as far as humans are concerned. Biogeochemical and biogeophysical cycles are not fully understood; for example, there is much to learn about the cycling of carbon, phosphorus, sulphur, and many other elements. Without better insight, accurate modelling and prediction of global change is difficult. Cycles may be classified as (1) natural, (2) upset by humans, and (3) recycling (managed by man and sustainable) (Chadwick and Goodman, 1975: 4). Many of group (1) have already been converted to (2); conversion of some type (2) to (3) is an important goal for environmental managers.

Environmental limits

Von Liebig's Law of the Minimum states that whichever resource or factor necessary for survival is in short supply is the critical or limiting one which restricts population growth (e.g. water, space, nutrients, recurrent fires, or a predator). The population reaching a limit may suffer gradual or sudden, limited or catastrophic collapse in numbers, a vacillation, or a cyclic boom-and-bust growth pattern. Solar energy drives most of the Earth's ecosystems; a very few exceptions include deep ocean hydrothermal vent communities and bacteria deep below ground level (Cann and Walker, 1993). Solar radiation receipts are thus the main limiting factor, and, in theory, it is possible to crudely estimate maximum global food production by mapping available surfaces and factoring in photosynthetic potential. Given that few of the world's agricultural strategies function at anything like potential maximum photosynthetic efficiency, some improvement of food and commodity production without further expansion of farmland could be possible. Factors limiting human development (e.g. key resources, ability to dispose of waste) may be modelled, and this has prompted debate as to what the comfortable and sustainable maximum global human population is. Miller (1991: 138) suggested that, with likely technology and foreseeable economic development, global population might reach ten or even thirty billion. Humankind is already more than halfway towards the lower end of these two estimates, so it is advisable to treat the problem with urgency. People also require reassurances such as a sense of security, adequate space, and law and order; these are more likely to be available at population levels below the possible maximum. If a population can be held a reasonable way below maximum population, there is more chance of sustaining a reasonable standard of living, and probably better adaptation in the event of a problem.

Caution is necessary when dealing with estimates of the population the Earth might support, as they are to some degree speculative. It may be possible to produce 40 tonnes of food per person for the 1990 global population, but will there be investment, environmental and social conditions allowing that productivity in the future, let alone improvement? In addition, disaster for one group of people may occur under very different circumstances than for others (Diamond, 2005). Meadows *et al.* (1992) and many others have argued that the limits have already been exceeded, but that there is still hope if appropriate development is pursued soon enough to convert 'overshoot' to a lower sustainable population.

Resources

A resource may be defined as: 'something which meets perceived needs or wants', so it is the expression of appraisal, a subjective concept (Zimmermann, 1993). Resources become available through a combination of increased knowledge, improving technology, and changing individual and social objectives. Mitchell (1997: 2) noted: 'In summary, natural resources are defined by . . . perceptions and attitudes, wants, technical skills, legal, functional, and institutional arrangements, as well as by political customs.' Economic and non-economic criteria determine utility. Non-economic criteria include: aesthetic quality, sense of moral duty to conserve wildlife, cultural importance, religious beliefs, and many others. An economist might subdivide resources into those with actual value, those with option value (possible use perceived), and those with intrinsic value (no obvious practical value, but there is a will to maintain them). In each development situation the environmental manager can recognise inputs (which include food, water and energy) and outputs (including sewage, garbage and heat), and resources such as a sense of security or attachment to the land which may also prove limiting. Von Liebig's law (stated above) recognises critical resources limiting ecosystem function and the survival of organisms. When dealing with human development, recognition of key resources and critical thresholds can be difficult due to the 'interface' of technology, culture and trade.

Resource demand changes as human perceptions alter, new technology is developed, fashions vary, environment alters, and new materials are substituted. Opponents of those warning that humans will exceed environmental limits and suffer have argued that economic forces (the 'invisible hand of the market') will intervene. Other optimists feel that demographic transition (to marked slowing or even negative population growth) is happening more rapidly in developing countries than it did in the past in nations such as France or Russia. It is unwise to wait and see what business as usual will bring (i.e. little or no significant change in development behaviour), and better to seek stronger controls on resource use. There is a case to be made for environmental management to place more stress on non-utilitarian goals, so that resources are valued for their own sake and, if need be, utilisation is forgone; how to effectively promulgate such a change is not clear.

A rough classification of resources useful for environmental management is as follows:

- those that can be safely and easily stretched by humans;
- those that can be stretched with care;
- those that cannot or should not be stretched.

Stretching of resources might be achieved through strategies such as the alteration of natural vegetation to agriculture; the conversion of slow-growing woodland to fast-growing plantation; farming of fish rather than fishing wild stocks, and so on.

Many natural resources are unevenly distributed and it is possible to miscalculate what is actually available – a major oil company recently admitted making gross over-estimates. The amount of a particular resource believed to exist is the total resource; the term 'identified resource' is applied to that which has actually been mapped and assessed. A reserve or economic resource is that which is judged extractable, given current technology, economic conditions and civil order. Undiscovered resources are those which specialists think are likely to exist but are unproven. Resources vital to a country are termed critical resources and those needed to ensure national security are strategic resources. There is often pressure to relax environmental controls and to alter other rules if strategic resources are to be developed. A comparison of known resource supplies and rates of use yields a depletion rate, typically the time it takes for 80 per cent of known reserves to be used.

Resources may be crudely grouped as: *non-renewable* (finite or exhaustible and can be used only once); *renewable* (if well managed, and there is no natural disaster, these can be used indefinitely); *inexhaustible* (resources such as sunlight, gravity, wind power and wave power, which it is virtually impossible to damage or over-exploit). Excessive use of a renewable resource or a disaster can alter it to a non-renewable. The process can be insidious; for example, carefully managed grazing may allow indefinite use, but a bushfire at a critical moment or even light grazing when plants are vulnerable could prevent regeneration and initiate soil erosion. In obvious cases of renewable resource overuse the term 'mining' is often applied to indicate usage in excess of the rate of recovery.

Part of the role of environmental management should be to exercise sound steward-ship over natural, human and economic resources. Specialist natural resources managers are usually employed to deal with minerals, water, forests, fisheries and so on; some-times the agencies and companies managing the various resources do not communicate, let alone co-operate. So, environmental managers may have to act as intermediaries, or somehow co-ordinate resource management. Large profits may be associated with exploitation, as well as the aforementioned strategic values, and this is likely to mean powerful challenges to environmental management efforts. Resource extraction may take place in remote areas where the big business frequently involved is difficult to monitor. In the real world many natural resource managers are unlikely to place environ-mental stewardship or social welfare as high on the agenda unless compelled to do so.

Monitoring, getting co-operation or enforcing environmentally sound stewardship is challenging, even more so when the resources are in common ownership, for example in international waters. Diamond (2005) felt that there was a better chance of sound resource management when powerful decision makers were not remote from other people and the 'grassroots' environment. Problems may be more likely when resource exploiters reside at a distance from the exploited resource. However, poor people may damage resources they are in contact with, because they have no choice but to do so to survive; also, people can use resources in a damaging manner through ignorance or unwillingness to change their outlook.

Resource exploitation usually depends on know-how. Environmental managers may therefore have to deal with human resources, technical skills, organisational abilities and knowledge. Some traditional knowledge may be useful worldwide, raising issues of ownership and royalties. Similarly, corporate knowledge gained by costly research may be needed by countries which cannot afford to pay back through royalties or market prices. There are also situations where resources lie on or under land occupied by indigenous peoples who may have very different values from those of the national government or world community. In the past such people were generally driven out, ignored or exterminated; now resource and environmental issues may require co-operation with them or learning from them. Some resource usage can thus present legal and ethical challenges, but law and ethics are still developing and may be inadequate to meet such needs.

Over the past half-century, as people have over-stressed the land, congregated in urban areas, demanded manufactured goods, and have been fed with the produce of modern farming, concern for outputs (pollution and waste) has grown. Ecosystems can each render a certain amount of contamination harmless – their assimilative capacity. The time needed for an ecosystem to deal with the pollutants varies, being affected by the types and quantities of pollutant received, the season, and other factors. This capacity may be seen as a renewable resource. However, the sudden arrival of a very toxic compound, large quantities of the usual pollutants, unusual weather conditions, or some other environmental variation, may cause a breakdown of assimilative capacity that is difficult or even impossible to restore. Before the spread of the 'polluter-pays' concept, outputs were rarely allocated economic value and were often ignored. During the past forty years there has been some progress in addressing pollution and waste problems; there is a growing awareness of the need to monitor and control carbon emissions, stratospheric ozone-scavenging compounds, pesticides, volatile organic compounds, polychlorinated biphenyls, nuclear waste, and many other compounds. There has been huge, but still inadequate, progress with measuring techniques, legal controls, and the establishment of international standards.

The Gaia hypothesis

Since the 1860s Darwin's concept of evolution – adaptation of organisms to the environment – has held sway (Goldsmith, 1990). However, the Gaia hypothesis proposed in 1969 by James Lovelock calls for some modification of evolutionary theory. James Hutton expressed similar views as early as 1785: he, and later Pierre Teilhard de Chardin, and Lovelock suggested that the biosphere acts as a self-evolving homeostatic system. The Gaia hypothesis received little support before the late 1980s, but acceptance has since been growing. If the Gaia hypothesis is proven, it would be a strong argument for a holistic approach to environmental management (Hunt, 1998).

There are several variants of the Gaia hypothesis (Lovelock and Margulis, 1973; Schneider, 1990: 8) but, whichever variant is accepted, it runs counter to the prevailing attitude in the West that humans can freely exercise controls over the Earth (Lovelock,

1979, 1988, 1992; Watson, 1991). Whether or not they accept the hypothesis, many have been stimulated by it to think carefully about environment and development issues. For example, it has helped provoke valuable research into the global carbon cycle. The Gaia hypothesis also provides a framework for people–environment study that is holistic (Levine, 1993).

Broadly, the hypothesis suggests that life on Earth has not simply adapted to the conditions it encountered, but has altered, and controls the global environment to keep it habitable in spite of disruption from factors such as changes in solar radiation, occasional asteroid strike, or large volcanic outpourings. The hypothesis seeks to explain the survival of life on Earth by treating the organic and physical environment as parts of a single system ('Gaia') in which biotic components act as regulators enabling control and repair (this is not a conscious process, nor is there implied a design or purpose). Temperature and composition of the Earth's atmosphere, according to the hypothesis, are regulated by its biota, the evolution of which is influenced by the factors regulated. Without Gaian regulation, the suggestion is that average global temperatures would be inhospitable to higher lifeforms, and atmospheric oxygen would probably be locked up in rocks.

In effect, the Earth is seen as a superorganism, a single homeostatic system with feedback controls maintaining global temperature, atmospheric gases and availability of nutrients. The controls involve a number of biogeochemical cycles, notably those of carbon dioxide, nitrogen, oxygen, sulphur, carbon and phosphorus. The system functions in the 'interests' of the physical environment and biota: the whole is greater than the sum of the parts. If so, humans are part of a complex system and must fit in, obey the limits or be cut out. Upset Gaian mechanisms, and there could be sudden, possibly catastrophic, runaway environmental changes.

Environmental crisis?

Warnings that the Earth faces a 'crisis' or is already in crisis have blossomed since the 1960s, some predicting disaster before 2000 (Ehrlich, 1970; Eckholm, 1976; White, 1993). 'Crisis' is a turning point, a last chance to avoid, mitigate or adapt. The cause is usually identified as one or a combination of the following: people's cavalier use of nature; over-population; misapplication of technology; faulty development ethics. What is perceived to be a crisis is subject to changing beliefs, fashion, technological ability and so on. One may recognise several categories of perceived crisis (the following are not arranged in order of importance, do not represent a comprehensive list, nor are they all wholly separate and discrete):

1 renewable resource depletion and degradation (especially shortfall in food production, problems with water, and energy supplies);
2 global environmental change;
3 pollution;
4 nuclear or biological warfare;
5 biodiversity loss;
6 increasing hunger and poverty;
7 increasing human repression, marginalisation and disempowerment;
8 rapid, often poorly planned, urban growth;
9 increasing population – this caused more concern in the 1970s than now;
10 debt burden – some regions may have problems due to debt repayment or structural adjustment measures introduced to counter it.

'Crisis' has become an overworked word which affects how people respond to warnings. People's circumstances and perceptions differ, so not all agree on what constitutes a crisis – 'crisis' for some may just be normal to others, and an opportunity to yet others. The term is also prone to emotive, journalistic usage (Blaikie, 1988). Some, mainly on the political left, suggest that the idea of a crisis may serve as a 'liberal cover-up' to divert attention from doing anything about 'real problems' such as social injustice and poverty (Young, 1990: 142–143). Other crisis supporters feel that environmental problems are mainly due to unsound concepts of development and modernisation – a social or ethical fault lies behind environmental crises (e.g. Weston, 1986: 4; Caldwell, 1990; Merchant, 1992: 17; Castro, 1993; Lomborg, 2001). In a few countries (e.g. Rwanda, Burundi, Bangladesh or Haiti) population densities are so great that 'Malthusian disaster' manifest as environmental degradation and genocide have resulted (Diamond, 2005: 307). At the time of writing (late 2005), fears of an oil crisis were being expressed, with the danger point likely within twenty years. Peak oil supply had been passed a few years previously and exploration companies did not seem to be finding enough new reserves.

While there are undoubtedly serious local or regional environmental and socioeconomic problems there is no immediate global crisis, although many would agree that current rates of population growth and consumption trends will cause one within a generation. A crisis-fighting, short-term focus approach to development planning is not a good idea. However, it may be necessary on occasion to conjure up a fear of crisis to get results. Environmental management issues often need a 'ginger group' to prompt action and a follow-up with sound research to establish what is happening and what is needed.

Identification of a large-scale crisis may be a mistaken response to a patchy, localised problem, reflecting inadequate observation. Careful research is vital, providing the environmental manager with the means for objective and careful monitoring, which helps prevent such errors (Thompson *et al.*, 1986; Blaikie and Unwin, 1988: 7; Blaikie, 1989). Writing on 'rural poverty unperceived', Chambers (1983: 13–27) noted a range of social science research errors which led to false impressions (physical science can make similar errors). For example, a researcher's tendency to view roadside areas and miss the 'interior'; the fact that the majority of studies are made during dry seasons; interviews with unrepresentative groups of people; research that is too short term; researcher bias. Ives and Messerli (1989) discuss areas of the Himalayas, which have often been identified as in environmental crisis, noting that there is little evidence that this is so. Indeed, some records show that conditions were markedly worse several decades ago. Blaikie and Unwin (1988: 13) cited an example of gully erosion in Zimbabwe identified as constituting a crisis, where careful study revealed that only about 13 per cent of total soil loss was from the spectacular gullies, while 87 per cent was from insidious inter-gully sheet erosion. Funds could easily have been spent treating gullying (a symptom of the problem) rather than sheet erosion (the actual problem). Another danger in adopting a crisis orientation is that decision makers suddenly respond to a problem (crisis management or 'fire-fighting') rather than make sustained efforts to avoid or solve it (Henning and Mangun, 1989: 3). It should be stressed that sudden crisis events are possible; so, rapid avoidance and mitigation responses are needed – a new form of civil defence is called for.

The world's growing number of environmental problems has often been interpreted as indicating a 'progressive loss of ecological stability' (Simonis, 1990: 26) – it may also reflect more research and awareness. There may be a risk of cumulative causation leading to a crisis; Sir Crispin Tickell noted: 'We can remove one, two, or ten rivets. But at a certain point – it could be the eleventh or the thousandth rivet . . . things fall apart' (*The Times*, 27 April 1991: 4). With any complex system there may be a failure of component parts, the breakdown of one of which is relatively insignificant, but sooner or later one breakdown might, alone or in combination with other factors, contribute to overall collapse. Environmental managers need to recognise significant thresholds and try to monitor whether these are being approached. An area of mathematics, catastrophe theory, which is concerned with the way in which systems can suddenly change by passing a crisis point, may aid the identification of critical environmental thresholds before they are reached. Threshold identification may also be assisted by ultimate environmental threshold assessment. This is derived from threshold analysis, which is based on the assumption that there are final boundaries which may be broken by direct or indirect, including cumulative, impacts. Kozlowski (1986: 146) defined these thresholds as 'stress limits beyond which a given ecosystem becomes incapable of returning to its original condition and balance'. It is possible to recognise temporal, quantitative, qualitative and spatial dimensions of these thresholds, and to assess their present and future status. There have been regional catastrophes which ultimate environmental thresholds assessment might have helped avoid, such as the 1970s to 1990s ruination of the Aral Sea, or the late 1990s recent forest fires in Brazil, Venezuela, Mexico and South East Asia.

The Brundtland Report rekindled global crisis warnings made in the 1960s and 1970s, and suggested a rough timescale: 'Most of today's decision-makers will be dead before the planet feels the heavier effects of acid precipitation, global warming, ozone depletion. . . . Most of the young voters of today will still be alive' (World Commission on Environment and Development, 1987: 8). At present global warming is generally seen to pose the greatest threat, yet as Stott (*The Times*, 4 September 2004) pointed out, it is a 'politico (pseudo) scientific construct' rather than scientifically proven. There are many other factors that could cause a crisis, including: population growth (compared with projected per capita availability of key resources such as land, water, food and fuelwood); pollution; soil degradation; and loss of biodiversity. In roughly one generation from now human population will probably have doubled, and might use 80 per cent of primary production. Even if climatic change and pollution do not depress photosynthesis, and if agricultural productivity improves, the limits are getting close, and living close to the limits is dangerous (Holmberg, 1992: 27).

Once potential causes of crisis are identified, those seeking to reduce the threat must apportion blame and achieve controls. Questions such as the following must be explored:

- Is there a global or a Southern crisis developing?
- Are the environmental problems faced by developed countries and developing countries the same?
- Are some or all of the developing countries' problems caused by the developed countries (or vice versa)?
- Are developing countries suffering more environmental damage than the developed countries?
- Are the developing countries or the developed countries more vulnerable to problems?
- With limited resources and political constraints, what deserves priority attention?

Countries have a tremendous diversity of environment, government, administration, historical background and so on. However, two things do seem to be widely shared by developing countries: poverty and environmental degradation. Whether this reflects accidents of history or special handicaps associated with the tropics has been debated

(Huntington, 1915; Adams, 1990: 6–8; Kates and Haarman, 1992). Developing countries' populations are currently growing more than those of the developed countries. However, they consume far less per capita of the world's resources than do developed countries. India and China are rapidly increasing consumption, which will place severe stress on the environment. In an interdependent world both developed and developing countries will have to co-operate, or conflict and failure to resolve problems will follow.

Africa is frequently singled out as having or being close to an environmental or poverty crisis, or both, especially in the Maghreb and south of the Sahara (excluding South Africa). The UK decided recently to focus more aid on African problems (Watts, 1989; Davidson et al., 1992). However, does Africa actually have a crisis compared with other parts of the South? Blaikie and Unwin (1988: 20) noted that Africa's soil erosion was not serious enough to call a crisis, although things are serious, and getting worse. Those who identify a crisis in sub-Saharan Africa (Harrison, 1987: 17–26, 56) argue that it is caused by:

1   a decline in per capita food production;
2   increasing poverty;
3   a debt crisis;
4   civil unrest (Africa, with less than 10 per cent of the world's population generated almost 50 per cent of the world's refugees in the late 1980s);
5   poor governance and corruption;
6   social factors (tribalism, greed, corruption, communal land use and so on).

Drought is often cited as cause of a sub-Saharan African difficulties, yet in most countries there is no conclusive evidence that rainfall receipts have diminished or become more variable in recent decades (Holmberg, 1992: 225). It is more likely that drought in Africa reflects or exposes other weaknesses – a 'litmus of development' (for a comprehensive report on the African environment see AEO, 2005; also available online at http://www.grida.no/aeo/ – accessed March 2005).

Often it is possible to recognise what might cause a problem, but tracing why these things happen is less easy. Western ethics are commonly blamed as the root cause of environmental problems – a cancer that colonialism has spread (metastasis) around the world. However, non-Western, non-colonised countries, remote areas not penetrated by capitalism and the former communist bloc also have serious environmental problems. Population growth cannot always be blamed, for there are situations where, despite very low settlement density, there has been severe damage. In various countries there are densely settled regions with unfavourable environments where people have sustained themselves for centuries (in the case of interior Papua New Guinea, with simple stone and wood tools). Population growth projections are therefore not a certain indicator that environmental problems will occur, although it makes it more likely.

Livelihood strategies, which long served people, often in harsh environments, have often broken down in recent years frequently causing environmental degradation. The reasons are diverse, including population increase; structural adjustment; social changes; spread of commercial agriculture; adoption of new crops; and restrictions on movement of people or livestock. It is also valuable to examine past crises to see what threats have materialised, and to try and unravel how society reacted. Lessons from the past may prove invaluable to modern environmental managers, but caution is needed because history rarely repeats itself exactly (Barrow, 2003; Diamond, 2005). It is also useful to look at developments in similar current environments, but there is no guarantee of the same outcomes.

The *Millennium Ecosystem Assessment* (MEA – see above, p. 55), backed by the UN, the World Bank, the World Resources Institute and scientists in ninety-five countries at a cost of US$24 million, is the first thorough worldwide assessment. The MEA indicates serious degradation of the Earth's life-support systems, lack of sustainable resources use, and a growing risk of abrupt and drastic environmental change. Whether governments will heed the warnings is uncertain; it is difficult at the time of writing to assess the impact this publication may have. Hopefully it will prompt more serious thinking and it does offer some hope that with proactive approaches disaster may be averted.

Summary

- Environmental management should consider the threat of infrequent but severe events and, whenever possible, steer development to reduce human vulnerability, conserve biodiversity and cultural riches, and enhance adaptability.
- Environmental management must look carefully at the physical, social and economic factors involved in each situation before drawing conclusions – false impressions are easily gained.
- In recent decades there has been a spread of interdisciplinary approaches.
- Ecosystems are widely used as study, planning and management units.
- Few ecosystems are wholly natural; many have altered drastically and must therefore be managed to avoid degradation – nature cannot regain control.

Further reading

Diamond, J. (2005) *Collapse: how societies choose to fail or survive.* Penguin/Allen Lane, London.
Explores how past civilisations failed to withstand environmental, social or economic crises; Diamond asks what modern societies can do to enhance their chances.

Dickenson, G. and Murphy, K. (1998) *Ecosystems: a functional approach.* Routledge, London.
Coverage of the ecosystems approach.

Jackson, A.R.W. and Jackson, J.M. (1996) *Environmental Science: the natural environment and human impact.* Longman Scientific and Technical, Harlow.
Introduction to environmental concepts and issues.

Lovelock, J. (2006) *The Revenge of Gaia: why the earth is fighting back – and how we can still save humanity.* Allen Lane, London.
Call for action and a rexiew of what is being done.

O'Riordan, T. (ed.) (1995) *Environmental Science for Environmental Management.* Longman Scientific and Technical, Harlow.
Widely used interdisciplinary introduction to environmental science.

4 Environmentalism, social sciences, economics and environmental management

> When the history of the twentieth century is finally written, the single most important social movement of the period will be environmentalism.
>
> (Nisbet, 1982: 10)

Before the 1940s only a few individuals expressed environmental concern; by the 1950s there were some environmental lobby groups and NGOs. National governments had passed pieces of environmental protection legislation since the eighteenth century or earlier and by the 1940s there had been a few international agreements. Richer nations had environmental management professionals (mainly trained as scientists with limited social or economic skills or political experience) concerned with pollution control, conservation, agriculture and fisheries by the 1930s. However, it should be stressed that before 1970 very few citizens knew the words 'environment' or 'ecology', environmental problems were seldom important political issues and economics wrote of environmental costs as 'intangibles' or unimportant. Environmental concern and management have come a long way since the 1960s.

Growing environmental concern (1750 to 1960)

Some societies protect certain plants and animals for reasons of religion or local economy (e.g. baobab trees are protected by people in many parts of Africa, and here and there rulers established reserves for hunting and recreation in parts of India before the fifteenth century). From the late seventeenth century European and American geographers, explorers and naturalists popularised natural history among the leisured classes, stimulated academics to seek better understanding of it, and encouraged policy makers to legislate for better treatment of nature. By the 1760s colonial powers were enacting legislation to try to protect forests on Tobago, Mauritius, St Helena, and other islands (Grove, 1992, 1995).

Two broad groupings of environmentalists (see discussion of environmentalism below) had evolved in Europe and America by the late nineteenth century.

Utilitarian environmentalists

In the late nineteenth century the British sought assistance from German foresters to sustain timber production in Burma and India. Political theorists (e.g. Pyotr Kropotkin in Russia) professed forms of 'utilitarian environmentalism' by the 1890s, which aimed to improve man through better working and living conditions (Kropotkin, 1974). Kropotkin, an anarcho-communist, argued for small, decentralised communities close to nature and avoiding industrialisation and the division of labour – something quite similar to what many environmentalists seek nowadays. Europe had similar 'utopian liberals' and proto-socialists such as William Morris (1891), and social reformers such as John Ruskin and Robert Owen (the latter founded utopian colonies, with limited success, in the UK, Ireland and the USA in the 1820s). In South Africa, other African colonies and India, legislation was passed to try and reduce soil erosion, control hunting and conserve forests and areas of outstanding natural beauty. By 1900, reserves had been established in Kenya and South Africa, often by hunters or ex-hunters (Fitter and Scott, 1978; Dalton, 1994). In North America by the 1850s, damage to forests, wildlife and soil was marked. Some already feared frontiers were closing and that limitless land and resources were a thing of the past. One of those who were concerned was George Perkins Marsh, who in 1864 published an influential, if somewhat deterministic, book: *Man and Nature*. This and publications by others prompted action – essentially two groups concerned for the American environment formed in the late nineteenth century: 'preservationists' and 'conservationists'. The former included John Muir, who wished to maintain unspoilt wilderness areas; the latter included Gifford Pinchot, and both were prepared to see environmental protection combined with careful land use (McCormick, 1989). Environmental managers still face this preservation or conservation choice today.

During the 1860s the US National Parks Service and the US Forest Service were established. Pinchot, Chief of the US Forest Service between 1890 and 1908, was a major force in establishing parks and reserves and is one of the founders of 'conservation', although the British already had conservancies in India. John Muir has also been hailed as 'Father of the US conservation movement'. In 1892 he founded the Sierra Club in California – still an influential NGO, it played an active role in promoting popular environmental concern between the mid 1960s and mid 1970s; it also gave rise to Friends of the Earth, one of today's foremost environmental NGOs (for a history of the American conservation movement, see Kuzmiak, 1991).

After 1917 divergence of development paths between Russia, and later other socialist economies, and the West made little difference – both had and have serious environmental problems (Gerasimov *et al.*, 1971; Komarov, 1981; Smil, 1983; De Bardeleben, 1986). The eastern bloc has played an active part in international conservation and environmental protection activities, and the former USSR, China and Cuba have established many national parks and reserves.

Romantic environmentalists

Eighteenth- and nineteenth-century industrial revolution led, especially in Europe and North America, to overcrowded, filthy cities, damaged countryside, loss of commons, disease and misery. Various intellectuals questioned capitalism, agricultural modernisation and industrial growth. Some were dubbed 'romantics', saw nature as a source of inspiration, and advocated a less damaging relationship with the environment. They include poets like Wordsworth, Blake and Coleridge, writers like Henry Thoreau (1854), and artists like Holman Hunt and John Turner. They inspired twentieth-century environmentalists, but their contribution is 'more escapist than visionary' (for a review of romantic environmentalism see Bate, 1991).

Drought in the USA midwest Dust Bowl, especially between 1932 and 1938, caused crop loss and soil erosion. The wind-blown dust was apparent as far away as Chicago and Washington, DC. Large numbers were ruined and displaced. The folksinger Woody Guthrie and novelist John Steinbeck commented on the degradation and misery; at first seen as subversives, they helped provoke public and government concern. To counter these problems President Franklin D. Roosevelt promoted integrated development of natural resources, and in 1933 he established the US Soil Erosion Service and in 1935 its successor, the US Soil Conservation Service.

The Second World War hindered the growth of concern for the environment, accelerated the development of resources and led to the production of new threats such as DDT and atomic weapons. During the first decade or so after 1945 efforts focused on economic and industrial reconstruction, on raising agricultural production, and on the Cold War. A few publications on the environment began to appear from the late 1940s (Osborn, 1948; Vogt, 1948; Leopold, 1949; Dale and Carter, 1954; Thomas, 1956). Of these it was especially Aldo Leopold (1949) who stimulated many of the 1960s to 1970s environmentalists. In 1949 the UN held one of the first international environmental meetings, the Conservation Conference at Lake Success (USA), and during the early 1950s helped establish the International Union for the Protection of Nature, which in 1956 changed its name to the International Union for Conservation of Nature and Natural Resources (IUCN).

Environmental concern from the 1960s to the 1980s

NGOs began to speak out on environmental issues in the late 1950s. By the mid 1960s there had developed what has been variously called an environmental(ist) movement, environmentalism, the ecology movement, an environmental revolution, and the conservation movement. In the 1960s and 1970s, particularly in California, public-interest law firms (such as the Environmental Defense Fund or the Natural Resources Defense Fund), supported by grants or foundations, acted on behalf of citizens or groups of citizens (previously action had to be undertaken by individuals) to protect the environment (Harvey and Hallett, 1977: 62). Understanding of the structure and function of the environment was improved by initiatives such as the International Geophysical Year (1957–1958), the International Biological Program (1964–1975) and the International

Hydrological Decade (1965–1974), plus expanding research. The USA Civil Rights movement, hippies, the anti-Vietnam War movement, European anti-nuclear weapons protests and the 1960s to 1970s 'pop culture' in general encouraged people to ask awkward questions about environment and development (Maddox, 1972; Ward and Dubos, 1972; McCormick, 1989). After a peak of interest in the early 1970s media coverage and public interest declined from 1974 until the mid 1980s, then climbed (Sandbach, 1980: 2–6; Simmons, 1989: 6; Atkinson, 1991b).

In the 1970s, the environmentally concerned, although active in publication, litigation and protest, were relatively non-political (in New Zealand, Germany and the UK politically active Green Movements were developing) (McEvoy, 1971; Morrison, 1986; Dunlap and Mertig, 1992). The focus was on over-population (Ehrlich, 1970), conservation of wildlife, and problems associated with technology (Farvar and Milton, 1972). Many of the publications between the mid 1960s and mid 1970s were dogmatic: warning of coming crisis, so that some environmentalists became known as 'prophets of doom' or 'ecocatastrophists' (White, 1967; Commoner, 1972). Miscalculation, hyperbole and other biases frequently clouded environmental campaigning in the 1970s, and a great deal of what happened was apocalyptic advocacy short on practical solutions.

In 1965 the US Ambassador to the UN, Adlai Stevenson, used Buckminster Fuller's metaphor Spaceship Earth in a speech; Boulding (1971) also used it, and the catchphrase spread the idea that the world was a vulnerable, effectively closed system. The International Biological Program, and later the UNESCO Man and Biosphere Program, helped establish an awareness that global-scale problems were real and the Earth's resources were finite. By 1970 some identified population growth as the primary cause of environment and development problems – neo-Malthusians (Ehrlich, 1970; Ehrlich et al., 1970). The more extreme neo-Malthusians went so far as to discuss the possibility of triage (withholding assistance from over-populated countries with little chance of improvement, to concentrate resources on recipients who might with help achieve control). Neo-Malthusian views have been criticised as simplistic and invalid (Boserüp, 1990: 41; Todaro, 1994: 339).

Hardin (1968, 1974a, 1974b) published an essay on the fate of common property resources in the face of population growth. His 'tragedy of the commons' argument was that people will tend to overuse commonly owned resources, in all probability destroying them, because without overall agreement each user seeks to maximise short-term interests and does not assume sufficient responsibility for stewardship (Box 4.1). Hardin's views have been widely attacked on several grounds, one being that he was describing more of an open-access resource situation than most common property resource exploitation. Harrison (1993) noted that seldom is use of commons a free-for-all; communities do generally have some controls and manage things.

Two early 1970s publications helped shake the West's complacency: *Blueprint for Survival* (Goldsmith et al., 1972) and *The Limits to Growth* (Meadows et al., 1972). The latter was intended to promote concern and further research, and explored a range of possible future scenarios which depended on how population and other key development parameters were managed (McCormick, 1989: 75). A second Club of Rome report was published by Mesarovic and Pestel (1975) and a heated futures debate developed between those advocating slow or even zero economic growth and others, such as Kahn et al. (1976) or Simon (1981), of the view that an open-access free market and human ingenuity would overcome environmental difficulties before limits were met, making it unnecessary for zero growth (Freeman and Jahoda, 1978; Hughes, 1980). Those with excessive optimism about limits have been called 'cornucopian' and the over-pessimistic 'Cassandras' (Cotgrove, 1982). Critics of the warnings find data and

Box 4.1

Common property resource

The relationship between the returns to labour on a given resource (e.g. cropland or a fishery) and the number of labourers exploiting it

Under private ownership For any additional employee hired beyond N*, the cost to the producer W will be greater than the employees' marginal product, and the difference will represent a net loss to the owner. To maximise profit requires the hire of N* workers, with a total output equal to AP* multiplied by the number of workers, N*.

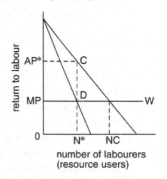

Under a system of common property Each worker is able to appropriate the entire product of their work, which is equivalent to the average product of all workers. Worker income will continue to exceed the wage (W) until enough workers are attracted to cause the average product to fall to the level of the wage, at which point the labour force = Nc. The implication is that aggregate welfare will fall and resource use is inefficient (and causes degradation).

Sources: Drawn from several sources, including Todaro (1994: 338–339)

modelling faulty; the Club of Rome was accused of 'crying wolf', and it is held up as a case of a situation where applying the precautionary principle in 1972 and freezing growth would have caused huge poverty and probably famine.

The 1972 warnings have been rechecked, new data have been substituted, and computer modelling has been improved – sequels have been published (Meadows *et al.*, 1992; Van Dieren, 1995; Meadows *et al.*, 2004), with the message that severe problems are virtually certain within fifty years; however, catastrophe could be avoided, provided the right measures are adopted soon. Economists and politicians have been slow to heed the warnings.

There have been a number in the 1970s who blamed industrialisation and capitalism for environmental and social problems. One was Schumacher, who in 1973 warned that giant organisations led to specialisation, economic inefficiency, environmental damage and inhuman working conditions. The remedies he offered included 'Buddhist economics', 'intermediate technology' (technology using smaller working units, local labour and resources) and respect for renewable resources. Starting in 1970 the USA passed a number of key pieces of environmental legislation, notably the National Environmental Policy Act (NEPA). NEPA established environmental impact assessment (EIA) and proactive environmental legislation in the USA; within ten years EIA had spread worldwide to become an important input to environmental managment.

Fashions shift; in the 1960s and 1970s technology was generally welcomed and technocratic approaches were often adopted. Perhaps as part of a process of winning respect, social scientists often used quantitative data and complex statistical techniques; unfortunately this 'quantitative revolution' sometimes resulted in approaches which were not transparent and sometimes proved inflexible.

Environmental concern from the 1980s to the present

Two seminal publications of the early 1980s were the *World Conservation Strategy* (IUCN, UNEP, WWF, 1980) and the Brandt Report (Independent Commission on International Development Issues, 1980). The Brandt Report stressed that many world problems would be solved only if it was recognised that rich and poor countries had a mutual interest – the solution of developing countries' problems was not just a question of charity but of global interdependence. The *World Conservation Strategy* promoted conservation for 'sustainable development' (the first time the latter phrase was widely publicised). The World Commission on Environment and Development set out in 1984 to re-examine environment and development problems and to formulate proposals for solutions. The Commission's findings (the Brundtland Report – World Commission on Environment and Development (1987) *Our Common Future*) highlighted the need for sustainable development and urged a marriage of economics and ecology. The Brundtland Report may be said to have initiated a new relationship between social science, natural science, economics and policy making, and is one of the most important publications of the twentieth century.

By the late 1980s the World Bank had adjusted its policies to give greater support to environmental management (Warford and Partow, 1989), oil prices had fallen, and a Green Movement had emerged, particularly in Europe, embarked on policy advocacy and made politicians of all persuasions aware of environmental issues. By 1988 environmental matters were on the agendas of politicians and decision makers with a higher public profile than ever before. Although green activity in politics has declined from a peak in the early and mid 1980s (Bramwell, 1994), it is by no means on the wane. By the 1990s fashion had shifted towards transparent and participatory approaches. Since the mid 1980s fashion seems to be focusing on integration and holistic approaches. In the past few years in a growing number of countries, 'popular environmentalism' and 'environmental justice movements' have been appearing as marginalised groups in urban and rural situations look for environmental improvement, better livelihoods and sustainable development. These goals demand better environmental valuation and understanding of environmental conflicts, and political ecology and ecological economics are seen to be of potential help (Martinez-Alier, 2003).

Environmentalism, ecologism and the Green Movement

By the mid 1980s, many environmentalist groups had developed, the members of which were willing to alter their lifestyles and encourage or force others to do so, in order to try and halt environmental damage (Buttel, 1978). It is difficult to give a precise coverage: what follows is intended to serve as a brief introduction. The expansion of green (environmentally concerned) thinking has coincided with the weakening of communism and socialism; indeed, some claim it played a significant role in that decline. Many greens see economic growth and consumerism as tainting both Western capitalist economies and those developed from socialist states. Less radical green philosophy can be embraced by existing politics, from liberal to dictatorship; radical deep green beliefs demand fundamental changes in politics, worldviews and ethics.

Environmentalism

'Environmentalist' was not used prior to the 1970s, but has been applied retrospectively to those involved in environmental matters long before that (Pepper, 1984; Grove, 1990,

1992). Environmentalism has been described as: 'a moral code or a set of mediating values to manage human conduct' (O'Riordan, 1976: viii); 'activism aimed at improving the environment' (http://www.en.wikipedia.org/wiki/Environmentalism – accessed March 2005); 'concern for environment elevated to a political pursuit' (McCormick, 1989: ix; Fox, 1995). Environmentalism calls for a managerial approach to environmental problems, secure in the belief that they can be solved without fundamental changes in present values or patterns of production and consumption. Dobson (1995: 1) argued that it was not an ideology, but rather a diverse group of people who all share a concern for the environment and seek sustainable development, even if their ideologies and exact objectives differ (O'Riordan, 1991) (see Box 4.2).

From the 1960s to mid 1970s environmentalists operated with what Rees (1985: 2) called 'messianic fervour'; they stimulated popular interest that has seldom ventured from advocacy to real solutions or political activism (Lewis, 1992). That had changed by 1980, with environmentalism increasingly involved with politics, commerce, law and business (Wilson, 1994). Some environmentalists are willing to embrace technology, biotechnology and the free market; many will not (Anderson, 1993; Narveson, 1995). There are also those on the side of science and rationalism who challenge environmentalism and the Green Movement (Brick, 1995).

Box 4.2

Some common green characteristics

The 'four pillars of green':

1   ecology
2   social responsibility
3   grassroots democracy
4   non-violence

The 'six values of green':

1   decentralisation
2   community-based economics
3   post-patriarchal principles
4   respect for diversity
5   global responsibility
6   future focus

Green characteristics:

• holistic approach
• disillusionment with modern unsustainable development paths
• non-violence
• a shift in emphasis away from philosophy of means to ends
• a shift away from growth economics
• a shift towards human development goals
• a shift from quantitative to qualitative values and goods
• a shift from impersonal and organisational to interpersonal and personal
• commonly a feminist interest
• a decentralised approach – 'think globally, act locally'

Sources: Spretnak and Capra (1985: xx); Porritt (1984: 10, 15); Merchant (1992: 15)

Environmentalism, it has been suggested, is a rejection of modernism (Pepper, 1996). Modernism may be defined roughly as 'seeking to fulfil human needs through the development of technology and the creation of wealth'. This has caused problems, and led to calls for postmodern alternatives (for a discussion of modernity see Giddens, 1991; and for postmodernism, Harvey, 1989). While 'postmodern' is widely used, the concept is confused (Funtowicz and Ravetz, 1991). Some recognise an ongoing postmodern period, beginning during the early 1960s (Frankel, 1987; Cosgrove, 1990: 355), characterised by the collapse of 'normality' and increasingly post-industrial or post-material activity and a holistic worldview (Bell, 1975; Roszak, 1972, 1979).

A postmodern and holistic approach might offer ways of understanding cultural and environmental phenomena, especially when circumstances demand multidisciplinary study of problems (Capra, 1982; Cheney, 1989; Warford and Partow, 1989; Kirkpatrick, 1990; Young, 1990). There are also signs that maths and fundamental physics are moving from Cartesian order (the systematic, reductionist approach to understanding chaotic complexity) towards postmodern holism, for example, by embracing chaos theory and fractals (Peat, 1988: 341; Lewin, 1993). Some have gone beyond postmodernism to advocate a post-environmentalism approach to environmental management (environmentalism is a reformist philosophy which tends to maintain a distinction between human affairs and nature; post-environmentalism seeks to reduce that separation when developing environmental ethics) (Pearce et al., 1989, 1990, 1991; Barde and Pearce, 1990; Pearce and Turner, 1990; Gare, 1995). The postmodern concept may prove useful, given that it is increasingly difficult to maintain a separation between science and politics and so on. 'Ecologism' is a generic term for an ideology that argues for care of the environment and a radical change in the human relationship with nature to get it. Put crudely, ecology is the science and ecologism is a worldview that draws upon it (Dobson, 1994; Kirkman, 1997). Dobson (1990: 36) described ecologism as 'the ideology of political ecology'.

Ecologism

Ecologism lies at the radical or fundamentalist end of the environmentalist spectrum. It is a political ideology or philosophy for relating society to nature with a strong spiritual component. Dobson (1990, 1995) noted that adherents hold that a sustainable, fulfilling existence 'requires radical changes in the human relationship with the natural world, and in the mode of social and political life' (most deep greens would support this). Other environmentalists are usually willing to manipulate and alter the environment if human needs are pressing enough (Smith, 1998).

Green spirituality

Spiritual ecologists include those who focus on established Western religion; for example, Pierre Teilhard de Chardin (1959, 1964) and Matthew Fox (Fox, 1983, 1989; Spretnack, 1986; Nollman, 1990; Merchant, 1992: 124; Kimmins, 1993; Gottlieb, 1996; Kearns, 1996); and those who look to pre-Christian religions of Europe, America or the Orient for inspiration to transform human consciousness so that it will have reverence for nature. Environmentalists often blame problems on the Western dominant Judaeo-Christian worldview (Cooper and Palmer, 1990, 1992; Barkey, 2000). In 1986 the World Wide Fund for Nature held its twenty-fifth annual meeting at Assisi, where leaders of Buddhist, Christian, Hindu, Islamic, Judaic and other faiths established an International Network on Conservation and Religion, and published the Assisi Declarations on Man and Nature (WWF, 1986). Batchelor and Brown (1992) explored Buddhism and ecology. In some countries sacred groves and other religious sites are often important biodiversity conservation sites (Singh, 1997). A prominent American environmentalist (S. Clark, Executive Director of the Colombia Foundation) observed: 'when we use the term "environment" it makes it seem as if the problem is "out there" . . . the problem is not external to us; it's us.'

The Green Movement

The Green Movement is a very diverse social or cultural movement that shares a common environmental concern and which often embarks on political action, mainly of a reformist or radical nature. 'Green' roughly means 'environmentally friendly'; 'greening' roughly means 'environmental improvement'. Reich (1970), writing about the possibilities for a new development ethic after the demise of the corporate state, was probably the first to use 'greening'. The use of green terminology increased after the mid 1980s in politics and as a popular alternative to 'environment', soon becoming common in media discussions.

There is little about green philosophy that is wholly new (Hill, 1972; Weston, 1986). Although 'green' often implies politicised environmentalism, some groups are not politically active, and even eschew politics. Greens are essentially mounting a cultural attack on the ills of modern society and economics, a sort of parallel to the economic attack by socialists (Redclift, 1984; Adams, 1990: 71). What would probably have been called Gandhian in the mid 1970s is now likely to be called green. Greens may be socialists, conservatives, intellectuals, poor or rich people, Buddhists, Christians, Muslims or humanists. Most share a fear that industrial nations are pursuing an unsustainable, dangerous development path (Porritt, 1984: 15). Greens may be roughly subdivided into romantic, anarchistic and utopian; or simply into 'light' or 'dark'. They may be said to have grown from partially American roots and to draw upon the writings of Henry Thoreau, Theodor Roszak, Ivan Illich, Aldo Leopold, Martin Luther King and others (Roszak, 1979; Spretnak and Capra, 1985: xvii; Devall and Sessions, 1985) (see Box 4.2).

The Green Movement has tended to develop a schism between light-green (or shallow) and deep-green (or deep) ecology. The division was largely initiated by the Norwegian philosopher and founder of deep ecology Arne Naess (Naess, 1973, 1988, 1989). It may be more accurate to talk of deep and shallow ecologies, as there is a wide spectrum of interpretation of what 'ecology' means.

European environmentalists became politically active in the 1970s; a leading role was played by Hamburg greens (die Grunen). By the 1980s greens had won a number of parliamentary seats in Germany and some other European countries. Early 1980s popularity in Europe faltered from the mid 1980s, partly because established political parties partially hijacked the green cause. In the USA environmentalists concentrated more on getting supportive legislation and advocacy, and green politicisation barely took off.

Deep green or deep ecology seeks to replace the existing social, political and economic status quo with new environmentally appropriate bioethics and supportive politics. Supporters blame many environmental problems on the anthropocentric nature of modern development, and adopt a biocentric (ecocentric) outlook, granting all life (human and non-human) intrinsic value (Evernden, 1985; Grey, 1986; Devall, 1988; Sessions, 1994, 1995). In general, deep ecology is synonymous with radical ecology and extends beyond the approach proposed by Naess, to include perspectives such as social ecology and eco-feminism, and some incorporate Taoist or Gandhian philosophy. It may be argued that deep ecology gives non-scientific input similar importance to (if not greater than) scientific, and is sometimes hostile to science. North (1995: 3) noted that greens often overlook the fact that they are the 'flowering of a science-based industrial society', and that unreasoning opposition to scientific progress, business and so on may not achieve useful environmental progress. Some of the most radical groups of deep ecologists such as 'Earth First!' may use violent methods – 'monkeywrenching' or 'ecotaging' (forms of sabotage) (Abbey, 1975; Davis, 1991a).

Social ecology is generally seen as a deep-green stance; it was largely initiated (in the USA) by the anarcho-socialist Murray Bookchin, who was critical of deep ecology (Bookchin, 1980, 1990; Light, 1998). In some camps it is seen as separate from deep ecology; others view it as an offshoot. Social ecology supporters see environmental problems as basically the result of social problems, and adopt an anthropocentric, decentralised, co-operative approach – a sort of eco-anarchy (Tokar, 1988; Devall, 1991). Another difference from mainstream deep ecology is that social ecology is humanist rather than ecocentric.

Light-green or shallow ecology seeks to apply ecological principles to ensure better management and control of the environment for human benefit – it is usually anthropocentric. There is far less of the rejection of established science characteristic of most deep ecologists. Shallow ecology is more inclined to try to work with existing economics and ethics (Fox, 1984, 1995); it is more likely to be concerned with solutions than with efforts to avoid problems in the first place. If decisions have to be made to protect the environment without adequate proof, deep ecologists are more likely to give support because they require no obvious human advantage. Jacob (1994) explores the potential of deep and shallow ecology as routes to sustainable development. It seems unlikely that extreme stances – deep or shallow – can effectively serve environmental management; some blend of their ethics is required, which places adequate emphasis on science (Norton, 1991).

Ways in which social sciences and environmentalism support environmental management

The social sciences provide information for one side of the human–environment interrelationship (Burch *et al.*, 1972; Sutton, 2004). The potential inputs to environmental management from the social sciences are:

- to provide information on social development needs and aspirations to explain, present and predict future human attitudes, ethics and behaviour;
- to study and develop ways of focusing the activities of social institutions, nongovernmental organisations, groups of consumers and so on to achieve better environmental management;
- to show the environmental manager social constraints and opportunities;
- to unravel the often complex and indirect social causes of environmental problems;
- the articulation and fulfilment of the shared interests of people (so far mainly at the local, regional or national level). National governments have mainly been reactive rather than forward-looking: social science will be needed to clarify how people think, nations relate to each other and institutions behave if a more proactive approach is the goal;
- to cut through 'technological determinism' so that the voice of social science may be heard (Redclift and Benton, 1994).

Environmentalism plays a vital role in the evolution of better environmental ethics but some of it is of limited value. Adams (1990: 83) warned: 'it is necessary to move outside environmental disciplines, and outside environmentalism, to approach the problem from political economy and not environmental science . . . to the understanding of environmental aspects of development which uses both natural science and social insights.' An example of such an integrated approach is that applied by Blaikie (1985) to the problem of soil erosion (Box 4.3). There has been a huge increase in the interest taken by social scientists in the environment since the late 1970s, with a shift from mainly enlightened activists at first to more widespread interest since 1992 (Chappell, 1993). There has been borrowing of concepts and jargon by social science from the environmental sciences, but sometimes things become distorted because some environmentalists and agencies do not derive their concepts by a process of logic, but bolt on scientific justification to values they already hold. For example, the sustainable rural livelihoods approach is widely used by development agencies and researchers (having originated in the early 1990s – Chambers and Conway, 1992). This appears to support sustainable development; however, users vary in their interpretation of sustainability, some virtually ignoring the need to avoid environmental damage.

There is inconsistency and imprecision of terminology, so it is often a good idea to try to understand the stance of those involved (Moghissi, 1995) (Box 4.4). There has been progress in recognising ongoing problems and predicting future impacts, and in exploring causes of problems (Albrecht and Murdock, 1986; Yearley, 1991; McDonagh and Prothero, 1997: 21). Social science has contributed in a practical way to social forestry management; agricultural development (e.g. advice on extension and project implementation); irrigation extension and management; pastoral development and range management; involvement of indigenous peoples in conservation; fisheries management and conservation; human resources management; risk perception; hazard avoidance; consumerism; property rights, and much more (Shankar, 1986). Historians explore past attitudes and approaches to environment; political studies specialists and economists consider the politics and economics of environmental usage; and theologians and philosophers probe the human–environment relationship. Anthropology and human resources management are increasingly used to inform the environmental manager about human behaviour, attitudes and beliefs, institutions, and organisational capacity (Wehrmeyer, 1996). Environmental management has also been much influenced in recent years by the development of *participatory* research, management, monitoring and appraisal (Burton *et al.*, 1986; Brokensha, 1987; Montgomery, 1990a).

Anthropologists have worked more with environmental managers than sociologists: working with indigenous peoples, livelihood strategies, archaeologists, palaeoecologists and ecologists, they often help clarify human–environment interrelationships. Anthropological input has been especially strong in the fields of relocation and resettlement, pre-development appraisal, impact assessment, conservation area management planning, and in studies of resource use, hazard perception and survival strategies adopted by land users (Jull, 1994; Blackburn and Anderson, 1995). Ethnobotany involves anthropologists and ethnographers assessing indigenous peoples' use of plant and animal resources in the hope of identifying useful crops, pharmaceuticals and so on. Anthropologists have also played a role in helping governments and environmental managers to understand and reach working arrangements with indigenous peoples, and in assessing social and cultural impacts of development on them (Snipp, 1986; Dale, 1992). The development of environmental sociology and ecosociology explores the interactions between society and the environment (Barry, 1999), and has rather eclipsed human ecology (Hannigan, 1995; Irwin, 2001).

Box 4.3

Concepts dealing with human–environment relations which may have discouraged social scientists from taking an interest in environmental management

Environmental determinism

From the 1870s a number of environmental determinists argued that the human–nature relationship was such that physical factors (e.g. climate) influence, even substantially control, behaviour, and thus society and development. For the past half-century these views have attracted condemnation. Some (e.g. Pepper, 1984: 111–112) recognise 'crude' and 'scientific' environmental determinists. Crude environmental determinism, and associated concepts, like comparative advantage, were expressed by intellectuals (e.g. Richter, Kant, Ritter, Ratzel, Semple (1911) and List). Scientific environmental determinists (e.g. Ellsworth Huntington, 1915) were a little more objective (Simmons, 1989: 3).

There can be no doubt that human fortunes often reflect natural events. However, much of what has been written by environmental determinists ignores the fact that humans can make different choices under similar environmental conditions, and often modify the environment. Nevertheless, environmental determinism is not dead and debate about its value continues, especially among social scientists, geneticists and psychologists concerned with inheritance of traits, deviant behaviour and upbringing, culture and anthropology (Milton, 1993, 1996). Supporters of the Gaia hypothesis could be said to accept a type of neo-determinism, and the interpretations of human development history put forward by Diamond (1998) are distinctly deterministic (Stout, 1992; Frenkel, 1994; Mannion, 1996).

Social Darwinism

Closely allied to environmental determinism is the concept of social Darwinism. At its core was the idea that humans are fundamentally controlled by nature – competition and struggle, rather than co-operation and mutual aid, were seen as natural and justifiable ways to behave, and the group best able to adapt to the environment would become dominant (Pepper, 1984: 134; Chappell, 1993). By the 1920s eugenics was supported by many as a way of improving a particular human group's genetics and thus their long-term survival and achievements. Eugenicists encouraged the breeding of 'desirable' people and suppressed 'undesirables' – an approach embraced in Nazi Germany. By the 1950s it was accepted in most quarters that social and economic development could overcome environmental factors and determine evolution, so social Darwinism fell out of fashion.

Environmental possibilism

A concept put forward by Vidal de la Blanche, and later by Febvre (1924) – environmental possibilism – holds that the environment constrains human endeavour and sets limits, but that choices between courses of action for man are possible within those limits; the same environmental opportunities may be used differently by the various cultures.

Box 4.4

Broad groupings of greens (avoiding deep and shallow categorisation)

Conservationists/traditionalists Heirs to the nineteenth-century romantic liberal rejection of industry and materialism. Less interested in drastic change of attitudes and lifestyle than some greens. Includes traditional conservationists such as members of the UK Royal Society for the Protection of Birds or the Council for the Protection of Rural England, and in the USA of the National Audubon Society or Sierra Club.

Reformists No particular tradition, midway between the previous and following groupings. Tend to be single-issue groups with problem-orientated aims (e.g. a group opposed to construction of a new airport or road or rail route).

Formal political parties and political groupings For example, Die Grunen, UK Green Party, Greens in the European Parliament, SERA. These produce regularly revised manifestos of wide-ranging policies. Green thinking has also been incorporated into the policies of a range of political institutions and has prompted new perspectives. Academic responses to green issues – Marxist/structuralist and market (mainstream) economics – tend to be hostile to or dismissive of many green paradigms (including new economics and some aspects of sustainable development).

Radical environmentalists Draw ideas from sources such as Kropotkin, Henry Thoreau, Theodor Roszak, Aldo Leopold, Godwin and so on. Recognise the need for considerable change in attitudes and lifestyles because environmental problems arise. They seek to alter other people's outlook, the economic system, social inequalities and so on. Often take a holistic, multi-issue approach. Considerable range, from moderates (e.g. Friends of the Earth) to extremists (e.g. Earth First!), who espouse militant tactics such as 'ecotage' (sabotage of things and people they see as a threat to the environment), 'ecovangelists' (who profess reverence for environment, not just stewardship) and even shamanists. (A schism has opened up between practical and spiritual factions of Earth First!)

Eco-feminists Believe women need to organise to achieve sustainable development and blame male-centred approaches to development rather than anthropocentric approaches, so can be hostile to deep ecology.

Cornucopians Place faith in technology and science as a solution for environment and development problems (e.g. Fuller, 1969).

Rational Seek to use science, social science and technology with care to achieve sustainable development. For example, non-cornucopian techno-fixers (e.g. work by the Rocky Mountain Institute – http://www.rmi.org/newsletter/97fwn/index.html).

Mystics A wide diversity, who turn to their inner voices for inspiration and guidance. This grouping would include those who derive their inspiration from Teilhard de Chardin, Buckminster Fuller, Taoism, Zen and paganism. The label 'New Age' was coined in the late 1960s by journalists to incorporate a hotchpotch of greens who rely on astrology, the occult, Gaianism, non-mainstream religions and so on as a guide to their relationship with the environment – in effect those with a postmodern spiritualist worldview. Many New Age supporters look towards the change from the present solar age of Pisces to Aquarius early in the twenty-first century as a moment of opportunity and possibly crisis (Henderson, 1981b). Certainly, there are greens who might be dismissed as 'cranky'.

Sources: Porritt (1984: 4–5); Weston (1986: 20); Taylor (1991)

A late twentieth-century paradigm shift?

Many recognise an ongoing worldwide paradigm shift, whereby a wide diversity of political groups, religious persuasions, old and young, share concern for the environment to a far greater extent than has been the case in the past. What were desirable goals in the past are being questioned; the way forward is far from clear and the environmental manager is charged with finding the best path. Social science must warn of changing attitudes, advise on human institutions that will work for ecologically sound development, and help identify supportive policies. It seems unlikely that development, as practised so far, will enable the world's poor to reach and sustain standards of living achieved in rich countries. It may also be difficult to maintain the quality of life in rich countries. Changed attitudes and new approaches are needed, and humankind probably has limited time to acquire them. Social science will play a vital part in managing the stresses which societies will probably undergo in the coming decades. Recognition of problems and reactions to them depend on what individuals and communities think of themselves and how they relate to their environment. At the roots of many of the world's environmental problems lie unsound concepts of development and modernisation. A widespread problem is that people tend to make Faustian bargains – decisions that sacrifice long-term well-being for short-term gain. Another is that people can react in an emotive way to questions which require careful investigation. Environmental managers must weed out unreliable advocacy and ensure that rational enquiry is not discouraged.

Ethics for environmental management

Ethics are the non-legal rules and principles which order human existence. Ethics are related to values, things which people hold dear and wish to support. Worldviews, the perceptions a person or group have of their surroundings, overlap with ethics and values (Kalof, 2005). There is unlikely to be a single worldview, even within a single family, although one may be reasonably dominant – generalisation must be cautious. The ethics embraced by individuals, professions and societies, like legislation, can change with time. Currently both environmental ethics and environmental laws are evolving to meet needs but that process is incomplete and there are often inadequacies. The development of environmental ethics is moving quite fast (see International Society for Environmental Ethics for a bibliography – http://www.cep.unt.edu/bib/ – accessed March 2005; there is a journal, *Ethics and the Environment*, published by Elsevier).

One can recognise an 'ethical spectrum' ranging from vague eco-friendly utilitarianism to aggressive and draconian, even eco-fascist. Environmental management has to operate within that spectrum; when problems are critical and results have to be obtained it may be necessary to move towards the draconian ethics.

Women and the environment

Environment–human interrelationships are often gender-sensitive (gender being a set of roles) (Shiva and Mies, 1994). Women have played a key, if not the major, part in establishing environmentalism and green politics (Petra Kelly) and many of the world's conservation bodies. Rachel Carson's book *Silent Spring* (1962) helped prompt environmental concern. A few societies are matriarchal and have female-inheritance systems. It has been argued that women view environment and development differently from men because of their reproductive role; there are also gender differences in employment, income, freedom, and perception of resources. It must also be noted that women play a large part in training children – influencing future opinions and behaviour to support environmental care lies very much in the hands of women. Changes in women's attitudes affect population increase. Rough statistics gleaned from the literature suggest that about 50 per cent of the global population are women, that they do around 75 per cent of all work, they receive about 10 per cent of total income, own approximately 1 per cent of property and undertake most childcare. Women and children frequently have different diets and exposure to pollution and other threats. Whether rural or urban, women and men are likely to have different livelihood activities and dissimilar access to resources; there may be some overlap and sometimes these roles are mutually supportive, but any development must examine female and male sides carefully (Rodda, 1991; Sontheimer, 1991; Seager, 1993).

Rural societies have often experienced the out-migration of men to work in cities or mines or overseas, leaving female-rich settlements. Women are often sidelined from inheritance, burdened with bringing up children, and form the poorest sector of society. Where men are dissolute and lazy, women often initiate change like conservation activities, community forestry/tree planting, and improvement of water supplies. In the past it was not uncommon for men to be consulted on development initiatives, yet it was the women who actually worked the fields or collected fuelwood. In rich and poor countries it is often women who are activists; for example, in the Love Canal pollution disaster in the USA; in Europe in protesting against the stationing of nuclear cruise missiles; in The Netherlands women fought the Lekkerkirk pollution disaster; in India women protested against the Narmada Dams; also in India they formed the Chipko Movement to protect forests; in North Africa they acted to counter desertification.

Whether rural or urban, women and men are likely to have different livelihood activities and dissimilar access to resources; sometimes these roles are mutually supportive, but often they are not. Where males migrate to find employment, women are often left to cope with families and farms, and are commonly resilient and inventive because they have to be when absentee wage-earners fail to send back remittances. Debates on women and development are relevant to environmental management, and can be subdivided into Women in Development (WID), Women and Development (WAD), and Gender and Development (GAD). WID focuses more on improvement of women's welfare and their role in economic development. WAD explores relationships between women and the development process, not just strategies to improve the integration of women. GAD looks more at the roles of sexes, their needs and interests, and ways in which each can actively participate in development. Anyone exploring environment and gender issues should familiarise themselves with these concepts and the writings of authors such as Ester Boserüp (Rathgeber, 1990; Wallace and March, 1991; Kabeer, 1994).

Ecofeminism sees parallels between the oppression of women and exploitation of the natural world – a gender-neutral approach is inadequate and masculine control has to be opposed (Merchant, 1980). More romantic environmentalists flag the earth goddess, sensitivity-to-nature aspects of femininity, and see development as too often the 'rape of nature' (ecofeminism bibliography – http://www.ecofem.org.biblio/ – accessed March 2005). There is also highly practical eco-feminism; for example, in the 1980s Anita Roddick used her *Body Shop*® chain of stores to support fair trade and environmentally sound marketing. Women are often concerned with local issues and are consumers, so they can play a crucial part in sustainable development efforts. Donella Meadows was one of *The Limits to Growth* team; Dame Barbara Ward helped initiate interest in sustainable development in the early 1970s; and Gro Harlem Brundtland placed sustainable development on the world's political and business agenda (World Commission on Environment and Development, 1987; Braiddoti *et al.*, 1994; Harcourt, 1994).

Social science and environmental management in practice

Environmentalism, green politics and environmental management are largely creations of Western democracies; their influence and usage has spread worldwide, but there often needs to be a considerable degree of adaptation to or substitution for locally appropriate approaches (Selin, 1995). Another consequence of the spread of environmentalism, the Green Movement and environmental management is that other cultures and ecological conditions are influencing their evolution. For example, Western law may be unsuitable for managing water in tropical environments and where there are traditions such as Islam. New insights and tools are developing outside the West and in some cases are valuable worldwide. Fields in which there has been considerable non-Western influence are social aspects of resource use, and indigenous peoples and environments.

Social aspects of resource use

An understanding of people's attitudes, capacities and needs is often vital for managing fisheries, forest resources, biodiversity conservation, pastoral development and so on. Anthropological botanists or botanical anthropologists (ethnobotanists) can discover from local people what plants have potentially useful properties. Political ecologists are often invaluable in unravelling the way in which communities relate to nature and to other humans and the economy. In the past, sustainable use of natural resources was often assisted by local rules, taboos and superstition; the past fifty years or so have witnessed the breakdown of these controls in many places as development takes place, and often nothing satisfactory takes over. The result is breakdown of sustainable resource use and environmental damage. Such changes have affected fisheries and forest use in a number of parts of the world. New socially appropriate and workable ways have to be found, and new institutions often have to be built and maintained.

Social forestry deals with the establishment and management of forest, woodlots and hedges by or for local people. The focus is on establishing tree cover, where it is needed, in the most appropriate manner, with minimal dependency on outside help (Lee *et al.*, 1990). The social forester may also be interested in why people destroy trees and in ways of countering such behaviour; for example, finding substitute fuels, or establishing alternative livelihoods. In some regions there has been NGO activity, some of it more or less spontaneous, whereby local people have come together to improve forest conservation or to support reforestation and woodlot planting (Tiwari, 1983; Arnold, 1990; Chatterjee, 1995). Whether the approach is farm-based, community-based or focused on women's groups, the key feature is people's participation in planting and management. Getting effective participation may require careful encouragement, perhaps manipulation of people, which may be assisted by guidance from applied social scientists.

Indigenous people can make ideal guides, managers and police for areas of managed forest and conservation areas, and they may also derive an adequate livelihood in their traditional environment in doing so (Wesche, 1996). Conservation efforts have often been insensitive to local people, which has alienated them and sometimes triggered poaching and other destructive activities. The best route to conservation is likely to be to avoid alienation and get effective local involvement. However, simply promoting participatory approaches (as has rather been the fashion recently) does not guarantee effective conservation or resource management; there are strong criticisms of community-based conservation. Oates (1999) is one conservationist who warns against politically correct but ineffective conservation; others have also commented on effective authoritarian approaches (Diamond, 2005: 440). When the cost of failure is the extinction of biodiversity, a 'top-down' approach may be excusable until workable alternatives can be found.

There is nothing new in the idea that some less sensitive conservation may, with care, be combined with livelihood activities. National Parks in the UK and a number of other countries allow some agricultural, recreational and other activity, and as the world becomes more crowded similar arrangements will become more common. The late Chico Mendes and others promoted the concept of the extractive reserve in Brazil in the 1990s as a way of protecting flora and fauna and allowing local people to extract products such as rubber. A similar result is obtained from tolerant forest management strategies – the thinning of understorey species and encouragement of useful tree species while removing the minimum of wild species. The latter has been developed in several parts of Amazonia (A.B. Anderson, 1992), although utilisation is more intensive. Similar strategies are to be found in South East Asia, and elsewhere in the tropics. Extractive reserves have been supported by the Brazilian state environmental agency (SEMA), making locals 'guardians of the forest', or in the case of marine reserves, such as those near Cabo Frio, responsible for reefs and fisheries.

The 'greening' of economics

Regardless of the stance adopted, whether utopian, utilitarian, libertarian, Malthusian or whatever, for almost as long as economics has existed, economists have invoked the 'invisible hand of the market' as a mechanism which supposedly ensures that it becomes uneconomic to exploit a potentially renewable resource before it is badly damaged. The pareto optimum theorem of welfare economics states that through market exchange, with each person pursuing their private interests, there are effective controls over resources exploitation and use of the environment. It also states that, except in inefficient market situations, it is not possible to make anyone better off without making at least one other person worse off. Unfortunately, the market has not been an effective control: there are plenty of examples of ruined fisheries and lost forests to prove it.

The reason for this is that '"the free market" does not provide consumers with proper information, because the social and environmental costs of production are not part of current economic models. Private profits are being made at public costs in the deterioration of the environment and the general quality of life, and at the expense of future generations' (Capra, 1997: 291). Thus, currently, the market often fails to control exploitation for various reasons. One is the difficulty in valuing many resources; for example, it is not easy to assign a value simply because a species is rare, and some things are valueless since a use has yet to be found for them. Resources and environment may be used to give outputs (such as crops) or benefits (such as recreational use) or there may be non-use (intrinsic) value (e.g. conservation provides material for future pharmaceutical use or crop breeding). When a resource or the environment has current utility (i.e. can give 'satisfaction'), this may be gained directly, say by the use of land for recreation or tourism, or indirectly through manufacturing (Perman *et al.*, 1996) (Figure 4.1).

Many of the attempts at a concise definition of economics mention 'resources', 'the Earth', 'the environment'; for example, 'economics is essentially the stewardship of resources' (Hanson, 1977); or 'economics offers a framework within which to analyse the problems which we face in making choices about the environment in which we live' (Hodge, 1995: 3); or 'economics is concerned with the allocation, distribution and use of environmental resources' (Perman *et al.*, 1996: 24). It is thus puzzling why, before

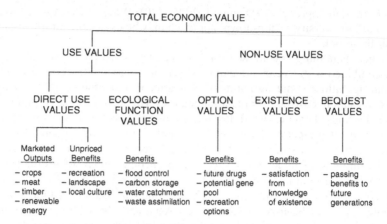

Figure 4.1 The elements of total economic value
Source: Hodge (1995: 7, Fig. 1.2)

the last decade, there was little contact between economics and environmental studies. The failure to weave environmental sensitivity into economics has been flagged as a cause of many of the world's problems. Given the difficulties involved in effectively valuing nature, and in dealing with human use of the environment and resources, such criticisms are perhaps unfair. Nevertheless, before the 1980s few economists recognised that the Earth was finite and most encouraged expansion, and there was little effort to remedy things.

One of the first to publish on resource and conservation economics was Ciracy-Wantrup (1952). Fourteen years later Boulding (1966) inspired many with his writings on the economics of Spaceship Earth which acknowledged that the world was finite and vulnerable. Further impetus to 'greening' was given by the publications of Meadows *et al.* (1972) and Schumacher (1973). By the late 1970s some economists were concerned about growth and environment (Krutilla and Fisher, 1975; Hanson, 1977; Kneese, 1977; Cooper, 1981; Ekins, 1992a; Buarque, 1993). Work on environmental economics expanded after the mid 1980s (Lowe and Lewis, 1980). Considerable effort has gone into seeking alternatives to reliance upon market controls (Redclift, 1992). A particular boost was given by the publication of *Blueprint for a Green Economy* (Pearce *et al.*, 1989), and related texts have regularly appeared ever since, the majority from the London Environmental Economics Centre (Pearce *et al.*, 1990, 1991; Pearce, 1995).

There has been an increasing interest in economics and environmental management (Redclift, 1992; Barbier, 1993; Turner *et al.*, 1994; Funtowicz and Ravetz, 1994; Mikesell, 1995; McGillivray, 1996). Since about 2000 there has been increasing focus on sustainable development (Pearce *et al.*, 1990; Tisdell, 1993), pollution control economics (e.g. Forsund and Strom, 1988), and economic development and environmental management (Schramm and Warford, 1990), and on environmental taxes. As discussed in Chapter 2, some economists adopt the concept of sustainable development, without acknowledging that the Earth and its resources are finite, and then talk of 'sustained growth' and ways of achieving it. Today, two widely stated goals of green economics/environmental economics are: (1) to cut extravagant resource exploitation; (2) to seek sustainable development. Some economists argue that environmental care should stimulate economic growth by improving the health of the workforce, making it more productive, and creating employment in the green sector (pollution control and environmental remediation and so on).

Box 4.5

An example which may have widespread promise: Curitiba City, Brazil

Much of the success of Curitiba's greening since the mid 1970s has been through the efforts of its Mayor, Jaime Lerner, and has not depended on much outside funding. He established an effective refuse-collecting system for Curitiba's slums (*favelas*), where narrow alleyways make it impossible to use lorries. Recycling bins were placed around the *favelas* and the people were paid in city transport system tokens or welfare tokens for sorted, recyclable trash. Organic waste went to farmers for composting, and people collecting this were rewarded with food stamps. The approach provides a sort of social security system for the poor, who in return scour the city for refuse. The travel tokens offer better access to employment and boost the use of public transport, there are no costs for running garbage trucks and less need for street construction. Numerous other, largely self-help, innovations have made Curitiba a landmark in green urban development. Curitiba has improved living conditions for the poor and upgraded its infrastructure in spite of having had one of Brazil's most rapid growth rates (Rabinovitch, 1996). The city has been able to become self-funding and no longer seeks aid from state government, is comparatively clean and relatively prosperous. The city has an improved bus transport system and crime rates have been kept low compared with other Brazilian cities.

Large sections of the world's population are still not directly affected by economics. The reality is that many of those suffering from environmental problems are 'marginal' – economically, they often have subsistence lifestyles, and may live in remote situations. In addition, one must not assume universal co-operation between nations to protect the environment: some companies, countries, power groups and individuals try to gain from global challenges, and exploit situations.

There are economists who are keen to avoid the 'commodity fetishism of mainstream economics', and develop workable 'green economics' and 'barefoot economics' (Scitovsky, 1976; Henderson, 1981a, 1981b; Max-Neef, 1986, 1992a, 1992b; McBurney, 1990; Dodds, 1997). Innovation does not just happen in rich countries: India and other developing nations have world-class economists, so some ideas do flow South to North. For example, the Brazilian city of Curitiba has evolved novel transport and waste disposal systems and ways of paying for them (Box 4.5). One suggestion is that steady-state economics be developed to ensure that growth does not lead to serious environmental degradation. How much inroads into mainstream economics this makes remains to be seen (Booth, 1997). There has also been some greening of mainstream accountancy (Gray, 1990).

By the 1980s environmental economics was expanding (Costanza, 1991; Common, 1996) and environmental issues had gained a much higher profile (Tietenberg and Folmer, 1998). Even so, mainstream economics still has far to go to become adequately green. Issues such as shadow pricing may be used more often, but macro-economics is still generally reluctant to include environmental costs in calculations of things such as gross national product (GNP). Economic growth is still a major goal – little change from the 1970s when an environmentalist observed that 'growth for the sake of growth is the ideology of the cancer-cell'. Most companies still seek to maximise profits for their shareholders, and governments are overwhelmingly driven by short-term goals, which do little to encourage investment in sustainable development and environmental quality. While green economics stresses quality of life and is less concerned with capital accumulation, mainstream economics still puts little effort into meeting environmental and social needs. However, well-publicised major disasters such as Bhopal have frightened some companies and governments into more cautious and sensitive strategies.

Various economists and political economists have tried to encourage more concern for the environment and 'invisible' sectors such as the poor (ignored because they do not 'appear' to contribute to economic growth or have any impact). Calls to radically rethink have come from advocates of local focus, post-industrial economics, some seeking reduction of human material demands, and others keen to stress that the problem is not growth, but *how* it occurs (Henderson, 1981, 1981b, 1996; Ekins, 1992b); some might be described as 'barefoot economists' (Schumacher, 1973; Max-Neef, 1982).

Economists have tried to improve the environmental sensitivity of cost–benefit analysis, and there have been attempts to incorporate economic evaluation into environmental impact assessment (James, 1994). There is still a lot of improvement needed (Georgiou *et al.*, 1997). For an introduction to the economic theory involved in policy making for environmental management, see Baumol and Oates (1988). Some economic tools are discussed in Chapter 8.

Global environmental problems and economics

A number of transboundary issues have become apparent in recent years and, in order to address them, there is a need to know the likely costs, discover ways of funding solutions, and where appropriate to develop economic controls. Pollution, and in particular the threat of global climate change, is attracting attention – some would say too much attention, at too high a cost. Considerable efforts are being directed towards trying to resolve the apportionment of blame, estimation of costs and development of controls (Agarwal and Narain, 1991; Funkhauser, 1995; Tietenberg, 1997; Proost and Braden, 1998). Another area of interest is the cost of technology change: Farvar and Milton (1972) suggested that careless application of technology caused serious problems. It makes sense to try to forecast the economic impacts of proposed innovations (Tylecote and Van der Straaten, 1997). The process of globalisation (see also Chapter 11) can affect the environment and society. As well as prompting challenges, globalisation could also offer opportunities for better environmental management. The literature on globalisation and the environment is expanding at a rapid pace (Haas, 2003; Tisdell and Sen, 2004).

Environmental accounts

There are a number of environmental auditing approaches: eco-audits, environmental stocktaking, eco-review, eco-survey, eco-footprinting, and more. State-of-the-environment accounts, environmental quality evaluations and environmental accounts systems collect data on the environment and resources to try and show the state of a land area or sea such as the Baltic or Aegean. Most of these accounting procedures treat the environment as natural capital and try to measure its depletion or enhancement. Techniques such as eco-footprinting seek to trace and value flows of resources and activities associated with discrete areas or activities.

The foundation for these procedures has often been the UN model of Standard National Accounts, usually with 'satellite accounts' added for environmental items – some call these 'environmentally adjusted national accounts' (UN, 1993). Such accounts seek to establish the stocks of resources, value of environmental features and their use over time (Newson, 1992: 92). National environmental accounts systems (new systems of national accounts, green accounts, patrimonial accounts or state-of-the-environment accounts) have been developed to assist with data gathering and storage and to value environment and natural resources (Pearce *et al.*, 1989: 93–119). Canada, Denmark, Norway, France, Japan, USA, The Netherlands and the World Bank have developed national state-of-the-environment accounts since the early 1970s and the UNEP has been promoting this type of accounting in developing countries (Alfsen and Bye, 1990; Hartwick, 1990; Schramm and Warford, 1990: 30; Common and Norton, 1994; McGillivray, 1994). Most follow the Dutch model, comparing output of each sector of the economy with how much it depletes finite resources such as fossil fuel. Some countries are moving to include water pollution, radioactive waste and other factors in their accounting.

These accounting systems seek to set out a region's environmental, social and economic assets, and may be used to assess whether economic development is consistent with sustainable development, or help ensure optimal use of natural resources and environment (Ahmad *et al.*, 1989; Hamilton *et al.*, 1994). For example, a natural resource accounting system can help a manager establish what percentage of, say, mineral exploitation profits to invest in long-term sustainable development so that a region or country does not suffer boom and decline. In practice, being able to make such investments depends on the type of government, people's attitudes and the persuasiveness of environmental management. Natural resource accounts can show the linkages between the environment and the economy, may be useful for forecasting, and can establish which habitats are of importance. They should make land use more rational, and are an improvement on the use of indicators such as gross national product (GNP) (Thompson and Wilson, 1994: 613), but stop short of encouraging a crucial change in people's and administrators' attitudes towards environmentally sound development.

In the mid 1990s the UK Office of National Statistics produced national environmental accounts to try to measure the country's economic performance, assessing the environmental impact of each industry, using 1993 statistics. These accounts show for various economic sectors the percentage contribution to the national economy against percentage of total: greenhouse gas emission; responsibility for acid deposition; and smoke emission (*New Scientist*, 4 September 1996: 11).

Estimating the value of the environment and natural resources

Ever since Ciracy-Wantrup (1952), resource inputs have been divided by assessors into: *renewable* (also called 'stock resources') which are robust enough to withstand poor management; *potentially renewable* (dependent on effective management); and *non-renewable* (Figure 4.2). Some renewable resources can be converted to non-renewable through poor management or natural disaster. Certain resources cannot be remade if damaged or exhausted (e.g. biodiversity). The absorptive capacity of the environment, its ability to absorb and neutralise damaging compounds or activities, should be assessed by economists. There may be opportunities to substitute for a given resource, using labour, capital or alternative materials.

The following techniques for valuing environmental/natural resources are widely applied.

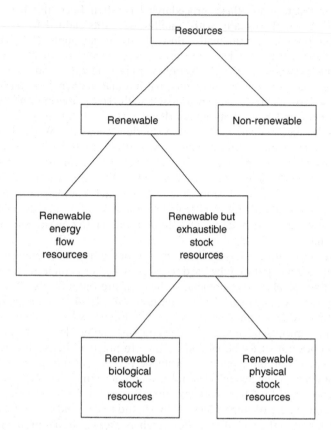

Figure 4.2 A classification of environmental resources
Source: Perman *et al.* (1996: 4, Fig. 1.1)

Cost–benefit analysis

Cost–benefit analysis (CBA) seeks to identify the impact of development on each person affected at various points in time, and so estimate the aggregate value which each person gains or loses. There is a huge literature on CBA, its shortcomings, modifications and alternatives (Brent, 1997). Widespread dissatisfaction with CBA's effectiveness in valuing environmental issues has led to many suggested improvements or alternatives, some favouring quantitative approaches, and others qualitative (Munda *et al.*, 1994). One focus has been to try to improve its consideration of environmental issues, but this is still far from adequately solved (Hanley and Spash, 1994). A development since the early 1990s is the concept of best available techniques not entailing excessive costs (BATNEEC). This places the onus on developers to adopt the best techniques available, with only 'excessive cost' as a viable excuse for not doing so (Pearce and Brisson, 1993).

Shadow prices

The difficulty of establishing the value of 'externalities', including environmental factors, in monetary units has been addressed in several ways: one is to use shadow prices. A shadow price is a value that reflects the true opportunity cost of a resource or service. The real value of something reflects the most desirable alternative use for it. For example (in the case of industrial production), the opportunity cost of producing an extra unit of manufactured goods is the lost output of childcare, food production or whatever, forgone as a result of transferring resources to manufacturing activities. In consumption, opportunity cost is the amount of one commodity that must be forgone in order to consume more of another (Todaro, 1994).

Paying for and encouraging environmental management

Environmental management may be funded and encouraged by national and international taxation. Some of the funding raised may be made available to poor countries for environmental management tasks via provisions such as the Global Environmental Facility (see below) or the International Finance Facility recently proposed by the UK government. In addition, charitable funds may be available and grants from profits on some recreational activity such as an international lottery (some cultures would be unhappy to receive money generated by gambling or usury). Any tax can be 'green' if it is spent on environmental matters (and has not generated environmental impacts). There are many, as yet untapped, international sources; suggestions include: a levy on a global body from all major international gold transactions; a tax on use of gesynchronous orbits; a tax on weapons sales; a tax on air travel. There are also possibilities for generating funds through investment. Already, a number of life assurance companies and pensions schemes support only environmentally or socially beneficial investment; international bonds similar to those issued in wartime in the UK and USA may also be a possibility.

Fair trade

Often small farmers in developing countries have made the transition from subsistence cropping to specialist production of export crops. Such farmers become vulnerable to market trends, controls on transportation, marketing and so on. For example, coffee provided a useful income for many until the late 1990s when world prices started falling; the consequence in countries such as Peru has been small farmer migration to cities, a shift to narcotics production, and in some cases a resort to shifting cultivation on easily degraded land. Attempts to even out market fluctuations through coffee cartels such as the International Coffee Agreement (set up in the 1960s) have not been enough. Escape from poverty and avoidance of land degradation in areas with smallholder coffee producers now lies with fair trade initiatives through companies such as Cafédirect. Fair trade arrangements seek to improve the revenue going to actual producers in poor countries by cutting out the middleman. The improved income and support has prompted some local co-operatives to purchase Internet equipment so that they can track market prices – in time, like wine growers trading on production region, they may be able to market their produce as something with local character and to carve out part of the market.

Contract farming

Small producers can be very vulnerable to swings in market processes; they may also have difficulties acquiring inputs such as improved seeds and fertilisers. They also face problems in packing and transporting produce to market, and in advertising and selling it. A number of large companies, especially large Western supermarkets, have developed contract farming approaches. The supermarket finds a dedicated and reliable supply which it can control, and it knows the price will not fluctuate too much. The farmer is insulated from very low prices, misses out on high prices, but enjoys stable medium prices in return for signing a contract. The seller provides input and advice, transport and marketing, and insists on quality control and work practices (which may include environmental care and employee welfare measures). In some cases production would be unlikely to start without the support of company patronage. The alternative might be degenerating subsistence agriculture, land damage and poverty. However, there are downsides: farmers become dependent on others; reliance on a specific distant market may entail risks; an unscrupulous company could in the long term drive down producer prices once they become locked into the system – rather like sharecroppers. Or some sort of globalisation debt-peonage might appear. The development of contract farming needs to be monitored to ensure that it improves livelihoods and does not cause environmental damage.

Organic farming

Strategies which reduce and ultimately eliminate the use of harmful chemical inputs certainly promise to reduce environmental pollution and could have far-reaching effects. In many parts of the world the thrust of modern agricultural development has resulted in environmental degradation which is increasingly manifest and will raise demands for a shift to organic-type production (organic farming or permaculture methods). There are other possible benefits: these alternative farming approaches may lock up more carbon in the soil than approaches relying on agro-chemicals; they may aid soil and water conservation; help promote sustainable development; and by reducing the need for expensive outside inputs they cut dependency and make access to the strategy easier for poor people and those in remote areas. New techniques are appearing, some suitable for tropical environments with poor soils – and some of these could open up little-farmed areas to production.

Here organic production is aiming at marketing produce to offer the consumer reduced exposure to harmful chemicals, and a sense of supporting environmental care. One consequence is that health-conscious people will pay more for organic produce – perhaps over 30 per cent more. The demand for such produce has been spreading from rich countries to developing country cities and demand is growing fast. Farmers face disadvantages in making a shift to organic production: it usually takes time and expense to ensure the land is free of chemicals and to get accreditation; yields may decline and labour inputs increase when chemicals are unavailable for weed and pest control, and chemical fertilisers are prohibited. Ultimately, changing consumer demands may force the change and the price premium may at least compensate.

Integrated area development

Development agencies and governments appear to be more interested in integrated approaches. This may make it possible to identify and promote activities which can be 'dovetailed'; thus farming may aid tourism (agritourism activities attract visitors), tourism may aid farming (visitor levies can be channelled to develop agriculture), and stable and environmentally aware farming and tourism support conservation and improved environmental management. Bioregional approaches such as watershed, river basin or coastal zone development could offer units for integrated development.

Green taxes

Taxation is an important tool for seeking environmental management goals. It may be used to discourage undesirable activities; reward beneficial activities through rebates or reduced taxation, and make issues public through the release of accounts and profit data. Pearce (1995) urged environmental management to seek a balance between using economic command and control, largely through taxes, and incentives. Capra (1997: 292) suggested that one of the most effective ways of countering environmental damage and supporting sustainable development would be to shift the tax burden from income to 'eco-taxes'. These could be added to products, energy, services and materials to reflect true environmental costs. These measures mean the *consumer* pays. While there have been national measures for some time, interest in green taxation on an international scale is recent and still mainly theoretical, triggered by increasing transboundary pollution, competition for internationally shared resources, and the threat of global environmental change.

The function of green taxes is not to raise revenue for government but rather to provide participants in the marketplace with accurate information about true costs. For example, a tax on CFCs reflects their impact on ozone (Farber *et al.*, 1995). Green taxes counter the pursuit of lower prices by externalising the true costs. It is important that attempts to integrate external costs of production into prices do not burden the poor or 'punish' the middle classes. The aim should be to give people and companies incentives to invent, innovate and respond to environmental challenges (Repetto *et al.*, 1992). Green taxation ideally encourages manufacturers to seek to reduce waste and other environmental damage to keep down their costs and thus prices to the purchaser – there is incentive to improve environmental practice. Taxation is also becoming an important tool in the quest for sustainable development (Von Weisäcker and Jesinghaus, 1992). One problem associated with attempts to agree international green taxation is that it may come into conflict with sovereignty (Nellor, 1987).

Pigouvian taxes

The idea of 1920s' UK economist Arthur Pigou, these are intended to be levied on external costs such as pollution, or activities it is desirable to discourage to achieve sound environmental management. Essentially it seeks to use market forces to effect an efficient allocation of resources. The polluter-pays a tax equal to the value of the external cost.

Carbon emissions taxes

There are a number of taxation approaches that have potential for controlling global climate change: tradable emission quotas; carbon (emissions) tax; energy use tax; taxation associated with technology transfers; reduced taxation for providing carbon sinks. A number of countries have already taken steps to adopt these (Cornwall, 1997).

Tradable emissions quotas

Tradable emissions quotas/credits (TEQs), also called marketable or auctionable permits or tradable emissions permits, have been adopted by a number of countries. In the USA they have been used for control of air pollution emissions for over a decade, and in France to control water-borne effluent since 1969 (Owen and Unwin, 1997: 402). There has been considerable interest in the use of TEQs for dealing with transboundary atmospheric pollution, especially carbon dioxide emissions (Koutstaal, 1997).

The 1997 Kyoto Treaty (Kyoto Protocol to the United Nations Framework Convention on Climate Change) established a TEQ 'club', which trades emissions permits among its members (see also Chapter 11). The Treaty commits the industrialised nations to reducing greenhouse gas emissions by around 5.2 per cent below the 1990 levels by 2012 (at the time of writing – late 2005 – a number of signatory countries were unlikely to meet their goals). Modified at Bonn in 2002 the Treaty waited a few years for Russia and some nations to sign. The Protocol came into force in February 2005, and its key elements are:

- a system of TEQs;
- tough verification of emissions;
- financing arrangements to aid poorer countries to comply.

Concern has been voiced that there will be inadequate controls over the future emissions off China, India and Brazil. There is still much disagreement about the Protocol; broadly middle-class people see the global warming threat as real, and poor people are more worried that TEQs will cut jobs. Some radical environmentalists have voiced suspicions that the Treaty will give benefits to bureaucrats, and that funds may be siphoned off by governments. Many feel that global warming is attracting attention away from other threats, and the Treaty will give poor returns.

Energy use taxes

Energy taxes (in most cases carbon taxes) seek to discourage pollution by increasing costs: for example, burning low-grade coal would attract a higher tax per Btu than the use of oil or gas – which emit less carbon. Tax on vehicle fuel, domestic power supplies and household heating fuel can be used to discourage excessive consumption. Energy taxes encourage efficient use and change to non-polluting alternatives, but may be unfair to countries with less scope for the latter – such as those lacking hydroelectricity or already committed to coal or oil.

Green funding

Environmental care is increasingly a condition of aid (Keohane and Levy, 1996). Since 1978 USAID has helped spread precautionary planning by requiring EIA whenever US foreign aid is likely to significantly affect the environment or people in developing nations. Funding and aid agencies are increasingly focusing on environmental management and sustainable development (Rich, 1986; Turnham, 1991; Feitelson, 1992), and they also check for risks, such as contaminated land, before supporting developments (Kopitsky and Betzenberger, 1987). The growth in green and socially appropriate insurance, shares and pensions management was mentioned earlier; the likelihood is that in the future more and more funding, whether aid or investment, will seek to promote better environmental management.

The problem of assisting poor countries to fund environmental care is being addressed; one initiative is the Global Environmental Facility (GEF), launched in 1990 as a corporate venture between governments of industrialised and developing countries. The GEF is jointly managed by the World Bank, UNDP and UNEP to assist developing countries to tackle globally relevant environmental problems such as climate change; loss of biodiversity; management of international waters; and stratospheric ozone depletion. The GEF is targeted at poorer countries and involves NGOs in identifying, monitoring and implementing projects. There were efforts in 1992 at the Earth Summit to increase the profile of the GEF. Criticisms include the complaint that donors to the GEF have simply cut back on other aid to finance it; that participation is not open enough or wide enough; and that some developing countries want poverty alleviation included.

Aid and the environment

There is a wide diversity of aid approaches: recipients may be governments, bodies, and groups of people or individuals. Aid may be in the form of grants, loans, equipment, training, secondment of skilled staff and so on. Donors can be international agencies, NGOs, individuals, groupings of governments, or national governments. Sometimes donors contribute aid directly to recipients, or it can be via an intermediary such as an NGO or a UN body. When aid is government to government it is termed bilateral aid; when several governments or an international organisation have contributed it is multilateral aid. Frequently aid is tied – that is, conditional: a recipient may have to behave in a particular way or a percentage of the provision must be used to buy goods and services from the donor nation. The latter arrangement includes 'aid for trade provision', and it is not unknown for obsolete, overpriced or unsuitable goods or services to be traded (Hayter, 1989: 21, 92). Aid may be in the form of funding, foodstuffs or other supplies, sometimes training or secondment of skilled manpower rather than donation of goods or funds. Green aid has conditionality – it depends on the exercise of environmental care or seeks environmental improvement. A risk may be that it is perceived as neo-protectionism or neo-colonialism, or an extra cost, or a sign that there is a risk that support could be diverted. Aid can help the environment without actually being focused on green goals if it seeks to ensure that it reduces impacts on the environment (Dinham, 1991; Hildyard, 1991).

Voicing environmental concern is not enough; worse, some aid may hide behind an environmentally friendly façade – a form of 'greenwashing'. At the Earth Summit in 1992, Japan offered more aid for the environment than any other nation, but some of that aid appears to be tied to her export or resources import policy. Japan has, according to Forrest (1991), tended to support large super-projects that have sometimes caused serious environmental impacts. Aid may be well intended, but even providing something 'harmless' like better roads or wells can cause problems. Environmentally benign aid is not easy to achieve and problems are often not intentional. Avoiding impacts may not be easy. What to a donor seems like sensible safeguards to avoid unwanted environmental and socio-economic impacts, may appear to a recipient to be excuses for conditionality, delay and perhaps loss of a portion of funding to pay for appraisals, safeguards and remedial measures, and intrusion into sovereignty (Linear, 1982, 1985; Adams and Solomon, 1985; Hayter, 1989).

To combat global environmental problems will require considerable aid to poor countries. At the Earth Summit richer nations were clearly reluctant to commit themselves to the GEF, either for fear it would slow their economies or because they wished to ensure tight control over how the aid was spent. The 'democratisation' of the USSR and its allies has meant less spending on arms and propaganda in both the East and West but it may capture aid which would previously have gone to developing countries. Most agencies have developed environmental guidelines and have staff to assess impacts prior to granting assistance. For example, the World Bank established an Office of Environmental Affairs in 1970 (Warford and Partow, 1989), and the UK Overseas Development Administration (now the DFID) established environmental appraisal procedures (ODA, 1984, 1989a, 1989b).

Debt, structural adjustment and the environment

When countries acquire debts they are bound into interest repayments which restrict funds available for environmental care; prompt efforts to generate foreign exchange – which often means insensitive exploitation of natural resources, and may lead to funding bodies insisting on austerity measures – which cause breakdown of established land use and land degradation. If the impacts on the world environment are serious enough it may make sense for those owed debts to write them off.

Debt and the environment

During the 1970s many developing countries financed their economies by taking loans. Falling prices for exports of primary produce, rising costs for oil imports, and in some cases disorder and maladministration led to escalating debt. The 1973 to 1974 OPEC oil price increase caused further recession, driving down export prices and making debt repayment difficult. The 'debt crisis' broke in 1982, and soon claims were made that it resulted in environmental degradation, although there was little clear proof (George, 1988, 1992; Adams, 1991; Reed, 1992a: 143; UN, 1992). Various impacts of debt have been recognised: (1) money diverted to servicing debt is unavailable for environmental management; (2) resources are put under pressure to earn foreign exchange for interest or to pay off debt; (3) means to combat debt cause difficulty, notably structural adjustment measures.

By the 1990s Latin American and sub-Saharan countries were spending about 25 per cent of their total foreign exchange each year servicing debt (Davidson *et al.*, 1992: 161). In spite of paying US$6,500 million a month interest between 1982 and 1990, debtor countries were still 61 per cent more indebted in 1991. Debt problems, by reducing biodiversity and degrading the global environment, and by causing poverty and conflict, also affect richer nations indirectly (George, 1992: 1–33). In 1991 under the Trinidad Terms the Paris Club of creditors agreed to cancel some debts.

Linkages between economics and environment are often complex, and caution should be exercised when debt–damage relationships are recognised. Debt servicing is not the only reason countries exploit resources: it might be to support urban facilities, industrialisation or special-interest groups.

Structural adjustment and the environment

When recession began to take hold in the developed countries, the World Bank and the International Monetary Fund began to impose structural adjustment programmes to try to stabilise the economies of debtor nations, protect creditors and generally shore up the international economy (Bello and Cunningham, 1994). The tool used to try and stimulate growth, and ensure debt repayment, fight various inefficiencies and improve the flow of traded goods, was the structural adjustment loan. Structural adjustment began in Turkey in 1980, and by 1990 another sixty-four countries had adopted measures. These measures varied in detail from country to country, but were always granted on condition the recipient deregulated their economy, reduced state expenditure and freed exchange rates. The goal was to give priority to export earnings, make the economy more efficient by cutting spending on wages and welfare, and reduce state controls to boost productive sectors.

There was limited success, but in some countries there were significant or marked ill-effects: reduced household incomes, increased unemployment, inflation, cut-back in support for welfare and public services, and less spending on environmental management. Limited impacts have been felt in developed countries, but in some poorer nations there has been significant increase in poverty, greater childhood mortality (George, 1988), land abandonment, riots and rural to urban migration. Structural adjustment may impact on the environment through progressive disempowerment of the poor (Redclift, 1995). At its worst, structural adjustment can oblige people to sacrifice environmental assets for short-term survival. Since the mid 1990s there has been interest in the effects of structural adjustment policies on the environment (Reed, 1992a, 1992b, 1996; World Bank, 1994). The World Bank and some other agencies involved in formulating structural adjustment policies try increasingly to support better environmental management and to seek sustainable development.

Debt-for-nature/environment swaps

The value of conservation in economic terms is considerable, but not adequately acknowledged. An assessment in the late 1970s suggested the contribution of plant and animal species to the USA economy was about 4.5 per cent of GDP – in the region of US$87,000 million (McNeely et al., 1990: 18). If the value were better known, conservation might be given more funding. Provided it is compatible with conservation, forest extraction, tourism and sport may generate supporting funds. However, some species will withstand little disturbance and need to be kept in isolation. So far, there has been little progress with taxing biotechnology or pharmaceutical industries to support conservation, yet both draw upon biodiversity. However, ways have been found of trading off debt for conservation or other forms of environmental care. Since the 1980s debt-for-nature/environment swaps (Thomas Lovejoy has been credited with their invention in 1984 – nature swaps usually focus on conservation, environment swaps can address other issues) have been negotiated in a number of countries, and are widely seen to provide a way for the recipient to pay off some or all foreign debt with less loss of face than would be caused by defaulting. The debtor country avoids defaulting and retains control over conservation or environmental activities, and banks should be able to write off some of the expense against developed country taxes (Simons, 1988; Pearce, 1989, 1995: 35, 47–48; Cartwright, 1989: 124; McNeely, 1989; Shiva et al., 1991: 63). The earliest debt swaps were negotiated by Ecuador and Bolivia in 1987, and subsequently in Costa Rica, the Philippines, the Dominican Republic, Mexico, Malagasy, Jamaica, Cameroon, and other countries (George, 1988: 168; Patterson, 1990; Yearley, 1991: 182, 258).

Debt swaps take a variety of forms, but most involve conversion of hard currency debt to local currency debt. When a lender realises it will probably never recoup, it sells the debt, at a discount, to another who then releases cash to the debtor country in local currency, and the donation supports environmental management or conservation. Some debt swaps are bond-based (a central bank pays interest on a bond created, usually for an NGO, over a period typically of five to seven years); others are government policy programmes (under which the recipient government pledges to implement a policy or initiatives aimed at improving the environment or conservation). Carbon sequestration deals are sometimes linked to debt-for-nature swaps. Through these a developed country or company establishes tree plantations to lock up carbon dioxide to compensate for emissions elsewhere. The developing country land is usually cheap, is not settled by people with legal tenure, and has better tree-growing conditions than colder climates.

There are opponents of debt swaps, especially in the recipient countries and on the staff of some NGOs (Hayter, 1989: 258; Sarkar and Ebbs, 1992). Criticisms are that:

- they offer limited potential to pay off debts (because they are tiny compared with typical national indebtedness);
- they may be used to 'smear' indigenous environmental groups' efforts, i.e. opponents of environmental protection spread rumours of foreign interference to divert attention from other issues;
- there may be difficulties in adopting them in some countries due to different accounting and regulatory systems;
- there is no guarantee of ongoing protection or care;
- they may be seen as an erosion of a developing country's sovereignty;
- if operated through NGOs, swaps may not assist or train local agencies;
- they do little to change commercial forces that damage the environment;
- they have so far been applied to a limited range of activities, mainly park and reserve establishment and maintenance;
- the main beneficiaries, it has been argued, are the debt-seller banks (Mahony, 1992).

Trade and environmental management

Trade impacts upon environment; for example, it affects:

- rates of deforestation;
- demand for animal and plant products, and may be a major reason why a species is endangered;
- global carbon dioxide levels;
- extraction of mineral resources, production of food and commodity crops;
- levels of pollution in developed countries;
- pollution controls in developing countries.

Some forms of trade may be less damaging than others: export of renewable forest products should be less damaging than logging, and may discourage deforestation if it is carefully controlled and local people benefit (Buckley, 1993). To combat logging the Body Shop® store chain has tried to encourage environmentally benign forest product trade by minimising middleman profits. However, such products may have limited markets, which restricts what can be achieved. Falling commodity prices on the world market mean farmers get poor returns on crops, yet, committed to purchasing inputs, they are forced to expand the area farmed, or intensify production, or practise shifting cultivation and the extraction of other resources to supplement their farming activities, leading to environmental degradation. Going back to a pre-cashcrop economy cannot solve the problem. Through trade, countries can obtain materials and continue to expand production. It may also mean that production impacts (pollution due to manufacture and problems associated with consumption of goods) are felt over a wider area. International agreements on issues such as carbon emissions control can have a considerable effect on trade and industry (and possibly agriculture) (Maxwell and Reuveny, 2005).

In the early 1990s probably over 80 per cent of the world's trade was in the hands of MNCs and TNCs (Anon., 1993: 220). In 1974 the Group of 77 (G77) – a coalition of 100, mainly developing, countries – demanded a New International Economic Order (NIEO) at the UN General Assembly. The NIEO included plans for new commodity agreements, alteration of what were seen as unfair patent laws and general North–South economic reform, especially expanded free trade as a way of creating employment and wealth. These demands have received considerable support, and TNCs and MNCs can benefit from better access to world markets. Some are less keen, and advocate a new protectionism – a reduction in the volume of trade, as an alternative to free trade, to cure the market problems that led to demands for NIEO (Lang and Hines, 1993).

Recent decades have witnessed the growing globalisation of trade, media, law, NGOs, and much more) (globalisation is also discussed in Chapter 11).The main vehicle for reform has been the General Agreement on Tariffs and Trade (GATT) (Morris, 1990; Shrybman, 1990; Davidson *et al.*, 1992: 174). There are other multilateral trade agreements (e.g. the North American Free Trade Agreement (NAFTA) between the USA, Canada and Mexico in late 1993) (Ritchie, 1992); the Asia Pacific Economic Cooperation (APEC) (founded in 1989 as a loose grouping of fifteen nations); the Common Agricultural Policy (CAP), which seeks to promote production and effective use of agricultural resources to maintain food supplies and give EC farmers a fair standard of living. The CAP uses price supports and has had significant effects on the environment of Europe and other countries which trade with Europe.

The GATT, which became the World Trade Organisation (WTO) in 1994, is a multilateral agreement covering about 90 per cent of the world's trade, first drafted in 1947 to establish rules for the conduct of international trade, the hope being to lower tariff barriers erected in the 1930s that were held to be a hindrance to world development. There were eight rounds of meetings to discuss GATT before 1985 to 1986; the last, the Uruguay Round, should have run from 1986 to 1990 but failed to reach agreement until some years later (Raghavan, 1990; Anon., 1992). Matters had stalled over cutting subsidies to agriculture: in particular the French farming lobby was opposed to the 1992 Blair House Agreements to reduce farm subsidies. In 1993 in Tokyo the Quad Group of GATT (Japan, USA, Canada and the EC) agreed to abolish or reduce many tariffs, effectively agreeing New World trade rules. Debate on the global and more local environmental impacts of WTO policies can be fierce (Sampson and Whalley, 2005). Ideally, the WTO could be harnessed to support sustainable development (Sampson, 2005).

Trade affects the environment and socio-economic conditions in diverse ways (Gallagher and Werksman, 2002). Trade conflicts and sanctions can radically alter people's welfare, affecting the way they use natural resources (in either negative or positive ways).

Free trade can lead to environmental damage: when the Roman Empire adopted free trade grain prices seem to have fallen, prompting large landowners with many slaves to practise more ruthless commercial farming that caused soil degradation, and smaller farmers were forced out of business. Richard Cobden was aware of the environmental implications for the UK of freeing up trade by the repeal of the Corn Laws (1846) (legislation which had protected farmers from falling wheat prices) – with free trade landowners drained and cleared more land, intensified land use and damaged farmland. Boxes 4.6 and 4.7 outline some of the impacts. The WTO had a long-drawn-out disagreement over the USA restricting imports of shrimps caught by countries like Mexico with nets that endanger wildlife (*The Times*, 28 April 1998). One problem is that signatory countries have less control over their imports because quotas and restrictions can be interpreted as trade barriers, which are outlawed (Bown, 1990; Westlund, 1994). There are also worries that free trade could favour developed countries' biotechnology (Acharya, 1991).

GATT established a Disputes Panel to resolve problems but so far it has not been effective enough at dealing with environmental issues. Interest in further greening free trade has resulted in a growing literature (Sorsa, 1992; Esty, 1994; Marsh, 1994; Rugman

Box 4.6

The positive and negative effects of free trade on environmental management

Free trade might help environmental management through:

- ending tariff barriers that raise produce prices, causing farmers to overstress land for profit;
- reducing the dumping of cheap US and European food surpluses, which, by making it difficult for developing country producers to get a fair price, discourage them from leaving land fallow or investing in land improvement, erosion control and so on;
- removing restrictions that make it difficult for developing countries to produce and sell finished wood products to other developing countries. This should yield much better profits and reduce logging;
- harmonising standards and co-ordinating trade impacts on the environment on a global scale.

Free trade might harm environmental management because:

- much existing or proposed environmental legislation could be interpreted as illegal non-tariff trade barriers. There is thus a reduction in controls which discouraged logging, trade in endangered species, use of cattle growth hormones such as BST and so on;
- trade liberalisation may lead to increased specialisation of production that may over-stress a resource or environment;
- the struggle to keep down costs to be competitive may mean exports are expanded to compensate and resources or the environment are put under stress;
- reduced import restrictions will remove opportunities to counter trade in hardwoods, endangered species and so on;
- there may be increased opportunities to sell commodities (e.g. beef, sugar), and this might encourage increased forest clearance and poor land management in countries that are keen to boost production;
- producers may think twice about spending money on pollution control or other forms of environmental management if another country does not, and they are competing with it to sell similar goods, on otherwise equal terms (Ritchie, 1992);
- it may be less easy, without the threat of trade restrictions, to get countries to reduce carbon dioxide emissions or other pollution;
- poor countries may reduce domestic food prices, import grain, and raise more export crops such as soya (e.g. as in Brazil);
- any domestic support for the peasantry in developing countries or poorer farmers in developed countries could be interpreted as unfair protection. Small farmers might become marginalised and then damage the land trying to survive;
- larger farmers, encouraged by free trade to practise industrial (agrochemical-using) agriculture to produce export crops, may damage the land;
- there is a risk that foreign inputs and MNC controls will increase, leading to more dependency;
- if free trade leads to reduced home production there is a risk of problems if overseas supplies fail;
- it could be difficult to pass and enforce national environment and resource management or health protection laws.

Box 4.7

Clash between free trade and environmental management: the yellow-fin tuna case

From 1972 the USA had restrictions on its own tuna fishing. Between 1988 and 1991 conflict arose because America felt the Mexicans were using purse-seine netting techniques that killed marine mammals and other wildlife. In 1992 the USA enacted the International Dolphin Conservation Act, which prohibited the import of fish or fish products from countries which were deemed to have inadequate measures for protecting marine mammals.

The 1992 Act led the USA to place an embargo on imports of Mexican canned yellow-fin tuna. Mexico complained (before the Act was passed) in 1991 that it would violate free trade rules. GATT found in favour of Mexico (the ban being a violation of its Article XI) (Anderson and Blackhurst, 1992; Charnovitz, 1993; Musgrave, 1993).

Similar problems have been caused by the USA's (1987) Driftnet Enforcement Act and (1992) Wild Bird Conservation Act. Problems with the former are continuing at the time of writing. The difficulty is deciding whether this sort of restriction is justified under free trade rules.

---

and Kirton, 1998). GATT set up a group on Environmental Means and International Trade, and bodies such as the OECD are keen to harmonise free trade and environment (De Miraman and Stevens, 1992; Zarsky, 1994). It would also be wise to seek greater co-ordination between the various free trade organisations and the UN Commission on Trade and Development (UNCTAD). In 1985 the eighty-five signatories of GATT undertook to try to restrict the export of hazardous materials. However, pollution control activities are not easy due to difficulties in disseminating information on pesticides and other compounds and their effects, and because monitoring and enforcing controls in the real world are often problematic. Measures were taken to improve controls; for example, in 1986 the FAO issued an International Code of Conduct on the Distribution and Use of Pesticides, and by 1990 about 100 countries were signatories. The FAO and WHO set up the Codex Alimentarius Commission to establish food standards, including acceptable pesticide levels, and this publishes standards annually. Under GATT the Codex seems likely to have increased powers. However, it has been argued that Codex decisions are determined too much by developed countries and MNCs or TNCs (Avery et al., 1993). Any nation that already has, or is setting, standards higher than the Codex may well be deemed to be putting up trade barriers and could suffer sanctions (for a recent study of WTO agreements and the environment, focusing on how to solve the difficulties, see Cameron and Fijalkowski, 1998; *Environment and Trade: a handbook* produced by IISD and UNEP – http://www.iissdl.iisd.ca/trade/handbook – accessed March 2005).

Summary

- In 1965 the US Ambassador to the UN, Adlai Stevenson, used Buckminster Fuller's metaphor 'Spaceship Earth' in a speech; the catch-phrase spread the idea that the world was a vulnerable and effectively closed system. In 1972 *The Limits to Growth* publication prompted debate about environmental limits and how growth/development should proceed – a debate that is ongoing.

- When the history of the twentieth century is finally written, the single most important social movement of the period will be environmentalism. Environmentalism appeared after the mid 1960s and is still evolving; whereas ecologism seeks philosophical changes, environmentalism is more managerial in approach.
- A diversity of political groups, religious persuasions, old and young, share concern for the environment to a far greater extent than has been the case in the past. What were desirable goals in the past are being questioned; the way forward is far from clear and the environmental manager is charged with finding the best path.
- The failure to weave environmental sensitivity into economics has been flagged as a cause of many of the world's problems. Given the difficulties involved in effectively valuing nature, such criticisms are perhaps unfair. Nevertheless, before the 1980s few economists recognised that the Earth was finite, most encouraged expansion, and there was little effort to remedy matters. For the past twenty years or so there has been an increasingly energetic effort to 'green' economics.

Further reading

Barry, J. (1999) *Environment and Social Theory.* Routledge, London.
Comprehensive introduction to social theory and environment looking at theorists and 'green' social theory.

Dobson, A. (1990) *Green Political Thought: an introduction.* HarperCollins, London.
Excellent introduction to green politics, although with a mainly UK focus.

Gilpin, A. (2000) *Environmental Economics: a critical review.* Wiley, Chichester.
Good introduction.

Group of Green Economists (1992) *Ecological Economics.* Zed Press, London.
Radical coverage of green economics.

Jacobs, M. (1991) *The Green Economy.* Pluto Press, London.
Radical introduction to green economics.

Scaltegger, S. and Burrit, R. (2002) *Contemporary Environmental Accounting: issues, concepts and practice.* Greenleaf Publishing, Sheffield.
Introduction to environmental accounting.

Tietenberg, T. (1992) *Environmental and Natural Resource Economics* (3rd edn). HarperCollins, London.
Good introduction which is widely available.

www sources

Environmentalism – Wikipedia bibliography http://www.en.wikipedia.org/wiki/Environmentalism – accessed February 2005.

Green economics site: http://www.greeneconomics.net/what2f.htm – accessed March 2005.

Green Economics Resource Center: http://www.progress,org/baneker/home.htm – accessed April 2004.

Social Science Information Gateway (SOSIG) environment section [Comprehensive route into many social science/environment websites]: http://www.sosig.ac.uk/environmental_sciences – accessed March 2005.

5 Environmental management, business and law

- Environmental management and business
- Corporate environmental management in the 1990s
- Corporate environmental management since 2000
- Corporate visions of stewardship – a paradigm shift to environmental management ethics?
- Approaches adopted to promote environmental management in business
- Greenwashing
- Environmental management and business: the current situation
- Environmental management and law
- The 1969 US National Environmental Policy Act (NEPA) – environmental 'Magna Carta'?
- European law and environmental management
- International law and environmental management
- Indigenous peoples and environmental law
- International conferences and agreements
- Alternative dispute resolution
- Prompting and controlling environmental management
- Summary
- Further reading

> Our products reflect our philosophy . . . respect for other cultures, the past, the natural world, and our customers. It's a partnership of profits with principles.
> (Anita Roddick – The Body Shop®
> promotional literature 1990)

While only a few companies have embraced environmental ethics as much as the Body Shop®, business and legal aspects of environmental management have developed greatly and generated huge interest in recent decades. In many respects business and legal aspects are the cutting-edge of environmental management. Business drives a lot of human activity, and can degrade people and the environment, or offer routes to new development ethics and sustainable development. Often business wields more influence than a poorer nation can, and often has the flexibility to fund innovation in ways governments often cannot match. Law should provide guidelines and rules for arbitration, without which chaos and destruction ensue. Both business and law must evolve rapidly to face challenges such as competition for limited and degrading natural resources, globalisation and transboundary problems adequately.

Environmental management and business

By the late seventeenth century in the Caribbean, Mauritius and several other places, the trade in sugar, timber and other commodities by bodies such as the Dutch East Indies Company, the (British) East India Company and clearances by numerous smaller producers were causing deforestation and soil erosion (Grove, 1995). By the mid nineteenth century English romantic liberal and socialist intellectuals like William Morris, and in Russia the proto-anarchist Pyotr Kropotkin began to criticise industrialisation for its pollution, human degradation and shoddy products (MacCarthy, 1994). But there was little popular protest until the 1960s, by which time people in developed countries had improved standards of living, and enjoyed sufficient free time and access to a more or less democratic media, to become aware of and lobby for environmental issues. Accidents like the *Torrey Canyon* oil-tanker spillage and pollution disasters such as Three Mile Island, Love Canal and Seveso had raised public awareness in the USA and Europe by the mid 1970s. In addition, environmental NGOs, consumer protection groups and popular writers were fanning public interest.

Accidents helped prompt environmental controls. From the 1970s American NGOs and groups of lawyers interested in environmental issues (notably the Environmental Defense Fund and the Natural Resources Defense Fund) began to fight group court actions against those damaging nature and lobbied for environmental legislation. In Europe and New Zealand green politics began to emerge. Research and contact between scientists increased after the 1957 to 1958 International Geophysical Year, leading to improving awareness of environmental issues, better understanding of the Earth's structure and function, the development of international standards, and sharing of data. The USA passed the 1969 National Environmental Policy Act (NEPA) in 1970, and established an Environmental Protection Agency (Seldner and Cottrel, 1994: 61–96). The UN held the 1972 Conference on the Human Environment in Stockholm, and in 1973 established the UN Environmental Programme (UNEP).

NEPA required US developers to meet environmental standards, and effectively promoted the precautionary principle. Business was also being prompted by other legislation, international bodies, NGOs, public opinion and self-interest to pay attention to the environment. It was not enough to obey the law; there was also a need to appear concerned for public relations reasons and to avoid negligence charges – what Brenton called 'defensive greenness' (1994: 148). Some companies saw opportunities for commercial gain – building a green image and marketing environmentally friendly products or providing services for environmental management; in some cases this has been a charade (see below on 'greenwashing'). There was a realisation that 'end-of-pipe' solutions, cleaning up rather than prevention, were more costly, gave a bad public image, and that environmental management could be a way of cutting costs to gain a 'competitive edge' (Beaumont, 1992: 201; Taylor, 1992; Winter, 1994).

Other factors have prompted business interest in environmental management:

● globalisation (i.e. media, finance and so on becoming global);
● 'glasnost' (i.e. increasing public demand for access to information);
● activity of green business groups, especially since the 1992 UN Conference on Environment and Development;
● trade union and NGO concern for environmental issues;
● a wish by companies to reduce inspection by regulatory bodies;
● insistence by funding, insurance and licensing bodies that required environmental impact assessment (EIA) and eco-audit be conducted;

- ethical (green) investment policies adopted by some companies (in the USA a group of powerful investors now apply a set of environmental policy principles – the 'Valdez Principles') (North, 1992);
- genuine sense of responsibility (some companies have been founded by people with a strong sense of moral duty);
- avoidance of litigation;
- the establishment since the 1970s of increasingly powerful environmental ministries in most countries;
- formation of bodies such as the Institute of Environmental Management (UK);
- promotion of the Integrated Systems for Environmental Management and the Business Charter for Sustainable Development (International Chamber of Commerce, 1991);
- provision of courses on environmental management at university business schools;
- the UN Center on Transnational Corporations has promoted sustainable development.

After 1978 the US Agency for International Development (USAID) insisted on EIA during planning when development was likely to affect the environment significantly.

Corporate environmental management in the 1990s

Business interacts with a wide range of parties (Figure 5.1). Satisfying the investors and shareholders is currently the driving force; the adoption of environmental management implies concern for a wider range of stakeholders: the public, bystanders, employees, consumers, and the regional and global environment. Environmental management must address its objectives within the context of company practices; if at all possible it should not slow down completion schedules (Seldner and Cottrel, 1994). As its value is proven, those practices may be modified to help environmental management. The larger company, firm or business is the focus of this chapter (corporations), but government departments, cities and institutions also increasingly adopt corporate approaches to environmental management.

The tasks of a 'corporate' environmental manager include:

- education of employees to be aware of environmental issues;
- updating management on relevant environmental regulation, laws and issues;
- selecting specialists and checking that environmental management tasks contracted out to consultants have been satisfactorily conducted and are properly acted upon;
- ensuring waste management is satisfactory;
- avoiding legal costs, reducing insurance premiums, risk and hazard assessment;
- if necessary, correcting mistakes of the past.

A typical definition of business environmental management is 'efforts to minimise the negative environmental impact of the firm's products throughout their life cycle' (Klassen and McLaughlin, 1996: 119). The range of tasks is so wide, and involves working with so many within and outside a corporation, that co-ordination is a key skill for an environmental manger. To summarise, businesses are adopting environmental management: (1) because it helps identify *opportunities*; (2) because it can improve *efficiency* (for example, identifying waste recycling potential); (3) because there is *fear* engendered by disasters and a wish to avoid such problems, and to cut liability and insurance costs; (4) for *public relations*; (5) out of genuine *ethical concern*.

If business fails to adopt environmental management in a serious fashion there will be little progress, for, as Hawken (1993) noted, corporations are the Earth's dominant

Figure 5.1 Corporate environmental management: the parties involved
Sources: Partly based on Royston (1978a: 7, Fig. 3); Hunt and Johnson (1995: 69, Fig. 4.1)

institutions – many corporations have earnings in excess of those of most developing countries, and some command more riches than some developed nations. Governments are often lobbied and prevailed upon to do what national business, MNCs or TNCs want. Big business often has better access to information, resources and skills than do poor nations, and may have greater stability for year-to-year planning than do some governments. Since about the 1980s there has been an increasing flow of books on environmental management and sustainable development for business (e.g. Elkington and Burke, 1989; Davis, 1991b; Sandgrove, 1992; Schmidheiny, 1992; Smith, 1992; Allenby and Richards, 1994; Hutchinson and Hutchinson, 1997). This literature may be subdivided into:

● greening of business (often focused on a particular sector such as petroleum production);
● environmental management for sustainable development of business;
● green corporate environmental management;
● total quality management/environmental management systems;
● eco-audit;
● impact assessment, hazard and risk assessment;
● green business ethics;
● green marketing, labelling and life-cycle assessment;
● recycling and waste disposal;
● health and safety;
● environmentally sound investment and funding;
● environmental law and business.

By 1992 the chemical industry in developed countries was spending an estimated 3 to 4 per cent of its sales income per annum on environment, health and safety in the

USA alone: that constituted about US$10 billion a year (Greeno and Robinson, 1992: 231). Spending has increased markedly since the late 1980s, and with accidents like Bhopal in 1984 and the *Exxon Valdez* in 1989 (the latter cost Exxon over US$2 billion) it is easy to see why.

Corporate environmental management since 2000

The growth of corporations and the emergence of globalisation mean that increasingly businesses operate a company-wide environmental management policy and set of principles worldwide. They can help spread improvements to countries which would otherwise be slow to adopt anything similar, and once established, competitor companies and suppliers, joint venture partners, and contractors are prompted to comply. The US Environmental Protection Agency began to revise its environmental management regulations in the 1990s, and by 2005 had achieved a mix of public regulation, government-supported self-regulation and mandatory information disclosure.

Along with the development of environmental management since the late 1990s there has also been a spread of support for corporate social responsibility (concern for the relationship which a company has with government and wider society). While it was proposed as early as 1970 and attracted growing attention after 1987, sustainable development since the late 1990s has moved increasingly beyond conceptual discussion and advocacy, to strategy formulation, workable measurement and governance issues. Since the late 1990s there has been development of approaches that explore business ecosystem function – flows of energy and materials, evolution of production, symbiosis between industry and environment, and so on. Concepts growing out of this include industrial ecology, life-cycle assessment, eco-footprinting, farming systems approaches, sustainable rural livelihoods approaches, and many others, some of which may effectively support sustainable development.

Reaching international agreements on environmental issues is a new art little practised before the late 1990s; progress is difficult because each country has different population structures, energy consumption patterns and natural resource endowment, which make it difficult to decide measures fairly.

In 2002 the Conference on Sustainable Development was held in Johannesburg. This produced the 2003 Johannesburg Declaration on Sustainable Development (http://www. johannesburgsummit.org/html/ – accessed September 2003), which has helped maintain widespread interest in sustainable development.

Market-based approaches to environmental management have great potential in a world where lack of public funding means initiatives must pay for themselves (Swingland, 2003). It is important to build partnerships between locals, NGOs, business, international agencies and so on. But caution is needed; for example, the 1997 Kyoto Protocol included agreements for rewarding those who sponsor reforestation of deforested areas – but this could lead to speculators clearing unspoilt land and then restoring it for a profit if safeguards are not implemented.

Corporate visions of stewardship – a paradigm shift to environmental management ethics?

'Fordism' of the 1920s to 1960s emphasised mass production, mass consumption, corporate control and resource exploitation (Amin, 1994: 2). Businesses in the main see economic growth as the route to development; however, that has not well served large numbers of the world's people. Economic growth has frequently failed to improve infra-structure and services, has done little to improve law and order or access to human rights, and has so far failed to maintain environmental quality. The goal of progress has to be redrawn to emphasise environmental quality and improved human well-being. After the 1960s various thinkers, 'barefoot economists' and environmentalists have questioned growing consumerism (i.e. excessive consumption) stimulated through marketing (Elkington and Hailes, 1988; Adams *et al.*, 1991). The problem is how people (consumers) and business (supplying the consumers) will shift to something more supportive of environmental goals. Hawken (1993) in *The Ecology of Commerce* argued that free market capitalism, the economic and social credo of most of the world, must rapidly shift to a 'restorative economy' based on 'industrial ecology' (see below). Only business, he argued, and no other human institution, has the power to make adequate changes. Allenby and Richards (1994) saw industrial ecology as a means of integrating environmental concern with economic activity. Whether it is termed post-Fordian, post-modern or post-industrial, what Hawken and others argue is that the world's future economy should be organised with guiding principles coming from industrial ecology. These post-Fordians seem convinced that the profit motive will be replaced by a more environmentally sensitive approach. Some even suggest that environmental management values are supplanting shareholder interests and a paradigm shift is beginning. However, there is a risk that 'greening' of business is appearance rather than substance, simply the adoption of environmental management tools to improve profits and public relations (Garrod and Chadwick, 1996). In Western nations (e.g. the UK and USA), environ-mental groups with strong ethical beliefs, such as some animal rights supporters and groups willing to sabotage what they see as environmentally damaging activities, have a marked impact on business. In some cases companies have moved activities overseas to try and evade the attention of such groups, while others have met with and agreed acceptable compromise solutions.

While there may seem to be few incentives at present to encourage a shift to better environmental management, there have been efforts to promote it (Greeno and Robertson, 1992: 224; Welford, 1996, 1997). One of the more significant moves has been the publication in 1991 of a Business Charter for Sustainable Development by the International Chamber of Commerce (ICC) at the 1991 World Industry Conference on Environmental Management (Box 5.1). One of the first questions asked by business of such proposals is: 'Can they improve financial performance as well as lead to sustain-able development?' Klassen and McLaughlin (1996) put this to the test, and concluded from studies of firms' performances that the adoption of environmental management did increase profits.

Applying standards and regulations to, and monitoring thousands of households and millions of individuals is a challenge; it is easier to seek environmental goals through medium and large companies serving those millions (Cairncross, 1991: 95). At the other extreme are those who attach less value to commercial efficiency and seek post-indus-trial alternatives to corporate globalisation as a route to sustainable development (Milani, 2000). Businesses vary a great deal in their impact on the environment, the resources they have available for environmental management, and their outlook.

More extreme environmentalists tend to write off all business as exploitative; however, people who do care for nature run some companies and there are advantages in going 'green'. Many companies have huge resources, both financial and in terms of expertise and ability to lobby governments, well in excess of anything that can be mustered by developing countries. Companies involved with potentially damaging activ-ities can no longer afford to risk legal action, bad publicity, disillusioned investors, refusal of cover by insurers, or loss of government licences – environmental manage-

Box 5.1

Business Charter for Sustainable Development: principles for environmental management

1 *Corporate priority* To recognise environmental management as among the highest corporate priorities and as a key determinant to sustainable development; to establish policies, programmes and practices for conducting operations in an environmentally sound manner.

2 *Integrated management* To integrate these policies, programmes and practices fully into each business as an essential element of management in all its functions.

3 *Process of improvement* To continue to improve corporate policies, programmes and environmental performance, taking into account technical developments, scientific understanding, consumer needs and community expectations, with legal regulations as a starting point; and to apply the same environmental criteria internationally.

4 *Employee education* To educate, train and motivate employees to conduct their activities in an environmentally responsible manner.

5 *Prior assessment* To assess environmental impacts before starting a new activity or project, and before decommissioning a facility or leaving a site.

6 *Products and services* To develop and provide products or services that have no undue environmental impact and are safe in their intended use, that are efficient in their consumption of energy and natural resources, and that may be recycled, reused or disposed of safely.

7 *Customer advice* To advise, and where relevant educate, customers, distributors and the public in the safe use, transportation, storage and disposal of products provided; and to apply similar considerations to the provision of services.

8 *Facilities and operations* To develop, design and operate facilities and conduct activities, taking into consideration the efficient use of energy and materials, the sustainable use of renewable resources, the minimisation of adverse environmental impact and waste generation, and the safe and responsible disposal of residual waste.

9 *Research* To conduct or support research on the environmental impacts of raw materials, products, processes, emissions and wastes associated with the enterprise, and on the means of minimising any adverse impacts.

10 *Precautionary approach* To modify the manufacture, marketing or use of products or services or the conduct of activities, consistent with scientific and technical understanding, to prevent serious or irreversible environmental degradation. The 1991 Second World Industry Conference on Environmental Management (Rotterdam) promoted the 'precautionary principle'. One problem for those proposing a development is how much proof of a risk they need before taking possibly expensive precautions – what seems to be widely followed is to establish whether there is a 'reasonably foreseeable risk' or a 'significant risk' (Birnie and Boyle, 1992: 95–96).

11 *Contractors and suppliers* To promote the adoption of these principles by contractors acting on behalf of the enterprise, encouraging and, where appropriate, requiring improvements in their practices to make them consistent with those of the enterprise; and to encourage the widest adoption of these principles by suppliers.

12 *Emergency preparedness* To develop and maintain, where significant hazards exist, emergency preparedness plans in conjunction with the emergency services, relevant authorities and the local community, recognising potential transboundary impacts.

13 *Transfer of technology* To contribute to the transfer of environmentally sound technology and management methods throughout the industrial and public sectors.

14 *Contributing to the common effort* To contribute to the development of public policy and to business, governmental and intergovernmental programmes and educational initiatives that will enhance environmental awareness and protection.

15 *Openness of concerns* To foster openness and dialogue with employees and the public, anticipating and responding to their concerns about the potential hazards and impacts of operations, products, wastes or services, including those of transboundary or global significance.

16 *Compliance and reporting* To measure environmental performance; to conduct regular environmental audits and assessments of compliance with company requirements, legal requirements and these principles; and periodically to provide appropriate information to the Board of Directors, shareholders, employees, the authorities and the public.

Note: The International Chamber of Commerce established a task force of business representatives to create this Business Charter for Sustainable Development – it was launched in April 1991.

Source: International Chamber of Commerce (1993)

ment has become something they cannot ignore. In the past, business was often keen to oppose, side-step, pay lip-service to or reluctantly comply with environmental controls.

There have been warnings that where conflicts arise between economic growth and the environment, 'corporate expertise' may seek ways to sideline environmental, social and ethical issues, and stakeholders' interests. The goal of business may be 'eco-efficiency', which has been defined as 'adding maximum value with minimum resource input and minimum environmental damage'. Companies embracing eco-efficiency have to meet a number of demands, which have been listed recently by one authority as:

● reducing material demands of goods and services;
● reducing the energy demands of goods and services;
● reducing pollution;
● improving recycling;
● maximising sustainable use of renewable resources;
● making products/services more durable;
● improving the intensity of service of goods and services.

Some argue that companies may adopt bureaucratic, poorly transparent approaches that do not support the best practices or have scope for ongoing improvement. Increasingly popular corporate greening tools like environmental management systems may not be as beneficial as many claim. Some businesses are self-deluding, well-meaning but ineffectual and, as discussed above, try greenwashing. However, a growing number genuinely try and usually have some degree of success.

Approaches adopted to promote environmental management in business

Some of the change towards environmental responsibility in business is being driven from 'within' by various stakeholders: sometimes it is management which has become enlightened and seeks greening; staff may take the initiative and it can boost their company pride and morale; consumers may welcome or demand it; insurers may promote better environmental awareness to reduce the risk of accidents and costly claims; other companies, retailers or consumers may force it by refusing components or products from environmentally unsound companies; government or international regulations may prompt greening, and NGOs, funding bodies and governments also encourage the shift (Buchholz, 1998). Already, some large companies insist that their component suppliers and other support subsidiaries meet strict environmental criteria. The crucial question is: Do businesses just seek to comply with regulations, and avoid legal liability, taxation or insurance claims – or does it go beyond compliance? A tax per unit of pollution, or other environmental damage, does not discourage sudden discharges that may be difficult to monitor – it is better to adopt regulation which seeks to ensure that the capacity of the environment to cope with damage is not exceeded.

Voluntary codes of environmental management include the ICC Business Charter for Sustainable Development (see end of this chapter for website), which has been endorsed or used by many companies. The Charter is used increasingly in developing countries. Various industry associations have developed their own codes of environmental management conduct, often with the support of the Charter. Money can be saved when businesses or other bodies practise recovery of waste products, and use less energy or raw materials. In practice, savings may be less clear-cut, possibly recouped over very long periods or in ways that are not easy to measure (Schramm and Warford, 1990; Brown et al., 1992; Beaumont et al., 1993; UNCTAD, 1993). The likelihood is that businesses will seek win–win approaches to try to reduce environmental damage and improve their competitive edge through increased productivity and/or lowered costs.

Klassen and McLaughlin (1996) noted: 'the long-term goal of environmental management is to move toward . . . considering environmental aspects in an integrated fashion in product design, the entire manufacturing process, marketing, product delivery and use, consumer service, and post-consumer product disposition.' Already, several fields are well developed, including: industrial ecology, green marketing, consumer protection bodies, eco-labelling, total quality management, covenants, and life-cycle analysis. Business may embrace environmental management with any one or more of a range of motives, including:

● precautionary principle – seeking to avoid problems and litigation costs;
● eco-efficiency – green to be efficient/profitable;
● proactive – forward planning and possibly a wish to promote new environmental standards;
● compliance – simply doing what state and/or public opinion asks for.

Industrial ecology

This is an approach which examines industrial, economic and resource activities from a biological and environmental, rather than a monetary point of view (Frosch and Gallopoulos, 1989; Graedel and Allenby, 1995; Ayres, 1996; Ayres and Ayres, 1996; Green and Randers, 2006). Allenby and Richards (1994) saw it as integrating environmental concern with economic activity. Industrial ecology regards waste and pollution as uneconomic and harmful, and seeks to 'dovetail' them with demands for raw materials. This 'industrial symbiosis' means that wherever possible industry should use by-products, and go beyond the reduction of wastes to make use of what remains from the producer or other bodies. The product does not cause damage, and leads to a system of commerce where each and every act is inherently sustainable and restorative (Hawken, 1993: xvi). Effectively, the environmental price of a product is included in its retail price. Supporters see it as a practical way of guiding business towards sustainable development, a way of shifting 'commercial metabolism' to better fit with nature.

Industrial ecology seeks to ensure that industrial activity is not viewed in isolation from its surrounding environment, but rather is seen to be integrated with it (Socolow et al., 1994; Graedel and Allenby, 1995). It seeks to understand how the industrial system works and how it interacts with the biosphere, and then to use this knowledge to develop a systems approach aimed at making industrial activity compatible with healthy ecosystem function. Industrial ecology is probably most easily practised in systems with clear boundaries: a geographical region; a specific process related to some material or energy (e.g. oil production); a sectoral grouping of organisations; or a cluster of industrial and service facilities (Boons and Baas, 1997). There has been interest in industrial ecology for at least forty years but until recently little has been achieved, apart from in Japan. Since the early 1990s the approach has received more attention, and there is a growing literature linking industrial ecology with environmental management, occupational health and safety, and planning (for further information see Journal of Industrial Ecology; Lowe et al., 2000).

This application of the ecosystem concept to industry means linking the 'metabolism' of one company or body to that of others. This is not far-fetched: some groupings of companies and settlements do it already. For example, Kalundborg (Denmark) has a coal-fired power station, oil refinery, pharmaceutical companies, concrete producer, sulphuric acid producer, fish farms, horticultural greenhouses and district heating which are well integrated. Kalundborg's industrial ecology has happened more or less spontaneously, as companies seek to minimise costs of energy and raw materials and cut the output of waste. Finland has embraced industrial ecology, and there are a number of examples in The Netherlands, Sweden (Hawken, 1993: 62) and Denmark, where sewage, agricultural waste and household refuse disposal are often integrated with district heating and electricity generation. One day the huge problem of urban sewage may become a valuable resource. It has received active promotion from bodies such as the International Society for Industrial Ecology (http://www.umich.edu/~nppcpub/resources/compendia/ind.ecol.html – accessed April 2005).

Industrial ecology has so far mainly been pursued in two ways: using a product-based approach, or adopting a regional industrial ecosystem approach. The Netherlands has established eco-industrial parks to encourage utility sharing and dovetailing. Industrial ecology can be a way to find innovative solutions to complicated industrial/environmental problems, and a way to integrate technical, ecological and economic expertise; it also links with environmental management systems and ecological engineering (Allenby, 1998).

Ecological engineering

Ecological engineering is the design, creation and management of ecosystems that process by-products and waste or recover minerals from effluent or mine spoil (often using biotechnology tools). The goal is to create sustainable ecosystems which integrate human society and natural environment for the benefit of both (http://www.aeesociety. org/ – accessed April 2005).

Pigouvian taxes

Some people advocate going beyond the type of industrial ecology-based strategy adopted by Kalundborg to a fully cyclic economy, i.e. one which yields virtually no waste because recycling and by-product recovery are complete. Making manufacturers responsible for some or all of the costs of recycling or waste disposal is one way of encouraging waste reduction and industrial ecology. There are various ways of doing this: one is to levy Pigouvian taxes; these aim to ensure that a manufacturer pays all costs from raw material and energy provision to final collection and recycling.

Pigouvian taxes may present problems: large companies may make sufficient profits to afford fines, but small companies could be crippled. Thus the 'polluter-pays' principle can be a virtual licence to pollute if the fines are not set high enough, and that can damage small businesses (Beaumont, 1992). One way of avoiding such problems is to use licences; for example, in Germany manufacturers pay a fee to the government to display a green dot on packaging which authorises (compulsory) recycling. As the costs have to be passed on to the customer, this encourages companies to reduce expensive packaging and use cheap, recyclable materials.

Green marketing

Some companies and public bodies had recognised by the early 1980s that a satisfactory green image could improve public relations, and perhaps provide a marketing niche (Charter, 1992; Coddington, 1993; Peattie, 1995). There are manufacturers that have gained from this, and offer genuinely improved products (e.g. refrigerators that use less electricity, do not leak CFCs and which are easier to recycle), and firms which manufacture equipment for monitoring and managing environmental quality. Less enlightened companies may sell goods because of public fears about the environment (e.g. sunblock creams and sunglasses for those afraid of increased UV). AEG reputedly increased sales of electrical goods by c. 30 per cent in a static market by running a marketing campaign on its green strengths. In America in the 1980s McDonald's commissioned an environmental audit and acted on it to shift from plastic packaging foamed with CFCs to environmentally friendly cardboard. This proved good for public relations and was much cheaper (Elkington and Hailes, 1988).

As well as trade agreements, the world's citizens are increasingly demanding material possessions; this 'consumerism' is being fuelled by advertising, and by the media presenting 'lifestyle' images to which people aspire. Many environmental activists are deeply concerned that globalisation and consumerism conspire to threaten any hope of sustainable development and the maintenance of adequate environmental quality. Welford (2000: 56) summed up the present situation succinctly: 'There is now . . . a dominant corporate culture which believes that natural resources are there for the taking and that environmental and social problems will be resolved through growth, scientific advancement, technology transfers via private capital flows, free trade and the odd charitable hand-out.' Some environmentalists urge a robust response – to establish a 'postmodern' and green worldview/culture to replace the current globalisation plus-consumerism-militarism.

Where an environmental asset is not being bought or sold it is not easy to give it a value (Haab and McConnel, 2002). People tend to be unwilling to pay if they do not directly benefit. Yet sometimes there are situations where people accept a cost without much immediate personal benefit. For example, they contribute to wildlife conservation yet few will set foot in the tropics; and the recent donations for the Indian Ocean tsunami also demonstrate considerable altruism. On the other hand, people spend large sums of money to support football, which is not crucial to human survival, but balk at taxes to conserve biodiversity, control soil erosion or pollution. Marketing has a crucial role to play in influencing support for environmental management.

Consumer protection bodies

Alongside the growth in green marketing there has been a spread of green consumerism (The Council of Economic Priorities of the United States, 1989; Irvine, 1989; Mintel, 1990). Consumer protection bodies have been active since the 1960s, and have not been restricted to the developed countries (e.g. one Malaysian body has been active in its own country and works for consumers elsewhere – the Consumers Association Penang).

Eco-labelling

The marking of goods to indicate that they are environmentally friendly (eco-labelling) has been adopted in many countries, including Canada, the USA, Germany and Sweden (Figure 5.2). In most cases the product is judged against similar goods by an independent agency to establish whether it has less environmental impact (without formal eco-auditing). Germany was one of the first countries to introduce eco-labelling in 1978, with its Umweltzeichen or Blaue Engel system (Hemmelskamp and Brockmann, 1997). This relies on a jury of experts supervised by the Federal Environment Ministry to award the right to display a mark on packaging or in adverts. This is a way of influencing the behaviour of consumers, helping them identify the environmental impacts of products, and encourages manufacturers to reduce these impacts.

Eco-labelling assesses environmental impact and communicates this to the consumer or middle merchant. The focus is on the product and often nothing is said about the process of production or distribution. So, an 'environmentally friendly' product may come from a factory which causes pollution, or present a disposal problem after use. There is also a need for standardisation and policing of eco-labelling. However, under current World Trade Organisation (WTO) rules this may not be easy. West (1995) warned that without better legal enforcement it tended to become a marketing gimmick. It is important that eco-labelling is accredited by an independent body to reduce the risk of company 'greenwashes' (see below).

Total quality management and environmental management systems

Total quality management (TQM) (also called company-wide quality management) aims to provide assurance of adherence to policy and specifications through a structured management system, and to enable demonstration of it to third parties through documentation and record keeping. TQM was first formulated in the USA, and largely developed in Japan in the early post-war period to try to improve industrial competitiveness. Environmental management systems (EMSs) show adherence to a suitable environmental policy, the meeting of appropriate environmental objectives (equivalent

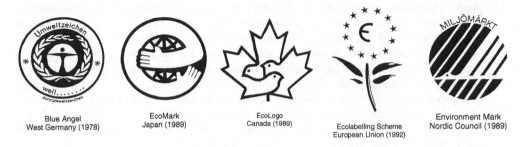

Figure 5.2 Eco-labelling

Note: Date of introduction in parenthesis.

to specifications in quality management) and the ability to demonstrate to a wide range of interested parties ('customers' in TQM) that the system requirements and objectives are met. EMSs usually require that a company or body publishes and regularly updates an environmental policy statement. An EMS provides an organisational structure, procedures and resources for implementing environmental policy. It also provides a language of performance and quality that may be understood by management (Willig, 1994; British Standards Institution, 1996). So far, adoption of EMS has been mainly voluntary with rapid growth of interest and continuing modification and improvement. Hunt and Johnson (1995: 4) suggested that this indicates business has shifted from 'defensive environmental management' to accepting the need for probity.

There are critics of EMSs, who argue it is possible to rig them by setting easy-to-achieve targets; that it is more important (and difficult) to nurture satisfactory environmental ethics; and that EMS is still being developed and tested (for a critique see Welford, 1996: 52).

Covenants

A government or other regulatory body can provide companies with a more stable regulatory environment and encourage development of better pollution control plans or adoption of an EMS (Beardsley *et al.*, 1997: 33) through a covenant. This is a written, voluntary agreement signed by the company or other body and the government or agency seeking regulation. The Netherlands has made extensive use of covenants as part of an integrated approach to national environmental management policy. A Dutch company undertaking a covenant would be expected to produce a development plan every four years, to be reviewed by local authorising bodies. The plan coverage includes pollution control and energy conservation, and is seen as a way of getting national policies implemented at local level. Measures were initiated by the National Environmental Policy Plans (adopted by the Dutch Parliament in 1989), and by 1997 over 1,200 companies had signed covenants. The covenanting approach can be quite effective, particularly in cutting pollution. However, some NGOs are not keen on the approach, viewing it as closed or cosy and not sufficiently open to third parties to check. There are also some worries that it may lead to a softening of enforcement controls. Nevertheless, it is an approach which encourages company self-regulation.

Life-cycle assessment

Many development activities are processes which have different stages (e.g. manufacturing a car or running a power station involve raw materials and energy provision, plant construction, manufacturing, distribution, use and disposal or decommissioning). Equipment is usually subject to wear and tear, and so varies in performance and presents different risks as it ages and as management acquires experience (or becomes complacent). Industrial and power generation sites, for example, often accumulate contamination, and so the environmental threat is not constant. It is therefore undesirable to assess impacts or develop environmental management policies by simply taking a snapshot view. Life-cycle assessment has been developed to try to consider the whole of an activity, which may extend beyond the time horizon of a single owner. It is a cradle-to-grave study of an activity or company (British Standards Institution, 1994a; Fava, 1994; Pidgeon and Brown, 1994; Franklin, 1995).

Life-cycle assessment seeks to determine the environmental (and in theory economic and social) impacts of a product or service through its entire life, including any recycling or decommissioning. It can suggest efficiencies and problems that need to be avoided or mitigated. There are currently efforts to improve the detail and spatial coverage of life-cycle assessment and to apply it to the quest for sustainable development.

Greenwashing

There are companies which see environmental management as a cost and a burden, as do some developing country administrators. Thus there is a temptation for some to lie about their environmental performance and hide behind a false façade of green publicity, which has been termed 'corporate greenwashing' (Greer and Bruno, 1997; Welford, 1997). Without effective and transparent environmental accounting such greenwashing disinformation is easier. Improving media and Internet communications also helps counteract greenwashing by making it easier for environmentalists to find and exchange information and attack offenders. Various NGOs seek to discourage greenwashing by exposure and public ridicule; for example, making regular greenwash awards: http://www.corpwatch.org/campaigns/PCC.jsp?topicid=102 (accessed March 2004).

Environmental management and business: the current situation

Since the 1980s it has increasingly been accepted that economic growth does not have to be at the expense of the environment. In a world where money is usually the key to achieving goals, the role of business in environmental management is important (GEMI, 1998). With growing globalisation, multinational and transnational companies commonly play a central role in activities which affect the environment, and many – perhaps half of the world's richest institutions are businesses not governments – generate much more revenue than any developing country can hope to muster. Some companies have changed to become environmentally sensitive – Jansen and Vellema (2004) presented case studies showing how agribusiness has responded to environmental issues and food production demands, and how it may be harnessed for better environmental management. However, there are still many businesses which strive to maximise profits to show economic growth and pay shareholders – many of the latter in developed countries. Only in the past few years has the idea of ethical investment become practical – allowing investors to direct their money to green activities, or those acceptable to pacifists, supporters of Islam and so on. There are signs that shareholders may shoulder more of the costs of environmental damage and insist on greener activities. Shareholders have started to ask questions on environmental policies at company board meetings.

Welford (2000:12) observed: 'The drive for economic growth and hunger for Western levels of consumption in the newly industrialising countries and the ex-Communist world are developing precisely at a point at which consumerism in the West is beginning to appear socially self-defeating and ecologically unsustainable.' Environmental controls are still often viewed by businesses and some politicians as 'green tape', slowing down development and raising costs. But the market alone is unlikely to deliver a less degraded environment, although business could play an increasingly green role (Barlow and Clarke, 2002).

Some of the commitment by business to environmental management in the past few decades has been rhetoric, and the question increasingly being asked by environmentalists is: 'Will sustainable development be adequately pursued before the collapse of established market economics – which many predict is going to be caused by environmental degradation?' With the previous question in mind, it is wise to nurture business to seek economic growth *and* support environmental management and sustainable development. This will be especially important in developing countries, where there is pressure for economic growth to counter poverty, and limited tax revenue.

The argument is often made that environmental resources are in least danger of degradation when they are privately owned, rather than being common resources or in public ownership, which is often the case in developing countries. When environmental resources in private ownership become scarce their price may be assumed to rise so that owners take better care of them and nurture them, and consumers are encouraged to seek alternatives – 'the invisible hand of the market' hopefully comes into play and counters over-exploitation. For elements of the environment not in private ownership there is no regulation, although there may be traditional controls.

There has been concern voiced about common resources for over thirty years. Since the 'tragedy of the commons' arguments of the late 1960s (Hardin, 1968; http://www.dieoff.org/page95.htm – accessed 2003) environmentalists and natural resource planners have continued to ask whether collective management of common resources can work (Berkes, 1989; Orstrom, 1990). In Europe, common resources such as land had started to shift to private ownership before the sixteenth century. In the UK by the eighteenth century enclosure and privatisation of common land was a driving force for the economy and caused thousands of rural folk to relocate to urban areas; these were developments which helped support the Industrial Revolution. Similar 'enclosure' is currently under way in many developing countries; the state or senior administrators sell licences for land, minerals, logging, fisheries and so on. Traditional users usually have no legally enforceable claim, even though they may have a long history of use, so that displacement and marginalisation often result (The Ecologist, 1993). The MNC or large national businesses which acquire such licences provide profits for shareholders, foreign income for the developing country, cash for ruling elites, and exploit resources which the host nation would otherwise probably be unable to develop. Resource exploitation may also be used to strengthen national claims over territory, as a display of sovereignty and progress.

One may summarise the present situation (see Beaumont, 1992: 202) as:

- the majority of businesses are aware that environmental issues are important;
- some businesses are doing something – it may be from genuine concern, but often it is for public relations or profit motives;
- too often businesses adopt a 'react and repair' approach, rather than following precautionary principles;
- only a few businesses are acting at a strategic level;
- businesses are in need of strategies such as industrial ecology, but will need to be encouraged or forced to adopt them.

Environmental management and law

Law is adapting to meet the demands of environmental management; many Western law schools now offer environmental law courses. Law should provide a framework for regulating the use of the environment (Harte, 1992; McEldowney, 1996; Bell, 1997) (Box 5.2). Law is crucial for environmental management in a number of ways, aiding:

● regulation of resource use;
● protection of the environment and biodiversity;
● mediation, conflict resolution and conciliation;
● formulation of stable, unambiguous undertakings and agreements.

Environmental management may involve a number of resource development situations (e.g. individually owned (private) resources; national resources; shared resources; open-access resources; common property resources; global resources). Law covers some of these better than others (Berkes, 1989; The Ecologist, 1993). There are different legal systems – for example, based on Roman Law, on customary laws, Islamic Law, the Code Napoléon, to name but a few. Environmental managers may have to work with unfamiliar national or local laws, or they may have to seek agreement between parties with different legal systems. Some countries have legal systems that combine more than one of these; say, indigenous and colonial era legislation, plus Islamic Law. Areas may be subject to state and federal laws and to secular and religious laws. In most countries statutory law is written by politicians and passed by national legislature; and judges compile common law (with reference to past cases and prior statutory law).

Most legislation evolves in response to problems, so there is often delay between need and the establishment of satisfactory law. Some sectors are relatively new and are developing at a rapid rate, making it difficult for legislation to keep pace. Biotechnology is one such case, particularly genetic engineering – rich and poor countries are trying to pass laws to prevent accidents and misuse, and to ensure 'biosafety' (see http://www.twnside.org.sg/bio.htm – accessed February 2005). The Cartagena Biosafety Protocol was signed in 2000 to help control risks associated with genetically modified organisms (GMOs) and the sustainable use of biodiversity (elements of the agreement invoke the precautionary principle – for example, requiring assessment of GMOs before there is any risk of escapes). There are gaps in the 2000 agreement and some non-signatory nations.

Without effective legislation, resource use, pollution control, conservation, and most fields of human activity are likely to fall into chaos and conflict. Law can encourage satisfactory performance, enable authorities to punish those who infringe environmental management legislation, or confiscate equipment that is misused or faulty, or close a company; it may also be possible for employees, bystanders and product or service users to sue for damages if they are harmed. Existing laws are predominantly anthropocentric – putting human needs before environment – rather than ecocentric.

Some countries have been active in developing environmental management law, notably Sweden, The Netherlands, the USA, Canada, Australia and New Zealand. Some environmental laws are ancient: Indian rulers promulgated controls on hunting and forest felling centuries ago; the UK had local pollution control laws as early as the twelfth century AD, and passed nationally enforced pollution control legislation such as the Alkali Act (1863) over a century ago. Environmental management increasingly involves transboundary problems that reach beyond traditional sovereignty limits, issues of negligence and the need for nations to co-operate. International law is evolving to address such issues, although it is difficult to develop and enforce (McAuslan, 1991). Often

BOX 5.2

Forms of regulation or legislation

Principles, standards, guidelines, etc., which are not firm laws, but which help lawmakers (definitions are not rigidly fixed)

Principle Broadly, a step towards establishing a law. Once established, tested and working, it can be incorporated into law.

Standard Levels of pollution, energy efficiency and so on that are desirable or required. They provide a benchmark so that different individuals, bodies and countries are as far as possible dealing with the same values. A treaty may incorporate standards.

Guideline Suggestions as to how to proceed, usually without real force of law.

Directive Documents that set out a desired outcome, but to some extent leave the ways of reaching it to companies, states or countries.

Licence A right granted to a body, which agrees to terms or pays, which requires adherence to strict practice and does not give any guarantee of permanent ownership or usufruct.

Law Laws and statutes that require certain actions or standards, and may punish failure to achieve them.

Treaty A solemn, binding agreement between international entities – especially states. Treaties can lay down rules or treaty constraints. Stricter, more precise treaties are likely to involve fewer states, and the process of drafting, adopting and ratifying means that this can be a slow process, and environmental management often needs rapid action. Vague treaties are quicker and easier to get signed. Few multilateral treaties are adopted in less than five years: the UN Law of the Sea Convention took nine years (1973–1982), and some take much more. Treaties can be difficult to enforce – often enforcement is attempted by an international organisation (e.g. the International Whaling Commission). Treaties should bind states that sign and ratify them to accept terms as customary law, but in practice they do not always get transformed into customary law, and some are largely ignored.

Declaration A general statement of intent or drafting of guidelines to follow. 'Softer' than the obligations of a treaty.

Convention Multilateral instrument signed by many states or international institutions. Conventions can be vague, which ensures that countries are not afraid of signing, but this can undermine effectiveness.

Protocol Less formal agreement, often subsidiary or ancillary to a convention.

Contingency agreement A good way of dealing with uncertainty surrounding many global environmental management issues. Agreement of what to do if something happens.

powerful MNCs or TNCs are involved in issues, and these may prompt and drive forward innovation, not necessarily to the benefit of the environment or the public. Walker (1989: 30) likened them to seventeenth-century city states that had insufficient public account-ability. The problem is to ensure that changes are for the good of the environment and the greater common good, rather than just suiting a large company or more countries. Most laws, whether civil or criminal, are corrective – punishing wrongdoers and deter-ring others from infringing rules and agreements, or from causing nuisance or injury. In the main, therefore, legislation has not been very pre-emptive. Environmental managers must also be aware that there is little point in passing laws or making inter-national agreements if there cannot be adequate enforcement. Three developments have had particular impact upon environmental legislation:

1   The *precautionary principle* – has evolved to deal with risks and uncertainties faced by environmental management (Rogers *et al.*, 1997). The meaning is still not firmly established by law. The principle implies that a little prevention is worth a lot of cure – it does not prevent problems but may reduce their occurrence, and helps to ensure contingency plans are made (Mitchell, 1997: 80). The application of this principle requires either cautious progress until a development may be judged 'inno-cent', or avoiding development until research indicates exactly what the risks are, and then proceeding to minimise them. Once a threat is identified, action should be taken to prevent or control damage even if there is uncertainty about whether the threat is real. Some environmental problems become impossible or costly to solve if there is delay, so waiting for research and legal proof is not costless. Much Western law demands that a misdemeanour has actually been committed, or is clearly planned; adopting the precautionary principle may demand legal action *before* something happens. There are also fears that laws adopting the precautionary principle will delay and raise the costs of development (see also Chapter 2).

     Some hold that the principle should be applied in situations where both the prob-ability and cost of impacts are unknown. The principle was stressed in many of the decisions reached at the Rio Earth Summit in 1992. For example, it was endorsed by Article 15 of the 1992 Rio Declaration on Environment and Development (Freestone, 1994: 209–211). Article 130r of the Maastricht Treaty (Treaty on European Union) of February 1992 states that EU policy on the environment shall be based on the precautionary principle.

2   The *polluter-pays principle* – in addition to the obvious – that the polluter-pays for damage caused by a development – this principle also implies that a polluter-pays for monitoring and policing. A problem with this approach is that fines may bank-rupt small businesses, yet be low enough for a large company to write them off as an occasional overhead, which does little for pollution control. There is debate as to whether the principle should be retrospective (e.g. today, a purchaser who acquires contaminated land in good faith is often forced to clean up the mess others have left – if the polluter-pays, how long back does liability stretch?). Developing countries are seeking to have developed countries pay more for carbon dioxide controls, arguing that they polluted the world during the Industrial Revolution, yet enjoy the fruits of invention from that era (see also Chapter 2).

     The polluter-pays principle is more a way of allocating costs to the polluter than a legal principle. OECD member countries adopted the principle in 1972, at least in theory (OECD, 1975).

3   *Freedom of information* – if the public, NGOs or official bodies are unable to obtain information, environmental planning and management may be hindered. Democ-racies have begun to release more information – the USA led with a Freedom of

Information Act, followed by the EU in early 2004. Few countries have such well-developed disclosure as the USA, which requires public registers of development activities, publication of environmental impact statements, hazard warnings on products, and so on. Some governments and multinational corporations fear industrial secrets will leak to competitors if there is too much disclosure, and there are situations where authorities declare 'strategic' needs and suspend disclosure. There have also been a few situations where authoritarian and relatively secretive governments have cared for the environment (e.g. the Dominican Republic).

In many countries, court actions, even if they were fought in the public interest, had to be brought by an individual, who, if they lost, paid costs. This acted as a deterrent to anyone tackling government, a large company or powerful individual wrongdoers, because they lacked equivalent resources. It is desirable that NGOs and individuals be allowed to bring legal actions to protect the environment, if necessary as group cases (class actions). In the USA the Environmental Defense Fund, the Sierra Club, and environmental lawyers such as Joseph Sax managed to achieve the right to bring class actions (or group actions) in the 1970s. Subsequently Canada, the UK and several other countries saw similar legal changes. Class action in the USA forced the US Aid Agency (USAID) to apply EIA to all developments which it funded overseas if they looked likely to significantly affect the environment. This has been a key trigger for precautionary environmental management in developing countries. Since the early 1990s the Internet has also been a way to spread information and counter secretive development.

The 1969 US National Environmental Policy Act (NEPA) – environmental 'Magna Carta'?

Discussions leading to NEPA began in the early 1960s, when the need was perceived for the USA to make a basic declaration of national environmental policy and an action-forcing provision. The US government was largely reacting to public opinion that conventional planning did not adequately take account of the environment; it already had responsibility to steward resources and protect the environment under the Public Trust Doctrine. However, before NEPA the USA had little effective federal control over the environment and lacked land-use regulations, which some other countries had. NEPA was signed into US law on 1 January 1970, to reform federal policy making and influence the private sector to reorientate values (Barrow, 1997: 168). It was originally intended that NEPA would change the nature of federal decision making. However, it has become more of a procedural requirement. Caldwell (1989) – one of the architects of NEPA – felt that, had it not happened in the USA, something similar would have appeared elsewhere.

NEPA required environmental impact assessment (EIA) prior to federally funded projects that might 'significantly' affect the environment – a message to officials to 'look before you leap'. NEPA Section 101 set regulations to protect the environment, Section 102 (2c) ensured they were pursued, and Section 103 included provision for EIA statements to be challenged in court. This happened a lot at first because NEPA was untested and used expressions such as 'significant' and 'human environment' that were poorly defined. There was also some need to clarify which developments required EIA, and how and by whom it was to be conducted. Virtually the first use of the expression 'EIA' occurs in Section 102 (2)c of NEPA, which requires US federal agencies to prepare an environmental impact statement (EIS) (bearing the costs against taxes, and sending copies to federal and state agencies and to the public) using EIA, prior to taking action.

There were three main elements in NEPA:

1   It announced a US national policy for the environment.
2   It outlined procedures for achieving the objectives of that policy.
3   Before supporting or funding any development likely to significantly affect the environment NEPA required federal agencies to conduct an environmental impact assessment (EIA).

Provision was made for the establishment of a US Council on Environmental Quality (CEQ) which was to advise the US President on the environment, review the EIA process, review draft EISs, and see that NEPA was followed. In addition, in 1970 the US government created the US Environmental Protection Agency (EPA), its brief to co-ordinate the attack on environmental pollution and to be responsible for the EIA process (the EPA is in effect 'overseer' of impact assessment in the USA).

NEPA was the first time US law had really allowed for development to be delayed or abandoned for the long-term good of the environment, and for efforts to be made to co-ordinate public, state, federal and local activities. Effectively, NEPA put environmental quality on a level with economic growth, a revolution in values in a country where state intrusion was anathema – for this reason, many see it as a sort of Magna Carta, although it stopped short of making a healthy environment a constitutional right. Public participation is written into NEPA to the extent that it might be described as a corner-stone.

NEPA is statutory law: it was written after deliberation, and did not evolve from custom, practice or tradition. Consequently, like a charter, it was imperfect; there were problems, especially delay, as litigation took place over various issues. Many felt NEPA had been abducted by lawyers and could become a bureaucratic delaying tactic. These teething problems have largely been resolved, although some feel NEPA should be strengthened, possibly leading to changes in the US Constitution to better manage the environment (Caldwell, 1989). NEPA has been a seminal concept and catalyst for EIA in other countries, although bodies such as the Canadian Environmental Assessment Research Council and the International Association for Impact Assessment also deserve credit for spreading and improving mandatory development review processes.

Effective implementation of EIA demands legislation and law enforcement to ensure that:

● there are no loopholes, so that no activity likely to cause impacts escapes EIA;
● the assessment is adequate;
● the assessment is heeded;
● the public are kept sufficiently informed or, ideally, involved in assessment.

European law and environmental management

The European Community (EC), European Economic Community (EEC) or European Union (EU) grew from the six original states which signed the Treaty of Rome in 1957 to form a closer union of fifteen nations in 1995; by 2005 this had expanded to twenty-two countries. The Community has grown, and will continue to do so, and is likely in the future to exert more influence in international fora on environmental matters. EC members such as Sweden and The Netherlands have long-established traditions of environmental concern but others have given the environment far less attention. Growing EC integration should prompt and support better policies more widely. It will also ensure common rules and ways of monitoring, setting standards and so on. In 1992 the EC established a European Environmental Agency as a clearing-house for environmental information. Its role is also to evaluate and disseminate information and develop means for applying the precautionary principle, but not enforcement of environmental policy.

The Council of Europe (established in 1949) had thirty-five EC and other member states in 1995 (many had been former colonies or trading partners), and by 2005 this had expanded further, and countries like Turkey were looking to join. The Council is active in advocacy, cultural relations and raising awareness of issues including conservation and environmental protection. A UN agency that acts as a pan-European forum is the UN Commission for Europe (UNECE), which supports sustainable development, environmental research, and has launched or serviced several agreements dealing with issues such as pollution (e.g. the 1992 Convention on Transboundary Effects of Industrial Accidents) (Hewett, 1995). Environmental legislation is an important part of the emerging pan-European legal system (with the European Court of Justice as an overall arbitrator). The European Environmental Agency has not got as much enforcement power as the US Environmental Protection Agency, and serves mainly to gather information on the state of the European environment. The EC has also established a European Environmental Information and Observation Network; a European Economic Community (EEC) Directive on Environmental Impact Assessment (Directive 85/337) – which requires environmental assessment to be undertaken by developers; an EEC Directive on Freedom of Access to Information on the Environment (Directive 90/313/EEC) – which requires authorities to ensure public access to relevant environmental information; and an EC Regulation on Eco-Management and Auditing (EMAS) (Regulation 1836/93).

One could make the broad generalisation that EC/EU environmental law has focused on co-ordination, codification and integration (Ball and Bell, 1991; Vaughn, 1991; Lister, 1996). Since about 1973 there has been more interest in integrating wider environmental issues into politics alongside concern for achieving economic growth. In 1985 the European Commission (governing assembly of the EC) decided environmental protection should be an integral part of economic and social policies at macroeconomic level and by sector. This was incorporated into the Treaty of Rome in 1987 (Article 130r) and was strengthened by the Maastricht Treaty (1993), which included a statement of concern for sustainable growth (Winter, 1996: 7, 271). EC legislation seems to be aligning itself increasingly with global conventions such as those relating to global warming or waste management. Since 1993 EC law has been enacted to support more freedom of environmental information, better standard setting, the precautionary principle, and the polluter-pays principle; Winter (1996: 277) has listed the core objectives:

● to preserve, protect and improve the quality of the environment;
● to protect human health;
● to prudent and rational use of natural resources;
● to promote measures at international level to deal with regional or worldwide environmental problems.

Hughes (1992: 86) has noted that environmental management law should be 'vertically integrated' between regional, national and international systems. The EEC system allows this to some extent. Efforts to develop an overall EEC environmental policy resulted in the publication in 1973 of the First Programme of Action on the Environment; the Second, Third, Fourth and Fifth Action Programmes appeared in 1977, 1983, 1987 and 1992 (reviewed in 1995), and lay down principles to which EEC environmental legislation should adhere (Hughes, 1992: 89). The Fifth Programme of Action on the European Environment seeks to incorporate good environmental policy into all

Community policies (CEC, 1992). In the UK the 1995 Environment Act created a powerful, wide-ranging Environmental Agency for England and Wales (also a Scottish Environmental Protection Agency), which brought together the functions previously spread among many agencies (pollution control, fisheries management, flood defence and so on) (Lane and Peto, 1995).

International law and environmental management

International law governs relations between states, and has no direct effect on domestic law or individuals. It is often difficult to force a sovereign state to sign, and then honour, a treaty or similar agreement. International law must thus depend a great deal on voluntary agreements by governments and international bodies (the Brussels and Lugano Conventions on Environmental Law cover this issue of ensuring compliance) (Székely, 1990a, 1990b). When negotiation fails a possibility is to refer the case to the International Court of Justice (The Hague) (not a very friendly process), or set up an International Joint Commission. International law has tended to be *laissez-faire* and *ad hoc* (Birnie and Boyle, 1992).

From the mid nineteenth century until the 1950s co-operation, exchange of information, agreement and international guidelines or rules were often initiated by international public unions (e.g. the International Postal Union, or the International Telegraphic Union). Nowadays, the UN and its fifteen specialist agencies (including the FAO, WHO, UNESCO and UNEP) often initiate the development of international environmental law. For example, the UNEP has published guidelines on principles of conduct over shared natural resources (1978) and, more recently, on exchange of information on chemicals in international trade. NGOs such as Greenpeace, Friends of the Earth and the World Wide Fund for Nature also lobby for environmental legislation.

Various observers note the UN-supported system of environmental treaty making is valuable, although it needs strengthening (e.g. the UN General Assembly can only recommend, not insist, that law be made). Developing countries have complained that international law is too US- or Eurocentric and there is a wish in some countries to see more application of Islamic Law. Since the 1972 UN Conference on the Human Environment (Stockholm), most of the UN-prompted multilateral treaties have been developed by a two-step process: a relatively vague framework convention which acknowledges a problem is presented (most countries are happy to sign such a non-binding agreement); that step prompts action, especially data collection, discussion and propaganda, which reduces opposition and raises interest so that a protocol may be introduced and agreed to (Susskind, 1992: 67).

International law faces a number of challenges. One of the greatest is the management of 'global commons': oceans and their resources; world weather and climate; atmosphere; stratospheric ozone; space and so on (Cleveland, 1990). Many resources, and also pests, migrate or move, so that effective management of ocean fisheries, migratory fish in rivers, whaling, disease or locust control and so on needs to be through multilateral agreement.

In the late 1970s a class action by an NGO forced the US Agency for International Development (USAID) to insist on pre-development environmental assessments before granting funds. In effect the precautionary principle embodied in NEPA was extended to the Third World with respect to aid. Within a few years most aid agencies had adopted environmental guidelines and rules (Wirth, 1986). The end of the Cold War may mean more opportunities and resources for international environmental law to develop (Walker, 1989).

International law and sovereignty issues

Sovereignty affects access to data and monitoring, and can be a major constraint on environmental management. Countries are usually reluctant to sign any agreement which affects their sovereign powers. Yet growing transboundary and global environmental problems make it important to get co-operation. There are transnational and multinational corporations sufficiently powerful to threaten and bribe their way around sovereignty and other controls. Terrorism can have a transnational or global impact, so there should be better international controls and co-operation to counter it. Unfortunately for many environmental management issues, obtaining multi-state agreements is a slow process.

The 1977 Stockholm Declaration on the Human Environment affirmed the sovereign right of states to exploit their own resources and their responsibility to ensure that activities within their jurisdiction or control do not cause damage to the environment beyond the limits of their national jurisdiction (Stockholm Principle 21). This affirmation has had considerable influence on subsequent international environmental law making (Birnie and Boyle, 1992: 90). International trade agreements, notably the GATT/WTO provisions, mean that if a country has environmental protection laws, say, controlling the import of pesticide-contaminated produce, timber cut in an environmentally unsound fashion, or fish caught using nets that kill dolphins, these measures may be unenforceable because they impair 'free trade' (Sinner, 1994). The level playing field demanded by trade agreements may make it difficult to control importation of food and commodities produced by means of genetic engineering and growth- or lactation-enhancing hormones. Conversely, there may be situations where globalisation helps countries adopt and enforce better standards (care must be taken to ensure that the motive is to improve environmental quality and not an attempt to make production costs uniform or create a global market for standardised products that enjoy economies of scale). It may be argued that the present is the era of post-nationalism and globalisation. Globalisation of patent rights has generated concern; MNCs and TNCs seek to recoup research costs and control markets; poor countries fear bio-piracy with corporations patenting and claiming intellectual rights on genetic resources and ideas derived from such resources. The patenting and control of sales of crop seeds (modern varieties) and pharmaceutical products have also caused much friction.

Protection and extension of sovereignty can lead to wars; the testing and storage of weapons; and territorial claims. These affect the environment and need to be more firmly addressed by international agreements and law (Shaw, 1993). The pollution associated with the Gulf War underlines the importance of negotiation. Hostile environmental modification is covered by the 1977 Environmental Modification Convention (invoked to hold Iran to reparations for damage to Kuwait), and there are controls on nuclear, chemical and biological weapons.

Box 5.3 presents some of the treaties and agreements relevant to environmental management. A number of trends are apparent here. There has been a move towards the precautionary principle – since about 1972 countries have been guided to try to prevent pollution accidents and misdemeanours. Obtaining damages for, or penalising, transnational pollution has been patchy (e.g. there were no adjudications over Chernobyl, Amoco Cadiz and many similar disasters). There has been little progress in establishing 'environmental rights' (i.e. rights of natural objects or organisms), although in some Western countries there is a vociferous animal rights lobby. Various agreements and conventions have reaffirmed and extended state sovereignty over natural resources (especially apparent in respect to ocean territorial limits). There has been some progress (e.g. the EC is developing a form of supranational legislation, and the UNEP argues that international law should deal with protecting the world's life-support systems).

Box 5.3

A selection of treaties, agreements and so on relating to environmental management

Internationally shared resources
In 1972 the USA and Canada signed the Great Lakes Transboundary Agreement for the comprehensive management of the water quality of the Great Lakes.

Protection of endangered species
The 1946 International Convention for the Regulation of Whaling; the 1973 Convention on International Trade in Endangered Species (CITES); the 1979 Bern Convention on the Conservation of European Wildlife.

Protection of environmentally important areas
There are many areas agreed by scientists, social scientists and other specialists to be in need of formal protection. Protection may be supported by a state; privately funded by a group or individual; or by an international body or bodies. For example, there is a world-wide scatter of Biosphere Reserves; the UK has state-protected Sites of Special Scientific Interest; many countries have reserves and national parks. Some conservation areas are established and watched over by international treaty – the 1971 Ramsar Convention (Convention on Wetlands of International Importance) provides a framework for protection of wetland habitats, especially those used by migrating birds. The UN Educational, Scientific and Cultural Organization (UNESCO) supports and oversees many sites of special cultural value.

The Antarctic
In Antarctica territorial claims have been set aside (but not eliminated) under the Antarctic Treaty which came into force in 1960 (signed 1959) (Theutenberg, 1984) (see Figure 5.3). Basically this is an international treaty by which signatories have agreed to keep Antarctica and its surrounding seas open for scientific research by all nations deemed to be pursuing scientific exploration south of 60°S. The treaty requires demilitarisation, no nuclear weapons and a commitment to conservation (Triggs, 1988; Holdgate, 1990) (for a review of Antarctic law see Auburn, 1982; Beck, 1986).

While it has been quite a flexible treaty, modified as need arose, it has been put under some pressure as interest in resource development (notably oil, minerals, krill, squid and fish) comes into conflict with its conservation requirements. There are also demands from non-treaty nations (basically those which have not maintained a significant research presence there) and some NGOs for there to be changes to give the whole world (probably through the UN), not just signatory nations, control of Antarctica (a coalition of over 200 NGOs and non-treaty nations – the Atlantic and Southern Ocean Coalition – has been seeking such a goal). There have been some moves which in theory could allow mineral resources to be used – the 1988 Convention on the Regulation of Antarctic Mineral Resource Activities allows exploitation only if very stringent environmental assessments are made and accepted by treaty nations. The Falklands conflict is a warning that if potentially attractive mineral resources are identified territorial claims may reappear in Antarctica.

Transboundary pollution
In 1965 Canada and the USA became involved in the Trail Smelter pollution case. The outcome was acceptance that no state has the right to permit use of its territory in such

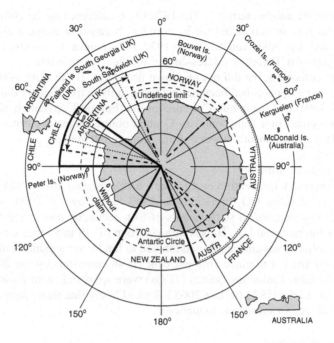

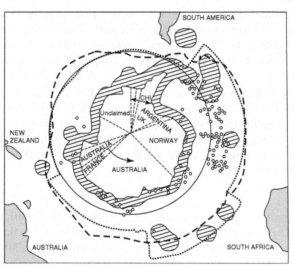

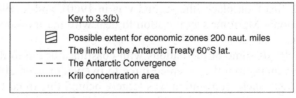

Figure 5.3 The Antarctic: (a) territorial claims; (b) possible economic zones to 200 nautical miles, and limit of Antarctic Treaty (60ºS lat)

a way as to injure another territory. The 1972 UN Conference on the Human Environment in Stockholm was in part called for by Sweden, because of concern about acid deposition generated by other countries. In 1979 the Geneva Convention on Long-range Transboundary Air Pollution addressed the problem of transboundary sulphur dioxide atmospheric emissions, but did not lay down firm rules. By the late 1980s the resolution of transboundary impacts had become an increasingly active field of diplomacy (Carroll, 1988). The 1991 UN Economic Commission for Europe Convention on Environmental Impact Assessment in a Transboundary Context obliged signatory states to act to protect transboundary environmental impacts from proposed activities.

Controls on global warming

The UN Framework Convention on Climate Change (signed at UNCED in 1992) obliged signatories to stabilise CO_2 emissions at 1990 levels by AD 2000. The 1997 Kyoto Conference was intended to settle details of CO_2 reduction and to see that targets were enforced by international law. However, a coalition of US industries was opposed to any limit on greenhouse gas emissions, and lobbied to hinder agreements. Finally, agreements were made by the EU to make an 8 per cent cut in emissions by AD 2010 and arrangements for Tradable Emissions Quotas (TEQs) were approved (with Russia able to sell its unused quotas to the USA) (in late 2005 it looked likely that many signatories would not meet their targets for emissions reduction).

Ozone damage controls

Efforts to phase out and if possible ban the use of CFCs were made at the 1985 Vienna Convention for the Protection of the Ozone Layer. The 1987 Montreal Protocol on Substances that Deplete the Ozone Layer – revised 1990 – derives from the Vienna Convention. The Protocol aimed for a 50 per cent cut in CFCs over a short period (twenty-four, mainly developed nations signed – by 1994 this had increased to seventy-four, including some developing countries) and was signed in the face of considerable uncertainty about ozone damage. The Protocol is a landmark in that for the first time nations agreed to impose significant costs on their economies in order to protect the global environment. India and China held out, seeking agreement for funding to assist with ozone controls. In 2005 ozone thinning, especially the Southern Hemisphere 'hole', was still a matter for concern.

The Law of the Sea

In 1954 the International Convention for the Prevention of Pollution from Ships was established to try to reduce the discharge of waste oil from oil-tankers and other ship-related discharges (with limited success). For ocean pollution control to be effective, agreements that cover rivers, effluent outfalls, air pollution and so on are required, because pollutants arrive in the sea from such sources (Boyle, 1992). In 1958 the First Conference on the Law of the Sea took place (the second was in 1960), and in 1959 the UN established the International Maritime Organization to deal with marine safety, law, pollution control and so on.

From the early 1970s some of the nations with coastlines began to declare extensions of their territorial waters from the accepted three to twelve, or even 200 nautical miles. The 1950 Continental Shelf Convention was largely behind this trend towards extension of exclusive sovereign rights to continental shelf or seabed resources. To try to formalise these trends the Third Conference on the Law of the Sea was held in 1974. The UNEP's Regional Seas Programme has brought together coastal states of a number of marine regions, resulting in several Regional Seas Treaties, covering: the Mediterranean; the Gulf; West Africa; Southeast Pacific; Red Sea; Caribbean; East Africa; and the South Pacific. These treaties led to the development of Environmental Action Plans and then co-ordination to fight pollution and so on. In 1977 the North Sea ceased to be 'high seas' as far as fish and mineral exploitation were concerned, when the EC established zones laying claim to the continental shelf. A number of the regional seas (e.g. the North Sea, Japan's Inland Sea, the Baltic and the Mediterranean) have been the subject of convention or treaty agreements in addition to the efforts of the UNEP to try to control pollution more effectively.

Meeting in Jamaica in 1982, the UN launched the Convention on the Law of the Sea (with agreements effective to 2500-m depth from the shore). Some developing countries are keen to see the oceans (e.g. Antarctica) declared common heritage, rather than becoming de facto possessions of those countries with the wealth and technology to exploit the resources. Controls over damage to ocean fisheries were not adequate at the time of writing (2005).

Natural resources are often under no clear and enforceable single ownership or even national sovereignty. There have been recent claims that sea fishing is so poorly controlled that 'the final roundup' is currently taking place (i.e. over-exploitation is unchecked). There are indications that nine of the world's seventeen largest ocean fisheries were being over-harvested in 2003 (New Scientist 178–2395, May 2003: 3). World fish stocks are in a poor state and time for developing workable agreements and enforcement is scarce. Unilever, Europe's largest fish trader, recently established a Marine Stewardship Council to support an approved trader label which indicates that fishing companies catch their products in a reasonably sustainable way. It is interesting to see the initiative coming from the business sector; unfortunately, it is nowhere near enough, and what has appeared has been judged ineffective and open to abuse. Soon, it seems, staple fish such as Atlantic cod will be on the CITES Endangered Species List (http:// www.cites.org / – accessed February 2005).

Currently there is a sort of 'enclosure' or privatisation of common resources (e.g. genetic material via patent law and claims of ownership of intellectual property rights). This could accelerate and disadvantage developing countries. There have been many attempts by business to uncover and control traditional knowledge (in common ownership) using ethnobotanists and social scientists (Berkes, 1999). For example, there have been protests over attempts by companies to patent products clearly based on folk remedies associated with India's neem tree (Azadirachta indica); the raw material and ideas for its use are 'stolen', a product is synthesised and then sold at a profit, and attempts are made to protect the trade by patent. As biotechnology develops, similar issues are likely to increase.

Indigenous peoples and environmental law

IUCN (1997: 27) estimates that there are over 250 million indigenous peoples who interact with environmental law with respect to:

1   protection of natural environment together with indigenous people;
2   rights of indigenous people over natural resources;

3   rights over traditional knowledge (e.g. to prevent ethnobotany becoming 'bio-piracy' (gathering indigenous knowledge which is patented and sold));
4   damages to indigenous people for past environmental wrongs by 'outsiders';
5   views of indigenous people, which could be fed into environmental law making.

Indigenous people often retain knowledge, skills and beliefs that relate closely to the natural environment. The protection of the environment is often vital to their physical and cultural survival, and they have insights which may aid environmental management and law making. The rights of indigenous peoples are recognised by the UN Commission on Economic Development (UNCED) 1992 Convention on Biological Diversity and by the 1994 Draft UN Declaration of the Rights of Indigenous Peoples. Nevertheless, indigenous people often still have no written land tenure, making them vulnerable to abuse or resettlement if there are natural resources to be exploited.

Worldwide, governments and companies involved in resource exploitation come into contact with local people. In the past the relationship was seldom beneficial for the latter. Nowadays, laws insist that there is fairer treatment, NGOs and the media are watching, and the locals are increasingly vociferous, aware of their rights, likely to hire lawyers, and better organised. Indigenous peoples are also networking with similar groups and exchanging experiences. In a number of countries indigenous people have helped put in place more effective environmental and social impact assessment. The values placed on environmental resources by modern society may be very different from those of indigenous peoples; for example, in New Zealand and Australia landscape features may have very important significance – state government and foreign companies may not share those views. In some countries local tradition and values are scorned. Where people have no firm, legally recognised title to land it is easy to license exploitation by outsiders.

In recent decades several countries have made changes to improve indigenous peoples' control of their environment and natural resources. Whether this will lead to better environmental management is debated. In Australia, New Zealand, the USA, Canada and Amazonian Brazil aboriginal people have fought for their sovereign rights to control and manage, or at least share in, resources (Shutkin, 1991; Dale, 1992). In Australia debate about aboriginal territorial rights has become heated recently. The Australian High Court has ruled that Australia's indigenous people enjoy native title and access rights to land leased by Euro-Australian farmers, which means two land users should legally coexist. An Aboriginal claim to coastal waters and the Great Barrier Reef, if awarded, would have considerable impact on fishing and coastal resorts (*The Times* (UK), 30 December 1997: 11).

A question increasingly asked is: Who should bear the cost of rehabilitation after resources exploitation? For example, the Pacific island of Nauru, now independent, provided phosphates for some ninety years. Does it have any claim on past colonial powers to remedy damage? Nauru claimed through the International Court of Justice for damage done before its independence in 1967 (Anderson, 1992). Similar retrospective actions have arisen in Australia and in other Pacific islands, over nuclear weapons test sites, and in Papua New Guinea concerning mining.

International conferences and agreements

International conferences and agreements on environmental issues mainly developed after the 1940s, with the first broad environment and development focus being the UN (Stockholm) Conference on the Human Environment in 1972. Business and law have

---

Box 5.4

Agreements made at the 'Earth Summit', 1992

- *Rio Declaration on Environment and Development* Updated version of the Stockholm Declaration (of 1972); published general principles for future international action on environment and development.
- *Framework Convention on Climate Change* Framework for negotiation of detailed protocols to deal with control of greenhouse gas emissions, deforestation, sea-level change and so on.
- *Convention on Biological Diversity* Intended to arrest alarming rate of species loss (criticised for having been poorly and hurriedly drafted).
- *Declaration on Forests* A principle, not legally binding, this was substituted for the original idea of a Forest Convention.
- *Agenda 21* An action plan for the rest of the century and a framework for dealing with environmental and developmental issues. Consists of forty chapters (not a legally binding instrument). Has proved a seminal document, prompting many local and regional initiatives.
- *Global Environmental Facility* A fund established for global problem-solving. Under the auspices of the World Bank, UNEP and UNDP. Designed to be 'democratic and transparent' and helpful to poor nations. Among other things, intended to support Biodiversity and Climate Change Conventions.

---

played a significant part in international conferences and negotiations, especially since the UN Conference on Environment and Development (1992 Rio 'Earth Summit' or UNCED). The Rio Conference was a test of the ability of the international political and legal order to reach a consensus for the good of the whole world (Tromans, 1992; Grubb *et al.*, 1993). Originally there were hopes that UNCED would agree an Earth Charter, but this was not achieved, although several new declarations were made and conventions were established (see Box 5.4) (Johnson, 1993; Freestone, 1994). Freestone (1994) reviewed the implications of UNCED, stressing that it did crystallise principles which contribute to the development of international environmental law. However, some feel the Earth Summit tended to weaken international environmental law by focusing on development issues (see Sands, 1993). A follow-up meeting to UNCED, the Rio II Conference, was held in New York in 1997. Another summit, which may be singled out as a key environmental management event, is the 2002 UN World Summit on Sustainable Development (the 'Johannesburg Summit'). Essentially a sequel to UNCED it aimed to see better implementation of *Agenda 21*. Other conventions and agreements which may be singled out are: the Kyoto Protocol, which seeks to improve global agreement of how to counter climate change; and the Cartagena Protocol, aimed at reducing risks and negative impacts associated with biotechnology and genetic engineering. These agreements are binding signatory countries to restrictions and international taxation, to a level unheard of a few decades ago and, not surprisingly, some nations have been reluctant to ratify, and some environmentalists argue that the effort and expenditure is misplaced.

There have been many other environmental agreements in the past decade, which are better dealt with when discussing particular issues such as control of global warming (for a review see BambooWeb – http://www.bambooweb.com/articles/e/n/Environmental_ agreements.html – accessed April 2005). Some of these are legally binding, some not; in many cases not all countries have ratified agreements.

Alternative dispute resolution

Disputes about resource exploitation and environmental management can be addressed in a number of ways (see Mitchell (1997: 218–239) for an overview; Napier (1998) for more detailed coverage of environmental conflict resolution):

1 through legal measures (judicial);
2 through political measures;
3 through administrative measures;
4 through alternative dispute resolution measures (which may not use law).

Legal measures rely upon courts, litigation, protocols and procedures. Political measures rely upon elected or established representatives to decide. Administrative measures may be used to improve resources and environmental management. Alternative dispute resolution may be through a range of measures, including:

● negotiation;
● mediation;
● arbitration;
● public consultation.

  Environmental management legislation may specify the use of some of these measures, or they may be adopted voluntarily. Negotiation is a process whereby two or more groups agree to meet to explore solutions, in the hope of reaching consensus. Mediation is similar to negotiation, but involves a mutually accepted neutral third party that finds facts and tries to facilitate discussion. The mediator may act with groups that are unwilling to meet face to face, if necessary 'filtering' the exchanges to help reach agreement. Arbitration involves a third party like mediation, but at the outset the parties involved agree to give the arbitrator power to make decisions (which may or may not be binding).

Prompting and controlling environmental management

Business is often seen as the 'motor' for change. Certainly considerable sums of money are being expended on training some of the brightest young adults in business schools and these people are likely to play a major role in shaping future development. It remains to be seen whether large multinational businesses like national governments tend towards conservative habits and slow gradual change. Business can develop fast and adapt swiftly; most would accept that Ford played a key role in rapidly developing modern production and consumption patterns. Private companies and joint state/commercial ventures have quickly exploited marine petroleum resources, spawning formidable new technology. In the past the roots of the British Empire lay in the East India Company. Breakthroughs in new energy sources will probably come from companies, rather than from state bodies. Large businesses can grow quickly from tiny 'garden-shed' origins and huge corporations can seed specialist semi-autonomous companies which are good vehicles for adaptation (Fuller, 2000).

Law may be used to both control and encourage actions, and must be able to deal with business, transboundary issues and controlling situations where many individuals play a part in determining environmental quality. Law has so far been less adaptable and slower to evolve than business. Lawyers and company executives tend to expect a future that is broadly like today – a 'business-as-usual' scenario. Technical innovation holds out the main hope of environmental improvement; people will rarely change their behaviour or accept a reduction in living standards (Cairncross, 1995: viii). There are unwise assumptions: that the environment is stable and generally benign, and that peace and Western democracy are here to stay. Expectation of limited change will probably mean poor preparedness for future disasters and too slow progress towards halting environmental degradation and working towards sustainable development. Since the 9/11 World Trade Center tragedy and the 2004 Indian Ocean tsunami it has become more apparent how interdependent societies are. Growing Chinese and Indian development will mean greater non-Western influence. There are signs in 2005 of an awareness that the poor, especially in Africa, have to be assisted by richer nations and at a scale not accepted before.

Summary

- The world is increasingly globalised and affected more and more by business and consumerism. The greening of business will play a key role in future development. Indeed, with much of the globe's economic power in the hands of business, progress in environmental management will tend to be backed by commercial bodies.
- Some businesses have genuinely embraced green approaches, some have made half-hearted efforts, and others have cynically exploited greening, practising corporate 'greenwash'. Somehow, environmental managers must catalyse the change to green economics and green business practice – through legislation, taxation, controls, propaganda and education.
- Business shapes the world and is increasingly a nurturing ground for new environmental management ideas and tools.
- Both business and law have to redefine their goals.
- Law is crucial for environmental management in a number of ways, aiding: regulation of resource use; protection of the environment and biodiversity; mediation, conflict resolution and conciliation; formulation of stable, unambiguous undertakings and agreements.
- Law tends to lag behind in meeting environmental management needs.
- Law and business will increasingly have to embrace non-Western ways.

Further reading

Capra, F. and Pauli, G. (eds) (1995) *Steering Business Towards Sustainability*. Earthscan/UNU Press, London.
Greening business.

Elkins, P. and Max-Neef, M. (eds) (1992) *Real Life Economics: understanding wealth creation*. Routledge, London.
Alternative ways for business to explore.

Wade, R. (2003) *Governing the Environment: the World Bank and the struggle to redefine development*. The World Bank, Washington, DC.
Review of the way forward, through World Bank eyes.

Welford, R. (2000) *Corporate Environmental Management: towards sustainable development.* Earthscan, London.
The third of three volumes on corporate environmental management which focuses on development issues and sustainability.

www sources

Business for Social Responsibility http://www.bsr.org/ (accessed May 2005).

Coalition for Environmentally Responsible Economics (USA) http://www.ceres.org/ (accessed May 2005).

Green Business Net http://www.greenbusiness.net/ (accessed May 2005).

Green Ethics Investors – numerous pension funds, investments, banks, and so forth which profess to be environmentally friendly. For further information see *Green Money Journal* http://www. greenmoneyjournal.com (accessed May 2005).

Greening of Industry Network (UK) http://www.greeningofindustry.org (accessed October 2005).

Guidelines for Multinational Companies – *OECD Guidelines for Multinational Enterprises* (2000) http://www.oecd.org/ (accessed October 2005).

Industrial Ecology Compendium – information on industrial ecology. North American focus (bibliography/case studies) http://www.unimich.edu/~nppepub/resources/compendium/ind.ecol. html (accessed May 2005).

International Business Ethics Institute http://www.business-ethics.org/ (accessed May 2005).

International Chamber of Commerce (ICC) – a global business organisation which promotes sustainable development through such initiatives as the *ICC Business Charter for Sustainable Development* http://www.iccwbo.org/ (accessed May 2005).

International Network for Environmental Management – seeks to develop and apply principles of environmental management. Over 500 member companies in 1994; non-profit organisation established in 1991 by Austrian, Swedish and German businesses http://www.inem.org/ (accessed May 2005).

⬤6 Participants in environmental management

The point is often made that people cause many environmental problems; therefore they must be involved in efforts to resolve and avoid them (Naess, 1989). This chapter explores the stakeholders involved in environmental management. Participants may be 'players' or 'bystanders' but all have an interest in the environment. Stakeholders may be individuals, groups, institutions, organisations or nations – they may cause change for the better or worse, and they can benefit or be harmed by their alterations and by natural shifts. It can be very useful to conduct a stakeholder analysis before undertaking a project or programme to help establish who is involved and how environmental management can harness support and reduce opposition (see below).

Adams (1990) singled out two voiceless groups affected by or causing environmental change: 'the blind' and 'the dumb'. The 'dumb' may include people or governments who are uninformed of the implications of development, or who are unable adequately to promote their views and affect change. The 'blind' may include consultants, scientists, economists, bankers and those bent on riches or blinkered by concern for sovereignty, religion or national security. The 'dumb' are often marginalised people, victims of disaster or unrest, or the underclasses – those without enough influence or power to lobby effectively when they feel change is needed. The environmental manager has to try to disseminate information to the 'dumb', possibly protect or empower them, and, if necessary, inform and control the 'blind'. It may take some effort to identify the marginalised. There is a further group, 'the unaware', which may include scientists who fail to perceive a problem or opportunity, not due to prejudice or greed, but because the issue is unfamiliar to them, and/or it happens too slowly to register (a creeping situation), or too fast and unexpectedly. Research and monitoring and adaptive environmental management approaches can help reduce these problems.

Modern development has focused on yield increase, often for the benefit of individuals or special-interest groups. It is only in the past few decades that appropriateness, sustainability, equity, participation and security have also become common goals. Traditional resource users often seek sustainability, equity and security, and much can be learned from them, although such traditions are often in decline through various development pressures. Before the 1970s many development agencies failed to ask whether a proposal was 'appropriate', or made any effort to seek indigenous knowledge, or involve local people in decision making and management. Increasingly developers must show some degree of environmental and social responsibility and consult locals. In any given environmental management situation there are likely to be a number of different views, and hence various possible responses. The environmental manager has to try to avoid conflicts between stakeholders and minimise damage to the environment (Box 6.1) (Bowander, 1987). In this there are parallels with the role of the state: environmental management similarly deals with policy, planning, legislation, control, implementation and management (Cooper, 1995). Environmental managers must relate to and steer people, build teams, establish supportive institutions, influence opinions, inform and educate – seldom can one individual do all this, and it is a team, consultancy or company effort (Litke and Day, 1998). Innovation depends a great deal on there being supportive institutions and networks of relations; the environmental manager must strengthen those already present and form new networks if necessary.

Learning from past peoples

Modern environmental managers can benefit from information about past peoples, for example:

- *Environmental history* – provides hindsight knowledge on environmental issues in the past, how stakeholders reacted to challenges and opportunities, and much more. Before the late 1980s there was some suspicion that the field overlapped crude environmental determinism. Recently, however, a number of researchers and more popular writers have been exploring environmental history; for example, ENSO effects and the impact of the Little Ice Age (Fagan, 1999, 2000; Diamond, 2005). This has helped prompt long-range forecasting of future ENSO impacts, gives insight into how present landscapes and traditions have evolved, and highlights threats which recur over a long time span.
- *Archaeology and palaeoecology* – provide information from prehistory that supplements historical data. The benefits include those just listed for environmental history plus information on techniques which people may have forgotten. Some useful agricultural techniques have been gained from archaeology and these plus other information may prove valuable for countering land degradation or developing sustainable agriculture in marginal areas.

Millennium development goals

The international community is committed to development which promotes economic progress, and sustainable development through the agreement and ratification of the Millennium Development Goals 1990–2025. In 2000 member states of the United Nations undertook to try and achieve these, which focus on eradication of poverty and hunger, improvement of human health, and attainment of environmental stability (for details see World Bank (2005) *Miniatlas of Millennium Development Goals*. World Bank, Washington, DC; http://www.developmentgoals.org/About_the_goals.htm and

Box 6.1

Participants in environmental management

- *Existing users*: Land or resource users (males and females may make different demands); there may well be multiple users.

- *Groups seeking change*: Government (may be conflicting demands from various ministries or policy makers); commerce (national, MNCs/TNCs), individuals seeking personal gain or to change the situation, international agencies, NGOs, media, academics, 'utopians').

- *Groups pressed into making changes*: The poor with no option but to over-exploit what is available without investing in improvement; refugees, migrants, relocatees, eco-refugees (forced to move or marginalised so that they change the environment to survive), workers in industry, mining and so on, who face health and safety challenges while carrying out changes.

- *Public* (may not be directly involved): May be affected as bystanders; may wish to develop, conserve or change practices (if aware of what is happening); expatriate or global concern.

- *Facilitators*: Funding bodies, consultants, planners, workers, migrant workers (latter two groups affected by health and safety issues), Internet exchanges of environmental data.

- *Controllers*: Government and international agencies, traditional rulers and religions, planners, law, consumer protection bodies and NGOs (including various green/environmentalist bodies), trade organisations, media, concerned individuals, academics, global opinion, and the environmental manager.

Note: For a given issue there is often more than one participant, some involved at different points in time and with varying degrees of involvement. As time progresses a group may become more aware of developments and/or empowered and act more effectively. There are subtle differences between 'involvement' and 'participation': the former may imply simply telling people what is happening or what will happen. Participation means that there is some degree of consultation and involvement (often far short of influencing whether a development takes place).

http://www.un.org/millenniumgoals/ – accessed April 2005). However, only one of the eight goals, number 7, actually pays much attention to environmental management. Thus there is a possibility that funding may be diverted from purely environmental management activities to those with a social development component.

Global change and people

Human health and food supplies are easily affected by environmental factors so there is interest in assessing the impacts of any change of climate, UV receipts, pollution and so on. Diseases may spread to new areas if there is global environmental change, and modern transport and lifestyles can alter transmission patterns. Disease transmission and infection usually depend on many variables, not just climate. Human habits and innovations are also important. Thus attempts to forecast future patterns for malaria and other diseases transmitted by insects or rodents must be cautious.

New patterns of disease can have marked impacts; for example, if malaria, sleeping sickness or yellow fever spread, people may alter their habits or even move, causing environmental changes. One response to new disease patterns might be increased use of pesticides, including those recently abandoned like DDT; the environmental impact would be considerable. Some diseases debilitate or kill off large enough numbers to affect the labour supply and land-use practices. The introduction of Old World diseases like measles and smallpox into the New World by the *conquistadores*, where people possessed little or no immunity, had huge effects on societies and their land use (Diamond, 1998). In a number of poor nations, especially in sub-Saharan Africa, HIV/AIDS is killing and weakening enough people to alter land use and food supply.

Stakeholders

For thousands of years people have evolved rights, taboos and skills for managing natural resources (Ewert *et al.*, 2004). Problems arise when traditional strategies and rights break down, are usurped, or do not cope with changing conditions. The problems may be caused by competing groups, urban elites, speculators, powerful commercial organisations and so on; but breakdown is often blamed on the inadequacies of local people or 'acts of nature', rather than on central government policies and weak administration. Worldwide, the expropriation of common resources from traditional users has become a problem (Jodha, 1991; The Ecologist, 1993), and this exclusion is to blame for a significant portion of all those forced into marginal situations. The politics of exclusion appears to be expanding – states license companies to exploit an area or resource used by people without documented rights and they are evicted to degraded marginal land or to settle in urban slums. Marginalisation usually means the weakening of livelihood, diminished standards of living, and more vulnerability to damage; in short, a poorer and more precarious position. Some groups are more prone to marginalisation: the poor, women, landless, indigenous peoples, unemployed, and the elderly. Marginalised people commonly come into conflict with the environment because they have no alternative and must use resources to try and survive; the official response may be to offer aid for social development, legislate against them, hound them on to somewhere else, or resettlement.

Stakeholder analysis and stakeholder management

It is important for environmental managers to engage with the right people in an appropriate way to achieve results. Stakeholder analysis seeks to identify all the stakeholders, and to work out their power, interests, capabilities, needs and so on. It then focuses on key individuals or groups and categorises them in terms of likelihood to support or oppose. Stakeholder analysis and management are widely used in business to assist in achieving goals (http://www.mindtools.com/ – accessed March 2005). Using these techniques, the environmental manager can target key individuals or groups and should gain some insight into their views and abilities, interests and relationships. It is not only important to win over powerful key stakeholders; knowing all those involved and keeping them informed in the right way ensures that they are more likely to cooperate and support efforts. Ideally, stakeholder analysis is conducted early in any development.

Indigenous groups

In the past, dominant societies usually ignored, exploited or persecuted indigenous people. Since the 1970s indigenous peoples have been strengthening their control over their lives and access to resources; many now hire lawyers and other advisers, and once isolated groups network with other peoples, some on the opposite side of the world. There has been a growing practice of seeking to consult and involve local people in environmental management, and to understand and make wider use of indigenous knowledge (Klee, 1980). Environmental management can learn a lot from study of people's livelihood strategies; for example, Geertz (1971) tried to understand the process of exploitation and ecological change in the real world, focusing on Indonesia. There is now a growing field of study of traditional knowledge (see IUCN Inter-Commission Task Force on Indigenous Peoples, 1997). Not all forest dwellers, pastoralists, hunter-gatherers or other indigenous groups support environmental management; it is a myth that pre-modern folk are always 'in balance' with nature and care for it. There are cases where peoples have granted mining rights to outsiders, accept fees for hazardous wash disposal on their lands, or build casinos. Nevertheless, some indigenous people have established sustainable and environmentally friendly livelihoods, such as ecotourism (Stalton and Dudley, 1999). Diamond (2005) examined a number of past and existing societies to try to assess what might be done to enhance chances of achieving and maintaining sustainable development.

Women

There has been a growth of interest in the role of women in development and environmental management, especially since the 1975 to 1985 UN Decade for Women. Some have attempted to subdivide the rapidly expanding literature and activism according to the perspective adopted:

1 Women, environment and development (WED) – focusing on women having a special relationship with the environment as its users and managers.
2 Gender and development (GAD) – with gender seen as a key dimension of social difference affecting people's experiences, concerns and capabilities. Gender may be defined as a set of roles; for a review of gender and development from an environmental management standpoint see Mitchell (1997: 199–217).
3 Women in development (WID) – focusing on reasons for women's exclusion or marginalisation from decision making and receipt of the benefits of development (Rao, 1991; Leach et al., 1995; Ngwa, 1995).

Women are relatively more adversely affected by environmental degradation; they tend to be the poorest sector of society and often depend more on common resources, loss of access to which hits them harder than the menfolk. Women must often survive with little support from men, who may have migrated to find employment, abandoned their partners, or are too impoverished or feckless to offer support. Female heads of household may manage farms or small businesses, and are commonly very inventive and adaptable. There are situations where environmental management efforts are best directed towards women in order to obtain results. In a number of countries women have taken the initiative; for example, starting reforestation projects. Women and children are commonly gatherers of fuelwood, food and water, so environmental damage means more work for them. Women are more exposed to hazards like insect pests while gathering, so the gender division of labour and routines further disadvantages them (Sachs, 1997). Commonly female diets, educational opportunities and levels of freedom are poorer than those of males. The two sexes are likely to respond to opportunities in different ways – so to think of even a single citizen social group as uniform is mistaken (A. Agarwal, 1992; B. Agarwal, 1997). In Burkina Faso studies revealed that productivity was better if men and women were given separate plots of land, rather than having women work on men's land (Zwarteveen, 1996). Gender differences in ownership can be important; if women are seen by men to be improving their crop yields or tree cover and they do not own the land, they may have it taken from them. To get their participation in soil conservation, biodiversity conservation, tree planting and other environmental improvements it is necessary to ensure that women enjoy the fruits of their labour.

There have been suggestions that women are more likely than men to be concerned about local environmental issues. There are many examples: in the USA Love Canal pollution case, women recognised the problem and campaigned for a solution; in India, the Chipko and related forest protection movements started with largely female memberships; and in the UK many of the protesters against new highways and so forth are women. Women benefit more from environmental improvements because they are often the fuel and water collectors – afforestation and improved water sources reduce the distance they have to walk and the risks they face (Dankelman and Davidson, 1988; Shiva, 1988; Momsen, 1991; Sontheimer, 1991; Jackson, 1993). In a number of peri-urban areas it has been the women who have organised gardening and tree planting.

Eco-feminism (ecological feminism) is a broad field, but in the main it recognises parallels between oppression of women and oppression of the natural world. The argument is that men dominate both, so 'greening of the Earth can only begin with the empowerment of women' (Diamond and Drenstein, 1990; Spretnak, 1990; Rodda, 1991; Mies and Shiva, 1993; Wells-Howe and Warren, 1994). Eco-feminism has made attacks upon other sections of radical environmentalism, including the deep greens and social ecologists, arguing that these gender-neutral attitudes are not enough to control male domination of women and nature (Mies, 1986; Merchant, 1992, 1996; Warren, 1997). Braidotti *et al.* (1994) and Harcourt (1994) have explored the role of women in attempts to achieve sustainable development.

There is also a more romantic debate on the contribution of women to environmental care, based on the perception that women, through reproduction and the nurture of children, are more closely attuned to nature. Women are certainly in a position to influence future behaviour by virtue of being first educators of the young. In developed countries women have long been at the forefront of raising environmental awareness: various pioneering conservation NGOs were founded by women; permaculture/organic farming was initiated by a woman. Rachel Carson and Barbara Ward were among the first to raise public awareness of pesticide pollution and sustainable development in the 1960s and 1970s. Women played a central role in the formation of green politics in Germany and elsewhere in Europe from the 1970s, notably the late Petra Kelly (Seager, 1993). A move towards establishing new environmentally friendly and more socially appropriate producer-to-market networks was taken by Anita Roddick's Body Shop® chain of stores in the 1980s. In some Western countries women are now more successful academically than men and play an important role in the consumption of manufactured goods; they are targeted by advertising, and can set trends, vote and alter buying patterns, all of which have significant environmental implications.

Individuals and groups seeking change

Powerful individuals or special-interest groups generally seek to control policy making and development, although fewer do so with the aim of improving environmental care rather than for personal profit. The lobbying may be subtle or open, directed at the public, or more focused on key political figures or departments.

In free enterprise countries rich individuals who fund institutions, finance university chairs or purchase land for conservation may further environmental management. There is commonly a pecking order among ministries, with some exercising more power and influence to try to influence change or resist it. The environmental manager should be vigilant for such influence, and seek to mitigate it if it acts against environmental quality and human welfare. This may necessitate alliance, covert or open opposition, or appearing to remain neutral. Diplomacy and politics play a major role in environmental management, and when there is more than one stakeholder, which is often the case, negotiation skills are at least as important as access to technology and knowledge (Vogler and Imber, 1995).

Individuals and groups with little power

The poor

Many identify two key challenges for those in charge of development at the beginning of the twenty-first century: poverty alleviation and environmental care. The two issues are sometimes closely related, although linkages are often unclear and complex. The poor, it is often claimed, degrade their environment in their effort to survive – a trap of poverty. Poor people are usually more vulnerable to environmental problems and hazards. In reality these hazards are usually part of a process not the cause, and may lie with trade, government policies, faulty land rights, marginalisation and so on. For example, the causes of environmental degradation in urban areas may lie with policies affecting agriculturists hundreds of kilometres away, causing people to migrate and swell city populations.

The 'Malthusian' concept is that demographic growth leads to a spiral of poverty, further population increase, environmental degradation and so on (Reardon and Vosti, 1995; Scherr, 1997). However, there are also situations where population increase has prompted intensification and improvement of land use, has cut environmental damage, and may ultimately halt population increase (Tiffen et al., 1994). There are affluent countries with very low population density suffering severe environmental degradation. Hotspots may be found where there is poverty–environment stress, including cities where population growth is outstripping employment and infrastructure; vulnerable land where marginalised people have congregated; and areas where traditional livelihood strategies are degenerating (Leonard et al., 1989: 19). There is also national or institutional poverty: nations may be unable to afford adequate environmental management or funds have been misspent. Aid may assist efforts to improve environmental management. For example, debt-for-nature swaps, or richer countries may support poorer countries' efforts in environmental management. The Montreal Protocol set up funds to assist with stratospheric ozone protection, and the UN Conference on Environment and Development tried to establish a Global Facility (initially involving the World Bank in 1990). The Earth Increment (established 1992) is supposed to support developing countries seeking to implement Agenda 21. So far, the Increment has been hindered by squabbles over allocations, and the failure of many funding nations to pay up enough (Holden, 1991; Patlis, 1992).

Displaced people

People relocate for a variety of reasons, some willingly, others reluctantly. The move may be expected and even gradual, or it may be sudden and unexpected – the latter usually means that relocates have no funds or tools, and are badly disorientated. Eco-refugees are people displaced by natural or human-induced environmental disaster or environmental degradation (El-Hinnawi, 1985; Ramlogan, 1996) (Figure 6.1). Eco-refugees (or environmental refugees) are not classed as true refugees (in 2005) – the UN High Commission for Refugees recognises only those who are persecuted. Others, including eco-refugees, are classed as displaced persons whom host countries have no legal obligation to settle, and they are expected to return to their origin at some point – difficult if it has been flooded by the ocean or buried by volcanic ash! Given the UN invented the term eco-refugee in 1985, it has been slow to address the legal difficulties.

Worldwide there were over fifteen million, possibly as many as fifty million actual refugees in 1998, a large proportion displaced by unrest and warfare. Eco-refugee displacement can be caused by dam construction, flooding, drought, tsunami, and many other natural disasters or environmental degradation. Other causes of displacement (apart from civil unrest) include: accidental pollution, like that of Chernobyl; market or communication changes which make cashcrop agriculture less viable; social or economic changes that trigger abandonment or neglect of traditional livelihoods; large-scale cropping or ranching development, political expediency or planners' desire to provide services for scattered populations (Parasuraman, 1994).

Those who are forced or tempted to move may relocate within national boundaries (relocatees) or move to another country. Migrants share some of the characteristics of the displaced but retain their roots, returning home seasonally, from time to time, or at the end of an extended period of employment, and in some cases regularly remit money back home. Migrants can cause environmental degradation in the areas they have left as a consequence of labour depletion, which then leads to unsustainable livelihood strategies. However, there are situations where migrants are able to earn enough funds to finance improved land husbandry in their homelands, or simply by leaving prevent excessive subdivision of landholdings and over-exploitation of resources. Government relocation and land development schemes sometimes support voluntary migration of those seeking employment or new land. In many regions of the world, the bulk of relocation is undertaken by unassisted migrants. Displaced people, even when officially aided, may have difficulty in establishing new or restarting their old livelihoods. Compensation and support, if provided, may be inappropriate or inadequate. Displaced people may face conflict with host populations in the areas they move to, may have problems with other refugee groups and frequently adopt short-termist strategies for survival which can damage the vegetation, soil and other resources (Black, 1994).

There can be beneficial effects through relocation, when, for example, the host country receives skilled and resourceful people (McGregor, 1994: 123). The depopulation of an area by relocation sometimes leads to conservation and tourism benefits. The Scottish Highlands (UK) is a region of scenic beauty in part due to eighteenth-century clearances (forced relocation of tenant farmers). However, negative impacts can occur where there is movement of people out of rural areas; those left cannot maintain traditional agriculture and resort to less laborious, damaging activities. Movements of people can spread diseases and organisms affecting humans, crops and wildlife, and may have a serious impact on food security in the host region (Döös, 1994; Prothero, 1994).

In Malawi in the early 1990s displaced persons outnumbered the host populations in some regions. Relocatees may return to their original homes once the reason for their move has been resolved, or if their hopes for better conditions have not materialised,

Figure 6.1 Tucuruí Dam, across the Tocantins River, Amazonian Brazil, *c.* three years before completion. The reservoir flooded about 2,300 km² and led to the forced relocation of a large number of smallholders and their families, some settled only a decade or so earlier by offical land development programmes!

but there are camps in some countries which have become virtually permanent features and a burden on the host. Refugee camps concentrate people, which can lead to dependency and could deter people from returning. Large numbers of people restricted to a camp and its surroundings are vulnerable to diseases and natural disasters, and can have a serious impact on the environment through collecting fuelwood, wild plants and game. Water bodies and streams may also become depleted and polluted. Pakistan's Northwest Frontier Province received over 3,500,000 refugees in the late 1980s, which caused serious river, forest and pasture damage (Young, 1985; Allan, 1987; Black, 1994). To reduce refugee camp impacts it may be possible to provide alternative fuel and stoves to discourage wood collection and sewerage systems to reduce pollution. Where camps seem likely to remain for a long time, their inmates should be educated and supported to establish tree cover and sustainable horticulture (UNHCR, 1992). Refugees and migrants often follow roads into less settled areas, and such spontaneous unofficial relocation is usually difficult to control.

One of the best-researched aspects of human displacement is dam- and reservoir-related resettlement (Figure 6.2). Numerous studies have been undertaken since the late 1950s and a number of agencies have developed resettlement guidelines (Cernea, 1988; Gutman, 1994). With the accumulated hindsight of decades it should be possible to reduce problems, but real-world situations make it difficult (Thukral, 1992). There is also a tendency to consider only the disruption to people in the area flooded by a reservoir, yet downstream of a dam many people may suffer changes in livelihood as river flows are altered and have to relocate, largely unassisted (Horowitz, 1991).

Various countries have used planned resettlement schemes to relocate people from areas of high population, environmental degradation, or other problem regions (sometimes

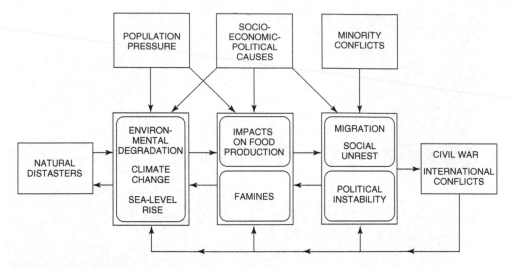

Figure 6.2 Schematic illustration of the links between the major factors that can have an influence upon, or reinforce, environmental degradation, resulting in an increased risk of environmental migration (eco-refugees)

Source: Döös (1997: 43, Fig. 1)

to strengthen their sovereignty over the relocation area). In Malaysia, the Federal Land Development Authority (FELDA) has opened up large areas of forest for resettling small farmers from land-hungry states (Fong, 1985; Sutton, 1989); Indonesia has an ongoing transmigration programme, settling people from Java on less populous islands. Similar state-supported land development and resettlement programmes may be found in Latin America (e.g. Bolivia, Ecuador), Kenya, the Sudan, Ethiopia, and several other countries (Collins, 1986; Pichon, 1992). These schemes often cost a lot, move relatively few people, may fail to sustain the settlers and prevent damage to the environment (Scholz, 1992). Studies have shown why settlers may fail to get established and resort to damaging the environment and re-migrating (Moran, 1981). Much depends on there being adequate incentives for sustainable land use (e.g. enough return for labour, secure landholdings, access to services like healthcare and so on) and upon the attitude of settlers, factors which environmental management may be able to control. Resettlement planning should benefit from the adoption of a participatory approach (Hall, 1994).

Eco-refugees could become a much greater challenge in the future (Sinclair, 1990). There is also a risk that large numbers may become a significant threat to global peace (Homer-Dixon, 1991; Westing, 1992; Myers, 1993; Ramlogan, 1996). In 2005 there were an estimated ten million eco-refugees, and predictions are that an estimated 150 million may be displaced by sea-level rise and agricultural changes caused by global warming by 2050 (http://www.archive.greenpeace.org/climate/database/records/ zgpz0401.html – accessed April 2005). Much of Bangladesh, parts of China, notably the Shanghai region, Egypt, India, and islands such as Tuvalu, the Maldives, Kiribati, the Marshalls, and some Caribbean islands are especially vulnerable to sea-level rise. Döös (1997) noted that the countries receiving most refugees so far have tended to respond by tightening border controls, and there was a need to look beyond this and to address causes. Some countries are seriously concerned about the possibility of an influx of eco-refugees in the future (Nolch, 1994).

There are claims that the world's poor are tending to be displaced to disaster-prone areas by the 'politics and economics of exclusion', and so could further swell numbers of refugees. There are regions prone to recurrent disasters, yet people still settle them; for example, coastal Bangladesh. Northeast Brazil suffers recurrent drought which is blamed for the hardship and relocation of poor people, although it is more likely to be lack of land reform (Hall, 1978). Some 'natural' catastrophes have an anthropogenic component – excessive grazing makes land vulnerable to droughts, global warming may raise the incidence of severe weather events, and tourism may encourage settlement in coastal sites vulnerable to storms and tsunami (Woehlcke, 1992). People displaced by 'elemental forces' may have little warning, and those shifted by gradual environmental degradation are likely to be better prepared, although the poor have less adaptability than richer people. Often environmental causes are merely a trigger for relocation because the poor have become more vulnerable or less able to recover from environmental problems. A number of international agencies, notably the UN High Commission for Refugees (UNHCR), have expertise on eco-refugees, and those concerned with global environmental change have tried to predict likely future scenarios (McGregor, 1993; Myers, 1993; Döös, 1997).

Public

The public usually consists of more than one group of stakeholders who may have different, perhaps conflicting, views and goals. Powerful groups tend to dominate and weaker people get marginalised, so the environmental manager has to establish the needs of all groups and try to ensure that none are ignored, yet if possible work with the influential. So far, mainly in developed countries, there has been legislation since the 1970s to ensure planning and development are more transparent. The environmental manager checks to see that public disclosure rules are followed, and, if needed, publishes impact assessment statements, environmental audit reports and so on.

Participatory environmental management

One way of ensuring that the weaker are heeded is to give them a say in what should be done and try to empower them so that they actively state their viewpoints. Participation and empowerment have become important for most Western nations, NGOs and many international agencies. However, there are some countries which prefer not to pass on too much control to the public: some are simply authoritarian regimes, but in others the people seem to prefer to have the state co-ordinate firmly, and sometimes authorities feel the public is not ready for participation. Increasingly, good environmental management is seen to be that which deals effectively and sensitively with people at the local or community level. This is not simply finding ways to control stakeholders; environmental managers need to gather local knowledge, understand feelings, in order to learn, alter practices and inspire people (Keen et al., 2005). Local knowledge is often crucial for resolving environmental disputes (Sidaway, 2005). Various aid agencies seek to engage poor people in such issues as conservation or improvement of the urban environment. There is usually a strong desire to reduce poverty, and sometimes this and environmental care act in symbiosis. But there may be situations where fighting poverty must take second place to environmental care, and this will not be politically correct. In addition, unpopular as the point may be, improved livelihoods can lead to more consumption, more refuse problems, and increased water and energy demand, all potentially causing greater environmental degradation (http://www.ann.sagepub.com/cgi/content/refs/590/1/73 –

accessed April 2005). Environmental managers must try to counter problems, but this may take away some of the development funding which others would like to devote to social development and generate opposition.

Participation is limited if there is not adequate access to information and transparency in decision making. The USA and most Western countries enjoy considerable freedom of information but many nations do not. This can mean that Western tools and techniques do not function well in non-Western situations. The media, consumer protection bodies and use of the Internet can help remedy this, but only to a point. How are citizens to enjoy improved participation? The state, NGOs or citizens themselves can prompt it. One route is the Deliberative Inclusionary Process, which seeks to help people shape environmental policy (Holmes and Scoones, 2000). Another approach is co-investment, whereby locals make efforts to improve environmental care, such as soil and water conservation works, and an aid agency or government provides help in the form of funds, materials, machinery, or whatever is not locally available. Environmental damage may be caused by insecurity of tenure or a weak legal claim to traditional resource use. These problems can discourage sustainable development because people will not invest in the future if they are unsure of benefiting. These problems also make it easier for government, business or private bodies to expropriate land and other resources. In such cases, simply providing better tenure and documentation may resolve problems.

Sustainable development strategies need to be designed to fit local conditions and to be co-ordinated to ensure that one locality or stakeholder group does not conflict with another. Better still, different regions and stakeholder groups should seek to develop integrative and mutually supportive strategies – 'dovetailing' waste from one activity to become a raw material for another. Environmental management should act as mediator and catalyst to develop collaborative approaches (Selin and Chavez, 1995). In this, public support can be crucial (Box 6.2). For example, it is pointless promoting tree planting if people later fail to take care of the growing saplings. Environmental problems are often a sum total of individuals' actions, so each person may have to change their attitudes to ensure a solution. Working with local people can inform environmental managers of threats, limits and opportunities they may otherwise have missed (Lise, 1995; Park, 1997).

Participatory approaches to data gathering, problem solving and development implementation have been progressed by a diversity of social sciences, agricultural extension agencies, public administration and development bodies, and NGOs, and have been adopted for environmental management (Messerschmidt, 1986; Cumberland, 1990; Chambers, 1994a, 1994b, 1994c). Since the 1980s it has become common to involve community members in participatory monitoring and evaluation of projects or programmes (community monitoring and evaluation or participatory monitoring and evaluation). The aim is to establish what stakeholders want (and even children may be consulted), need, do, and could adopt. There are potentially a number of benefits:

- The community is involved and can learn.
- It reduces the need for expensive experts.
- It reduces costs.
- It can engender support.
- There are opportunities to tap community creativity and traditional knowledge.

There is no single standard procedure for community/participatory monitoring and evaluation. Using a multidisciplinary team is important. For consultations, it is better to hold a number of small sessions than a few large ones. Typical methods include focus groups, group discussions, observation, and asking locals to draw maps or diagrams.

Box 6.2

Why the public should be involved in environmental management

- The public may be able to provide advice that would be missed otherwise.
- Open planning and management should be more accountable and more careful.
- Fears and opposition to management may be reduced if people are informed.
- If people identify with management they may well support it.
- It reduces risk of a communication gulf between 'experts' and 'locals'.

Note: The public is often a mixture of different groups: local people of differing age, sex and so on; regional, national or global groups.

'Involved' may mean minimal information; adequate information; active input to management before and during development; or involvement after management decisions.

Sources: Author; Wilkinson (1979)

It is important that methods and objectives are clearly explained to the people. Those consulted should be carefully selected and the data cautiously interpreted. Assessors must also be prepared for various viewpoints from a community – not all will support proposals. In addition, it is desirable to assess impacts upon neighbouring communities and to gather information on social capital. Social capital helps determine how vulnerable, resilient and innovative a group is. Ideally there should be an assessment to determine whether social capital is strong or in decline.

Participation can be invaluable, but it can also be complex and demanding (Ghaai and Vivian, 1992). If the environmental management team does not understand society and history as well as environment, serious difficulties can arise – Fairhead and Leach (1996) note the misinterpretation of the nature of forest 'islands' in the savannahs of Guinée. Similar warnings are given by Leach and Mearns (1996) that received wisdom is not enough, and that local knowledge and objective multidisciplinary or interdisciplinary study are needed. Local knowledge is often, but not always, valuable. Without local knowledge there may be misconceptions, prejudices, taboos and so on which hinder environmental management. Nevertheless, sometimes an 'outsider's' viewpoint is needed as well as that of locals.

Currently, participatory approaches are fashionable and may be promoted even if the environmental results fail to confirm their value. It makes sense to employ local people in a conservation area, rather than exclude them and cause resentment and possibly poaching (and it also makes use of their local knowledge and skills) but there can be disadvantages. Debate has raged over the effectiveness of participatory wildlife conservation as opposed to 'top-down and authoritarian approaches' (Oates, 1999; Stalton and Dudley, 1999). People can obtain livelihood from careful use of conservation area resources – extractive forest reserves, tolerant forest management, and employment as guides, rangers and ecotourism staff. Livelihood improvement must be balanced against environmental goals. There are cases where environmental management has improved under distinctly authoritarian regimes; for example, Dominica. Environmental managers, if necessary, should adopt approaches with limited public participation if there is no effective alternative, although doing so may generate hostility.

Facilitators

There are many bodies and individuals that promote and assist environmental management. Techniques such as stakeholder analysis may be used to try to identify promising contacts. This section examines some of the potential facilitators.

Funding and research bodies

Funding bodies can support environmentally desirable developments or withhold money until proposals are modified to meet environmental requirements. Many funding bodies have developed environmental management units, guidelines and manuals (Turnham, 1991). Regional development banks like the Asian Development Bank commission environmental management studies and training. There have been cases where failure to carry out environmental management measures has led to withdrawal of funding from large projects already well under way (e.g. the Narmada Dams in India). A huge diversity of bodies conduct research aimed at improving environmental management: universities, private research companies, independent international research institutes, and UN or other international agencies. Most research is applied and in response to perceived needs, but it is vital that some is anticipatory to warn of possible threats and potentially useful strategies, and enough non-applied study is undertaken to expose new and unexpected results and information that may be valuable but which offers little commercial gain. Commercial research and development is less likely to focus on untested and unprofitable fields; there is thus a risk that important environmental and social issues will be neglected. International bodies, governments and charitable foundations can therefore play a crucial role in funding what others will not.

Communications

From Victorian times newspapers have occasionally helped to prompt environmental action: opposition to feather and fur fashions, anti-whaling, support for smokeless fuel use, and much more. Communications, media, education and marketing overlap, and all have the potential to win support for environmental management. Cigarette, chewing gum and tea companies were publishing collectable cards, which sometimes dealt with wildlife issues well before 1945, and helped broaden popular awareness. The development of newspapers, radio, television, and magazines has played a key part in establishing environmentalism since the 1960s; today satellite broadcasting is bringing wildlife and environmental concern programmes to an ever-widening audience.

The Internet is playing an increasingly important role in environmental protection and management (Anon., 1995, 1996; McDavid, 1995; DeRoy, 1997). NGOs can exchange information, report problems beyond a national boundary (commonly before they can be prevented by a company or state), and are able to co-ordinate activities. For individuals involved in environmental management the Internet has become an important source of information and means for dissemination and discussion (Schuman, 1996). The media are improving public awareness of environmental issues, although, unfortunately, coverage is not always objective or accurate. Improved telecommunications make monitoring easier as instruments can radio information back (often in real time) via satellite and mobile phone links to research or administrative bodies. Development of computers, software and GISs make data handling and analysis far more powerful than was dreamed possible even ten years ago.

Controllers

There are various ways in which use of the environment is controlled; in most societies traditions evolve, and in some cases develop into laws. Moral and religious beliefs influence how people deal with the environment; sometimes the influence is positive, sometimes it is damaging. For example, Diamond (2005) examined how Easter Islanders engrossed in religious activities could literally cut down the last trees and cause environmental collapse. In Western-style democracies public opinion can be a significant control. In any society charismatic individuals and influential books can alter outlooks and establish organisations that function long after they have died. Sometimes this has worked to help environmental management – as in the case of Henry Thoreau, Peter Scott, Gerald Durrell or Jacques Cousteau (to name but a few).

Traditional controls and laws can break down or become outmoded by new challenges. These include the spread of transboundary problems; the globalisation of trade; genetically modified organisms; the penetration of capital. In parts of Amazonia traditional fishing conserved stocks through local laws and taboos; commercial fishing flouted these controls with no apparent ill-effect, so locals also started to ignore them – fish stocks crashed. However, some developments can work for the good or the bad; while capital penetration in one situation damages the environment, elsewhere it may bring funds and developments which lead to improvement.

Traditions and spirituality

Traditional societies commonly control resource use through local rulers who can exercise secular and religious power. These may allocate land for cultivation, decide whether to move a village, prohibit hunting and so on by reference to omens or magic (Hallman, 1994; Gottleib, 1998). The West, it has been claimed by a number of environmentalists, is influenced in its management of the environment by Judaeo-Christian ethics which place humans before nature. In Buddhist countries like Sri Lanka wildlife may be seen roaming unmolested, even in cities, when it would be hunted for the cooking pot in Africa or Latin America. In a number of countries religious bodies are in the front line of action to protect indigenous peoples and the environment, in promoting improvements to slum areas, and in poverty alleviation. Ethics and green spirituality alone are not enough: they do not guarantee adequate co-ordination, generate data or monitor situations. The skills of environmental managers are vital to determine the best strategies for the survival of fauna and flora and to organise sustainable development.

Accreditation

Those involved in environmental management have until recently been subject to limited professional policing. That has been altering, and there is growing adoption of quality assurance and control by various professional bodies. Currently there is a dwindling opportunity for consultants and environmental management professionals to practise if they are not accredited. Accreditation is the process whereby a state department or professional body registers practitioners if they meet set standards. The body then monitors their performance, demands improvements if needed, calls for updated skills when necessary, supports new skills acquisition, and warns or even 'strikes off' (i.e. removes the credentials for practice) any who do not adequately comply. There are still some non-accredited experts but these increasingly tend to be bypassed when it comes to hiring. Accreditation is a very important shift, which sets standards and improves accountability of environmental management staff.

International bodies and NGOs

International bodies and NGOs have become important watchdogs of corporate, govern-ment and special-interest group activities. They have a highly multifaceted role: lobbying meetings and governments; media campaigning to increase public awareness and empowerment; fund-raising for environmental management, conservation and environ-mental education; researching environmental issues; acting as ginger groups, and much more. Between 1909 and 1988 international bodies (e.g. IUCN or UNEP) increased from around thirty-seven to about 309, and NGOs (e.g. Oxfam, Friends of the Earth, Greenpeace) expanded from about 176 to about 4,518 (Princen and Finger (1994) provide a list of environmental NGOs).

An important role for NGOs is to act as a link between the local, national and inter-national. Many NGOs have a tiered local-to-international structure (e.g. Friends of the Earth), and command huge resources in terms of funding and expertise. There is growing networking by NGOs, and increasing numbers of coalitions and, with compact satellite telephones and the spread of the Internet, it is becoming increasingly difficult for govern-ments or other powerful groups to keep issues hidden or to subdue opposition.

Large NGOs may have been active for decades, have wide experience and global reach, and command considerable resources. NGOs involved in environmental issues are a very diverse group: some are catalysts, some key actors; they promote, condemn, empower, expose and monitor; some are politically orientated and some apolitical; there are also objective scientific NGOs (bodies such as the Scientific Committee on Antarctic Research (SCAR), which deals with Antarctica). NGOs have negotiated covenants between business and government to reduce environmental damage in some countries such as The Netherlands.

From humble origins NGOs can grow to become globally powerful bodies. Their staffs can be very dedicated, in some cases aggressive and undemocratic, and some-times bent on ill-advised crusades. When NGOs promote misguided policies or project their polarised perceptions and act in a careless or obstructionist manner, the environ-mental manager will have to work with it or subvert it. At the 1992 UN Conference on Environment and Development NGOs played both official and parallel unofficial roles (in the Global Forum).

Environmental management problems can be difficult to solve with existing inter-state regulatory and scientific approaches. It is in these situations that NGOs can perform a valuable role. Princen and Finger (1994: 221, 223) felt that NGOs are especially useful for linking knowledge from science with the grassroots (i.e. to people and real-world politics). Often they are swifter to respond to environmental problems and challenges than other organisations or governments, and in many cases grow from the grassroots in response to issues (Ekins, 1992b; Zeba, 1996). It should be noted that some grass-roots NGOs are somewhat ephemeral.

There is a risk that NGOs may be pressured to find neat, comprehensive solutions to complex problems; their supporters expect to see 'magic bullet' solutions and some-times lose interest or withdraw support if these are not quickly forthcoming. This limits the staying power of such NGOs (Vivian, 1994).

Unions

Unions in rich and poorer countries have become involved with environmental manage-ment in a number of ways. They may pressure governments to act on issues of health and safety, which can help prevent disasters like Bhopal. Large unions have influence over considerable investment funds and can steer their members towards ethical and green insurance and pension companies. Unions may oppose special-interest groups, crime and unsympathetic government. In Brazil co-operatives and unions of poor farmers and rubber tappers have opposed large ranchers bent on clearing land and evicting poorer people when the government has done little. Unions may channel environmental care advice to workers when governments fail to do so.

There can also be negative effects; loggers' unions in western North America opposed environmentalist resistance to clearance of old forests, and petrochemical, car manufacturing and coal-mining unions may lobby against carbon emission control agreements to protect members' jobs.

Summary

- Stakeholders can be individuals, groups, institutions, organisations or nations – they may cause change (for the better or worse), and they can benefit or be harmed by their alterations and natural shifts.
- Increasingly, developers must show some degree of environmental and social responsibility, which means informing and consulting with people.
- Environmental management can learn a lot from ordinary people, both past and present.
- Effective environmental management is seen as that which deals with people at the local or community level. Adoption of community/participatory approaches should not be automatic; the benefits and contribution to environmental goals must be assessed carefully.

Further reading

Black, R.J. (1998) *Refugees, Environment and Development.* Addison Wesley, Harlow.
  Good introduction to displaced people and environmental issues.

Ewert, A.W., Baker, D.C. and Bissix, G.C. (2004) *Integrated Resource and Environmental Management: human dimensions.* CABI Publishing, Wallingford.
  Comprehensive coverage of natural resources management with many case studies.

Gadgil, M. and Guha, R. (1995) *Ecology and Equity: the use and abuse of nature in contemporary India.* Routledge, London.
  Explores the complex way in which Indian society interacts with the environment.

Morse, S. and Stocking, M. (eds) (1995) *People and Environment.* UCL Press, London.
  Good overview of people–environment interactions.

Stalton, S. and Dudley, N. (eds) (1999) *Partnerships for Protection: new strategies for planning and management for protected areas.* Earthscan, London.
  Looks at partnerships for conservation, involving citizens, indigenous people, companies, communities and so on.

www sources

Nottingham University (UK), Planning Department Bibliography (environmental management bibliography – includes women/development/environment material) http://www.nottingham.ac.uk/sbe/planbiblio5/bibs/Greenis/A/27.html (accessed June 2005).
Women's Environmental Network (UK) (accessed June 2005).

PART

II

PRACTICE

7 Environmental management approaches

- Environmental management focus and stance
- Participatory environmental management
- Adaptive environmental management and adaptive environmental management and assessment
- Expert systems and environmental management
- Decision support for environmental management
- Systems or network approaches
- Local, community, regional and sectoral environmental management
- The state and environmental management
- Transboundary and global environmental management
- Integrated environmental management
- Strategic environmental management
- Stance and environmental management
- Political ecology approach to environmental management
- Political economy approach to environmental management
- Human ecology approach to environmental management
- The best approach?
- Summary
- Further reading

Environmental management involves the application of a mixture of objective scientific and more subjective, often qualitative approaches. It is a blend of policy making, planning and management, but there is no single widely adopted framework to shape its application, although there are guides to policy and procedures, and standards and systems (e.g. the widely adopted ISO14001 (Croner Publications Ltd, 1997)). Each situation faced by an environmental manager is to some extent unique, and the approach adopted reflects the attitudes and background of those involved, the particular situation, time and funding available, and many other factors. One can list the more common elements of environmental management approaches, recognising:

- top-down (authoritarian);
- bottom-up (inclusive/participatory);
- centralised;
- decentralised;
- socialist;
- free market;
- Western;
- company focus;

- non-business focus;
- non-Western;
- light-green (technology accepted);
- dark-green (technology opposed);
- giving priority to social development (poverty alleviation);
- giving priority to environment before human welfare.

Other criteria might be added, and the above may be combined in various ways with emphasis on differing elements. Whatever the combination, all seek to maintain and if possible improve environmental quality. So far most development of environmental management has been in Western nations, and currently business has a growing interest. In coming years it is likely that new approaches will be generated in non-Western countries such as China or India.

Environmental management is sometimes little more than a catch-phrase. When seriously undertaken it is a process of decision making about the allocation of natural and artificial resources that will make optimum use of the environment to satisfy at least basic human needs for an indefinite period of time and, where possible, to improve environmental quality. Newson (1992: 259) noted that a large part of environmental management was 'decision-making under uncertainty'. There is generally more than one route to a goal: perhaps one is the best all-round solution, one the best practical, one is that favoured by the government, another is that favoured by a company. Another core role of environmental management is environmental arbitration. This can be attempted by an individual acting as a 'csar', by a democratic body or through 'green anarchy'.

Environmental management focus and stance

In an ideal world ethics and laws would provide strong guidance – but in practice these are often inadequate. Furthermore, knowledge is often incomplete and data inadequate; thus problem solving and decision making are not straightforward.

Much of what has just been said is difficult to separate from environmental planning. In the past, planners often neglected environmental issues, were insufficiently aware of the dynamic nature of Earth processes, and failed to identify natural limits, hazards and potential. Today it is hard to comprehend that before the 1970s bodies such as the World Bank or the United Nations had few, if any, established environmental advisers, and that environmental quality was often seen as an optional extra by teams of decision makers dominated by economists and lawyers. Planners nowadays are much more aware of environmental issues. Environmental planning might be defined as efforts to strike a balance between resource use and the environment, the primary objective of planning being to make decisions about the use of resources.

Environmental management overlaps a number of other fields. Landscape planning has a long tradition and runs parallel with environmental management, focusing on aesthetic issues (Kivell et al., 1988; Ashworth and Kivell, 1989; Foder and Walker, 1994). Environmental planning overlaps with environmental management especially during implementation. Regional planning has links and so has impact assessment. The Netherlands adopted a National Environmental Policy Plan (NEPP) in 1989, the first serious attempt by a national government to develop an integrated environmental policy based on explicit control principles and clearly formulated long-term objectives (Bennett, 1991). This is in sharp contrast to the more usual incrementalist (step-by-step) approach by most environmental planning and management. The Netherlands' NEPP environmental planning and management approach gives serious consideration to the concept of sustainable development and the 'polluter-pays' principle. Although NEPP is behind schedule for its implementation, it has already influenced several other governments to develop similar approaches.

At one time the main, if not only, means of trying to consider environmental issues in planning and management was to use cost–benefit analysis (CBA). For environmental management to be a significant improvement on CBA or cost-effectiveness analysis (which are inadequate because they require monetary valuation, which can be difficult, and they fail to consider social and environmental issues adequately), it must view things from social, economic and environmental perspectives. To do that effectively demands a multidisciplinary (or interdisciplinary) approach (Spash, 1996). However, a functional grouping approach is often adopted in practice (e.g. a pollution control agency; a conservation body), and this may hinder multidisciplinarity.

As if it is not enough to have to deal with complexity, incomplete knowledge and poor data, the environmental manager often has to cope with situations where the development objectives and strategy have already been decided by politicians, special-interest groups, aid agencies, company directors and so on. Environmental management may also have to proceed in a piecemeal manner, with inadequate jurisdiction, insufficient time to act effectively, and public and administrative mood swings (Trudgill, 1990). Environmental managers may be faced with a crisis-management (reactive, short-term response) situation even though one of their principles is anticipatory planning (Scher, 1991).

Environmental management may be subdivided into the following components:

1   *Advisory*

- advice, leaflets, phone help-line;
- media information (which can be covert, i.e. hidden in entertainment or open);
- education;
- demonstration (e.g. model farm).

2   *Economic*

- taxes;
- grants, loans, aid;
- subsidies;
- quotas.

3   *Regulatory/Control*

- standards;
- restrictions;
- licensing of potentially damaging activities.

In a given situation a mix of these components will be undertaken. When the mix results in poor enforcement, and/or the people involved are not won over, results are likely to be limited.

Environmental management can adopt three distinct stances:

1   preventive management – which aims to preclude adverse environmental impacts;
2   reactive or punitive management – which aims at damage limitation or control;
3   compensatory management – mitigation of adverse impacts through trade-offs.

One example of the latter type of trade-off is to protect some habitats of conservation or aesthetic value, and to develop other localities. The goal is to prevent an overall slow decline in environmental quality. Montgomery (1995: 186) suggested the environmental manager might be best advised to focus on: (1) modifying anthropogenic inputs (input management – controlling use); (2) responding to ecosystem attributes (output management – driven by assessment of resources). Ideally an environmental management framework will integrate (1) and (2) to control environmental degradation most effectively.

While co-ordination of environmental management approaches is desirable, it is difficult to see how too rigid a framework can help, given that each situation is to some degree unique. Companies, funding agencies, NGOs and governments have developed codes, manuals and guidelines to guide environmental management (Forrest and Morison, 1991; Nash and Ehrenfeld, 1997); Europe is adopting codes which will shape practices in all member countries, and in the USA the Environmental Protection Agency sponsors new environmental management programmes. There are demands for environmental planning and environmental management to act to strengthen the drive for achieving sustainable development (Costanza, 1991; Blowers, 1993). One way a government can pursue environmental management is to use covenants. Covenants offer a means of providing companies with a stable regulatory environment, and act as incentives to encourage development of pollution control plans and environmental management systems, and the government can focus its attention on companies and bodies that have not signed covenants. The Netherlands has one of the most innovative and best-developed approaches to environmental management which relies upon two primary components: (1) National Environmental Policy Plans (NEPPs); (2) covenants (Beardsley et al., 1997). NEPPs were adopted by the Dutch Parliament in 1989 and 1994, set targets for pollution reduction, and are a relatively integrated approach. The covenants are voluntary agreements between the Dutch government and various sectors of industry to facilitate the improvement of environmental management objectives and to keep down enforcement costs. This strategy has apparently been quite effective in achieving environmental management goals – mainly pollution control, but also sustainable development initiatives.

Before long there should be environmental management system standards widely in use. These, together with eco-auditing and environmental management system standards, will provide some internationally recognised foundations for environmental management to draw upon in any given situation.

Participatory environmental management

Participation and environmental management have already been discussed in Chapter 6; in recent years it has become more common to inform citizens or involve them. Indeed, participation (or collaborative approaches) and empowerment are currently so fashionable that they are almost politically correct. Where environmental problems result from human attitudes, participation is crucial because resolution depends upon people altering their views. There are a number of well-established approaches, particularly participatory rural appraisal (PRA) and rapid rural appraisal (RRA). These have evolved since the 1960s, but more especially since the 1980s, to support agricultural development, healthcare, decentralised planning and democratic decision-making. Much of the development of these approaches has taken place in developing countries although, in richer nations, market research has helped shape some of the methods used. In non-democratic societies village communes or people's discussion groups have offered parallel routes to participation. Participatory approaches are widely seen to be valuable in any quest for sustainable development.

PRA and RRA rely on multidisciplinary study and close contact with people to get a full picture of their needs, capabilities, limiting factors, opportunities and threats. An example of a successful participatory approach is the LANDCARE programme launched in Australia in the 1990s; this provided government support for voluntary rural groups of farmers, and other folk who seek to counter land degradation.

Adaptive environmental management and adaptive environmental management and assessment

Environmental challenges are so diverse and complex that a single rigid approach is unlikely to work. Whenever possible an adaptable strategy should be adopted to cope with unforeseen problems and opportunities. Adaptive environmental management means different things to different people: it is seen by some as a tool or approach that can be quickly modified to suit a particular situation; systems modellers see it as meaning the ability to explore various 'what if?' scenarios; or it can be an approach that is flexible and able to cope with poor data availability, and to respond to new challenges as they arise. The latter is the most common interpretation and involves a continuous learning process that should not be separated from research and ongoing regulatory activities. This probably never reaches a state where there is fully satisfactory knowledge for environmental management (Walters, 1986; McLain and Lee, 1996). Adaptive environmental management is a learning-oriented approach, which is suitable for managing complex situations with high levels of uncertainty. Policies are monitored and adapted if necessary, a learning process. It seeks to integrate scientific, local and social studies knowledge. Adaptive environmental management stresses an integrated approach which considers social, economic, political and environmental; it makes use of science and social studies. This means it accepts uncertainty and can abandon experimentation to arrive at decisions; it is also more management focused. There is usually quite a lot of reliance on modelling (Walters, 1986; Trudgill, 1990; McLain and Lee, 1996). Adaptive approaches have been applied to rangeland management and conservation areas (Salafsky and Wallenberg, 2000; Salafsky et al., 2001), and with good effect in efforts to control pollution in the Great Lakes of North America, the Baltic and Chesapeake Bay (USA).

In the 1970s Canadian impact assessment specialists and environmental managers (Holling, 1978; Walters, 1986) developed a related field – adaptive environmental assessment and management (AEAM). This seeks to integrate and deal with economic, social, environmental and other issues; it recognises the existence of many diverse stakeholders; and it addresses uncertainty (http://www.geog.mcgill.ca/faculty/Peterson/susfut/AEAM/ and http://www.gse.edu.au/Research/adaptive/Adaptive2.htm – accessed April 2005). AEAM differs from mainstream environmental management, which is based on informed trial-and-error in using risk-adverse 'best-guess' management strategy. It also probes issues, identifies uncertainties, and weaves this into its strategy.

Adams (1990) complained of widespread 'juggernaut' development, which was too inflexible and clumsy, and so caused environmental and socio-economic problems. The best response to such development is an adaptive one which can alter to match challenges, a strategy championed by natural resource managers in the 1970s who borrowed ideas from operational management and management science. Mitchell (1997: 82–85) outlines 'hedging' and 'flexing' strategies for decision making where there is severe uncertainty. Hedging is a process of trying to avoid the worst consequences, and flexing is a continuing search for other possible options even after a decision has been taken. Adaptive management is far better than the disjointed incrementalist approach often adopted (i.e. just muddling through). However, it is not perfect – McLain and Lee (1996) reviewed three adaptive environmental management case studies and found 'serious flaws', mainly in relation to how environmental management decisions were made. They also noted a risk of ignoring non-scientific knowledge.

Expert systems and environmental management

Expert systems are computer programs that rely on a body of knowledge to cope with a difficult task usually performed only by a human expert. Expert systems are tools, but, like many tools, may be used to the guide approach, and they are increasingly used where there is a shortage of skilled specialists. Costly to establish, the systems should improve with use. For examples of applications to environmental management see Moffatt (1990), Fitzgerald (1993) and Warwick et al. (1993). Some EMSs make use of expert systems to aid bodies with the inital returns for assessment for certification.

Decision support for environmental management

One of the problems faced in environmental decision making is complexity. With limited time in which to develop solutions, difficult-to-trace webs of interrelationships are a problem. Clarity of linkages and aid in interpreting data are thus very useful (data visualisation for decision support is reviewed in *Landscape and Urban Planning* 21(4) – published 1992). Decision support systems are derived from operational research and management science; they deal with complexity by 'playing' to learn fast. Usually they take the form of interactive computer-based systems, which help the decision maker to model and solve problems. Some argue that anything which aids decision making is a decision support – even a cup of coffee (Janssen, 1995). Whether complicated aids like the multiple criteria method (Paruccini, 1995) are of practical value is unclear. There is also a need for approaches that can help the environmental manager weigh goals against costs and risks, and structure strategies in the best way possible. Operational research or management and multi-objective decision support methods can provide useful help for the environmental manager (Bloemhofruwaard et al., 1995).

The use of computer-based systems in support of decision making in environmental management has increased over the past decade. Some systems integrate the use of GIS and modelling as well as aiding decision making (Zhu et al., 1998).

Systems or network approaches

Systems analysis and network approaches have been applied to environmental management since the 1970s (Bennett, 1984; Carley et al., 1991). These can be demanding of research and slow to perfect, but are useful for ongoing management of particular situations, and as a way of making sense of complexity.

Local, community, regional and sectoral environmental management

Environmental management has frequently been tackled on a local, community, bioregional or regional scale (Kok et al., 1993; Welford, 1993). Wilson and Bryant (1997:

141) stress that environmental management is a multilayered process: there may well be different tiers involved from local up to state and international levels, all interrelating. Smith (1998) found in the USA that, for pollution discharge control, decentralisation did seem to improve environmental management performance, the key factor being access to local knowledge. The social sciences offer a pool of experience on community development aspects of natural resource management which the environmental manager may draw upon (Smith *et al.*, 1994), and regional planners have often worked closely with environmental managers. The bioregional concept may offer a framework, as may river basins (and sub-basins), watersheds, small islands, coastal zones, and other suitable functional landscape units.

A number of sectors have developed approaches, standards and pools of expertise – for example, the petrochemicals, paper-pulp production, mining, oil, cement, sewage treatment and power generation industries. Tourism is also beginning to develop environmental awareness and ecotourism standards; for example, for ski resorts (Williams and Todd, 1997) and beach resorts. This expertise can greatly assist and speed up further environmental management in the same sector.

The state and environmental management

Environmental management is a politicised process (Wilson and Bryant, 1997). Is the process of environmental management, then, controlled by the state, NGOs, international agencies, or what? Ultimately, with global interdependence and shared world systems, there has to be some element of international co-ordination and control. Below that, the majority of environmental management is in state hands but, like medicine or economics, the profession should be able to steer the state towards certain goals. Hopgood (1998) has examined US policy on international environmental issues since 1972, seeking to establish whether the state had retained or lost control of policy making to environmental groups and international agencies. The answer was not clear. A decentralised approach might prove less robust against special-interest groups, large companies and so on than a centralised, state-supported approach (Walker, 1989). It is not uncommon for states in a federal system to come into conflict among themselves or with central government over environmental issues. One reason for the formation of the EPA in the USA was to co-ordinate and integrate efforts under a federal system.

Transboundary and global environmental management

The need for transboundary and global environmental management is growing. Local, regional, national and corporate environmental management can draw upon established social institutions, the market, law and, ultimately, the power of the state to force a resolution of conflicts. However, transboundary and global environmental management must rely on building international co-operation for monitoring and problem resolution. In practice, honouring agreements is more difficult than achieving them. There is also the question of who or what body should foster international co-operation to search for solutions to transboundary and global problems, oversee implementation and, if environmental management is to be anticipatory, identify potential problems and conflicts (Davos, 1986; B. Agarwal, 1992). Some see UN bodies as able to fulfil these roles; others suggest it should lie with internationally respected research centres. At present both these types of institution play a part, but overall co-ordination and enforcement is too weak.

With the spread of free trade since the General Agreement on Tariffs and Trade (GATT) (which became the World Trade Organisation – WTO – a few years ago), the North American Free Trade Agreement (NAFTA), and similar undertakings, environmental management must cope with problems caused if controls are to be interpreted as a 'trade barrier'. Efforts are being made to improve environmental management provisions in free trade agreements, but there are still weaknesses – like those which led to the USA–Mexico yellow-fin tuna débâcle in the early 1990s (Mumme, 1992; Seda, 1993).

Integrated environmental management

Much environmental management and planning has been reactive, narrow in focus, piecemeal and poorly co-ordinated. Integrated approaches have been explored to try to counter these problems, and to ensure that environmental management yields socio-economic benefits. Environmental management seeks the integration of environmental concern into proactive development planning; it aims to guide and not hinder development. Environmental problems cross political borders and boundaries between air, water and land; they also involve different disciplines and actors, so can be difficult to deal with without integrated environmental management. There has been considerable interest in integrated environmental management in recent years, from industry, academics, politicians, professional planners and resource managers (Müeller and Ahmad, 1982; Cairns and Crawford, 1991; O'Callaghan, 1996). There is a lack of firm agreement as to what exactly it is (Barrett, 1994; Margerum and Born, 1995). Terminology is a little vague; for those involved in corporate environmental management 'integrated' means the development of an environmental management system that combines health, safety and environmental quality issues. Alternatively, the Dutch government, concerned with the environmental management of the North Sea, would see 'integrated' as implying the assessment of all relevant environmental factors: pollution, fisheries, erosion and so on, and resolving issues in an integrated way (Wolters, 1994). Yet another interpretation is that it integrates environmental management with environmental engineering.

With a number of other environmental management approaches, there is a risk that academics and professionals become too engrossed and forget that it is a means to an end: achieving sustainable development; better resource use for the general good; reduction of environmental problems and so on (Born and Sonzogni, 1995). In spite of these problems, improvements should make it easier to adopt effective integrated environmental management (Rabe, 1996; Ewert et al., 2004).

The roots of integrated environmental management lie in integrated area development approaches and comprehensive regional planning and management, including comprehensive river basin planning and management. There are also similarities shared with areas of management science, such as total quality management. The key elements of integrated environmental management are, according to Born and Sonzogni (1995: 168):

- co-ordinated control, direction or influence of all human activities in a defined environmental system (such as a river basin or a watershed) to achieve and balance the broadest possible range of long- and shorter term objectives;
- a process of formulating and implementing a course of action involving human and natural resources in an ecosystem, taking into account the social, political, economic and institutional factors operating within the ecosystem in order to achieve specific societal objectives;

- an inclusive approach that takes into account the scope and scale of environmental and human issues and their interconnections. A strategic and interactive process is used to identify key elements and goals which need attention.

Strategic environmental management

The formalised, proactive, systematic and comprehensive process of evaluating the environmental effects of a policy, programme or plan and alternatives is known as strategic environmental assessment. This has been applied to such issues as aid programmes, structural adjustment, changes in public transport policy and so on (Partidário, 1996). Strategic environmental management integrates with fields like health and safety and strategic business planning. Overlapping a little with strategic environmental assessment is strategic environmental management (SEM), which may be defined as the preparation and implementation of policies that seek sustainable development of the environment (Nijkamp and Soeteman, 1988). SEM should ensure a long-term view and adequate monitoring of local, regional and global issues. The Netherlands has gone further than most countries towards adopting SEM as part of national policy (Ministerie VROM, 1989), and Europe is committed to adopting it (Figure 7.1). It has been argued that there are situations where SEM may not be the best option, especially for some companies, in spite of pressure for its adoption (Vastag et al., 1996).

Stance and environmental management

Political and ethical stances play an important role in determining environmental management goals and the strategies used to achieve them. An environmental manager

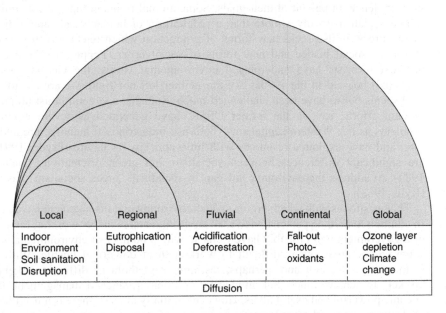

Figure 7.1 Linkages of levels in sustainable development tasks (based on the approach adopted by the National Environmental Policy Plan of The Netherlands)
Source: Carley and Christie (1992: 199, Fig. 9.2)

can follow a textbook-scientific approach, but it is unlikely to be completely uninfluenced by politics and his or her own outlook (Boehmer-Christiansen, 1994). Those who profess concern for the environment have a wide spectrum of viewpoints and may revise their ideas, so stance is usually rather elastic (Parkin (1989); O'Riordan and Turner (1983: 1–62) give an overview of environmentalist ideologies). There are light-greens, prepared to make use of science and technology to improve human well-being and environmental quality, and who are aware of limitations in doing this; there are 'cornucopians', who probably have excessive trust in the capacity of science and technology to cure all environment and development problems; there are deep-greens, who mistrust science and technology, some of whom adopt a romantic approach, and some favour spiritual development or New Age values (Naess, 1989). Some environmentalists are ecocentric and give nature greater priority than human needs, while others are anthropocentric and place human needs first, and some profess a 'holistic' approach; many greens adopt a decentralist, slightly anarchistic stance; others support established political parties (Pepper, 1984; Dalton, 1994; Dobson, 1995). Many deep-greens believe that ecological awareness is spiritual and that new ethics, vital for satisfactory environmental management, must be grounded in spirituality (Sessions, 1994: 21). Those who profess deep ecology also seek a paradigm shift, to a philosophy which aims at a sustainable society based on material simplicity and spiritual riches.

Supporters of social ecology advocate a decentralised, co-operative, anarcho-socialist lifestyle (claiming that if people are in harmony with one another they are more likely to be in harmony with nature – a far from established assumption) (Bookchin, 1972, 1982, 1986). Eco-feminism (see Chapter 6) has been critical of deep-green and social ecology viewpoints, arguing that gender neutrality is not enough, and anti-androcentric approaches are needed to end paternalistic behaviour which leads to exploitation of women and the environment (Cheney, 1987; Zimmerman, 1987; Merchant, 1992). Extreme eco-radicals or 'eco-warriors', such as the Earth First! groups, put environmental welfare before human welfare and may resort to eco-terrorism ('monkeywrenching'), even violence, in pursuit of their goals. Some animal rights groups take a similar line.

Eco-socialism involves more than a redefinition of human needs and redistribution of resources: it also seeks new forms of production which reject private ownership in favour of social justice and new forms of social order (Pepper, 1993). Marxist and socialist theorists have underplayed environmental issues; the German die Grünen ('greens') boasted in the 1970s: 'we are neither left nor right, we are ahead!' and the lead seems not to have been challenged much. Socialist and communist utopian development efforts, say, in the former USSR, have generated as severe environmental problems as has Western capitalism – both use industrialised manufacture and agriculture, and have exploited resources with little concern for nature (Pryde, 1991). There are significant differences between socialism and green orientations (Bahro, 1982, 1984); to address the environmental gap in socialism, 'green socialism' has appeared (Ryle, 1988).

The question is with which group(s) does environmental management have sympathy? In addition to holding personal views, an environmental manager working for a company or a government will probably have to liaise with a number of environmentalist groups, some co-operative, others difficult to work with or downright hostile. Without some form of co-ordination and, perhaps, restraint, a plethora of different stakeholders is unlikely to achieve much, but, guided by good environmental management, they may become powerful and useful allies. However, initially at least, there is a need for caution in dealings, to avoid misinformation, the risk of one group trying to gain advantages over another, over-powerful alliances and so on. Environmental managers are generally aware of these risks and have developed guidelines.

Political ecology approach to environmental management

Political ecology has its origins in the 1950s in the writings of Wallerstein and Gundir Frank. It has been defined as the study of the relationship between society and nature, and as the application of ecology to politics and study of political competition for control of natural resources (Bryant and Bailey, 1997; Bell et al., 1998; Low and Gleeson, 1998). It is an interdisciplinary area of research, which connects politics and economy to problems of environmental control and ecological change (Zimmerer and Bassett, 2003). Robbins (2004) observed that political ecology brings together cultural ecology and political economy. Cultural ecology is interpreted in different ways by geographers and anthropologists – the former view it as exploration of how society and humans affect the environment; the latter see it as study of how natural environment affects socially organised behaviour (http://www.answers.com/topic/political-ecology – accessed April 2005). Political ecology holds that radical changes in human social habits and practices are required in order to counter environmental degradation and achieve sustainable development (Dobson, 1995: 17). The political ecology approach implies an interest in cause–effect relationships, study of the different interest groups involved in using the environment, of their economy, habits and livelihoods (Chapman, 1989; Atkinson, 1991b; Hershkovitz, 1993; Oliver, 1994). Scientific study is not enough: social, economic and political issues must be considered; for example, the struggle against logging in Sarawak can only be understood in the light of the history of local peoples and present politics and economics (Colchester, 1993); and the process of environmental degradation in Honduras only becomes clear through studying the political ecology of poverty (Stonich and Browder, 1996).

Political economy approach to environmental management

An understanding of human–environment interactions may be gained through examination of how the social relations of power relate to the control and use of resources and nature – the political economy approach. There are likely to be different perceptions of environmental needs and problems between planners, policy makers, government ministers and various sections of the public. To deal effectively with environmental management demands an awareness of political economy. Blaikie (1985) adopted a political economy approach to examine soil erosion and its control in developing countries. Urging small farmers to control soil erosion for the national good is unlikely to have much effect if it brings them no significant benefit. It may require people far removed to pay and alter attitudes so that environmental management may be supported at local level.

Human ecology approach to environmental management

Human ecology is the study of relations between humans or society and nature through a multidisciplinary approach (Begossi, 1993). An alternative definition is the study of ecosystems that involve humans (Garlick and Keay, 1970; Hardin, 1985; Catton, 1994). There has been interest in human ecology since the 1920s or earlier (Barrows, 1926). The scale of approach may be local to global, and it supports holistic study (Sargeant, 1974; Steiner and Nauser, 1993).

The best approach?

There is no such thing as a single 'best approach'; each situation has unique demands. It is possible to say that whatever is selected a crucial element is to be *sensitive* to crucial issues, people's needs and fears, environmental limits and so on. To some extent the ends are more important than the means; the approach may not matter much when problems threaten human survival on Earth. However, for most challenges it should be possible to attain goals in a cost-effective way without resorting to draconian measures. Environmental management usually has two choices: (1) where time and funds are short, the 'quick-and-dirty' approach; (2) a more thorough, slower and usually more expensive approach. The former sacrifices depth of assessment and reliability for speed and cheapness. The latter is often too slow to be practical. The ideal is a quick, thorough, adaptable and transparent approach (i.e. the public and other onlookers can see what is being done) – and nothing quite fits that.

In the past a command-and-control (top-down) approach to environmental management was virtually universal, relying on regulations, fines, inspections and so on. That has been giving way to a more 'hands-off' voluntary approach, often bottom-up management, and relying on rewards rather than punishments to obtain results. However, the old ways are by no means extinct; indeed, they are probably still dominant. For dangerous activities there clearly still have to be strict controls. Where reliance is on an EMS the consequences of failure to meet standards may only result in warnings and finally being struck off (de-certified). That may not be a sufficient deterrent to ensure good environmental management.

Environmental management can be centralised or decentralised; technocratic or appropriate/human-based; sensitive to local needs (of people and environment) or insensitive. The level of activity is also diverse; environmental management may operate at:

- Local or even micro-level (involving individual stakeholders – farmers, as in LANDCARE in Australia or fishermen).
- Sectoral level (groups of villagers, farmers, or bodies).
- Regional level (watershed, river basin, island, or whatever).
- State or national level.
- Global level (bodies like the World Bank, OECD, and so on have established environmental management departments, policies and sets of guidelines).
- Alternatively a special-interest approach may be used – possibly combined with one of the above. This includes powerful groups, cartels, NGOs – often demanding adherence to a given agenda leading to polarised reactions.

Ideally there should be freedom to select the approach and tools which seem most suitable, but few, if any, environmental managers are wholly free of controls or bias.

Summary

- Each situation faced by an environmental manager is unique, and the approach adopted should reflect the attitudes and background of those involved, time and funding available, and many other factors.
- Environmental management is a process of decision making about the allocation of natural and artificial resources that will make optimum use of the environment to satisfy at least basic human needs for an indefinite period of time and, where possible, to improve environmental quality.

- There is generally more than one route to a goal: perhaps one is the best all-round solution, one the best practical, one is that favoured by the government, another is favoured by a company and so on.
- As if it is not enough to have to deal with complexity, incomplete knowledge and poor data, the environmental manager often has to cope with situations where the development objectives and strategy have already been decided by politicians, special-interest groups, aid agencies, company directors or others.
- Whenever possible an adaptable strategy should be adopted to cope with unforeseen problems and opportunities.
- There is no such thing as a single 'best approach'; each situation has unique demands. It is possible to say that whatever is selected an essential element is to be *sensitive* to crucial issues, people's needs and fears, environmental limits and so on.

Further reading

Holling, C.S. (1978) *Adaptive Environmental Assessment and Management.* Wiley, Chichester.
Advocates adaptable approaches to environmental management. There has been renewed interest, including that from social scientists involved in environmental management, and Holling is a key text.

Keil, R., Bell, D.V.J., Penz, P. and Fawcett, L. (eds) (1998) *Political Ecology: global and local.* Routledge, London.
A collection of papers exploring modern political ecology and issues such as sustainable development.

Robbins, P. (2004) *Political Ecology: a critical introduction.* Blackwell, Oxford.
Political ecology is an academic subject which strives to understand relationships between society and environment or resources – crucial for environmental management and the quest for sustainable development. Political ecology insight can help avoid misunderstanding and misinterpretation – the problem of 'myths' used in error by managers.

Walters, C.J. (1986) *Adaptive Management of Renewable Resources.* McGraw-Hill, New York.
Argues for a flexible and adaptive approach.

8 Standards, monitoring, modelling, auditing and co-ordination

This chapter and the following (Chapter 9) deal with methods, tools and techniques. These are frequently used terms but are seldom adequately defined and are frequently used interchangeably. My interpretation is that:

- a method is a general manner of approaching a problem;
- a technique is a specific application of a tool or tools;
- a tool is a means for collecting data, analysing data, presenting data, testing data, interpolating from data, helping select a course of action and so on.

There is considerable overlap and 'grey' areas. The choice of method, technique and tools is increasingly influenced by a trend towards a proactive approach. There is also a dominance of Western values, in particular: administrators consulting with the public and accountable for their actions; acceptance of the 'polluter-pays' principle; and a desire to pursue sustainable development.

Broadly, this chapter deals with the tasks of setting standards, deciding terms of reference, modelling, monitoring, audit of what data there is already, and co-ordination. Chapter 9 focuses on proactive identification of threats, limits, problems and opportunities, using hazard and risk assessment, impact assessment, modelling, forecasting and use of hindsight knowledge. The separation is not perfect: for example, tools such as eco-auditing and eco-footprinting are included in this chapter because they deal mainly with current status, but there is overlap with tools explored in Chapter 9.

Methods, techniques and tools are often borrowed and adapted from other disciplines. Sometimes tried and tested tools are available 'ready to use'. Sometimes a specialist consultant is hired, or the environmental manager may have to assemble a toolkit and devise an approach, and if the problem is novel, research is needed to develop something. Whether tried and tested or new, the method, technique or tool needs to be tested and focused. Pilot studies and test runs should be a key part of everyday environmental management – in practice they are often dispensed with due to the pressure to address issues fast and at reasonable cost. Sometimes things are best handled by specialist agencies, commercial or university consultants – a quick glance at the Internet or any journal shows a huge expansion in these services. However, there are disadvantages in using consultants:

- consultants tend to be hired for as short a time as possible and have moved on and are unavailable when some problem develops;
- sometimes they seek early completion bonuses and this can reduce caution and care;
- the specialist may report what he or she feels the commissioning body wants to hear to avoid friction and help secure future contracts;
- consultants may be 'outsiders' who may be unfamiliar with the environment or the socio-economic or cultural situation;
- there may be a focus on getting results which do not halt development, rather than best possible environmental practice.

Whenever possible environmental management methods, techniques and tools should be standardised so that the results can be easily checked and meaningfully compared with past studies or results from elsewhere. The following listing presents some of the common tasks of environmental management and a selection of the tools used for the functions:

- Developing terms of reference (scoping/setting limits to the exercise): brainstorming tools; desk research; pilot studies; test models; goal identification tools; participatory assessment (e.g. gather traditional/local knowledge, PRA, RRA); checking nearby and other relevant cases; cost–benefit analysis.
- Setting goals and objectives and strategic planning; brainstorming, SWOT assessment and so on.
- Selecting methods: consulting guidelines/benchmarks/standards; strategic management and planning tools (check how the proposed development fits in with others/nearby localities); selection of options tools.
- Collecting relevant data: surveys; focus groups providing local knowledge; GIS; establishing monitoring instruments/observers; desk research.
- Scenario development studies: hazard and risk assessment; environmental and social impact assessment; predictive modelling; visioning (scenario prediction techniques used in business) past study for indications of possible future developments; Delphi technique; life-cycle analysis; eco-footprnting.
- Providing support for environmental management and standardisation: environmental management systems (EMSs).
- Selecting enforcement methods: law, taxation, policing measures, incentives, education, propaganda/advertising.
- Assessing what funding is required, checking cost-effectiveness, reviewing spending.
- Pilot studies: test projects, trials, market research studies and so on.

- Informing the public and seeking reactions.
- Monitoring ongoing measurement to provide data on progress and to try and give early warning of problems.
- Modelling: checking situation (perhaps simplifying complexity to understand what is happening), seeking early warning of problems; trying to see how unexpected developments will probably proceed.
- Evaluation/auditing: eco-audit; sustainability audit; project/programme evaluation tools during the exercise or post-exercise appraisal.
- Public relations, public consultation, communication with company or government decision makers – used in a number of the above tasks: written reports/presentations; briefings; press reports; drop-in shops; websites, chat sites.

There is some overlap of tasks and tools, and in practice there may not be the funds, time or access to sites to allow thorough approaches.

In Chapter 1 the key principles of environmental management were noted: prudence and stewardship. Standards, monitoring, and the other approaches covered in this chapter are crucial for pursuing those principles. They are vital for meaningful evaluation on which to base forward planning and policy decisions, for law making and enforcement, for effective implementation, co-ordination, and for avoiding unwanted impacts. This chapter deals with some rapidly expanding fields of environmental management; in particular, environmental management systems (EMSs). Implementation of EMSs is now big business and is generating an expanding literature.

Data

The starting point for much environmental management is data acquisition and interpretation. Data may be crudely divided into quantitative and qualitative. These each come in two forms: reliable and unreliable. A statistician might prefer to divide data into parametric and non-parametric, basically that which may be further analysed with powerful statistical techniques, and that for which there are less powerful tools.

Whenever possible data should come from more than one source and more than one tool should be used, so that decisions are based on more than one line of evidence. The trend is also towards multidisciplinary teams dealing with environmental management problems. In the past environmental management was dominated by natural scientists who dealt mainly with quantitative data. Qualitative data was disparaged and seen as subjective, 'soft' and unreliable, and its use was largely restricted to social studies. Nowadays there are growing numbers of social scientists involved in key areas of environmental management and there is an expectation that development will as far as possible be socially beneficial and that the public will be involved to some degree.

Qualitative data is fine provided it is collected properly and the interpretation is careful. 'Hard' scientists often have to advise before they have adequate quantitative proof and may be unable to arrive objectively at a decision. Scientists routinely work on an equal footing with social scientists, often in multidisciplinary teams, possibly seeking a holistic approach. Even so, there is still a quantitative/qualitative divide – between the differing traditions, analytical theory and collection methods. Some social studies researchers are still suspicious of 'empirical' study and what they see as scientists' failure to engage with social reality. Environmental management seeks to make use of both quantitative and qualitative data, and in multidisciplinary teams to seek common ground and objective interpretation.

Standards, indicators and benchmarks

A standard may be defined as a widely accepted or approved example of something against which others may be measured. Standards allow meaningful evaluation, exchange and comparison of data, improve objectivity of judgement (so are important to science), aid recognition of crucial thresholds and limits, support negotiation, law making and comparison (between sites, between countries and between years). Environmental standards may be divided into three broad groups: those concerned with ensuring human health and safety; those concerned with maintaining environmental quality; and those concerned with the quality of consumer items. The establishment of widely applicable scientific standards has been one of the most important achievements of Western civilisation ('standards of behaviour' – ethics – have been discussed in Chapter 2). Monitoring, modelling, auditing and environmental management systems help ensure ongoing objectives are set and met, check progress and warn of problems and opportunities. Standards enable the establishment of benchmarks, and checking and stocktaking using those benchmarks.

Standards have existed from ancient times: some archaeologists recognise Mesolithic standard measurements, the Egyptians, Greeks and Romans had units of measurement and coinage, and medieval European craft guilds set standards for the quality of goods. By the nineteenth century, Britain, France and some other countries had set up institutes and observatories which developed, managed and regulated established standard units used to record data. Standards and benchmarks overlap a little; the latter are approved and standardised descriptions of best practice, procedures, and perhaps intents and goals. For some a benchmark is just a level that can be aimed for, or it can be a waymarker against which to judge standards, compliance or progress. For example, a country may have a benchmark for Masters-level courses in environmental management, indicating levels and scope of coverage, rigour of assessment and so on.

Unfortunately, in the past national standards collected in, say, a French colony would often have to be converted to units used in Britain. When similar indicators are used, conversion from one system to another may be simply done with arithmetic. However, sometimes the indicators are not easily comparable or the means of gathering data are more or less unique, so making even rough comparison may be difficult. There is another problem, namely that what is a useful standard in a temperate country may be meaningless when applied in the humid tropics, mountain or polar regions – there are still tropical countries which have building standards inherited from temperate colonial powers which specify roofs to cope with snowfall. Without worldwide standards it is difficult to research the structure and function of the environment and to monitor global conditions. Before the late 1950s international unions agreed standards for some fields, such as telegraphy and radio, but for the environmental sciences many improvements came only after 1960. One achievement of the International Geophysical Year (1957–1958) and subsequent global exchanges of hydrological, meteorological, geophysical and biological data was the development of better international environmental standards. Where there are no single agreed standards there is always a risk that someone will mistake one for another. Only a few years ago a NASA Mars probe was lost at huge cost because one of the teams involved had worked in metric and another in imperial, and a command intended to be in the former was sent in the latter.

As research into environmental issues progresses, new standards are needed, for example to assess 'safe' levels of chemical pollution, radioactivity, or to deal with genetically modified organisms. The process is ongoing, involving various national and international institutes and standards organisations such as the British Standards

Institution or the International Standards Organisation. Advances in medical knowledge, toxicology, ecology and so on, sometimes force the revision of certain established standards. Ozone-damaging CFCs were considered inert and safe in the late 1930s, and environmental levels of DDT caused little concern before the 1960s; today much stricter controls are applied. New standards are being developed which distinguish between groups of people; one example is pollution standards that may have to take into account the greater vulnerability of children to some compounds.

There are a number of ways of developing a standard, each with advantages and disadvantages; for example, a standard for checking that fruit does not exceed 'safe' levels of a pesticide might be based on a simple maximum residue level (MRL), or a sort of lump sum, or an acceptable daily intake (ADI) – which assumes that consumers all eat a given amount per day. It is consequently important that an environmental manager knows the characteristics of a standard as well as the levels measured by it, and the reliability of the measurements. The methods of data collection as well as the agreed units must be standardised. Taking the same meteorological measurements in the lee of a house and in open countryside or at various times of day gives quite different results, making comparison difficult. Collecting data is often expensive; it is therefore important to avoid poorly focused, encyclopaedic data collection, and it is a good idea to 'scope' first to assess what should be measured and how. Once a standard or benchmark is agreed it must be publicised and policed, and if necessary revised or replaced.

Standards play a crucial part in:

- monitoring;
- modelling to understand the environment and establish trends;
- negotiation;
- enforcement of rules;
- environmental auditing;
- maintaining environmental quality.

The fields of activity which make use of standards include:

- pollution control;
- the quest for sustainable development;
- health and safety;
- public hygiene and health (especially domestic water supplies, sewage and waste disposal);
- consumer goods (food standards, electrical safety, electromagnetic radiation safety);
- pharmaceutical products;
- transport safety and quality;
- disclosure of information to the public.

Standards are of little use if they are not effectively enforced. Another difficulty is that standards may sometimes be relaxed, usually for profit or strategic reasons. The expression REGNEG (renegotiating of regulations) has been applied to the situation where a developer succeeds in persuading the authorities to relax or modify regulations in its favour, making it easier to meet standards or avoid assessments.

Benchmarks provide reference points by which to measure something, set minimum targets, and are a means for sharing and promoting good practices. Benchmarking also assists the comparison of one situation with another. Various tools can be used in the task of benchmarking; one is trend analysis which consists of time-series tabulations of data which enable the pattern of change to be assessed and possibly some future forecasting. Trend analysis can also be useful in performance appraisal. Many bodies provide benchmarking documents and guides; for example, educational curricula or pollution control (e.g. GEMI http://www.gemmi.org/docs/bench/bench.htm – accessed February 2005).

Eco-labelling is a form of standard used increasingly on products to indicate how much they impact upon the environment. The consumer can judge one product against another and, hopefully, buy the greenest. Various independent assessors undertake the labelling so that it is objective. The focus is on product or service impact and says little about manufacturing or recycling impacts. Policing and standardisation need improvement.

Standards often rely upon indicators – things that can be relatively easily measured, and which have specific meaning and point out something: the stage reached, quality, stability, vulnerability. Indicators are widely used to try and assess whether things are getting better or worse, including sustainable development (Bell and Morse, 1999). A little desk research soon shows that there is a vast diversity of different indices; living species with known sensitivities may be used to show heavy metal pollution, acid deposition, frost occurrence, soil qualities, level of grazing, and much more. The chances are that if something needs measuring there is already at least one indicator for it which can be found easily with a little library research. Indicators have been developed by most if not all fields: ecology, economics, healthcare, pollution control, biodiversity assessment and conservation, social development, famine relief, and many more. Indicators measure and warn of a huge diversity of risks, threats, quality changes, degree of sustainability, aesthetic value and so on.

Some indicators are subjective and some objective, and, because there are so many, one problem is to establish them so that they are always measured the same way, and another is to validate them. Ideally an indicator should be sensitive, but not so much so that it triggers false alarms; it should respond fast, reliably and unambiguously, and if possible be cheap and easy to use. An indicator may be a single object or event (not a measurement on a scale), such as a distinctive tree which shows the soil is fertile, or lichen showing there is little acid deposition, a moss which is sensitive to heavy metals pollution, or the point when rural families sell gold jewellery because famine is approaching. A chemical or hormonal test kit may show a colour change in the presence of some compound – like a pregnancy test kit. Biodiversity may be shown by an index, possibly on a scale of 0–100. Bioindicators and biomarkers are often used for monitoring. These are plant or animal species with known sensitivity and preferences. Note: reliance must not be placed on a single bioindicator/biomarker, and more than one should be checked (Lagadic et al., 2000).

An index (plural indices) is a scale on which can be shown value, quantity or position. A standard may be a point on an index scale of pollution – so many ppm of a compound. Some indicators are precise and reliable, others less so. Ecologists and geographers have explored critical indicators, a single parameter, which determines whether an ecosystem or livelihood can flourish. Often this is something like water supply or degree of crowding. The concept of carrying capacity is based on the belief that an ecosystem can sustain only a certain density of particular organisms, and if that is exceeded predator–prey balance, nutrient supply or waste disposal will break down. While carrying capacity probably helped spawn the idea of sustainable development, it is risky to assume that because pressure on an ecosystem is below some threshold, all is well; a change of climate, arrival of a new species, or other unforeseen development may topple the balance (Postel, 1994; Pritchard, 1994). Environmental management has to build in margins of error around any indicator or critical threshold that is being monitored.

Sometimes when a broader focus is needed, or the process to be monitored is complex, a composite index may be devised which is the sum of a number of different measurements (e.g. the Human Development Index (OECD, 1991; UNDP, 1991)) or various sustainable development indicators.

Sustainable development indicators

In 1992 *Agenda 21* called for the establishment of indicators of sustainable development. Sustainability indicators, if they highlight the real underlying causes of environmental damage, will help prevent wasted efforts treating symptoms or pursuing 'cosmetic cures'. Because there is no single established definition of sustainable development, and there are different strategies for pursuing it, and the starting point and challenges differ from site to site, it is difficult to develop a universally accepted index to measure it (Kuik and Verbruggen, 1991; Victor, 1991; Pearce, 1992; Hanley and Spash, 1994; Trzyna, 1995; Van den Bergh, 1996; Friend, 1996; Atkinson, 1997; Jasch, 2000; Briassoulis, 2001; Velva, *et al.*, 2001; World Bank, 2003; Redclift, 2005). For information on sustainable development indicators and sustainability assessment (see below), visit: http://iisdl.iisd.ca/measure/compendium.htm (accessed September 2003); and http://www.sustainer.org/resources/index.html (accessed July 2003) (see Box 8.1). The OECD has put a great deal of effort into developing sustainable development indices.

Judging progress towards sustainable development demands prediction of the behaviour of complex socio-economic and physical systems, and using extensions of established economic, social and environmental indicators is unlikely to be adequate (Bell and Morse, 1999). The likelihood is that a number of indicators will become established based on different understandings of what is most important. In general, composite indicators have replaced single-dimension indicators; for example, the environmental sustainability index (ESI) (http://www.yale.edu/esi/ – accessed June 2005). Hanley *et al.* (1999: 59–62) critically reviewed and compared a number of indicators; for a bibliography of sustainability indicators and monitoring see http://www.nottingham. ac.uk/sbe/planbiblios/bibs/Greenis/A/24.html – accessed June 2005.

Setting goals and objectives

There are many tools for setting goals and objectives. Some have originally been developed by strategic planning, military strategists, policy research, public relations, business management, and many other fields. The problem is that goals and objectives are often set before environmental managers are consulted, rather than incorporating their input from the outset. When that is the case adequate solutions may be difficult. Most goals and objectives decision making starts with some form of contact between stakeholders, all or possibly only the powerful ones, and some type of 'brainstorming' usually follows. Simple brainstorming is cheap, quick, and is facilitated by e-mail and tele-conferencing. The brainstorming session is likely to consist of workshops with stakeholder representatives or experts, or focus groups may be consulted. A focus group is a relatively informal meeting with stakeholders where the observer prompts discussion in a limited way but essentially listens. A slightly more powerful tool for brainstorming (and future scenario prediction) is the Delphi technique. Its workshops of experts and other stakeholders are more orchestrated than focus groups, using controlled feedback, to get a pooling of various opinions. This tool is useful when there is less than optimal data and may be done via e-mail or tele-conferencing. The Delphi technique minimises

Box 8.1

Measuring sustainable development

- *Index of Sustainable Economic Welfare* – This is a socio-political measure first proposed in 1989.

- *Net Primary Productivity* – Derived from the ecological concept of carrying capacity, which is the maximum population of a given species that an area can support without reducing its ability to support the same species indefinitely.

- *Environmental Space* – Developed in 1992.

- *An Extension of the Human Development Index* – There have been efforts to modify or develop this to complement the Human Development Index (HDI) (Sagar and Najam, 1998). The HDI was proposed by the UN Development Programme in 1990 and has become a widely used multi-dimensional measure of development. Since then it has been considerably modified and now makes some provision for assessing sustainability. However, the HDI has a long way to go before it measures sustainability and environmental issues adequately, and analysts such as Neumayer (2003) suggest that consideration should be given to developing a 'green index' to complement the HDI, rather than further greening the existing HDI.

- *Factor X Concept* – This asks 'by what factor can/should the use of energy/resources be reduced and still have the same utility?' (Robért, 2000: 251). This is a flexible way of monitoring and modelling how to extract more from resources being used. It can be modified to ask 'By what factor must resource flows to affluent societies be reduced to allow the poorer societies to improve their living conditions?'

- *Composite Index of Intensity of Environmental Exploitation* – Similar to the HDI (Desai, 1995).

- *Less general, more focused Sustainability Indices* – These have also been developed for specific ecosystems and sectors of activity; for example, a *Sustainable Land Management Index*, and a *Sustainable Agriculture Index*. These have been prompted by doubts about the long-term viability of modern agriculture as a consequence of pollution by pesticides, herbicides, fertiliser runoff, and heavy use of petrochemical energy inputs (Rigby *et al.*, 2001: 465).

- *Indicators of farm level sustainability* – Might prove useful for highlighting the key inputs and practices, which hinder sustainability.

- *Eco-footprint* – Measures how much land is required to supply a particular city, region, country, sector, activity or individual with all needs usually expressed in hectares per capita. Effectively, it is a measure of 'load' imposed on the environment to sustain consumption and dispose of waste. The concept is based on the idea that each individual uses a share of the productive capacity of the Earth's biosphere both for resources and disposal of wastes. It is essentially a measure of human aggregate ecological demand and it can only be temporarily exceeded or the productive and assimilative capacity of the biosphere is weakened (Wakernagel and Yount, 2000). For example, the average North American had an eco-footprint in 1995 *c.* 4.5-times what it would have been in 1905 (http://www.sustainablemeasures.com/Indicators/ISEcologicalFootprint.html – accessed December 2003). Eco-footprinting is thus a useful ecological accounting method for assessing the demands made by humans on various productive areas (Palmer, 1999; Wackernaegel and Rees, 2003; http://www.redefiningprogress.org;http; http://www.ire.ubc.ca/ecoresearch/; http://www.iclei@iclei.org – all accessed August 2003 – give information on eco-footprinting). It can help show where human demands are problematic, aid evaluations of what could be done to improve sustainability, and provide a framework for sustainable development planning. However, it has been suggested that it underestimates human impacts and shows only the minimum needs for sustainable development, without a healthy margin for error, which the precautionary principle seeks (Ferguson, 1999).

Eco-footprinting may be pursued using a number of different models – for a company, island, region, sector of production, city, or whatever. A valuable tool for use in the quest for sustainable development, it allows comparison of the impact of different components on the same aggregated scale. It also aids the assessment of eco-efficiency (i.e. whether an organisation is making the most of its resources and waste disposal opportunities). Eco-footprinting has been used by regional planners seeking sustainable development (e.g. for cities, regions or small islands: Barrett, 2001). It may also be used by businesses, cities and other bodies to measure their environmental performance (Barrett and Scott, 2001). Roth *et al.* (2000) explored the value of an ecological footprint approach for aquaculture and found some serious faults, including its 'two-dimensional' interpretation of complex ecological and economic systems; and failure to recognise issues such as consumer preference or property rights; and it offers a temporally limited 'snapshot' view; natural systems are seldom stable – and it may fail to allow for this; consumption per capita, fashions and settlement patterns change and this may not be registered; also, it does not help to find the most appropriate path for human activities. However, it does seem to offer a graphic and easily communicated image; it is a relatively transparent tool; and it may stimulate creative thinking about environmental and developmental issues. Eco-footprinting should be used with care, and in combination with other tools.

squabbles and deference to powerful single opinions (see Chapter 9). When brainstorming, environmental managers should seek to be aware of the limitations of the data available. The search for goals and objectives may be initiated by a known threat or opportunity or by international or national agreements or guidelines; for example, the Millennium Development Goals or *Agenda 21*.

Once an initial set of goals and objectives has been determined it can be tested and refined, using simple tools such as SWOT analysis or cost–benefit analysis. SWOT analysis is a simple tabulation of strength – S, weakness – W, opportunity – O, and threat – T associated with a given choice. SWOT analysis is cheap, easy and fast. It offers a crude, slightly subjective overview of a situation or proposal so it is useful in brainstorming. Used to help set goals and for project or programme evaluation, the logical framework evaluation/analysis (Logframe approach) can help users to think widely about plans.

The tool was developed by management studies and provides a structure to describe a project or programme, and test the logic of the planning action in terms of means and ends. It focuses on how objectives will be achieved and what the implications of action will be.

Cost–benefit analysis (CBA) was being used by the 1930s, and possibly as early as the 1860s. It is applied to plans, projects, programmes and policies to try and calculate positive and negative impacts, in some cases in advance of a proposed development. It seeks to value impacts in economic terms, which can mean problems for assessing environmental and social items – efforts to do so are usually indirect, using techniques such as opportunity costs, shadow pricing and property values (Hanley *et al.*, 1999). CBA is a tool which is intended to help developers select from a set of defined development alternatives. It may be applied to projects, plans, programmes and policies. The results are given in monetary units (if necessary using techniques such as contingency valuation to try and estimate the worth of things that are difficult to value directly). In reality it is not as objective as many hope, and there can be difficulties valuing some things. There are ongoing efforts to update and modify CBA to improve its performance, none of which so far has cured all its faults. CBA is less useful in developing countries because people there are more likely to operate outside any formal market setting. In addition, poor people may largely exist 'outside' economics – they consume what they produce, and there are societies which value social or cultural things above material. Cost-effectiveness analysis seeks to select development alternatives on the basis of lowest monetary costs (i.e. best value for money). A goal is set, say an improved environmental standard, and assessors seek the least-cost way to achieve it.

Once these or other tools have selected a set of goals and objectives it may make sense to undertake a pilot study. A pilot study is a small-scale application and forerunner of the main development effort; it can identify problems, develop tools and train personnel – and offer some chance for adaptability. Unfortunately pilot studies are often neglected, resulting in difficulty adapting to the unforeseen. Pilot studies can raise costs, introduce delays, and small-scale results may be difficult to scale up and use at a wider scale.

Often environmental managers deal with processes – manufacturing, building and managing large projects such as dams and so on. Life-cycle assessment (or life-cycle analysis) (LCA) first appeared in the 1960s as a tool which seeks to identify impacts and demands at each stage of manufacturing, service provision and so on. Impacts do not cease when goods leave a factory; there may be pollution associated with their use and disposal, and LCA assesses impacts for the whole life cycle. It has been used increasingly by manufacturers since the adoption of legislation to require it in Europe and the USA. LCA can help identify stages in manufacturing or service provision where environmental measures are needed and are most effective (Frankel and Rubik, 2000). It is also a tool for assisting environmental managers to understand environmental problems. LCA may also be used to help evaluate the impacts and best practice at each stage in the provision of services, or in manufacturing or consumption (from raw materials to end-of-life disposal or recycling of products and decommissioning of a factory or other facility). Currently the UK is about to decommission many nuclear power stations – little thought was given to the challenges this would offer when they were built in the 1950s and 1960s, yet much could have been engineered in to help if the life cycle had been considered. Some organisations have considerable experience with LCA practices (e.g. ISO14040–14048 EMS standards). Heiskanen (2002) noted that the spread of LCA was encouraging practices such as design-for-environment, environmental labelling, and other practices which seek to integrate manufacturing and service provision with environmental concern.

A range of participatory tools is used for evaluating and monitoring needs, livelihood strategies, social capital, attitudes, useful traditional knowledge, vulnerability and much more (Scoones and Thompson, 1994; Nelson and Wright, 1995). Aid agencies and NGOs have developed these tools, with inputs from impact assessment specialists, anthropologists and sociologists (Save the Children, 1995). It has been fashionable for a couple of decades to involve people in data gathering, planning and decision making. In the past a failure to consult people commonly led to negative social impacts that could have been avoided, and valuable local knowledge and skills were missed. The approach is multidisciplinary or even holistic, and gender analysis is often an important component. Participatory appraisal may be employed in rural and urban situations.

Data are often wanted in a hurry and cheaply; thus tools which are relatively 'quick and dirty' (i.e. fast but not very accurate or detailed) are valued. So some techniques and tools seek to be rapid; rapid methods have been widely used in rural situations, but may be employed for assessing squatter settlements, urban populations and so on. Rapid rural appraisal is a methodology for rural development studies, developed since the early 1980s, which relies on researchers making in-depth and informal contact with people, observing local conditions and collecting other available data (Carruthers and Chambers, 1981). It is suited to investigating the numerous complex linkages involved in livelihoods. It places stress on relevance, comprehensiveness, multidiciplinarity, speed and low cost. It is much faster than most normal academic research (see: Rural Development Institute http://www.rdiland.org/RESEARCH/Research_RapidRural.html; United Nations University http://www.unu.edu/UNUpress/food2/UIN08E/; and http://www.developmentinpractice.org/abstracts/vol07v7n3a10.htm – all accessed April 2004).

Where data are needed fast in non-rural situations, the skills of market research may be tapped. The tools employed by market researchers include focus groups, question-naire surveys, observational studies, phone interviews and so on.

Monitoring

Monitoring aims to establish a system of continued observation, measurement and evaluation for defined purposes, 'continued' meaning constant or regularly repeated. Most assessment techniques give a spatially and temporally restricted 'snapshot' view which may not be representative soon after. Monitoring may use such techniques but repeatedly in order to build up a sequential set of observations. This is vital because things differ considerably between the start of a development, during implementation and after completion. There may also be changes in the economic, social, political and environmental conditions. Without monitoring, it can be difficult or impossible to establish how things are performing. Monitoring is the process of keeping the health of the environment (and with social monitoring, of society) in view (Spellerberg, 1991: xi). If sustainable development is a goal, monitoring is vital. Monitoring should be oper-ated to agreed schedules with comparable methods. The focus may be on biology, chemical pollution, air pollution, or any other aspect of the environment. It is seldom possible to obtain a precise, detailed picture of all environmental parameters (let alone social, economic, and so on). Monitoring is therefore often undertaken for a specific reason (or reasons), for the systematic measurement of selected variables (Mitchell, 1997: 261), to:

● improve understanding of environmental, social or economic processes;
● provide early warning;
● help optimise use of the environment and resources;
● assist in regulating environmental and resources usage (e.g. it may provide infor-mation for lawcourts);
● assess conditions;

- establish baseline data, trends and cumulative effects;
- check that required standards are being met, or see whether something of interest has changed;
- document sinks, sources and so on;
- test models, verify hypotheses or research;
- determine the effectiveness of measures or regulations;
- provide information for decision making;
- advise the public.

There has been increasing interest, spurred by transboundary problems, in developing international monitoring systems. These seek to monitor at the global level and ideally offer wide access to their information (those bodies involved include the UNEP, OECD, EEC; and the International Atomic Energy Commission). An independent international research unit was founded in 1975 to assist international organisations with monitoring – the Monitoring and Assessment Research Centre (MARC). This concentrates on biological and ecological monitoring, particularly pollution. The World Conservation Monitoring Centre was established in 1980 by upgrading an IUCN-run body, to monitor endangered plant and animal species. The UNEP has established the Global Environmental Monitoring System (GEMS), which is a co-ordinated programme for gathering data for use in environmental management and for early warning of disasters. The UNEP promoted global environmental monitoring at the 1972 UN Conference on the Human Environment. The US Food and Drugs Administration monitors pharmaceuticals and foods. International bodies monitor the spread and use of weapons, especially nuclear, chemical and biological devices. In most countries, doctors, vets and other professionals report observed effects to central monitoring bodies.

Monitoring may show how the environment, a society or economy changes, aiding understanding of structure and function and, hopefully, offering early-warning of problems. Monitoring, surveillance and screening (the checking of a specific thing, e.g. a particular disease in a population – not to be confused with impact assessment screening) are valuable development aids but they can generate problems over who should administer, enforce and pay for them.

One tool which may be used in monitoring to warn of the development of a critical situation is ultimate environmental threshold assessment. Derived from threshold analysis, this watches for a point at which it is known problems will start to develop. The thresholds may be global or local, environmental, social, economic, or whatever. The threshold has to be established by previous research (and ideally is a recognised standard). Much of the early development of the tool was by national park managers looking for a precautionary planning aid – it is no good learning of a conservation problem after the creature has been exterminated.

The geographical information system (GIS) has become an important surveillance, monitoring, planning, research and policing tool. Data acquired from a range of sources, some possibly updated in real time (i.e. constantly giving information on the current situation), are stored and updated in a computer system; the data may be retrieved and displayed in a huge diversity of ways. So, if an environmental management team want a map of, say the snowcover in March correlated with atmospheric pollution levels, it is possible provided the data has been stored.

Business management and project evaluation have developed a wide range of monitoring and evaluation techniques, many of which are used by environmental management. Some gather data (for example, from local people or different groups in a culture), there are tools to help simplify complex situations, and others can help establish best practice (e.g. logical framework assessment or Logframe approach). Project evaluation is often like impact assessment and provides a 'snapshot' (limited in time span and extent) of how a development is progressing and what may have gone well or wrong (Coleman, 1987; Bennett and James, 1999: 96–98).

Surveillance

Surveillance is repetitive measurement of selected variables over a period of time, but with a less clearly defined purpose than monitoring. It is more exploratory and can be undertaken to determine trends, calibrate or validate models, make short-term forecasts, ensure optimal development, and warn of the unexpected. Surveillance, like monitoring, can focus on the environment, people or an economy, and may:

● check whether statutory regulations are complied with (without monitoring and surveillance the setting of standards and rules is of little value);
● provide information for systems control or management;
● assess environmental quality to see whether it remains satisfactory;
● detect unexpected changes.

Where monitoring seeks to establish the ongoing picture, it may be important to examine past conditions and establish trends to understand the present and permit extrapolation of possible future scenarios. For example, studies of climate changes and ecological responses give clues to possible future conditions.

Environmental, social and economic monitoring have each generated their own practitioners and literature, which may focus at local, regional, national or global level or study 'pathways' (e.g. for pollution). Surveillance and monitoring may be done at source (where something is being generated), at selected sample points, at random, along transects, or by sampling some suitable material or organism. For example, pollution may be monitored by checking a smoke-stack, by a network of instruments, or by surveying lichen species diversity and growth. Regulatory monitoring checks its findings against set, in-house, national or international standards or stated objectives.

For the past few decades, and at a gathering pace, remote monitoring and surveillance have been possible. Data may be gathered by orbiting or geostationary satellites, reconnaissance aircraft, unmanned submarine craft, and automatic terrestrial land or marine data-gathering stations linked to the data collector by radio or phone link. Internet links and cheap data collection platforms are reducing the costs – rather than aircraft, balloons or unmanned drones, microlites or kites can be used. Sensing and transmitting equipment is now small and inexpensive enough to attach to fish, seals and other wildlife. The best data are of little use if poorly co-ordinated, so bodies have evolved to support surveillance and monitoring on an international scale and disseminate results to where they are useful. Remote sensing can give information where access is difficult or dangerous – not surprisingly, intelligence-gathering bodies initially developed much of the hardware and software. When a strategic overview is needed, remote sensing data collection can be very useful.

Modelling

There is a huge diversity of modelling methods; all seek to clarify without misreading the process under study, some allow forecasting, and some accept the input of alternative sets of variables to explore different scenario outcomes. The quality of data input and accuracy with which the process is understood are crucial – 'garbage in means garbage out'; even when data input is good and the model has been well tested the results should be interpreted with caution. Most models are a caricature or simplification of reality: often a set of equations used to predict the behaviour of a variable or variables. Some predictions can be imperfect, but good modelling should cope with change and inadequate data and give useful results. There are many types of model developed by various disciplines: computer models, analogue models, conceptual models, role-play exercises, and many others. Conceptual models are used to see what needs study, to help formulate and check hypotheses and to organise ideas. Simulation or predictive models can provide EIA with an indication of what may happen in the future, and can help environmental managers see how something is proceeding. Hydrologists may set up a scale model of river estuary and release flows of water to study tides, currents, flooding, scour and deposition. Climatologists are developing general circulation models, and using powerful computers to try to establish likely future climate change. Input–output models have been used by regional planners and environmental managers for integrated environmental management and strategic environmental management. Futures models or world models were used to produce *The Limits to Growth* and other predictions. Ecosystem simulation modelling is applied to specific ecosystems; social scientists use social modelling to predict socio-economic impact, and economists use economic models to try to establish micro- or macro-economic trends and to test ideas for manipulating an economy.

Environmental auditing, environmental accounting and eco-auditing

The environmental manager needs to have an idea of the state of the environment, and of any threats, future problems or opportunities. There has been some confusion over the use of the terms 'environmental auditing' and 'assessment'. Environmental auditing has been applied to stock-taking, eco-review, eco-survey, eco-audit, eco-evaluation, environmental assessment (another vague expression), state-of-the-environment assessment, the production of 'green charters' and the checking of impact assessments to determine their effectiveness (Cahill, 1989; Edwards, 1992; Grayson, 1992; Thomson, 1993; Buckley, 1995). A key step in any environmental activity is usually the acquisition of data, and often it must be expressed in economic terms (see: http://www.accg.mq. ecu.au/apcea; http://www.iucn.org/places/usa/literature.html; and International Environmental Management Accounting Research and Information Center http://www.emma website.org/about_ema.htm – all accessed February 2004).

Environmental accounting may be used to support eco-efficiency. This is an approach geared towards ensuring competitively priced goods or services that also satisfy environmental goals. It has been defined as the delivery of competitively priced goods or services that satisfy needs and bring quality of life, while progressively reducing ecological impacts. This might be shortened to 'doing more, better, with less', and it falls well short of sustainable development because it does not address poverty and is too limited in scope. Nor does it question the need for making a product or providing a service. Specialist auditing and assessment includes vulnerability audits (vulnerability to hazards), social capital audits (to assess whether social capital is adequate, is in decline, is improving, and whether it can be stabilised or uprated), and gender audits (to assess what the genders have, do and need). There is a need to improve the sustainable development aspects of eco-audits.

Environmental accounting is reasonably distinct from auditing; state-of-the-environment accounts and environmental quality evaluation use knowledge of how the ecosystem is structured and functions to collect data showing the state of an area (not only terrestrial; the Baltic and the North American Great Lakes have been assessed). The approaches discussed here seek mainly to establish the current status of an ecosystem – stock-taking; methods such as EIA (see Chapter 9) are predictive and focus more on the future effects of development. Environmental management accounting may be found at the IUCN Green Accounting Institute website: http://www.iucn.org/places/usa/literature.html and the International Environmental Management Accounting Research and Information Center website: http://www.emmawebsite.org/about_ema.htm – both accessed May 2005. Ecological evaluation seeks to establish what is of value. At first glance environmental auditing would seem to mean establishing the latest picture provided by monitoring; however, it is more complex. Environmental audits can be conducted at company, institution, state, national or global levels, and may include: (1) a stock-taking or inventory-focus approach to the environment which seeks to review conditions and evaluate impacts of development (e.g. new systems of national accounts); (2) studies aimed at avoiding or reducing environmental damage; (3) means by which a body systematically and holistically monitors the quality of the environment with which it interacts or is responsible for – something vital in any quest for sustainable development. The latter activity is now widely undertaken and is usually termed 'eco-audit'. Eco-audit is usually an internal review of the activities and plans of a business or other body; although these may be undertaken by potential buyers of a company and sometimes environmental enforcement bodies. Eco-audit is big business and is undergoing expansion.

State-of-the-environment accounts set out a region's or a nation's environmental, social and economic assets. Norway, France, The Netherlands, Canada and the World Bank have developed national state-of-the-environmental accounts systems – for example, France's Comptes du Patrimoine Naturel ('national heritage accounts'), developed since 1978. National state-of-the-environment accounts make use of environmental assessment and have been promoted as improvements on indicators like gross domestic product for documenting development status. They may prove important in future trade agreements to ensure that environmental effects are counted, and in the quest for sustainable development (Ahmad et al., 1989). However, there has been criticism of environmental accounting, mainly that it is just stock-taking and stops short of encouraging a change of attitude towards the environment and a precautionary approach.

In the UK environmental assessment has been used by government bodies to mean EIA; elsewhere the term is applied to pre-development stock-taking, for example site selection for a nuclear waste repository. In the USA an environmental assessment means a concise public document which should provide enough evidence for a decision to be made on whether or not to proceed to a full EIA. Environmental assessment has also been applied to surveillance or screening such as checking drugs or industrial activities, and it is used for studies which seek to establish the state of an environment with less focus on impacts than EIA. Environmental appraisal is a generic term used in the UK for the evaluation of the environmental implications of proposals. Environmental appraisal is sometimes used as an equivalent of environmental assessment or environmental evaluation. A number of agencies have published environmental appraisal guidelines (see Barrow, 1997: 23).

Overlapping environmental auditing and eco-auditing is the supply chain audit. A company or organisation conducts a comprehensive environmental (and sometimes also social) check on the products, materials and other inputs it purchases to manufacture or provide a service. Investigators may check that a supplier deals with waste properly, does not put the environment, workers or the public at risk, employs adults and pays adequately, and so on. Supply chain auditing puts medium and smaller companies and organisations under pressure to improve their environmental (and human resources) management – the larger company or body commissioning the audit may issue bench-marking or manuals to help support improvement.

Eco-audit

Eco-auditing (corporate environmental auditing or environmental management systems auditing) may be defined as a systematic multidisciplinary methodology used periodi-cally and objectively to assess the environmental performance of a company, public authority or, in some instances, a region. Eco-audits were originally developed in North America in the 1970s to evaluate whether business practices met legislative require-ments and good practice. There has since been considerable refinement, although adaptation to non-Western situations is still under way. The end-product of an eco-audit is an audit report for management, often released to the public, with an undertaking for ongoing repetition to improve future performance. Eco-audits may be done in-house by a company or government team, or by an independent, ideally accredited, specialist or team. The trend is for a consultancy company to undertake the audit and for an accred-iting body to oversee it – within a few years the likelihood is that there will be a single world accreditation body with standardised packages (this will be the International Standards Organisation (ISO)). So far, the decision to eco-audit has been mainly volun-tary (in relation to finance and company matters, 'auditing' more usually implies involuntary) and prompted by a desire to increase public awareness and aid the quest for sustainable development. Sometimes eco-audits are demanded: there have been cases where shareholders have asked for eco-audits at public meetings (*The Times*, 11 April 1997: 25); aid agencies often commission them before granting funding; NGOs may press for them; and insurance companies may require them before accepting a client or grant reduced premiums if there is a satisfactory audit. In the future governments may pass legislation requiring eco-auditing or EMSs, which include the approach.

Impact assessment deals with potential effects of proposed developments; eco-auditing focuses on actual effects of established activities. Both impact assessment and eco-audit can be valuable tools for environmental management provided that manage-ment is committed to adequate action on the findings. Eco-audit usage expanded in the 1980s as companies were held more responsible for the damage they caused and realised the need for a green image (EPA, 1988; Shillito, 1994; Buckley, 1995; Gilpin, 1995). It has been promoted in Europe by the International Chamber of Commerce and by some multinational corporations as a means of getting effective environmental manage-ment (International Chamber of Commerce, 1989, 1991). A significant step forward has been the development of eco-audit standards and environmental management and audit systems, the world's first being offered by the British Standards Institution (BS7750) in 1992. This soon developed beyond assessment of environmental effects to include an assurance to continual environmental improvement.

Eco-audit handbooks and guidebooks began to appear in the mid 1980s (Harrison, 1984; Blakeslee and Grabowski, 1985; Thompson and Therivel, 1991; Local Govern-ment Management Board, 1991, 1992; Grayson, 1992; Spedding *et al.*, 1993; McKenna & Co., 1993; Richards and Biddick, 1994). In 1986 the US Environmental Protection Agency issued an Environmental Auditing Policy Statement designed to encourage the use of eco-audits by US companies, and laid down guidelines.

Impetus was also given by the publication of *Agenda 21*, and by the European Commission's Fifth Environmental Action Programme (1992). The latter sought to

Figure 8.1 European Union Eco-Management and Audit Scheme (EMAS) eco-audit award logo

Note: This may be used on a company's brochures, letterhead, reports and advertisements, but must make no reference to specific products or services, and may not be used on product pacakages.

promote 'shared responsibility' by people, commerce and government for the environment, popular green awareness, and a move towards sustainable development. One of the tools it supplied to support these goals was eco-auditing (see Figure 8.1). In 1992 eighteen of the UK's top companies undertook eco-audits; by 1996 about half the country's firms had eco-audited – a rapid, voluntary spread. Eco-auditing is part of a growing shift from mere compliance with regulations to developing forward-looking environmental management strategies (Willig, 1994; Sunderland, 1996), so it supports the principle of prudence. There has been less progress in developing countries, although India has modified its Companies Act to include a requirement for eco-audits, and Indonesia has required companies to conduct eco-audits since 1995.

Eco-audits offer some or all of the following benefits:

- they generate valuable data for regional or national state-of-the-environment reports;
- they are a means for ensuring the continual improvement of environmental management;
- they may be a valuable way of monitoring;
- they can help establish an effective environmental protection scheme, which may reduce insurance premiums (Finsinger and Marx, 1996);
- they can assist efforts for sustainable development;
- they can inform the public about the body's environmental performance, which is good PR;
- they can help involve the public in environmental management;
- they help identify cost recovery through recycling, opportunities for sale of by-products and so on;
- they reduce risks of being accused of negligence and losing court cases;
- they may reduce the need for government inspections;
- they can ensure that often complex regulations are known about and followed, and that licences are obtained;
- they offer management more peace of mind.

There may also be risks associated with eco-audits:

- they may spot a problem that is costly to cure, which might otherwise have been overlooked without too much harm;
- they can be expensive;

Box 8.2

Types of eco-audit

- *Site or facility audit* – A company or body audits to see how it conforms to safety and other regulations, and care for the environment.

- *Compliance audit* – To assess whether regulations are being heeded and/or policy is being followed.

- *Issues audit* – Assessment of the impact of a company's or other body's activities on a specific environmental or social issue (e.g. rainforest loss) (Grayson, 1992: 40).

- *Minimisation audit* – To see whether it is possible to reduce: waste; inputs; emission of pollutants (including noise); energy consumption and so on.

- *Property transfer audits* (pre-acquisition audit, merger audit, disvestiture audit, transactional audit, liability audit) – A company or body audits prior to disvestiture, takeover, joint venture, alliance, altering a lease, sale of assets and so on to show whether there are any problems such as contaminated land.

- *Waste audits* – To see whether regulations are met, whether costs can be reduced by sale of by-products and so on (Ledgerwood *et al.*, 1992; Thompson and Wilson, 1994). The motivation to audit may be to comply with legislation or come from a desire to prevent problems.

- *Life-cycle assessment/analysis* – Evaluation that can extend beyond the time horizon of a single owner, company or government (it is cradle-to-grave) (e.g. impacts of something from manufacture through use to disposal) (British Standards Institution, 1994c; Fava, 1994).

Note: Even within a single company, eco-audit must check for variation from unit to unit and allow for change that takes place as plant ages. Eco-audit may extend to checking environmental impacts of suppliers, subsidiaries, use and disposal of products and packaging.

- a body may fear trade secrets will be exposed to competitors;
- smaller companies cannot conduct eco-auditing in-house and must use specialists from outside (costly, with a risk of loss of trade secrets).

There are two broad categories of eco-audit: (1) industrial – private sector corporate eco-audits; (2) local authority or higher level government eco-audits (sometimes called 'green charters') – these are more standardised than industrial (private sector) corporate eco-audits, and are commissioned by local authorities to show local environmental quality (Levett, 1993; Barrett, 1995; Leu *et al.*, 1995) (Box 8.2). Some local authorities produce state-of-the-environment reports, which are not the same as audits carried out as part of an environmental management system approach (see below). In the UK the first eco-audit by a local authority took place in 1989 (Kirklees District Council, assisted by Friends of the Earth). About 87 per cent of UK local government authorities had used eco-audit or planned to by 1991, encouraged by the UK 1990 Environmental

Box 8.3

Eco-audit–environmental management system standards

BS7750

In early 1992 the world's first eco-audit standard was published – British Standards Institute's BS7750 Specification for Environmental Management Systems (British Standards Institution, 1992, 1994a, 1994b; Hunt and Johnson, 1995: 89) – derived from an earlier Management Quality System BS5750. A number of countries adopted it, and it was revised in 1993 and 1994 to make it more compatible with the recently introduced Eco-Management and Audit Scheme (EMAS – which has drawn upon BS7750) (Bohoris and O'Mahoney, 1994; Willig, 1994: 33–42; Buckley, 1995; Sharratt, 1995: 41–53). BS7750 is a means by which an organisation can establish an EMS. To obtain BS7750 a body has to establish and maintain environmental procedures and an environmental protection system which meets BS7750 specifications and demonstrate compliance. It must also be committed to cycles of self-improvement through internal eco-audit. There are three elements to BS7750: (1) possession of an environmental policy; (2) a documented EMS; (3) a register of effects on the environment. Critics of BS7750 argue that it is possible to achieve the standard by promising to do better and then to release relatively little information to the public (it is not as open as, say, the US Toxic Releases Inventory). At the time of writing, BS7750 did not provide for a publicity logo and was rapidly being superseded by the ISO14001 series.

EMAS

The Eco-Management and Audit Scheme (EMAS) was launched in 1993 (EU Council Regulation 1836/93), although it was not until April 1995 that it came into force in the UK (Welford, 1992; EEC, 1993; Brown, 1995). EMAS goes beyond eco-audit to require an approved EMS and the production of an independently verified public statement. EMAS seeks to encourage industries in EU states to adopt a site-specific, proactive approach to environmental management and improve their performance. EMAS is in some ways similar to, and broadly compatible with, the already established BS7750, but it is much broader in scope and requires greater public reporting of audits. It is stronger than BS7750 on environmental protection, and is aimed more at industrial activities. EMAS is also stronger on ensuring that a body regulates its environmental impacts.

EMAS registration is voluntary (but is established in the EEC by regulation so that consistent rules are supposed to be set for all those participating). Participants write and adopt an environmental policy which includes commitments to: meeting all legislative requirements and ensuring continued improvement of performance; implementation of an environmental programme with objectives and targets derived from a comprehensive review process; establishing a management system (which includes future environmental audits) to deliver these objectives and targets; and issue public environmental statements (EMAS does not insist on full publication of audits). Originally it had been planned to make full public disclosure compulsory but this was abandoned. An accredited third party verifies all these measures (see *Journal of the Institution of Environmental Sciences* 4 (3): 4–7). If these terms are broken, the organisation may be suspended from EMAS, and so lose its right to a special logo (green credentials), which means loss of publicity advantage, and possibly increased insurance premiums or supplier, investment, or sales-outlet boycott.

Criticisms of EMAS include the charge that its auditing criteria are vague (Karl, 1994); that it disrupts the activities of an organisation; that it may reveal trade secrets, and perhaps cause public or workforce hostility. There are signs in the UK that small companies find the cost of BS7750 more of a challenge than do larger companies. There are also calls for EMAS to increase the focus on sustainable development (Spencer-Cooke, 1996). Europe is improving EMAS by introducing strategic environmental assessment to all plans, policies and programmes (Barton and Bruder, 1995: 11) (see Chapter 3). There were also plans at the time of writing to expand EMAS to make it more compatible with the ISO14001 series – which is effectively becoming the most used standard. If the EEC adopts an Environmental Charter, eco-audit will become more widespread, possibly even compulsory.

ISO14000/14001

The International Standards Organisation (ISO) has been seeking to develop a standard (or rather a series of standards, some advisory, some contractual) broadly compatible with EMAS and BS7750. ISO[DIS]14001 was introduced in 1996, and the series also incorporates ISO14004. ISO14001 provides information on the requirements for an EMS, and ISO14004 has the elements needed and guidance on implementation of an EMS. The ISO1400 series are roughly equivalent to BS7750 and EMAS, but more user-friendly and easier to understand, and seem likely to gain worldwide adoption (for details see Rothery, 1993; Baxter and Bacon, 1996; Jackson, 1997; Sheldon, 1997). These ISO standards are related to the ISO9000 series (roughly equivalent to BS5750) which are widely used by business world-wide and deal with quality systems (TQM) registration. The ISO14001 standard is taking over from BS7750 and is periodically updated (Knight, 1997).

Note: These standards, which deal with environmental management systems (EMSs), have evolved from total quality management (TQM), and are quality auditing systems. They must be widely applicable, effective at getting regulation, yet flexible. It is also desirable that they help integrate environmental management quality standards with commercial quality management (product/service quality) standards and occupational health and safety quality management standards (Young, 1994).

Protection Act (Grayson, 1992: 50). Unfortunately, some of the eco-audits produced little more than publicity documents.

There is considerable overlap between eco-audit and health and safety management. Some countries now audit the environmental quality of new buildings to ensure that they do not harm employees, that they use eco-friendly construction materials and do not waste energy. Energy efficiency and better employment conditions mean savings on power bills and less absenteeism as well as environmental benefits. Barton and Bruder (1995: xv) see local eco-audit as a key measure in the delivery of sustainable development and as 'a process for establishing what sustainable development means in practice – how to interpret it locally, how to test whether you are achieving it'. They recognised two components in eco-audit: (1) external – collation of available data to produce a state-of-the-environment report; and (2) internal – the state-of-the-environment report as a foundation for efforts to assess policies and practices (Barton and Bruder, 1995: 12).

Various bodies and companies publish eco-audit guidelines or manuals which can help other auditors, and there is also the use of computers, expert systems (see also Chapter 9) and information technology (retrieval systems such as LEXIS® conceptual, or hypertext searching). However, guidelines and computer aids are not enough: effective environmental management demands commitment. Some companies' authorities and educational establishments have tried to conduct eco-auditing on the cheap, which tends to give inadequate results. First-time audits are usually more complex than follow-up audits. Voluntary adoption of eco-auditing, especially if it is handled in-house, poses risks from institutional politics (e.g. ministries may compete; companies may be rivals; internal squabbles may distort things scientific) (Ludwig *et al.*, 1992; Rensvik, 1994; Reisenweber, 1995). The cost of eco-auditing varies, depending on the complexity, novelty, thoroughness and local circumstances. Poorer institutions and small businesses may need aid to be able to afford auditing.

Eco-auditing with recognised international standards is spreading; EMSs (like the ISO 14000 series) are spreading, and are constantly being updated and tuned to suit different situations. Better training and accreditation of auditors should reduce problems (Buckley, 1995: 292–293; Gleckman and Krut, 1996). Worries have been voiced that some standards are determined more by politics, special-interest groups and public opinion than by objective environmental managers. Box 8.3 presents some eco-audit–environmental management system standards.

Auditing, appraisal, assessment and evaluation are used in planning (pre-development) and to judge progress (during development or implementation), when implementation is finished (post-project), for ongoing monitoring, and if a development is decommissioned. The results do not just indicate development status; there may be clarification of what has happened which helps others in the future. Often there is a reluctance to undertake post-development or post-impact assessment appraisals; the reason may be that:

● money was made available only for implementation, and recurrent funds are scarce;
● expertise may have moved on and there is nobody to undertake it;
● interest has shifted to something new;
● nobody is keen to look for problems.

A range of social studies tools are available for determining stakeholder views, capabilities and needs. For example, there are various participatory appraisal techniques (Save the Children, 1995). Some of these participatory approaches seek speedy assessment, and usually accuracy or detail is compromised to some extent. Dealing with complex situations generally demands a multidisciplinary team approach.

Sustainability assessment

Sustainable development and sustainability indicators have been discussed earlier in this chapter; assessment demands indicators, techniques, and means of evaluating and presenting information (Gibson, 2005). Indicators may be used to police, identify opportunities and analyse situations – not just track progress towards sustainable development (Bennett and Jones, 1999).

Environmental assessment and evaluation

The emphasis in this chapter is on tools, techniques and methods that assess, evaluate or analyse mainly the existing situation, rather than one which might develop in the future (with or without a proposed development). In addition, many of the techniques and tools give a single snapshot view at each application. Assessment and evaluation tools are often derived from business management and project and programme planning for which a snapshot view may be sufficient. Environmental management may demand a longer term and wider view; some argue for a comprehensive, integrated or even holistic oversight.

Assessment and evaluation of resources is long established; most countries have survey bodies, and large oil and mining companies have evaluation/prospecting units. However, since the 1940s there has been a huge breakthrough in remote sensing. Evaluators can now obtain accurate data from aircraft overflights and satellites; towed sonar or geophysical instruments can probe the depths of the sea, and seismic or magnetic survey probes rocks below the surface of the land (Bishop and Romano, 1998). An example of the impact of evaluation improvements is offered by Amazonia, which stagnated economically until the early 1970s when Brazil's Projeto RADAM used airborne side-scan radar to map resources and helped trigger huge investment in mining, ranching and hydroelectric development.

Eco-footprinting

Eco-footprinting (or ecological footprinting) is a tool which seeks to measure ecological performance of a 'target' – an individual, group, a company, organisation, sector (such as an industrial process, service, transport network, supply of a commodity), or a region (such as an island, valley, highland, country, city, region). It tracks the impact of the target and compares it with what the environment can provide. It does this by calculating how much biologically productive land and water (as area or area per capita) is needed with current demand and technology to provide inputs and dispose of outputs (wastes) safely. The result is a footprint in something like km^2 for the target, which offers a visual image that is easy for people to grasp and which can be compared with other situations.

Eco-footprinting is a tool which may be used as an indicator by those pursuing sustainable development (Chambers et al., 2000). A number of international bodies and large NGOs have started to conduct regular eco-footprinting audits of nations or businesses. These show that some countries are using resources on an unsustainable basis and that there is considerable difference between targets. Eco-footprinting do-it-yourself kits for calculating the footprint of an individual, office, school or business are available; for example, see http://www.epa.vic.gov.au/Eco-footprint; http://www.bestfootforward.com – accessed May 2005. A slightly different form is the carbon footprint, which sets out to assess the amount of *carbon dioxide* emitted each year by a household, company or organisation. EU legislation requires organisations and businesses to trade in carbon if they exceed established quotas – it is thus vital to be able to assess accurately whether such a threshold is passed. Visit the British Petroleum site http://www.bp.com/carbon footprint to calculate household carbon footprint (accessed November 2005).

Integrated environmental assessment is an interdisciplinary process which seeks to collect, interpret and communicate the likely consequences of implementing a proposal (Van der Sluijs, 2002). It has been applied to global warming issues, acid deposition and other fields, and is similar to the Delphi technique, in that it draws on informed group opinion – but usually through focus groups (a group of informants who are interviewed in a relatively 'hands-off' manner, but who are asked set questions).

Environmental management decision making

Environmental managers have to be effective decision makers; guesswork is a last resort; funding bodies and those monitoring progress want a reasoned assessment backed by data, modelling and theoretical argument. The sorts of tools which may be used to support decisions include: life-cycle assessment, environmental risk assessment, hazard assessment, impact assessment, cost–benefit analysis, multi-criteria analysis, and many others (Guariso and Werthner, 1989; Wrisberg and Udo de Haes, 2002). Sometimes a systems approach is adopted which enables modelling to support decision making – it should be noted that a system is a mental construct. Usually decision making demands a guiding framework, tools to identify options, and to select from the options identified. Life-cycle assessment may be used for decision support (for example, to decide how best to dispose of municipal waste). There are other scenario-prediction or role-playing tools which may be used to arrive at decisions. Increasingly in the West there may be public involvement: the use of focus groups, questionnaire surveys, or even a referendum.

Seeking a strategic view

Strategic environmental assessment (SEA) tools seek to extend the focus from local/project to one which is broader/programme or policy or sector, even global (Therevel, 2004; Dalal-Clayton and Saddler, 2005; Wood, 2005). Environmental management decision making must be conducted when there are often inadequate data, incomplete knowledge, funding and time constraints, and lobbying by special-interest groups, citizens and politicians; it is also easy to lose the broader view when addressing a specific challenge. Companies, politicians and publics seldom adopt a long-term view, loans have to be repaid fast, developers want rapid results, and ministers look only ahead to the next election. Nowadays, materials are seldom tested for more than twenty years' durability. People do not worry much about the welfare of their children, let alone future generations beyond that, and they tend to adopt a local focus when it comes to investment (the worldwide response to the 2005 tsunami might signal a shift to a wider view). Environmental issues frequently demand a very long-term view and even a global focus. Tools to assist with more strategic decision making have been developed by strategic planning, strategic impact assessment and strategic environmental management (Pentreath, 2000; see also *Strategic Environmental Management*, a journal published by Elsevier). Interest in SEA is growing, in part prompted by the EU Directive (2001/42/EC) – the 'SEA Directive', and by the UNECE Protocol on SEA. The 'SEA Directive' came into effect in 2004 and requires EU Member States to implement SEA measures.

Environmental management systems

Single eco-audits give a snapshot view: they are more effective if they are part of a structured environmental management system (EMS). So, as well as achieving cost savings through environmental initiatives, an EMS allows an organisation to integrate environmental management into overall management (for an evaluation of EMS costs see Alberti *et al.*, 2000). An EMS may be defined as an organised approach to managing the environmental effects of an organisation's operations – it involves integrating environmental respect and awareness with economy and quality of production (Stuart, 2000). Adopting an EMS enables an organisation to set goals, monitor performance against them, and it shows when to take corrective action or make improvements; also, it supports the development of a reflective outlook which seeks to be environmentally sound (Hunt and Johnson, 1995: 5; Moxen and Strachan, 1995: 35). Sometimes the EMS is conducted 'in-house', but it is most likely to be undertaken by an accrediting body or a subcontractor – the ISO14001 series is becoming established as the most widespread in use. The EMS process should be one of continuous, ongoing improvement, with a cycle of goals set, checks conducted and results published. Thus, the process should take a body beyond mere compliance and encourage it to become proactive and stimulate good practice. Use of an EMS should also help keep a body aware of changes in knowledge, legislation and so on. Figures 8.2 and 8.3 illustrate the basic EMS approach. An EMS should help ensure a structured, standardised and balanced approach to environmental management, and improve the organisations' image, attractiveness to employees and so on. Regulators are more likely to treat bodies using EMS with a 'soft touch', and management may enjoy greater peace of mind and pride. Disasters such as the Bhopal incident and the *Exxon Valdez* oil spill which put large companies under severe financial stress and negative publicity have helped encourage the adoption of EMSs.

The first EMS was the British Standards BS7750 which evolved from a TQM standard (BS5750), and was first released in 1992 (see: http://www.bsi.gobal.com – accessed January 2003). Another EMS, the European Union Eco-management and Audit Scheme (EMAS), is a site-specific and proactive approach promoted since 1995 (http://europa.eu.int/comm/environment/emas/; http://www.quality.co.uk/emas.htm; http://www.inem.org/litdocs/inem_tools.html and http://www.eli.org/isopilots.htm – all accessed January 2005). The widely used ISO14000 series includes over sixty certification systems: ISO14001 to 14061 apply to eco-audit, life-cycle assessment and so on (International Organisation for Standardisation standards relating to environmental management – http://www.iso.ch/iso/en/stdsdevelopment/tc/tclist and http://www.iso14000-iso14001-environmental-management.com – accessed February 2004; and http://www.iso.ch/iso/en/iso9000–14000/pdf/iso14000.pdf – accessed October 2003). These

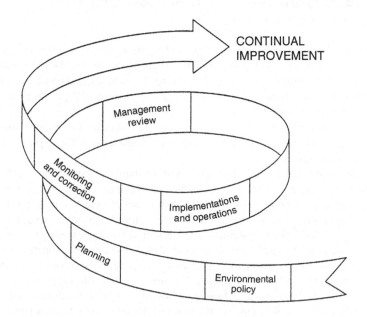

Figure 8.2 Basic environmental management system approach

Source: Based, with modifications, on Hunt and Johnson (1995: 6, Fig. 1.2)

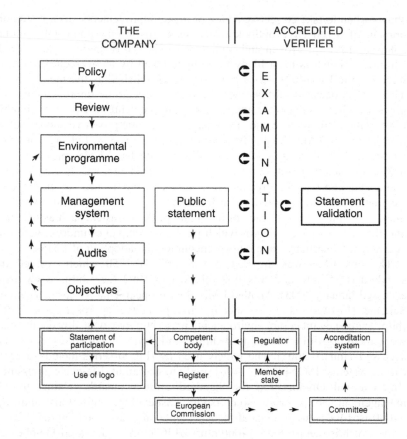

Figure 8.3 The basic provisions of the European Union's EMAS
Source: Hunt and Johnson (1995: 74, Fig. 4.2)

are derived from the ISO14000 series, and relate to the ISO9000 quality management series. Some clients use ISO9002 to certify their EMS, but ISO14001 (launched in 1996) provides what is virtually the 'world standard' framework and guidelines which support the voluntary development of assessment and environmental practices. It also indicates what is needed for an EMS (scoping): the format, objectives, targets and implementation, review procedures, correction and so on. Unlike the European Union's EMAS or eco-audit accreditation, most EMSs make no requirement for a public environmental management statement.

ISO14001 and its future derivative will probably become the world standard for environmental management (Schoffman and Tordini, 2000; Osuagwu, 2002). This means some degree of globalisation and uniformity of assessment criteria, and may help ensure that issues are investigated because an objective international assessor is involved. The ISO14001 system is often adopted by organisations with little environmental management experience in developed and developing countries. There have been some spectacular cost savings claimed as a consequence of adopting EMSs (Rondinelli and Vastag, 2000: 505). Increasingly businesses insist on suppliers or subcontractors having EMS certification, and failure to do so can be a hindrance for smaller firms. EMSs only certify each client, not the level of actual environmental performance. There is little point in an organisation or government developing an EMS if it has insufficient funds to address any problems revealed, which can happen in poor countries. Critics argue that EMSs may be a substitute for adequate environmental management, and are bureaucratic, mechanistic and insufficiently flexible; consequently they lead to mere compliance, not a will to improve. In addition, the cost of an EMS could deter those with limited funds and for some clients outweigh benefits. It is also difficult at present to de-certify a body if it is granted EMS standard status and then becomes sloppy.

Bodies adopting an EMS submit to a cycle of periodic audit and review. They are a management tool designed to help a business, region, country, city or institution improve its awareness of, and control over, environmental impacts. An EMS may be applied to a single site or a number of sites spread around the world. More and more 'ready-made' EMSs are adopted and adapted, with a handful of bodies now offering world standards. They are a means for helping industry, or other bodies, comply with environmental regulations, obtain technical and economic benefits, and are designed to ensure that an environmental policy and environmental objectives are adopted and followed (standards such as BS7750 or EMAS require an EMS to be established and maintained). EMSs can:

- help to develop a proactive environmental approach;
- encourage a balanced view across all functions;
- enable effective, directed environmental goal setting;
- control environmental impacts;
- involve all staff, including senior management, in environmental care;
- ensure that legal requirements are met (such as pollution control);
- develop objectives and targets;
- make the environmental auditing process effective.

An EMS usually requires a participating body to publish an environmental policy statement and regularly update it. The standards used to eco-audit typically test whether the body:

- has identified overall aims;
- understands constraints on achieving aims;
- identifies who is responsible for what;
- sets an overall timetable for achieving aims;
- has determined resource needs;
- has selected a project management approach;
- has a progress monitoring system.

EMSs rely on independent certification of compliance with set eco-audit standards to encourage more careful planning (Hunt and Johnson, 1995). EMSs can in practice be difficult to pursue effectively as a result of real-world institutional politics, funding problems, data shortages, need for industrial secrecy, health and safety issues. EMSs are a major and lucrative field of consultancy activity and the literature is also expanding at a rapid pace.

Most EMSs nowadays are undertaken by subcontracting to a team accredited by and using the approaches developed by one of a few global bodies, notably: ISO, BS, EN and EMAS (there are moves towards making these broadly equivalent).

Prompting environmental management

Environmental managers often have to encourage people, business managers, government ministers and NGO staff to adopt more environmentally desirable practices. This can demand keen skills of presentation and persuasion, if not actual cunning; there may be a need for forming alliances, covert negotiation and networking, and manipulation of others. A more gentle approach involves public education and propaganda or marketing. Since the 1970s most countries have developed environmental ministries, and the media have also established environmental reporting; there may be a reluctance to fund or to alter selfish habits, but in the last half-century there has been a sea-change in general awareness and interest in environmental matters.

Change may need the support of a suitable strategy, set of guidelines or manual. Treaties, agreements, protocols and international conferences play a crucial role in prompting and supporting environmental management. Support for environmental management is also provided by the Internet, which facilitates exchange of information, debate, lobbying, protest and a way for stakeholders to mobilise. To achieve sustainable development people will have to become more aware of the need for, and be willing to support, environmental management. Industrial ecology can help prompt such awareness. It has been described as 'an operational approach to sustainability' by which humanity can deliberately and rationally maintain environmental quality with continued economic, cultural and technical development (Frosch, 1995; Erkman, 1997; Dunn and Steinemann, 1998).

Environmental management guidelines began to be issues for powerful businesses and international agencies in the early 1980s, and within a decade bodies such as the World Bank, OECD, USAID, DFID, JICA, and many others had established them and insisted on their use before supporting developments. In addition to national and international legislation, and agreements, environmental management is supported by a growing number of academic, NGO, professional institute and trade journals which disseminate information; research and staffing foundations in universities and research institutes also help promote improvement. Prizes for good practice and ridicule for poor may help promote good practice. Some activities may drive a way forward, but are controversial and based on less than sound research – typically 'ginger groups' pursue them; just as important is slower and more cautious research and confirmation, left largely to academics and research institutes. Both of these are important, and the environmental management team may have to try and orchestrate them without getting too caught up in the process.

Summary

- Each situation demands the selection of a method, approach and set of tools (which often need refinement). The environmental manager(s) may be able to do this through simple desk research; however, new challenges often demand considerable research. Laws, socio-economic conditions and the environment may change, and new tools and approaches appear; often environmental management is more of a co-ordinating role and specialists will be gathered or hired to apply methods and tools.
- There is a huge diversity of tools, some unique to environmental management, but most borrowed from sciences and social sciences; some need 'fine-tuning', many are the product of Western, liberal, democratic countries and may be unsuitable for other political, cultural and economic environments. Some of the tools and approaches developed in non-Western nations are of value to Western countries.
- Environmental managers seldom find tools that are adequately reliable and precise; important decisions are ideally based on the application of more than one tool and

a number of sets of data. Some tools can generate a sense of false security (e.g. EIA), and some produce results which lack transparency.

● Objective decision making is seldom unhindered; developments are often piecemeal, and frequently hurried and constrained by funding and politics. Whenever it can, environmental management should seek to rise above such difficulties and adopt a more strategic view with a wider spatial focus and longer temporal span.

Further reading

Bryman, A. (2004) *Social Research Methods* (2nd edn). Oxford University Press, Oxford.
  Well-tested introduction to social research methods, including qualitative and quantitative tools.
Burke, G., Singh, B.R. and Theodore, L. (2005) *Handbook for Environmental Management and Technology* (2nd edn). Wiley, New York.
  Overview of environmental management for the non-technical – US focus.
Harrop, D.O. and Nixon, J.A. (1999) *Environmental Assessment in Practice*. Routledge, London.
  Introduces good practice for environmental assessment; includes a number of case studies.
Katz, M. and Thompson, D. (1997) *Environmental Management Tools on the Internet: accessing the world of environmental information*. CFC Press, Boca Raton, FL and St Lucie Press (100E Linton Blvd., Suit 403B, Delray Beach FL 33483, USA).
  Guide to finding environmental management information and tools on the Internet – US focus.
Manley, B.F.J. (2001) *Statistics for Environmental Science and Management*. Chapman and Hall, Boca Raton, FL.
  Statistical techniques for environmental management.
Save the Children (1995) *Toolkits: a practical guide to assessment, monitoring, review and evaluation* (Save the Children Development Manual No. 5 – prepared by L. Gosling and M. Edwards – revised and expanded edn published in 2004). Save the Children Fund, London.
  Draws on NGO expertise on assessment, monitoring and evaluation. The focus is mainly on projects and on social aspects, but also covered are EIA, Logframe analysis and other common tools.
Therevel, R. (2004) *Strategic Environmental Assessment in Action*. Earthscan, London.
  Practical guide to SEA with toolkit and case studies.
Thompson, D. (ed.) (2002) *Tools for Environmental Management: a practical introduction and guide*. New Society Publishers, Gabriola Island (Canada).
  Useful, easy-to-read compendium, although expensive and with a mainly North American focus.

www sites

Environmental Data Services Ltd (ENDS) – information journals (by subscription) covering environmental management, EMSs, environmental planning and so on (UK) http://www.ends.co.uk – accessed September 2005.
Environmental management tool box (free downloadable tools) http://www.gdre.org/uem/e-mgmt.html and http://www.inem.org/htdocs/inem_tools.html – accessed May 2005.
International Chamber of Commerce environmental management tools http://www.icewbo.org/home/state – accessed May 2005.
Sustainable development indicators: http://www.iisdl.ca/measure/compendium.htm; http://www.sustainer.org/resources/index.html – both accessed April 2005.
UNEP environmental management tools http://www.unepie.org/pe/pe/tools/ – accessed May 2005.

9 Proactive assessment, prediction and forecasting

- Environmental risk management
- Environmental impact assessment
- Social impact assessment
- Other tools for assessing the potential for development and impacts of development
- Livelihoods assessment
- Vulnerability studies
- Predicting future scenarios
- Hazard and risk assessment
- Technology assessment
- Health risk assessment
- Computers and expert systems
- Adaptive environmental assessment and adaptive environmental assessment and management
- Integrated, comprehensive and regional impact assessment, integrated and strategic environmental management
- Dealing with cumulative impacts
- Summary points
- Further reading

> Ecology is the science which warns people who won't listen, about ways they won't follow, to save an environment they don't appreciate.
>
> (Anon.)

One of the notable developments of the late twentieth and early twenty-first centuries is a growing willingness to predict potential problems and opportunities, rather than wait to see what happens and then possibly be unable to deal effectively with them. Hopefully, the above comment made in the 1970s is now less true than it was; however, there will be times when those involved in environmental management will agree.

This chapter examines approaches used to predict problems and opportunities and to suggest future scenarios, assess the impacts of development, and the risks and hazards posed by nature and human activity (there is some overlap with Chapter 8). Sustainable development is unlikely without effective prediction of threats, assessment of limits, and identification of opportunities to improve conditions, avoid or mitigate threats and enhance adaptation.

Environmental risk management

Environmental risk management incorporates a range of approaches (including risk assessment, discussed below) to:

- estimate risk;
- evaluate risk;
- respond to risk.

It deals with multidimensional risks, often involving interrelated physical and social impacts, and demands political judgement to improve the chances of optimum decision making (O'Riordan, 1979; Pollard *et al.*, 1995). There have been calls for these approaches to become more holistic (Harvey *et al.*, 1995), and some already overlap with eco-auditing (see Chapter 8). There is growing interest in assessing risks associated with global environmental change, including: biospheric catastrophe (unstoppable shift to conditions that threaten human and other life); climatic perturbation (natural or human-induced which threatens the well-being of people and wildlife); reduced provision of basic needs (threats to sustained production of food, access to adequate water and energy); and pollution (O'Riordan and Rayner, 1991). Burgman (2005) explores risk assessment for ecology, conservation, resource management and environmental management.

Environmental risk management and some of the approaches discussed in this chapter are imprecise, partly because the world is complex; a common cliché is that 'everything in the environment is connected to everything else'. The media often refer to the 'butterfly effect' – a concept from chaos theory, implying that in a delicately balanced world system a trivial event could lead to a vast cascade of changes that are impossible to predict accurately and have serious global consequences. Since the 1960s there has been a shift towards more appropriate development, and the right to damage the environment and people in the name of 'progress' is questioned. There is increased awareness that technology and biotechnology can pose threats and there is growing interest in sustainable development. This chapter looks at the approaches developed to identify and avoid problems or missed opportunities. In addition to warning of impacts, risks and opportunities, some of these approaches can help make planning and management more accountable to the public, and may encourage more careful decision making. They are often not the quantitative scientific approaches they seem; rather, they are ordered but *subjective* methods for improving judgement (Fairweather, 1993: 10). Futures studies focus on estimating whether development will stay within environmental limits and what physical and socio-economic changes there may be. The output is often based on rather subjective, even speculative, projections and suggests scenarios further into the future than more objective assessments.

Environmental impact assessment

There is no universal definition of what exactly environmental impact assessment (EIA) is, so it is best treated as a generic term for a process which seeks to blend administration, planning, analysis and public involvement in pre-decision assessment (Goodland and Edmundson, 1994). A shorter explanation might be 'an approach which seeks to improve development by *a priori* assessment' (Boxes 9.1 and 9.2). Figure 9.1 illustrates how EIA fits into planning, and Figures 9.2 and 9.3 show how it relates to

---

Box 9.1

An overview of EIA

The following observations describe EIA:

- It is a proactive assessment, and should be initiated pre-project/programme/policy, before development decisions are made. In-project/programme/policy and post-project/programme/policy assessments are common. While these may not allow much problem avoidance, they can advise on problem mitigation, gather data, feed into future impact assessment, improve damage control and the exploitation of unexpected benefits.
- It is a systematic evaluation of all significant environmental (including social and economic) consequences that an action is likely to have upon the environment.
- It is a process leading to a statement to guide decision-makers.
- It is a structured, systematic, comprehensive approach.
- It is a learning process and means to find the optimum development path.
- It is a process by which information is collected and assessed to determine whether it is wise to proceed with a proposed development.
- It is an activity designed to identify and predict the impacts of an action on the biogeophysical environment and on human health and well-being, and to interpret and communicate information about such impacts.
- It is a process which forces (or should force) developers to reconsider proposals.
- It is a process which has the potential to increase developers' accountability to the public.
- It usually involves initial screening and scoping (to determine what is to be subjected to EIA, and to decide what form the assessment should take).
- It should be subject to an independent, objective review of results.
- It should publish a clear statement of identified impacts with an indication of their significance (especially if any are irreversible).
- It should include a declaration of possible alternative development options, including nil-development, and their likely impacts.
- Ideally there should be public participation in EIA (it is often partial or avoided).
- There should be effective integration of EIA into the planning/legal process.

Source: Part-based on Barrow (1997: Box 1.1, p. 3)

---

other approaches. Impact may be defined as the difference between a forecast of the future with a development occurring and a forecast without the development (see http://www.uky.edu/Agriculture/Sociology/nre350htm – accessed June 2005).

Identifying consequences of a proposed activity is common sense, rather than a revolutionary idea. However, for much of history it has not been adequately conducted in the planning and management approach adopted. Impact assessment has been evolving for over quarter of a century, but it is still imperfect and is often misapplied or misused. The field has been dominated by EIA; however, there are a number of approaches running parallel, and sometimes overlapping, with broadly similar goals, frequently exchanging information, techniques and methods. These include social impact assessment (SIA) (see below), hazard assessment, risk assessment, technology impact

Box 9.2

The figure below illustrates the typical step-wise EIA process

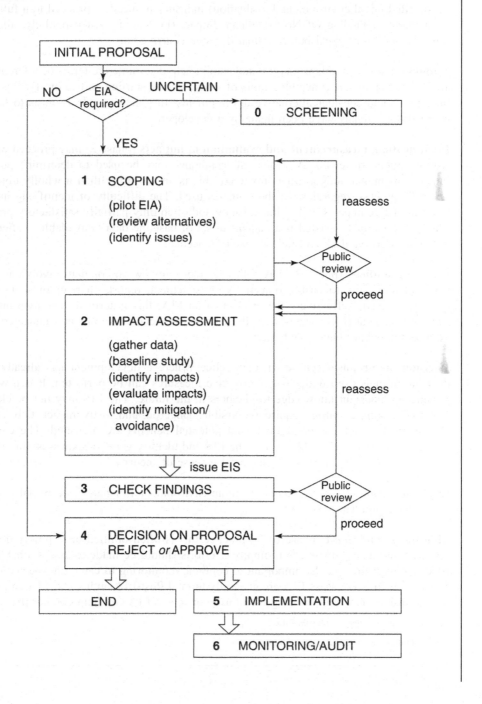

Screening (phase 0) is concerned with deciding which developments require an EIA. This should prevent unnecessary assessment, yet ensure that there is no escape when assessment is needed (in practice that is difficult). Screening may not be mandatory in some countries. Note that the term 'environmental assessment' is used for screening in the USA, but in the UK has been applied to EIA. In the USA if environmental assessment/screening (also called initial environmental evaluation) indicates no need to proceed to a full EIA, a statement of Finding Of No Significant Impact (FONSI) is issued publicly, allowing time for objection/appeal before a final decision is arrived at.

Scoping (phase 1) overlaps phase 0 and should help determine the terms of reference for an EIA, the approach, timetable, limits of study, tactics, staffing and so on. By this stage the EIA should consider alternative developments. In practice, a decision as to how to proceed may already have been made by a developer.

Identification, measurement and evaluation of impacts (phase 2) may proceed with or without public review(s). A variety of techniques may be used to determine possible impacts: as human judgement is involved, this is an art rather than a wholly objective scientific process, regardless of the statistics used. The difficulty of identifying indirect and cumulative impacts makes this a tricky and often only partially satisfactory process. This phase is much assisted if an adequate set of baseline data is available – often it is not, and extensive desk and field research is needed.

Checking findings (phase 3) may follow a public review and/or may involve an independent third party to ensure objectivity. A statement, report, chart or presentation is usually released – effectively the product of an EIA; this is termed the Environmental Impact Statement (EIS) and is what the decision makers, environmental managers (and perhaps the public) have to interpret.

Decision on proposal (phase 4): in practice, where a development has already been decided on or is even under way, corrective measures can be perfected. It is a way of passing on hindsight knowledge to planners in the future. The EIS may not be clear or easy to use: some countries require irreversible, dangerous and costly impacts to be clearly shown. It also useful if alternatives and potential benefits are indicated. The environmental manager must be able to read the EIS and identify gaps, weaknesses, limitations. An EIA must not be allowed to give a false sense of security.

Implementation (phase 5): this is where an environmental manager is especially active. Unexpected problems may arise.

Monitoring and audit (phase 6): in practice this is often omitted or is poorly done. If planning and management are to improve, efforts should be made to assess whether the EIA worked well. It is also important to continue monitoring to catch unexpected developments. Efforts to assess EIA are generally termed Post-EIA audits. An EIA can easily be a snapshot view, and ongoing monitoring or repeat EIA can help counter that.

Note: the idealised steps or phases 0 to 6.

Source: Redrawn from various sources by the author

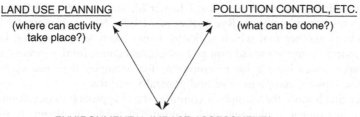

LAND USE PLANNING
(where can activity
take place?)

POLLUTION CONTROL, ETC.
(what can be done?)

ENVIRONMENTAL IMPACT ASSESSMENT/
SOCIAL IMPACT ASSESSMENT/
COST–BENEFIT ANALYSIS
(what are the effects/
who feels the effects/
what are the alternatives?)

Figure 9.1 How impact
assessment fits into
planning

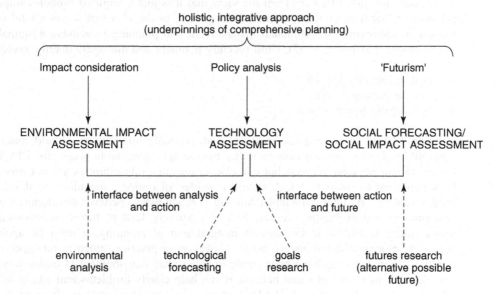

holistic, integrative approach
(underpinnings of comprehensive planning)

Impact consideration

Policy analysis

'Futurism'

ENVIRONMENTAL IMPACT
ASSESSMENT

TECHNOLOGY
ASSESSMENT

SOCIAL FORECASTING/
SOCIAL IMPACT ASSESSMENT

interface between analysis
and action

interface between action
and future

environmental
analysis

technological
forecasting

goals
research

futures research
(alternative possible
future)

Figure 9.2 Relationship of environmental impact assessment (EIA), technology assessment,
social forecasting and social impact assessment (SIA)

Source: Adapted from Vlachos (1985: 54, unnumbered figure)

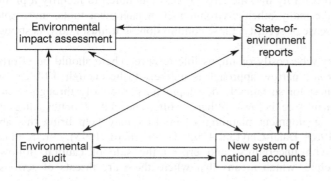

Environmental
impact assessment

State-of-
environment
reports

Environmental
audit

New system of
national accounts

Figure 9.3 Relationships,
possible exchanges of
information and
methodologies for
environmental impact
assessments, environmental
audits, new systems of
national accounts and state-
of-the-environment reports

Source: Thompson and Wilson
(1994: 612, Fig. 5)

assessment, eco-auditing and CBA (see also Chapter 8), and a range of forecasting or futures scenario-prediction methods (Ryecroft *et al.*, 1988). Net environmental benefit analysis (NEBA) is another tool which seeks to assess likely environmental impacts from a development (http://www.esd.ornl.gov/programs/ecorisk.html – accessed May 2005). These approaches have a lot in common: for example, they are systematic, focused, interdisciplinary, comprehensive and generally iterative.

EIA can offer much more than simply a common-sense approach to development: it can be a policy instrument, a planning tool, a means of public involvement and part of a pre-emptive framework crucial to environmental management and the drive for sustainable development. Some view impact assessment as a philosophy rather than just a technique; M. Smith (1993: 12) argued that it should be treated as 'a bridge that integrates the science of environmental analysis with the policies of resource management'. EIA was established in the USA in 1969, and by 1980 had spread widely. The 1992 Earth Summit stressed the value of impact assessment – seventeen of its twenty-seven principal declarations deal with some aspect of EIA.

Attitudes towards EIA vary from the view that it is just a required rubber-stamping activity, or that it determines optimal development, to the idea that it has a vital role to play in improving environmental management and planning to achieve sustainable development (Lawrence, 1997). Until recently planners and managers mainly asked:

- Is it technically feasible?
- Is it financially viable?
- Is it legally permissible?

Using EIA to consider goals, realities and available alternatives, should make it possible to identify the *best options* rather than simply acceptable proposals. EIA has tended to flag negative impacts but can also ensure that opportunities are not missed. It is important to stress that EIA should consider all options, including no development/no change. By improving understanding of relationships between development and environment and prompting studies, EIA can actively lead to better environmental management. If EIA is to become an integral part of planning, it must be applied before development decisions are made. However, in practice, much is retrospective, initialled after decisions have been made or even after development is under way or completed. This is still of value because it can help clarify problems and add to hindsight knowledge. Nevertheless, if EIA is done after key decisions have been made, it is unlikely to be able to force a change of plan to less damaging options. At worst it may simply be cosmetic – done to try to reduce opposition, with the results ignored, side-stepped and not enforced. In addition, while not a blatant cosmetic exercise, EIA is frequently an inflexible and devalued part of a development legitimisation process, added to which is the possibility that the EIA process has failed to identify a problem or opportunity. EIA, like many other environmental management tools, must not be allowed to give a false sense of security; those commissioning it must know its strengths and weaknesses.

Damaging impacts may be costly or impossible to cure. There should be efforts to avoid them, and an *ad hoc* narrow approach is unlikely to be enough. EIA can be a powerful tool in the quest for sustainable development, particularly through strategic environmental assessment (see below). What is often needed is better integration of environmental and development planning – and EIA may help bring this about (Jacobs and Saddler, 1989; Dalal-Clayton, 1992; D. Pritchard, 1993). Environmental managers must cope with uncertainty, and err on the side of caution, following the precautionary principle, which means that where there are threats of serious or irreversible environmental changes, lack of full scientific certainty should not be used as a reason for postponing measures to prevent environmental degradation (Dovers and Handmer, 1995: 92).

EIA should be more widely used early in planning and needs to be improved to consider more effectively indirect and cumulative impacts (Gardiner, 1989; Jacobs and Sadler, 1989; Anon., 1990; Jenkins, 1991; Wallington et al., 1994). A cumulative impact is the consequence of more than one direct or indirect impact acting together. Such impacts can be very difficult to predict. Thus a cumulative impact can be indirect, and often so far down the chain of causation that it manifests in unexpected places and possibly after considerable delay – and may be difficult to predict. Predicting direct (first-order) impacts is relatively easy – simple cause–effect relationships, second-, third- or higher order indirect impacts can be explored with some techniques, but it is likely to be expensive and slower than identifying first order. It is rather like driving fast with a choice of dim broad dip-beam headlights or very narrow-beam, not necessarily aimed where it is needed. Chemical and biological time bombs are forms of cumulative impact: a chemical accumulates, or a biological process continues, without causing a problem, perhaps hidden until a threshold is suddenly exceeded, either through continued accumulation or activity, or because some environmental or socio-economic change(s) triggers it (Stigliani et al., 1991). For example, pesticide gradually accumulating in the soil may suddenly be flushed out when acid deposition brings soil chemistry to a threshold; another example might be the gradual accumulation of a chemical in the environment, which suddenly reaches a threshold where it triggers infertility or gender distortions in a species. One vital role of environmental management is to recognise threats and warn if thresholds are approached. Cumulative effects assessment is also examined towards the end of this chapter.

Pre-development enquiries and evaluation are not new; for example, the *Report of the Volta Preparatory Commission* (HMSO, 1956a, 1956b) was proactive but it was not conducted in as systematic a way as EIA. Countries such as the UK and France have made use of commissions of inquiry to assess impacts and, to some extent, to keep the public informed since the sixteenth century or earlier. However, these generally took a great deal of time to deliver results, and were applied in an *ad hoc* manner only to some developments, in response to ministerial or popular concern. Their involvement of the public is in a very limited or controlled way, and they are not as systematic as EIA; an example of a UK commission of inquiry which conducted public meetings is the early 1970s Roskill Commission on the Third London Airport (HMSO, 1971). McHarg's (1969) *Design with Nature* stressed the value of anticipatory and systematic consideration of environmental limits, development impacts and alternatives, and is commonly seen as a forerunner of EIA. White also came close to proposing EIA in the 1960s (White, 1968). The first true EIA was probably in 1967, investigating copper mining in Puerto Rico (Mayda, 1993; Gilpin, 1995: 115). EIA has evolved from land-use planning, CBA, multiple objective analysis, modelling and simulation, and was established by the 1969 USA National Environmental Policy Act (NEPA) (Flamm, 1973; Ditton and Goodale, 1974: 145–151) (Box 9.3). NEPA has not been the only EIA initiative in the USA – sixteen states, plus the District of Columbia, had passed similar laws by 1991, but it was the most influential (Canter, 1996: 20).

What was new about EIA when it spread after 1970 was its systematic assessment and presentation of predicted impacts, available alternatives and mitigation possibilities. EIA evolved in an era dominated by a technocratic perspective on problem solving and with an emphasis on biophysical impacts. This may help explain why SIA has received less support, although activities similar to SIA pre-date EIA. NEPA was improved in 1971, 1973, 1976, 1977, and in 1978 the Council on Environmental Quality

Box 9.3

The 1969 US National Environmental Policy Act

President Theodore Roosevelt called for foresight in respect to pollution control during his 1908 Conference on Conservation, but it was not until 1969 that effective legislation was enacted. Preparations leading to the National Environmental Policy Act (NEPA) began in the early 1960s, when the need was perceived for the USA to have a declaration of national environmental policy and an action-forcing provision (Ditton and Goodale, 1974; Canter, 1996: 1–35). Before NEPA the USA had little effective federal control over the environment and lacked land-use regulations which countries such as the UK or France had (Wood, 1995: 16). NEPA became law on 1 January 1970. It was designed to reform federal policy making with the intent to influence the private sector – the hope being to transform and reorientate values (Heer and Hagerty, 1977; Caldwell, 1989). Originally it was intended that NEPA would change the nature of federal decision making. However, over the years it has become more of a procedural requirement (Wood, 1995: 75).

NEPA required an EIA prior to approval of federally funded projects that 'significantly' affected the environment – a message to federal officials to 'look before they leap' (Cheremisinoff and Morresi, 1977). NEPA Section 101 set regulations to protect the environment, Section 102(2) (c) ensured that they were pursued, and Section 103 included provision for inadequate EIA statements to be challenged in court (Wathern, 1988: 24; Hildebrand and Cannon, 1993). US federal agencies are required to prepare an environmental impact statement (EIS) (bearing the costs against taxes and sending copies to federal and state agencies and to the public) using EIA, prior to taking action (for a list of the federal agencies involved see Corwin *et al.*, 1975: 41). There was also some need to clarify what developments required assessment and how it was to be conducted.

There were three main elements in NEPA:

1   NEPA announced a US national policy for the environment.
2   It outlined procedures for achieving the objectives of that policy.
3   Provision was made for initiating the establishment of a US Council on Environmental Quality (CEQ) which was to advise the US President on the environment, review the EIA process, review draft EISs and see that NEPA was followed (i.e. recommendations and co-ordination). The CEQ effectively administers EIA legislation in the USA and issues the regulations which ensure that effective EISs are produced.

Also in 1970 the US government created the US Environmental Protection Agency (EPA), its role being to co-ordinate the attack on environmental pollution and to be responsible for the EIA process (the EPA is in effect 'overseer' of impact assessment in the USA).

NEPA created a more systematic, product-driven process of environmentally informed decision making. This was the first time US law had really allowed for development to be delayed or abandoned for the long-term good of the environment. Efforts were also made to co-ordinate public, state, federal and local activities. Overall, it was a revolution in values in a country where state intrusion was anathema – for this reason many see it as a sort of Magna Carta, although it stopped short of making a healthy environment a constitutional right, and some have been seeking to change that (Yost, 1990).

NEPA is statutory law (i.e. it was written after deliberation and did not evolve from custom, practice or tradition). Consequently, like a charter, it was not perfect.

By the late 1980s some of the initial weaknesses had been overcome and at least thirty other countries had adopted similar procedures (Manheim, 1994). Bodies like the Canadian Environmental Assessment Research Council and the International Association for Impact Assessment deserve credit for spreading and developing impact assessment. The results have been mixed: in some countries satisfactory, in others the NEPA approach is socio-economically and culturally inappropriate and needs further adaptation.

(CEQ) was established to give NEPA regulations more force. Although it had issued EIA guidelines in 1974, the US Agency for International Development (USAID) failed to apply them strongly enough. In 1975 a US public-interest group sued to force it to prepare environmental impact statements (EISs – the end-product of an EIA presenting the results) on its grants and loans. Consequently, by 1976 USAID and other bodies, notably lending banks and the USA State Department, were applying EIA to overseas investments and aid. In 1979 the Foreign Assistance Act effectively extended NEPA to the USA's foreign activities. By the 1980s there were calls for further reform, including better incorporation of social impacts into EIA procedures, and for NEPA to be more strongly written into the US Constitution (something yet to happen) (Renwick, 1988). Attempts to amend NEPA in 1990 to increase its overseas application to global change, biodiversity loss and transboundary pollution has had limited effect. Since the late 1980s, at least in Western countries, EIA has tended to become more participatory, making efforts to keep the public informed.

By 1995, about half the world's governments required EIA in some form (Robinson, 1992). Adoption has usually involved modification of techniques and procedures, because US experience may not be sufficiently relevant, and approaches and techniques are constantly evolving. The quality of impact assessment varies greatly; between 1970 and 1980 there was a tendency to develop technocratic and often not very transparent methods – perhaps seeking to appear 'scientific' and 'reputable'. Often inexperienced staff conducted the assessments poorly. After the late 1970s techniques came under academic scrutiny and have been generally improved; currently there is a move to improve techniques, and the conduct and qualifications of those conducting EIA as accreditation gradually enforce standards. So far, the greatest progress has probably been made in Australia, Canada, the USA, The Netherlands, Sweden and Norway (Coenen, 1993; McCormick, 1993; Prasad, 1993).

There are a number of ways in which EIA can be adopted:

● Adapt existing planning procedures to incorporate it (as in Germany, Sweden, Denmark and the UK).
● Create impact assessment legislation, like the USA, Australia and Canada.
● Develop global impact assessment regulations and supportive institutions.

EIA has mainly been applied at project, and to a lesser extent programme, levels. There is growing interest in extending assessment to a higher level: multi-project, programme and policy applications at national, transboundary and even global level.

Even though there were few binding agreements reached at Rio in 1992, nor much funding made available, the Earth Summit made clear that global and transboundary impact assessment were important. So far, attention has mainly focused on predicting impacts of global warming, ozone damage, world trade developments and structural adjustment policies. Less attention has been given to the impacts of soil degradation, ocean and atmospheric pollution, and loss of biodiversity, though these are very real threats (Barrow, 1997: 172–225). In 1991 the UN Economic Commission for Europe (UNECE) launched the Convention on Environmental Impact Assessment in a Transboundary Context at Espoo, Finland (signed by twenty-eight countries, including the USA and the European Community) (http://www.unece.org/env/eia/helptopics.htm; http://www.gdre.org/uem/eia/impactassess.html – both accessed June 2005).

This Espoo Convention was the first multilateral treaty on transboundary rights relating to *proposed activities*. The Convention provides for the notification of all affected parties likely to suffer an adverse transboundary impact from a proposed development. Signatories also undertake to give equal rights concerning impact assessment to all those affected by a development. Even if they are citizens of different countries, they can therefore be represented in the developer nation's public inquiries and so on. The EC Environmental Assessment Directive (85/337EEC of 1985) made similar but less wide-ranging provisions, which permit affected parties to participate in the developer's impact assessment if they so wish (Jorissen and Coenen, 1992). This directive goes beyond making provisions for project-level impact assessment to encourage programme- and policy-level assessments. This is effectively strategic environmental assessment (SEA) (see below) and offers a promising route for EIAs to be able to deal with transboundary impacts (Therevel *et al.*, 1992: 131).

Environmental managers should not be tempted to attempt their own EIA (or social impact assessment); there are plenty of handbooks, website guides and expert systems, but there is no adequate substitute for experience and specialist skills. However, it is important to be familiar with the strengths and weaknesses of impact assessment, to be in a position to commission subcontractors, to understand the findings, and to spot mistakes and abuse. There are many introductory sources (see e.g. the IAIA website; Gilpin, 1995). There is also a broad, readily accessible literature on specific applications of EIA, such as its application to developing countries, transport planning, health and so on.

EIA and SIA are sometimes applied to healthcare, and there is some overlap with health risk assessment (see below). Health impact assessment seeks to predict the impacts of a development (physical or social which have occurred or which are proposed) on health, seeking to establish opportunities and predict problems. This may allow the prediction of healthcare needs or help those seeking to improve health and safety measures. There is also interest in establishing the effect of disease on society, employment, taxation and agriculture; for example, establishing the impacts of HIV/AIDS transmission on farming. Health impact assessment information may be found by visiting http://www.hda-online.org.uk/downloads/pdfs.hia_review.pdf – accessed May 2005.

Social impact assessment

Social impact assessment (SIA) seeks to assess whether a proposed development alters quality of life and sense of well-being, and how well individuals, groups and communities adapt to change caused by development. SIA should also identify how people can contribute and show what they need (for an introduction and bibliographies see Vanclay and Bronstein, 1995; Barrow, 1997: 226–259; Goldman, 2000; Becker and Vanclay,

2003; http://www.iaia.org/Databases/SIA_Database/SIA_interface.asp; http://www.gsa.ene.com/factsheet/1098b/10_98b_1.htm; http://www.utas.edu.au/ruralcommunities/social-impact-assessment.html – all accessed May 2005). EIA and SIA deal with opposite ends of the same spectrum and blend together. There is also overlap with cultural impact assessment, concerned with effects on and influence of archaeological remains, holy places, traditions and so on. Freudenburg (1986: 452) saw mainstream SIA as part social science, part policy making, and part environmental sociology. SIA often uses qualitative data and may deal with more intangibles than EIA, and it has attracted the criticism that it is 'soft' and imprecise. Yet qualitative data, provided they are objectively gathered, can be as valuable for many purposes. Some of the issues SIA deals with are difficult to quantify; for example, sense of belonging, community cohesion (maintenance of functional and effective ties between a group), lifestyle, feelings of security, local pride, willingness to innovate, perception of threats and opportunities. However, these are things an environmental manager needs to know about.

According to Burdge and Vanclay (1996: 59), social impacts are alterations in the ways in which people live, work, play, relate to one another, organise to meet their needs, and generally cope as members of a society (and involve lifestyle, community cohesion, mental health and so on); while cultural impacts involve changes to the norms, values and beliefs of individuals that guide and rationalise their cognition of themselves and their society. SIA and cultural impact assessment consider how a proposed or actual activity affects way of life and attitudes. One may argue that socio-economic and biophysical aspects of the environment are so interconnected that impact assessment should not treat them separately. This is not a universally held view, and such a holistic total impact assessment is more of a goal than a reality. Yet there is often no distinct division between the EIA and SIA.

Social scientists and social historians were studying social impacts long before EIA and SIA appeared, but the emphasis was almost always on retrospective analysis. It is the focus on prediction, planning and decision making that separates SIA from other fields of social research, which tend to concentrate on causal analysis. Some claim that a French *philosophe*, Condorcet, carried out an SIA for a proposed canal in the late eighteenth century, but the first use of the term 'SIA' was probably in 1973 in connection with the impact of the Trans-Alaska Pipeline on the Inuit People (Burdge, 1994). In general SIA has remained underfunded and neglected compared with EIA, although attention in the USA increased following the CEQ's 1978 requirement that NEPA direct more attention to assessing socio-economic as well as physical impacts. Various disasters and legal actions by indigenous people (e.g. the Berger Inquiry in Canada in the early 1980s) around the world prompted the demand for SIA. The Three Mile Island incident (a US nuclear facility which suffered a near-meltdown that necessitated evacuation of householders in 1983) is seen by many as a landmark event because it was forced to use SIA to assess threats and public fears before restarting the reactor (Moss and Stills, 1981; Freudenburg, 1986: 454; Llewellyn and Freudenburg, 1989). The US Federal Highways Administration and the US Army Corps of Engineers have been active in developing SIA, mainly in relation to road developments; there has also been considerable activity in New Zealand from the early 1970s – prompted by the Environmental Protection and Enhancement Procedures (1973), the Town and Country Planning Act (1977) and the Resource Management Bill (1989). New Zealand had a Social Impact Assessment Working Group, established to develop and promote SIA by 1984 (and in 1990 an SIA Association was formed). It is probably fair to say that until the late 1980s there had been less interest in SIA in Europe than in the USA or Canada; for example, the physical effects of the Chernobyl disaster received attention but, apart from health impacts, the socio-economic effects received much less.

Methods and techniques used by SIA originate from a wide range of disciplines: social welfare, sociology, behavioural geography, social psychology, social anthropology and so on. This diversity, the complexity of SIA and the relative lack of funding have resulted in its becoming less standardised than EIA and it has spread more slowly. In addition to claiming that it is imprecise, critics of SIA argue that it is too theoretical; too descriptive (rather than analytical and explanatory); weak at prediction; *ad hoc*; mainly applied at the local scale; and likely to delay development (causing 'paralysis by analysis'). Another criticism is that few of the theories it uses are tightly defined so it is difficult to make comparisons between successive studies.

SIA, even with its imperfections, can help ensure that projects, programmes and policies generate fewer socio-economic problems. It can guide the management of social change in advance of the implementation of proposed developments, and has the potential to bring together various disciplines and types of decision maker (Soderstrom, 1981: v). The socio-economic (human) component of the environment differs from the biophysical in that it can react in anticipation of change; it can also be adapted if an adequate planning process is in place. It is also different in that reactions can be more fickle, because individuals or groups in a population are more often than not inconstant in response. There may also be a difference in timing as well as degree of impact on various sections of society, some of which may be especially vulnerable. For example, property owners will probably react differently from non-property owners. As with EIA, different socio-economic or socio-cultural impacts may be generated at various stages in a policy, programme or project cycle; for example, during construction, when the facility is functioning, and after it is closed down – too narrow a temporal focus and SIA may miss impacts (Gramling and Freudenburg, 1992). Spatially it is also important to adopt a wide enough view, as social impacts may be felt at the individual, family, community, regional, national or international level (or more than one level), not necessarily at the same time. Like EIA, SIA has been applied more at project rather than programme, plan or policy level. The crucial point is that SIA, like EIA, should identify undesirable and irreversible impacts.

Tools and methods used by SIA include: social surveys; questionnaires; interviews; use of available statistics such as census data, nutritional status data and findings from public hearings; operations research; systems analysis; social cost–benefit analysis; the Delphi technique (see below); marketing and consumer information; field research by social scientists, and so on. Behavioural psychologists are often involved in SIA to ascertain such issues as likely reactions, whether stress has been or will be suffered, what constitutes a sense of well-being and so on. The SIA equivalent of an EIA baseline study is the preparation of a social profile to establish what might be changed and what would probably happen if no development took place. Field research techniques can be divided into direct and indirect. Direct observation of human behaviour may be open or discreet (an example of the latter is the use of street videocameras), conducted during normal times or times of stress. Indirect observation includes study of: changes in social indicators, patterns of trampling, telephone enquiries directed at selected members of the public, historical records, property prices, suicide rates and so on. Communities are a unit which can be monitored for changes using demographic, employment and human well-being data, so SIA often adopts a community focus. Alternatively, especially when aid donors commission an SIA, the focus is on target groups, typically the people(s) investment is supposed to help. An issues-oriented approach is another possibility, or a regional approach, or it is possible to make use of rapid rural appraisal and participatory rural appraisal methods (Gow, 1990). Given the complexity of identifying and assessing direct socio-economic impacts, it is not surprising that much less progress has been made with cumulative impact assessment than is the case with EIA.

There are many variables of interest to SIA, including:

- assessment of who benefits and who suffers – locals, region, developer, urban elites, multinational company shareholders;
- assessment of the consequences of development actions on community structure, institutions, infrastructure;
- prediction of changes in behaviour of the various groups in a society or societies to be affected;
- prediction of changes in established social control mechanisms;
- prediction of alterations in behaviour, attitude, local norms and values, equity, psychological environment, social processes, activities;
- assessment of demographic impacts;
- assessment of whether there will be reduced or enhanced employment and other opportunities;
- prediction of alterations in mutual support patterns (coping strategies);
- assessment of mental and physical health impacts;
- gender impact assessment – a process that seeks to establish what effect development will have on gender relations in society.

The quest for sustainable development involves trade-offs that have adverse social and economic impacts; so it is desirable that these are forecast and avoided. It is also vital to assess whether there are any social institutions or movements which could support or hinder sustainable development. Without supportive social institutions, sustainable development will probably fail. SIA can help develop these (Ruivenkamp, 1987; Hindmarsh, 1990).

Sometimes a multidisciplinary team deals with both EIA and SIA, or there may be separate specialists, or SIA is a modest sub-component of EIA or environmental auditing. Whichever is selected, SIA should be conducted by competent, professional social scientists familiar with the local people. SIA has been most applied to road construction; boom towns; indigenous peoples impacted by development; large projects; land-use decision making; voluntary relocatees or refugees; and tourism development. In some countries SIA is required before gambling or drinking licences are granted (http://www.dgr.nsw.gov.au/html/GAMING/sia.html – accessed June 2005). Environmental management should make more use of SIA than is currently the case.

Other tools for assessing the potential for development and impacts of development

Ecological impact assessment considers how organisms, rather than people, will be affected by activities (Westman, 1985: 86; Duinker, 1989; *Guidelines for Ecological Impact Assessment*, from http://www.ieem.co.uk – accessed July 2005). Recently the expression has been applied to the description and evaluation of the ecological baseline used by EIA. More accurately, ecological impact assessment is concerned with establishing the state of the environment, whereas EIA focuses on predicted and actual effects of change. Treweek (1995a, 1995b) has reviewed ecological impact assessment and reported it was a valuable support for EIA. An aspect of ecological impact assessment which is growing in importance is its application to biodiversity loss (Hirsch, 1993). Ecological impact assessment may rely on selected ecosystem components as indicators or on ecosystem modelling. Ecosystem function can be complex and is often poorly understood, making accurate assessment difficult.

Habitat evaluation seeks to assess the suitability of an ecosystem for a species or the impact of development on a habitat (Suter, 1993: 8). There may be more than one habitat affected by a development, in which case each is dealt with separately. This approach has been used by the US Fish and Wildlife Service in assessments of the impacts of US federal water resource development projects, and by the US Army Corps of Engineers (Canter, 1996: 390).

Land-use planning is a process which may operate at local, regional or national scale; land capability assessment, land appraisal, land evaluation, land suitability assessment and terrain evaluation feed into that process. A land-use survey indicates the situation at the time of study, and is not the same as a capability classification, which looks to the future. There are various approaches and methods for land capability classification (e.g. the Ecological Series Classification or the Holdridge Life Zones System). Often the land-use planning approach adopted depends on a country's politics. It is widely felt that land-use planning is a valuable ingredient of EIA and in the quest for sustainable development, and that EIA can feed into land-use planning. In practice the two are often poorly integrated.

Land capability assessment, land evaluation and land appraisal generally follow a proactive approach similar to that of EIA (scoping, data collection, evaluation, presentation of decision) in the production of a land capability classification or land evaluation (Beek, 1978; Patricos, 1986). Some approaches consider a range of factors which might include the concept of carrying capacity, others just soil characteristics and slope. The end-product is a description of landscape units in terms of inherent capacity to produce a combination of plants, animals and so on; it is also likely to reflect government development goals, market opportunities, labour availability and public demands (e.g. terraced agriculture may be possible but labour is not available).

Simple inventories of land use and, to a limited extent, capability were made in medieval times – notably the Domesday Book. Modern land capability classification was developed by the US Soil Conservation Service in the 1930s in response to problems such as the US Dust Bowl. Linked to consideration of conservation and development, land capability classification can lead to a land suitability assessment (a rating of landscape units showing what development they might best support). Land suitability assessment may depend on overlay maps of various landscape or development attributes, or direct field observation of clues (something local people may traditionally do) (e.g. to seek distinctive plants indicative of good soil). Geographical information systems (GIS) and remote sensing are increasingly applied to land capability assessment.

The Universal Soil Loss Equation (USLE) is a predictive tool, which uses data on a wide range of parameters to estimate and predict average annual soil loss. It was developed in the 1930s by the US Soil Conservation Service, and was improved in 1954 and again in 1978 by the US Department of Agriculture. It is widely used by planners and consultants to check on existing and likely future soil loss, and to select appropriate agricultural practices and crops to sustain production. Developed in midwestern USA, it has been modified to make it suitable for other environments, so there are numerous revised versions (Hudson, 1981: 258). The USLE should be used with caution: problems arise when data are imprecise or unavailable, and it is best applied in situations where water rather than wind erosion occurs (although there are modified versions intended to cope with wind erosion). A typical form of the USLE is:

$$A = (0.224) \, RKLSCP$$

where:

A = soil loss;

R = rainfall erosivity factor (degree to which rainfall can erode soil);

K = soil erodability factor (soil vulnerability to erosion);

L = slope length factor;

S = slope gradient factor;

C = cropping management factor (what is grown and how);

P = erosion control practice factor.

The agroecosystem zones concept was promoted by the UN Food and Agricultural Organisation (FAO, 1978) to provide a framework for considering a range of parameters over a limited planning term with the aim of promoting sustainable development. An agroecosystem is an ecological system modified by humans to produce food or commodities, which generally means a reduction in diversity of wildlife. Agroecosystem assessment (or analysis) evolved in Thailand and attempts rapid multidisciplinary diagnosis that includes ecological and socio-economic concepts and parameters (Conway and Barbier, 1990: 162–193). It considers not only the farming system but also household characteristics, and regional, national, even global factors likely to affect the local community. The area under consideration is zoned – often making use of a land-use survey or land capability assessment. Agroecosystem assessment needs to be approached with some caution because it can lead to over-simple interpretation.

Farming systems research (FSR) is an open-ended, iterative, multidisciplinary, holistic, continuous, farmer-centred, dynamic process applied to agricultural research and development (it considers biophysical, social and economic factors, and seeks to integrate their study) (Shaner et al., 1982; Brush, 1986: 221). There is no single method but all approaches share five basic steps (Maxwell, 1986):

1 Classification – the identification of homogeneous groups ('target groups') of farmers.
2 Diagnosis – identification of limiting factors, opportunities, threats and so on for the target group.
3 Generation of recommendations – which may require field experiments, pilot studies and/or research station work.
4 Implementation – usually working with an agricultural extension service.
5 Evaluation – which may lead to revision of what is being done.

FSR is a systems approach applied to on-farm research, and is promoted as a way of increasing farmer participation in development, and of generating improved and appropriate approaches and technology. FSR includes study of factors which may be beyond the control of the farming community – world trade issues, global warming and so on. Unless some 'off-the-shelf' input is available, FSR usually takes time – often two years, sometimes from five to fifteen years or more.

There is considerable overlap between agroecosystem assessment, FSR and participatory assessment approaches, including rapid rural appraisal (see Chapter 8). These tools can assess the current situation and assist in the prediction of future opportunities and difficulties. PRA seeks rapport with and participation by those surveyed, and shares information with them. RRA simply aims to extract information as fast as possible (and may miss things by not liasing closely). Participatory assessment and monitoring may be defined as qualitative research or survey work, which seeks to gain an in-depth understanding of a community or situation (Yar, 1990). Rapid rural appraisal (RRA) is a family of approaches, focused mainly on land capability assessment, which seek to incorporate (or involve) local people in the process and to reduce the time and costs of preparation. It is a systematic, semi-structured activity carried out in the field by a multidisciplinary team and designed to quickly acquire new information on, and new hypotheses about, rural life. RRA has evolved rapidly since the late 1970s and there is no single standardised methodology – for an introduction see *Agricultural Administration* vol. 8(6), special issue (1981); *IDS Bulletin* vol. 12(4), special issue (1991); Conway and McCracken, 1990; Chambers, 1992). A central thesis of RRA is 'optimal ignorance', the idea that the amount of information required should be kept to the necessary minimum (something some EIA practitioners should also bear in mind). Another central thesis is 'diversity of analysis' – the use of different sources of data or means of data gathering, and a range of experts, if possible, familiar with every aspect of rural life.

RRA, according to Conway and Barbier (1990: 177–178), is: iterative (i.e. processes and goals are not fixed and can be modified as an exercise progresses); innovative (it is adapted to suit needs); interactive (team members work to gain interdisciplinary insight); informal (it often relies on unstructured interviews); and assessors should be in contact with the community. RRA can be of variable character: (1) exploratory – like agroecosystem analysis, it seeks information on a rural issue or agroecosystem; (2) topical – with a specific output expected, often a hypothesis that can be a basis for research or development.

Participatory rural appraisal (PRA) approaches seek to enable local people to share, enhance and assess their knowledge of life and conditions, to plan and to act. PRA differs from RRA, in that the latter extracts information, whereas the former shares it and seeks rapport. Multidisciplinary team studies and a stress on participatory public involvement also offer possibilities for better conduct of EIA. However, there has been a tendency to emphasise the strengths of RRA and PRA and to understate the problems which might be encountered. Speedy data collection is also needed for urban environments, and rapid urban environmental assessment has been developed to provide it; for a review see Leitmann (1993). Given the tremendous growth of cities, and the human and environmental problems that might arise, there has been relatively slow development of these techniques and tools.

Livelihoods assessment

The assessment of livelihoods is often a key element in improving environmental care, fighting poverty and reducing people's vulnerability. The focus has mostly been on rural livelihoods, although there is some interest in the urban poor. Assessment seeks to uncover how groups make a living, ways in which they can be disrupted, and the potential for improving livelihoods and making them more secure. The environmental manager is also interested in assessing how people regard and interact with their environment, and as livelihood and environmental quality generally interrelate, this is important (DFID sustainable rural livelihoods guidance sheets – http://www.livelihoods.org; UNDP livelihoods website http://www.undp.org/sl – both accessed June 2005). An element in the quest for sustainable development is almost always an understanding of livelihoods. Tools were first developed by social scientists and agricultural extension or healthcare staff, and seek to gather information on land use; tenure; access to crucial and useful inputs (credit, labour, seeds and so on); marketing of produce; and other relevant socio-economic and environmental issues (Carney, 1998; Scoones, 1998).

Vulnerability studies

There are signs of growing interest in predicting change in people's vulnerability to physical and socio-economic changes. Some of this interest has been generated by concern about global climate change and by eruptions (e.g. Martinique in the late 1990s), earthquakes, hurricanes, droughts and tsunamis during the past decade. The number of poor people is growing and poverty generally increases vulnerability. Urban growth and population increase mean that people are more and more concentrated, crowded and less able to evacuate or find food locally if normal conditions are disrupted. A number of key products are manufactured by a handful of factories which supply the world, and if these are disrupted the impact may be global – as was the case with microchips for some time following the Kyoto earthquake. A number of modern trends are making societies less adaptable and thus more vulnerable. A key element of any sustainable development strategy should be a thorough assessment of vulnerability together with ongoing monitoring. That should feed into efforts to reduce vulnerability and improve adaptability.

Predicting future scenarios

Forecasting is an essential part of planning and programme and policy formulation. There may also be ways in which prediction of future scenarios can help shift public opinion and prompt pre-emptive changes, as is the case with much of the current global warming debate. Forms of forecasting have been used by many peoples since prehistoric times to decide when, where and what to hunt, where to settle, to make agricultural decisions, embark on migrations or warfare and much more. Traditional forecasts may be good, but these techniques are frequently far from objective, and sometimes rely on soothsaying, magic and superstition. Since about the mid eighteenth century Western societies have based forecasting on rational observation, projection of trends and hindsight knowledge (Fortlage, 1990: 1). By the 1930s in Europe, the USSR and the USA post- and in-project assessments of development were being conducted, and cautionary guidebooks, checklists, procedural manuals and planning regulations (and, in the UK, occasional public inquiries) were in use to improve decision making (Caldwell, 1989).

The banking, investment and insurance industries had developed hazard and risk assessment methods by the 1940s, and military tacticians were trying to predict scenarios during the Second World War and the Cold War. Time-series data may be used as a basis for projections, and key indicators may be monitored for warning or modelled to predict. Models are used to understand complex situations and predict future scenarios for trend extrapolation and informed speculation. Models are also useful for assessing the impacts of a wide range of developments such as: altered land use, effluent discharges, global climatic change, modification of river channels, estuarine conditions, coastal erosion, agricultural chemicals impacts, acid deposition and so on. Models include physical models (e.g. laboratory tests, scale models of estuaries or catchments), statistical models (e.g. principal components analysis), computer models, systems models (for a review of ecosystem models used for environmental management see Jorgensen and Goda, 1986; for a handbook see Jorgensen et al., 1996).

Futures modelling, futures research and 'futurology' attracted attention in the early 1970s following the publication of The Limits to Growth (Meadows et al., 1972). A sequel reviewing how accurate the warnings had been appeared at the time of the 1992 Earth Summit (Meadows et al., 1992) and as recently as 2004. It should be stressed that futures research is often difficult, and frequently imprecise and unreliable. It has to allow for both gradual and sudden changes that are due to new inventions, attitude changes, environmental alterations and so on (Westman, 1985: 3). The further ahead one attempts to make predictions, the less accurate they are likely to be. The results of futures research are useful, but must be treated with caution.

The Delphi technique (see also Chapter 8) was developed by the RAND Corporation in the late 1940s to try to obtain a reliable consensus about future developments from multidisciplinary panels of experts (Stouth et al., 1993). These panel evaluations were used for Cold War purposes, and little was published for a decade until a report by Gordon and Helmer (1964). The approach is for expert assessors to be asked their views without communicating with each other. These are pooled, evaluated, and the assessors are allowed to see the result as a controlled feedback, and are given the chance to modify their opinions which are again fed back (a Gestalt approach); the feedback–pooled response process may be repeated three or more times to produce the final conclusions. The approach ensures anonymity for the assessors to prevent peer pressure or intimidation influencing results; and the controlled feedback helps to achieve a group viewpoint and an aggregate judgement.

The Delphi technique has been used in futures research on healthcare policy and innovations, gambling, tourism, marketing, management studies, resources allocation, technology innovation studies and war-games. The Environmental Evaluation Systems approach to EIA uses the Delphi technique; and the cross-impact matrix EIA approach has also been developed from it (Soderstrom, 1981: 20). It is useful for short-range and for longer range (over fifteen years into the future) forecasting, especially if high degrees of uncertainty are involved and where there is a need to predict impacts upon culture. The results are, of course, subjective and qualitative. Impact assessment asks what impacts may occur; the Delphi technique asks about the likelihood and date some impact will happen – it can thus complement impact assessment. It has become much easier to run with modern computers and may be done through a communications network such as the Internet, or even by mail without the need to gather expert assessors in one place. However, it can be slow.

Assessments of the technique suggest it is a valuable approach but one that has often been poorly applied. Careful selection of the experts is crucial to avoid gaps or bias, and it is also important to ensure that the questions they are asked are not too limited, or their expertise could be constrained and lost. Bias may be introduced if assessors are allowed to suggest other assessors.

Hazard and risk assessment

Environmental risk management demands accurate and up-to-date hazard and risk assessment. The importance of hazard assessment and risk assessment hardly needs stressing in the wake of accidents such as Seveso, Love Canal, Three Mile Island, Bhopal and Chernobyl, or natural disasters such as floods, earthquakes and the 2005 Indian Ocean tsunami. Until recently, one aspect of risk assessment, which has been neglected, is that posed by natural threats which have a long interval between occurrences. A number of very real hazards of huge magnitude have not occurred in historic times, and so attract little attention even if experts warn of them (Barrow, 2003). A hazard is a *perceived* event or source of danger which threatens life or property or both. A disaster is the realisation of a hazard; a catastrophe is a particularly serious disaster. Hazard assessment may be said to seek to recognise things which give rise to concern

(Clark *et al.*, 1984: 501). Hazard assessment tends to deal with natural hazards: flood, storm, tsunami, locust swarm and so forth. Human activities also pose threats (e.g. crime or technological innovation), and may initiate natural hazards and alter the vulnerability of the environment, wildlife or humans to them. Some people therefore divide hazards into natural, quasi-natural and anthropogenic.

One difficult problem faced by hazard or risk assessment, and to some extent EIA, is assessing what is 'acceptable'. Various groups, even within one society, may perceive and evaluate hazards and risks differently, and often vary in their vulnerability. The perception of risk is often not based on rational judgements: people have gut reactions to or dread of certain things and little fear of other, perhaps more real, threats. There are likely to be different risk perceptions from class to class, age group to age group, and for different religions and sexes: much depends on previous exposure or awareness through the media (Douglas and Wildavsky, 1990; Krimsky and Goulding, 1992). Perception also varies from individual to individual, and for any given group through time (e.g. a youthful person may be more tolerant of risk than someone older; and the poor face and have to accept more risks). In general, people are more concerned about the short term rather than the long term, and by 'concentrated' hazards – an air crash that kills 300 rather than the same number of fatalities from household accidents dispersed over a country, or in time. Some risks tend to attract more attention (e.g. radiation hazards compared with traffic accidents). If people think they are in control, as car drivers, for example, they are probably less worried than as passengers on a train; yet a motor accident is a much greater risk. Perception can be greatly affected by media and myth, and faced with a hazard or risk, people's responses are diverse.

Events such as the 2005 Indian Ocean tsunami, the ENSO events of the 1980s, recent bushfires, and increasing costs of hurricane and flood damage have prompted more willingness among politicians and citizens to consider proactive measures. Global warming impacts have rather dominated concern, especially the increased flood risk it may pose (Penning-Rowsell and Fordham, 1994). Some argue that there is a need to widen concern; for example, volcanic eruption risk and asteroid strikes pose a threat but attract little spending; and there is worryingly little concern for the relatively vulnerable world food security situation. More affluent countries have invested in earthquake and tsunami warning systems and have modified some land-use and building regulations to mitigate impacts (the US Geological Survey offers a seismic hazard map which may be checked using the ZIP code at http://www.eqhazmaps.usgs.gov/ – accessed June 2005). Some hazards become evident through research rather than disaster; for example, the problem of stratospheric CFCs and the global rise of CO_2 levels. A growing number of threats are human caused or human induced, so technology assessment is vital; these include: asbestos, fertility-damaging and gender-distorting pollutants, nuclear materials, bioengineering, and so on. Government agencies, international bodies and NGOs assess the safety of goods and services for consumers. Consumer protection in a globalising world must have a transnational overview. Employee health and safety must also look beyond national boundaries, and is monitored by governments, international agencies and NGOs.

Risk and hazard perception has generated a growing literature from behavioural psychologists, health and safety specialists, anthropologists and specialist risk or hazard assessors. The assessor can categorise hazard or risk according to such criteria as minor/severe; infrequent/frequent; localised/widespread, and may resort to estimating the value of a life to weigh against risk probability and risk avoidance costs (the Bhopal tragedy in India raised the question of higher life valuations awarded to citizens of rich nations). Involvement of the public in risk and hazard assessment can pose problems: predictions may involve companies that wish to avoid giving away the fruits of their experience or research, or a government that wants secrecy concerning strategic information or activity which they prefer the public or factions (e.g. terrorists) not to know about.

One can define risk as the expression of the chance or probability of a danger or hazard taking place, and risk assessment as going beyond predicting probability to identifying objectively the frequency, likelihood, causes, extent and severity of exposure of people or things or activities. Put simply, risk is 'probability x consequence' (Suter, 1993). Thus there is a threat of drowning involved in crossing the Atlantic, but the risk of doing it by row-boat is somewhat higher than going by commercial jet or cruise liner. Risk assessment (appraisal or analysis) is a loose term – it considers hazard and vulnerability: how people react to risk and their pattern of exposure. Risk assessment has been defined as 'the process of assigning magnitudes and probabilities to the adverse effects of human activities (including technical innovation) or natural catastrophes' (P. Pritchard, 1993). It involves identifying hazards; estimating the probability of their occurrence; evaluating the consequences; using these findings to assess risk; and presenting the conclusions, ideally with some indication of reliability of estimate. Hazard assessments and risk assessments are not precise arts: different assessments may assign different predictions to a risk, and some risks may be overlooked. In the end the environmental manager must exercise judgement.

Risk assessment may go on to identify coping strategies or establish what people will pay to avoid a risk. Some view risk appraisal as the assessment of communities' attitudes to risks. Risk assessment may be divided into that concerned with risks to the environment or biota and that concerned with risks to humans. Risk assessment typically consists of risk identification, risk estimation (establishment of character and levels) and risk evaluation (assessment of probability of occurrence, consequences, and so on (P. Pritchard, 1993). Risk assessment studies effects, pathways or factors involved (e.g. laboratory experiments into toxicity). Often, risk assessment involves weighing dangers against benefits (e.g. the threat of asbestos-related illness versus its value in protection against fire).

Risk assessment is an analytical tradition, not a legal definition, with centuries-old roots in the actuarial, investment and insurance professions, which has spread to engineering, development of new materials (especially chemicals, pharmaceuticals and biotechnology innovations), economics, healthcare and criminology. Risk assessment may also take the form of screening a new product or activity, to ensure that it is safe for user and environment, before releasing it for general use (i.e. laboratory or test-bed assessments). Insurance companies and bankers need to know risks before providing cover or loans. Administrators use risk assessment to reduce the likelihood that they could be accused of negligence if something goes wrong and for contingency planning. According to Suter (1993: 3), risk assessment can provide:

- a quantitative basis from which to compare and prioritise risks;
- a systematic means of improving understanding of risks;
- a means of making assessment more useful and credible by giving probabilities to predicted impacts.

Legislation such as the US Toxic Substances Control Act 1976 (which requires regulation if there is a risk to human health or environment through use or release of a harmful chemical or biological agent), or the UK Environment Act 1995 (which requires local authorities to carry out risk assessment and maintain registers of contaminated land) makes it increasingly important for the environmental manager to commission and interpret risk assessments (Asante-Duah, 1998). Separation of 'natural' from 'man-made', industrial or technical hazard assessment, and general risk assessment, is maintained by practitioners and literature rather than reflecting different concepts. Unlike EIA, risk assessment tends not to address development alternatives, and is at present less likely to be required by government policy or law (this is also true for technology risk assessment). Risk assessment is often better at estimating magnitude, certainty and timing of impacts than EIA. Risk assessment and hazard assessment are often applied where there is more uncertainty than that which EIA faces (Covello *et al.*, 1985: 16). EIA can increase planners' accountability to the public; risk assessment is likely to be more concerned with internal management or be applied by a regulatory agency.

Hazard and risk assessment usually use a template (to help order the process) to generate a statistical estimate of probability of occurrence of a certain level of impact (not a forecast but a statistical recurrence, e.g. a one in hundred-year chance of a serious flood). The probability estimate data may be used to produce a zoned map which can be used to determine land-use or building regulations, select transport routes or site infrastructure, prepare contingency or emergency procedures (e.g. provide hurricane shelters, tsunami protection walls and early-warning systems). Tsunami detection and warning systems, like the recently improved NOAA detection/warning arrangements, will probably spread to the Indian Ocean, Atlantic and other oceans, thanks to the 2005 disaster. Insurance companies often use risk assessment to determine premiums: for example, mapping physical risks and crime rates against postcodes/ZIP codes. Some threats appear suddenly, others creep up and may be obvious or easy to miss.

Well-developed areas of risk assessment include ecological risks, health risks, and technological and industrial risks. Like EIA, risk assessment is mainly applied at project level or to a particular process, although it is sometimes used at policy, plan and pro-gramme levels. Environmental risk assessment is a sub-field of risk assessment, which seeks to assess risks to the environment resulting from industrial activity and other devel-opments. Ecological risk assessment, another sub-field, seeks to define and quantify risks to non-human biota (i.e. to assess the likelihood of adverse change in an ecosystem as a result of human activity). The EPA has promoted it in the USA since 1990.

Technology assessment

Technology assessment (technical evaluation) explores the impacts of human techno-logical innovation; it seeks to establish whether equipment and techniques will work and what effect they may have. This may include assessment of use impacts to inform decision making and clarify problems and opportunities (*Impact Assessment Bulletin* vol. 53, special issue, 1987). Technology assessment follows a broadly parallel path to EIA, and may involve evaluation of indirect and cumulative impacts (Kates, 1978; Kates and Hohenemser, 1982). It involves systematic study of the effects on environment and society that occur when a technology is introduced, extended or modified.

Technology assessment was widespread in the USA by 1967, so it pre-dates EIA. In 1973 the US Congress created the Office of Technology Assessment to promote and oversee it. An International Society for Technology Assessment operated from the USA in the mid 1970s, developing into the International Association for Impact Assessment (IAIA), a body which promotes EIA, SIA, technology assessment, hazard assessment, risk assessment and related activities. The National Science Foundation in the USA also supports technology assessment, and Europe, Japan, Canada and Australia had estab-lished bodies to promote the field by the late 1980s. As well as having a warning function, technology assessment can, like EIA, aid decision making and planning in other ways. It may be initiated by government, international bodies, NGOs or the industries or agencies which plan to innovate.

Technology impacts can be a function of: technology failure; operator failure; poor maintenance; poor design; faulty installation; terrorism; natural or human accident; or adaptations prompted by the innovation. Not surprisingly, assessment practitioners are often engineers, so socio-economic issues may not be well covered. Thus the spread of mobile phones, which has had a huge socio-economic impact, has received little attention. The tendency has been to concentrate on morbidity and mortality – but there is now increasing interest in civil liberties and social aspects of technology innovation. Technology assessment has an important role to play in the quest for sustainable development, identifying threats and promising development paths.

Technology risk may be posed by a known potentially dangerous activity such as petrochemical processing or by new, untested technology, chemicals, biotechnology and pharmaceuticals (Ricci, 1981: 101). Technological innovation may relate to any aspect of life: attempts to improve agriculture, telecommunications, industry, transport and so on. Industrial hazard and risk assessment examines mainly established manufacturing practices, and is less likely to deal with unknowns arising from technical innovation than technology assessment proper. There is increasing interest in using technology assessment to 'tune' new technology, and it is being applied to biotechnology, including genetic engineering. Europe is applying it to long-term strategic policy making and as a means of early warning. There is a tendency for technological hazard to be 'exported' to countries where laws, monitoring and enforcement may be less stringent, and planners and regulators less well informed. Large sums of money may be involved in such exporting, making objective assessment a challenge. Awareness of the need for assessment has been enhanced by the activities of NGOs, professional bodies, trade unions and international agencies.

Health risk assessment

A rapidly expanding field is health risk assessment (Birley, 1995; BMA, 1998); it may involve assessing the risk posed by waste disposal, workplace environment, employment activities, noise, stress and so on. Alternatively, it may be much more health focused, and conducted by medical and related staff. With damage claims soaring in Western countries there is an incentive for pre-emptive assessment. Global warming fears have prompted attempts to assess future climate change-affected disease patterns; however, this is difficult because many factors are involved in transmission and in determining human or animal vulnerability. In 2005 fears of avian flu caused many governments and agencies such as the WHO to meet and exchange information, and to prepare contingency plans in advance of any pandemic. Some researchers were aware of the threat and had been studying the virus for decades.

Computers and expert systems

There have been attempts to computerise impact identification monitoring and assessment (see Geraghty, 1992; Guariso and Page, 1994; Benoît, 1995: 421–426). Canter (1996: 45) has argued that, as impact assessment gets more complex and laborious, in order to be more holistic and adaptive, computerisation becomes more important. Computer techniques have also been used for interpreting impacts (Baumwerd-Ahlmann *et al.*, 1991). The development of better microcomputers and software has made it possible to run impact assessments, expert systems, environmental information systems and models. There has been interest in integrating EIA, monitoring and GIS through computer use, and in applying computing to SIA (Leistritz *et al.*, 1995). Nevertheless, progress has still been limited by lack of user-friendly programs, and by the relatively low number of impact assessors who are skilled with computers (Guariso and Page, 1994). Once developed, software is often made available on shareware www sites.

The application of computing should be transparent, to reduce the risk of accidental or deliberate errors or unauthorised disclosure. Accidents such as the Chernobyl disaster have prompted a number of countries to co-operate and develop joint rapid impact assessment and data exchange systems. These are vital for coping with rapidly developing transboundary problems (e.g. airborne pollution). The European Community has gone partway to developing such a system for radioactive fallout by establishing the EC Urgent Radiological Information Exchange (ECURIE) in 1987.

Expert systems (or 'knowledge-based systems') can be valuable once perfected, as an aid (not replacement) for skilled assessors. However, they may take a lot of research and time to develop (Loehle and Osteen, 1990; Geraghty, 1992, 1993). They are particularly useful when there is a shortage of expertise to conduct assessment and may have potential for improving public involvement (Schibuola and Byer, 1991). They are also used for environmental planning (Wright *et al.*, 1993; Tucker and Richardson, 1995), eco-audit and environmental management (Benoît and Podesto, 1995), and to apply EIA to regional planning (Burde *et al.*, 1994). The approach involves developing a computer program that stores a body of knowledge and uses it to perform tasks usually done by a human expert (e.g. impact or risk assessment). These systems draw on heuristic (rule-of-thumb) reasoning to act as 'advisers', provide support for decision making, or aid data management.

Gray and Stokoe (1988) reviewed the potential and limitations of expert systems for impact assessment and environmental management, one of their hopes being that they could help achieve consistent quality of assessments. Mercer (1995), recognising that impact assessment increasingly uses qualitative methods of assessment, tried to develop an expert system capable of coping with this. For a bibliography on expert systems and EIA and an example of an expert system for EIA developed for application to water-related developments by the Mekong Secretariat, but adaptable to other situations see http://www.ess.co.at/EIA/ – accessed June 2005.

Adaptive environmental assessment and adaptive environmental assessment and management

Impact assessment generally adopts a static, 'snapshot' approach, but causal relationships are often not constant (e.g. monetary units may be devalued, the environment may alter, decision-making objectives change, attitudes of people shift); such an approach can therefore be ineffective. There is also a risk that a one-off impact assessment could discourage planners from adequate monitoring. The need is to ensure that assessment is continuous or repeated regularly (Holling, 1978; Gilmour and Walkerden, 1994). Two approaches have evolved: adaptive environmental assessment (AEA), and adaptive environmental assessment and management (AEAM). These are broader than mainstream EIA, and have a bias towards coping with uncertainty. In addition, AEAM seeks to integrate environmental, social and economic assessment with management. AEA has been used for managing the upper Mississippi River.

AEAM was pioneered by Holling and colleagues (Holling, 1978, 2005), Environment Canada, the University of British Colombia, Vancouver (Canada), and the Austrian-based International Institute for Applied Systems Analysis (IIASA). The Holling approach was applied to a mountain settlement in Austria, starting in 1974, by a UNESCO (Man and Biosphere Program)/IIASA/University of British Columbia team. It uses a series of carefully designed research periods followed by multidisciplinary modelling workshops which include science and social science experts, planners, managers, resource users and locals. The workshops develop alternative scenarios and management strategies, which are then compared to arrive at the best problem-solving approach. The workshops seek to ensure that the assessment team and participants continually review efforts to predict and model policy options for decision makers, and also provide a bridge for different disciplines and competing perceptions. The end result is a computer-based systems model that can be tested and tuned until it supports adaptive management and can help identify indirect impacts (Jones and Greig, 1985; http://www.geog.mcgill.ca/faculty/Peterson/susfut/AEAM/ – accessed June 2005).

AEAM can be useful where baseline data are poor. It also encourages and facilitates multidisciplinary assessment. However, it can be demanding in terms of research expertise and time for completion. Some see AEAM as particularly supportive of sustainable development (Grayson et al., 1994).

Integrated, comprehensive and regional impact assessment, integrated and strategic environmental management

Impact assessment should be better integrated into policy making, planning and administration (Htun, 1990; Hare, 1991; Jenkins, 1991; Pearce, 1992; Slocombe, 1993; Van Pelt, 1993: 99; Bowyer, 1994; Ortolano and Shepard, 1995: 16). The following approaches seek to cover more than just a restricted range of impacts, to do so over more than a snapshot of time, and at wider scales, and up through all project, programme and policy levels (from local up to international). Some of the approaches seek to cope better with indirect and cumulative impacts than mainstream impact assessment (Nijkamp, 1986). Integrated impact assessment is a generic term for the study of the full range of ecological and socio-economic consequences of an action (Lang, 1986; McDonald and Brown, 1990). It is difficult to predict the impacts of something if no account is taken of other current and planned developments. It also seeks to promote closer integration of impact assessment into planning, policy making and management, adopting a tiered approach (Parson, 1995).

To assess cumulative impacts a regional impact assessment approach may be adopted (e.g. where successive tourism developments lead to regional problems or a number of irrigation projects combine to cause difficulties). It is also useful for establishing planning objectives (e.g. the impacts of a new shopping centre (mall) were considered by Norris (1990) using such techniques). It makes sense to assess developments in their spatial setting rather than in isolation; it also allows the interfacing of planning and environmental management at the regional level and offers possibilities for assessing exogenous impacts on the region. Economists use econometrics and input–output analysis to explore economics and environmental linkages at regional level: for example, the impacts of an irrigation development on a region associated with Malaysia's Muda Scheme (Isard, 1972; Bell and Hazel, 1980; Bell et al., 1982; Solomon, 1985).

Integrated regional environmental assessment is similar to the approach discussed above, having the following objectives:

- To provide a broad, integrated perspective of a region about to undergo or undergoing developments.
- To identify cumulative impacts from multiple developments in the region.
- To help establish priorities for environmental protection.
- To assess policy options.
- To identify information gaps and research needs.

There is no single methodology for doing this, and the approach is more difficult than mainstream EIA. A solution might be to subdivide regions into smaller units for assessment (perhaps ecosystems or river basins, although there may be situations where administrative regions offer better possibilities). Integrated environmental management seeks to reconcile conflicting interests and concerns, minimise negative impacts, and enhance positive results. It is an approach which seeks to integrate impact assessment and evaluation into planning and decision making. For an example of an integrated environmental management procedure (proposed for South Africa), see Sowman *et al.* (1995).

While most EIA and SIA is applied at project level, it is also desirable to assess at programme and policy level, for example, to improve:

- overseas aid provision;
- structural adjustment programmes;
- free trade developments;
- public transport policies.

It is not easy to find an effective and flexible, integrated approach that may be applied to, say, national energy policy, an industrial development zone, or to an extensive area of scenic value. The greatest promise probably lies with tiered assessments (Lee, 1978, 1982: 73–75; Wood, 1988; Harvey *et al.*, 1995). These adopt a sequential approach with broad assessment at policy level (tier 1), e.g. impact assessment of national road policy; followed by more specific assessment at the programme level (tier 2), e.g. regional road programmes; and even more specific assessment of individual (road) project(s) (tier 3), e.g. local road construction. Efforts are made to cross-reference broad and specific assessments. Prior or parallel events in higher tiers condition events in tier 3, so it is unsatisfactory to look at a lower tier without also considering higher ones or vice versa. Tiered impact assessment can also adopt a multisectoral approach (horizontal tiers) – if sectors were considered in isolation cumulative impacts might be missed (or a sector might get missed). This requires a holistic approach to avoid missing interactive effects. Tiered impact assessment should acquire data that make subsequent or related impact assessments easier, faster and cheaper to conduct. Tiered impact assessments should complement each other and so avoid the duplication which might otherwise occur. It may be possible with some types of development to do broad impact assessments and dispense with a plethora of individual assessments (e.g. instead of factory-by-factory impact assessment it may be possible to conduct a single industrial estate assessment). The USA tries to encourage a tiered approach, and in other countries, such as The Netherlands, and more recently Europe as a whole, the trend is towards this approach.

Impact assessment experience at programme and policy level is more limited than for project level, but it is growing. Such assessment differs from mainstream project-focused assessments, in that it must allow for the fact that other programmes and policies, cultural and other forces have considerable effect on what is being assessed (projects can usually be studied in relative isolation). To cope with these challenges strategic environmental assessment (SEA) (or programmatic EIA) has been developed

(Partidário and Clark, 2000; Dalal-Clayton and Sadler, 2005; for a bibliography of SEA and other environmental management material see http://www.nottingham.ac.uk/sbe/planbiblios/bibs/Greenis/A/20html – accessed May 2005). SEA should be undertaken early in the development process, like EIA or SIA, at the start of draft planning and programme development. Strategic environmental assessment (SEA) offers means of viewing and co-ordinating development from policy and programme levels down to project level through a tiered approach (Hill *et al.*, 1994; Sadler, 1994). This is a form of tiered, nested or sequential environmental impact assessment that seeks to provide a framework within which project, programme and policy impact assessment can take place (EIA may be used at the project level, tiered with SEA to link it to programme and policy levels or, as is increasingly the case, SEA is applied to all levels) (Wood, 1992, 1995: 266–288; Therevel, 1993; Buckley, 1994; Partidário, 1996; Therevel and Partidário, 1996; Horton and Memon, 1997; *Project Appraisal* 7(3), 1992 – special issue on strategic environmental assessment) (Figures 9.4 and 9.5).

SEA may be applied:

● with a sectoral focus (e.g. to waste disposal, drainage and transport programmes);
● with a regional focus (e.g. to regional, rural and national plans);
● with an indirect focus (e.g. to technology, fiscal policies, justice and enforcement, sustainable development). SEA may be applied at a higher, earlier, more strategic tier of decision making than project EIA.

Provision for SEA was made by NEPA in 1970 and in California's Environmental Quality Act of 1985, and it is now in use in various countries, including Canada, The Netherlands, the USA (especially California), UK, Germany and New Zealand (http://www.ceaa.gc.ca/016/index_e.htm; http://www.scotland.gov.uk/consultations/environ-ment/seacpl-00.asp – both accessed June 2005). The Netherlands has had a statutory SEA system in force since 1987 for waste management, drinking-water supply, energy and electricity, and some land-use plans, and its formal requirements were strengthened in 1991 under the National Environmental Policy Act. New Zealand has had SEA laws since 1991, under Part V of the Resource Management Act (1995). The EEC and the UK published proposals for SEA in 1991 (although Therevel *et al.* (1992: 32) note that in the UK poor long-term strategic planning will probably make the adoption of SEA difficult). Since 1995 the EU has been moving towards requiring member states to adopt SEA procedures and the World Bank also supports it. In 2003, at Kiev, a Protocol on Strategic Environmental Assessment (the 'SEA Protocol') was signed. Once this is in force it will require parties to use SEA on draft plans and programmes and to involve the public in decision making (UN Economic Commission for Europe http://www.unece.org/ennv/eia/welcome.html – accessed June 2005). At the time of writing the EC 'SEA Directive' (Directive 2001/42 EC) was coming into force and effectively required SEA, although that term is not actually used, on a diversity of proposed plans and programmes (http://www.odpm.gov.uk/stellent/groups/odpm_planning/documents/page/odpm_plan; http://www.sea-info.net/ – both accessed June 2005).

SEA is useful for site selection, and by conducting such a 'higher order' assessment there may be less need for, and less depth required from, component project EIAs. The SEA approach may cope better with cumulative impacts, assessment of alternatives and mitigation measures than standard EIA. It is claimed that SEA can ensure that EIA is initiated at the correct point in the planning cycle and therefore makes it easier to pursue sustainable development by helping prevent problems that are difficult to reverse. Increasingly SEA is seen as a key approach for implementing the concept of sustain-able development, because it allows the principle of sustainability to be carried down

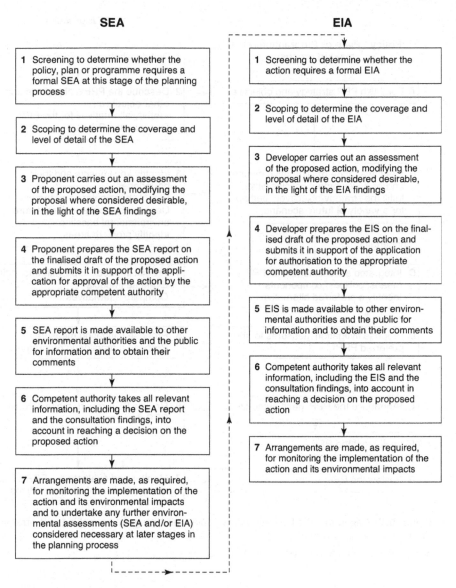

SEA

1 Screening to determine whether the policy, plan or programme requires a formal SEA at this stage of the planning process

2 Scoping to determine the coverage and level of detail of the SEA

3 Proponent carries out an assessment of the proposed action, modifying the proposal where considered desirable, in the light of the SEA findings

4 Proponent prepares the SEA report on the finalised draft of the proposed action and submits it in support of the application for approval of the action by the appropriate competent authority

5 SEA report is made available to other environmental authorities and the public for information and to obtain their comments

6 Competent authority takes all relevant information, including the SEA report and the consultation findings, into account in reaching a decision on the proposed action

7 Arrangements are made, as required, for monitoring the implementation of the action and its environmental impacts and to undertake any further environmental assessments (SEA and/or EIA) considered necessary at later stages in the planning process

EIA

1 Screening to determine whether the action requires a formal EIA

2 Scoping to determine the coverage and level of detail of the EIA

3 Developer carries out an assessment of the proposed action, modifying the proposal where considered desirable, in the light of the EIA findings

4 Developer prepares the EIS on the finalised draft of the proposed action and submits it in support of the application for authorisation to the appropriate competent authority

5 EIS is made available to other environmental authorities and the public for information and to obtain their comments

6 Competent authority takes all relevant information, including the EIS and the consultation findings, into account in reaching a decision on the proposed action

7 Arrangements are made, as required, for monitoring the implementation of the action and its environmental impacts

Figure 9.4 A comparison of strategic environmental assessment (SEA) and environmental impact assessment (EIA)

Note: EIS = environmental impact statement.

from policies to individual projects (Therevel *et al.*, 1992: 22, 126). SEA, at least in principle, can enable countries to work together on transboundary problems (see above discussion of transboundary EIA).

SEA is more demanding of data and expertise than mainstream EIA, but this is less of a problem if it overcomes many of the limitations of the latter. A difficulty faced by SEA is that programmes evolve in a subtle way, and at a given moment it may not be easy to see what actually constitutes a programme. Another problem is that policy makers may not want to give potential opponents or competitors a perspective on their strategy, so public involvement is a problem. Methodology is in need of development.

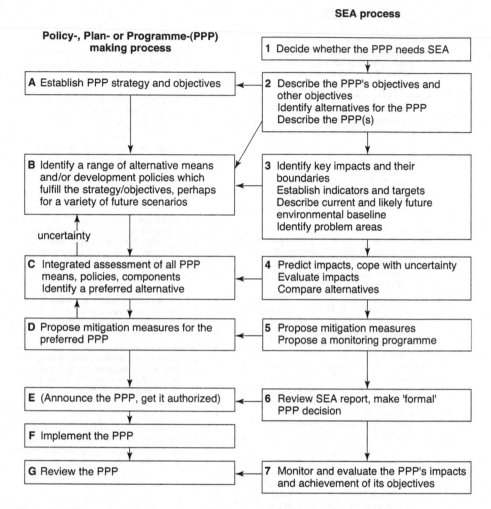

Figure 9.5 Stages in links between policy- , plan- and programme-making and strategic environmental assessment (SEA)

Source: Thereval and Partidário (1996: Fig. 1.1)

SEA must make accurate assessments in spite of often vague proposals and policies (compared with the project-level situation), and it must cope with often uncertain system boundaries; limited information on existing and future developments; a large number of possible alternatives to consider; the involvement of a number of, possibly uncooperative, bodies; and possibly more political pressures than are felt by EIA.

Dealing with cumulative impacts

The systematic and comprehensive identification and assessment of cumulative impacts – cumulative effects assessment (CEA) (cumulative impacts assessment) – is increasingly attracting the attention of researchers and practitioners. Mostly the focus has been on negative cumulative impacts. However, it can assess positive impacts as well. The

USA, Canada, New Zealand and a number of other nations now have regulations requiring assessment of cumulative impacts (in the USA, it has been part of EIA legislation since 1979, but in practice progress has been slow).

Spaling (1994: 243) observed that environmental changes accumulate through many different processes or pathways:

- incremental (additive) processes (repeated additions of a similar nature $a + a + a + a \ldots$);
- interactive processes ($a + b + c + n \ldots$);
- sequential effects;
- complex causation;
- synergistic impacts;
- impact which occurs when a threshold is passed as a consequence of some trigger effect (e.g. chemical or biological time bomb);
- irregular surprise effects;
- impacts triggered by a feedback process (antagonistic – positive feedback which reinforces a trend, as opposed to ameliorative – negative feedback which counters a trend).

In practice CEA is difficult. Nevertheless, there are methods which are at least partially effective (e.g. the component interaction matrix and the minimum link matrix). There are also specific CEA methods (see Spaling and Smit, 1993; Smit and Spaling, 1995). Some have tried to assess cumulative impacts by adopting a regional or strategic stance (see above discussion of SEA), and others have tried CEA at the project level.

There are signs that global stability and even some of the Earth's life-support systems are increasingly shaped by cumulative and global impacts that can affect local and regional systems. Cumulative impacts may result in a runaway process which exceeds some critical threshold and may be difficult to remedy (e.g. global warming leads to uncontrollable releases of greenhouse gases from various sinks, resulting in uncontrollable warming) – impact assessment has the potential to warn environmental managers of these.

Summary

- Environmental management has become increasingly proactive in approach. This is largely driven by a desire to establish sustainable development, and by disasters which have proved costly and which have generated widespread public concern. Tools and techniques are consequently required to place more emphasis on prediction and assessment of future scenarios, selection of optimum strategies and so on.
- There is a move away from piecemeal (*ad hoc*) approaches to more standardised and accredited measures, and from individual project and local focus to one that is more strategic and integrated (even holistic).
- Impact assessment and the forecasting of future scenarios is seldom accurate and 'gap'-free; the environmental manager must be cautious and seek multiple lines of evidence. Ideally the tools and strategy should be flexible and adaptive.
- Predictive tools can be problematic; EIA, SIA, hazard and risk assessments may give a false sense of security. This is because they are imperfect and mainly give a snapshot view (limited coverage in time and space). It is also difficult to reliably identify indirect and cumulative impacts. The unexpected must be expected.

Further reading

Becker, H.A. (1997) *Social Impact Assessment: method and experience in Europe, North America and the developing world.* University College London Press, London.
Good introduction.

Becker, H.A. and Vanclay, F. (eds) (2003) *The International Handbook of Social Impact Assessment: conceptual and methodological advances.* Edward Elgar, Cheltenham.
Good sourcebook on SIA.

Glasson, J., Therevel, R. and Chadwick, A. (1994) *Introduction to Environmental Impact Assessment.* University College London Press, London.
Although focused on developed countries, this is one of the best introductions to EIA.

Harrop, O. and Nixon, J.A. (1998) *Environmental Impact Assessment.* Routledge, London.
Mainly British focus; good introduction to practice, methods and techniques with numerous case studies.

Lawrence, D.P. (2003) *Environmental Impact Assessment: practical solutions to recurrent problems.* Wiley, Chichester.
Introduction to EIA with a practical focus.

Lee, N. and George, C. (eds) (2000) *Environmental Assessment in Developing and Transitional Countries: principles, methods, and practice.* Wiley, Chichester.
A comparative review of environmental assessment procedures and practice focusing on non-Western countries.

McGill, S. (1999) *Environmental Risk Management.* Routledge, London.
Good introductory text.

Marriott, B.B. (1997) *A Practical Guide to Environmental Impact Assessment.* McGraw-Hill, New York.
Introduction to EIA with a focus more on North America.

Modak, P. and Biswas, A.K. (1999) *Conducting Environmental Impact Assessment in Developing Countries.* United Nations University Press, Tokyo.
A more practical and developing country focus on EIA.

Petts, J. (ed.) (1999) *A Handbook of Environmental Impact Assessment* (vol. 1). Blackwell, Oxford. Useful source of further information on methods and processes; a second volume deals with practice and limitations.

Therevel, R. (2004) *Strategic Environmental Assessment in Action.* Earthscan, London.
Good introduction to a rapidly expanding field.

www sites

Ecological risk analysis (Oakridge National Lab. USA) http://www.esd.ornl.gov/programmes/ecorisk/ecorisk.html
International Association for Impact Assessment (IAIA) – a good introductory site for EIA and SIA: http://www.IAIA.extNodak.edu/IAIA
Risk assessment (USDA) http://www.nal.usda.gov/wqic/risk.html

All the above accessed June 2005.

⬤ 10 Key resources

This chapter examines key natural resources. While ongoing access to natural resources is vital for sustainable development, it is also important that wastes are disposed of satisfactorily, and that human and natural disasters are avoided. Some societies may have little in the way of natural resources; indeed, some of the most successful nations are poorly endowed or were unaware of useful minerals and so on when they were developing. Human resources – knowledge, culture and skills – may be used to obtain natural resources from elsewhere by trading or domination. A good example today is Singapore, a city state with little drinking-water, food supplies or energy resources within its territory. However, Singapore has invested in education, technology and the development of trading facilities. Some societies acquire their natural resources through cultural activities – Venice earns more today from its ancient buildings than merchant ventures.

Companies, governments and consumers far removed from the source area often determine natural resource exploitation, development and management. Natural resources are not uniformly distributed: some are essentially available anywhere, but many (some of critical importance) are restricted and the world's consumers have to live with this. This raises two questions: Who has decision-making power over resources? To whose advantage do they operate?

Natural resources management issues

A resource may be defined broadly as something which humans realise has actual or potential value. Resources may be subdivided into natural, technological, financial, organisational, cultural and aesthetic. There is overlap – natural features may have aesthetic value and influence culture, and without technology and organisational skills a mineral might not be exploited. Natural resources may be divided into:

1   finite no matter what;
2   renewable if well managed;
3   renewable if reasonably managed;
4   renewable no matter what.

Alternatively:

i   easy to damage and swift to recover;
ii   easy to damage and slow to recover;
iii difficult to damage and slow to recover;
iv   difficult to damage and swift to recover.

Some past civilisations seem to have collapsed through poor natural resource management rather than irresistible disaster – in various parts of the world water resources failed to match demand; in the Middle East salinisation took its toll. In East Asia land degradation and water supply problems probably damaged the Khmer (Figure 10.1); and Easter Island is generally cited as an example of poor forest and soil management destroying a culture (see Fagan, 2004; Diamond, 2005).

Figure 10.1 Ruins of Angkor (Cambodia). The huge temple complex was built between the ninth and fourteenth centuries AD by the Khmer civilisation. For over 500 years thousands of people sustained a living in the area, then something went wrong, townships and temples were abandoned and the jungle took over. Modern sustainable development must note history.

The following resource exploitation options are possible:

- Mining of non-renewable mineral resources – the source of income is exhausted, not sustained.
- Production of renewable mineral resources: sea-salts, sulphur from volcanic vents, nitrates.
- Sustainable yields from well-run plantations, farmland or natural vegetation – palm oil, natural rubber, rattan, bamboo, waxes, beverages, fibres, and many other materials. This can be a problem if it takes land from smallholders and drives them to marginal areas where they can damage the environment. Commodity production may take land out of food production and make countries more reliant on importing grain. It can displace and destroy biodiversity, and can be a source of pollution as a consequence of processing.
- Unsustainable production of natural commodities – excessive hunting of species, over-collecting wild plants, clumsy ranching, plantations which degrade the land (Eden and Parry, 1996). Recently, demand for medicinal plants has led to over-collecting of wild biodiversity in many countries. Animals such as tigers and rhino are suffering through hunting to supply traditional medicine. In Africa especially, wild animals are being hunted to extinction to provide growing cities with 'bush meat'.

The value of a resource may be fickle, sometimes depending on little more than fashion. New fashions, environmental changes and technological progress create resources. Those seeking sustainable development have to allow for such shifts – if all future options are to be maintained resource use must damage nothing, and currently useless items have to be given respect. Environmental managers are called upon to assess resources, identify resource-use impacts, develop management strategies, monitor and enforce controls, and if necessary try to rehabilitate or restore what has been damaged. There has been a trend away from *laissez-faire* resource development, to exploitation watched over by the state or others (e.g. NGOs, international bodies and so on).

Resources supply human needs and wants and generate outputs, which usually have to be dealt with (waste disposal and pollution control). Some resources are more important than they seem at first glance: the key resource may not be water or food, but cooking fuel, and people and wildlife need space, landscape may have aesthetic qualities or impart a sense of place and identity which drives an economy. Identifying the key resources may not be easy. In addition, natural resource use depends on human skills, knowledge, attitudes, funding and so on. There have always been resource shortages, which humans adapted to by being omnivores and mobile, practising infanticide, late and restricted marriage, developing taboos, common resource usage controls, technology and laws. Modern societies are probably less adaptable; improved technology and transportation have helped to compensate, but against this must be weighed the fact that modern populations are huge, livelihoods depend on a complex web of factors, opportunities to move to avoid problems are restricted, and we are currently enjoying more stable and pleasant environmental conditions than did most of our forebears. The environment is not as constant as most would like to think; a shift to something resembling the 'Little Ice Age' of the 1350s to 1830s would have serious implications. A climatic hiccup such as a diminution of the Gulf Stream which warms Europe or a large volcanic eruption shrouding the Earth with ash for a couple of years would cause severe resource difficulties. Environmental managers should provoke some consideration of realistic threats; at the time of writing the G8 were debating global warming – a little like contemplating one of a pack of playing cards.

Some natural resources are obviously important: water, food, energy and so on. Stratospheric ozone provides a vital shield, and is a global common resource that can be easily damaged – its vulnerability and importance were only seriously accepted after 1987. People living in a region and using natural resources may have no legal rights and may lose out to the state, powerful individuals or businesses. Economists have until recently assumed that market forces would prevent over-exploitation of natural resources. When individuals own or have access to a natural resource it may be difficult to control its usage. Large, powerful companies can be difficult to monitor and discipline. Common resources may be exploited with no adequate controls operating.

Some countries are natural resource 'poor' and others 'rich'; there are resource-rich countries which suffer poverty, and those which are resource poor, yet have a good standard of living. Many have argued that the key resource is human skill and initiative, pointing to states such as Hong Kong and Singapore as examples. Control of natural resources can be influenced by access to armed might, transport facilities and trading skills: companies and cartels wield considerable power. Many important resources are exploited, distributed and sold by a handful of companies. Commercial agriculture is dominated by a handful of large agroindustrial companies. Land in some parts of the world is not under individual or company ownership, and is allocated by traditional tribal rules. There are many who have insecure land tenure and who can easily be evicted. Both of the last two situations can discourage sustainable use and improvement. Security is also vital; if individuals, companies or a state feel unsafe they will not invest labour or money in sustaining or improving natural resources. Without law and order good resource management is difficult. Individuals, companies seeking to grow and satisfy shareholders, or states seeking power, often display greed. Effective sustainable natural resource management may not be compatible with greed or exploitation for strategic reasons.

Natural resource management can adopt a sector, ecosystem or bioregion (highland, coastal zone, river basin, dryland, wetland and so on) as a planning and management unit/focus, or apply political ecology, human ecology or sustainable development approaches. Recognising what is the key resource(s) in a given situation can be difficult; ecologists attempted to do this, using Von Liebig's Law to identify what was crucial to the survival of a species in a given ecosystem. Humans faced with natural resource shortage can sometimes resort to transporting material or substitution.

A key resource is something which:

● is important for survival and welfare;
● could trigger conflict if mismanaged;
● could catalyse development;
● is vital for maintaining an ecosystem, environmental process or biodiversity;
● inspires people or comforts them.

Prehistory and history are full of examples of potentially sustainable natural resources destroyed by human misuse: large mammals hunted to extinction, species lost through the introduction of rats, cats and dogs, fisheries over-fished. Sustainable development is still largely rhetoric; too few resource development schemes stress maintaining usage or recycling of waste – the bulk seek exploitation for short-term profit, rather than indefinite gains. Food and water are clearly key natural resources. The lack of any catastrophic food shortage (famine) is one indicator of development; few nations have maintained such a status for long.

Concern for natural resource management helped prompt a number of books in the 1970s and 1980s, which part-laid the foundations for sustainable development (Mitchell, 1979; Ruddle and Manshard, 1989; Omara-Ojungu, 1992). Allied fields are resource assessment or analysis (the establishment of how much there is), resource monitoring (checking whether it is degrading or improving), and resource law (developing and enforcing rules for optimum management). Resource control is power; it can generate foreign exchange, enslave groups of people and trigger wars. Some societies are vulnerable to the supply of one or a limited number of resources; any shortage, glut, substitution or alteration of demand can spell disaster. Ghost townships attest to the exhaustion of mineral resources, the banning and over-fishing of whales, the salinisation of soils, or the drying up of water supplies. The Millennium Ecosystem Assessment emitted strong warnings in 2005 that many natural resources and ecosystems were in difficulty and, without good management, could abruptly fail. Similar assessments are planned every five to ten years in the future (http://www.millenniumassessment.org//en/About. Overview.aspx – accessed June 2005).

Water

For much of human history people have been able to move and settle where water and other resources were good enough to support them; also, populations were small enough to limit over-exploitation and pollution. This is no longer the case: people are much less mobile, and growing human population, rising per capita demand for water and increasing pollution have led to concern that there is, or soon will be, a water crisis. Fresh water is vital for the survival of humans and most other organisms, and little development can take place without adequate supplies. Water rights are often more crucial than land rights. Water is frequently a sustainable resource if adequately managed, although some supplies are finite no matter how carefully used. Excessive exploitation and pollution, poor soil management, altered vegetation, building and stream channelisation can each or in combination lead to diminished and contaminated water supplies and altered stream flow. Poor land and channel management leads to siltation and flooding. Water resources management therefore needs to be integrated with the management of other resources.

Land drainage, soil compaction and urbanisation cause precipitation to run off quickly – the result can be severe erosion, flash flooding, reduced groundwater recharge and wildly fluctuating stream flow. Poor land use leads to nutrient over-enrichment and sediment-contaminated water, which damages plankton, fish, shellfish and aquatic plants, and silts up channels, lakes and reservoirs. Tillage, mining, overgrazing and bushfires also generate sediment. Streams, rivers, channels and lakes can silt up (siltation), causing flooding and destroying bridges and other infrastructure far away from the area generating the sediment. Reservoirs lose their storage capacity and large hydroelectric schemes can be ruined by silting, sometimes before they pay back construction loans. Siltation can destroy irrigation schemes, water filtration and sewerage systems. The sediment is a lost resource for the areas eroded, but where deposition occurs there may be benefits if it is fertile. Sediment transport is often episodic and can be very difficult to monitor: much is generated during storms, especially after tillage before crops provide ground cover. Soil and water conservation approaches which reduce run-off and retain soil and moisture can help sustain agriculture, recharge groundwater, and should help to reduce flooding and silting. Groundwater, surface water and soil management should be comprehensive and integrated (Thana and Biswas, 1990). Water demands often conflict unless there is careful integrated management.

Flood threat has been increasing because of: rising sea-levels; increased settlement in flood-prone areas; poor soil and water conservation practices (which increase and speed run-off); urbanisation (which also speeds/increases run-off); river channelisation which removes the flood moderation offered when floodlands are inundated, and in some areas subsidence caused by over-extracting groundwater or mineral exploitation. Flood damage is tending to rise because people increasingly replace traditional homes with dwellings made of materials which are vulnerable (electrical systems and fitted furniture), and because they rely on insurance, rather than wise location and removal/salvage of belongings. There is also a likelihood that storm patterns and associated flooding are altering through natural or man-induced climate change. Reliance on costly flood defences like those of New Orleans or Dacca (Bangladesh) may be unwise. The former city suffered badly in 2005 and the latter has attracted the criticism that its defences have shifted the threat to other unprotected areas and costs too much. Alternative flood mitigation strategies include: planned evacuation from flood areas and replacement with appropriate land use (recreational parkland, conservation areas and so on); better whole-catchment management; improved early warning and flood refuges (Ward, 1978). Recently a number of authorities have re-established flood regimes on regulated rivers – opening dam spillways to create artificial floods so as to flush away downstream silt and benefit riverine and floodland biodiversity. Some managers have created new or restored old riverside floodlands and water-meadows to alleviate flooding, aid ground-water recharge and offer conservation and recreation benefits.

Precipitation may fluctuate from year to year through natural climatic fluctuations, including El Niño/El Niña, and human-caused environmental change. Large parts of the world have seasonal shortages so supplies are manipulated through reservoir storage or water transfer schemes – engineering approaches, often on a large scale, and judged mainly by economic and technical criteria. Adaptability is important, especially where inadequate data makes it difficult to safely forecast, and also unforeseen changes could spell disaster for inflexible water development. Unfortunately, much water resources development is politicised, inflexible and inappropriate.

Many identify water supply as *the* challenge of the twenty-first century (see Chapter 11). Yet richer citizens generally take water for granted. Postel (1992: 24) noted that the world was entering a new era, one which contrasts with the past, when damming rivers and drilling wells was relatively straightforward, and supplies were fairly easily 'engineered'. Water will be in shorter supply and more likely to be competed for. Some claim that there is already a widespread 'water crisis' (Kobori and Glantz, 1998). In some countries the best supplies are already being used, so future demand will have to be met by less wasteful usage and from less than optimal sources. Water crosses borders above and below ground; rivers and lakes often form territorial boundaries, and the shared waters are vital. Water management has to effectively address multi-state sharing, recycling and reduction of wastage, and seek alternative sources (Barrow, 2005: 50).

Various authorities have expressed concern about the possible serious shortfall of water. In many parts of the world, if the current population, development and climatic change trends continue, there will be serious problems. There is clearly a wide need for improvement of management and, if available, development of new sources (*The China Daily* has shown concern http://www.chinadaily.com.en.english/doc/2005–04–20/content-43529.htm; see also http://www.ifpri.org/media/water205.htm; and http://www.idswater.com/water/Europe/Stockholm_International_Water – all accessed September 2005; Kobori and Glantz, 1998). Lobby groups have been trying to stimulate concern for the 'water crisis'. One of the most influential is the World Water Council established in 1996 as an international NGO, to raise awareness and act as an international 'think-tank'. Other groups include the Global Water Partnership (established 1996), and the World Commission for Water in the twenty-first century (Abu-Zeid, 1998). The latter body has called for:

- holistic approaches and integrated management of water;
- full-cost pricing for water, and targeted subsidies for the poor;
- institutional, technical and financial innovations in water management.

Many parts of the world have inadequate quantities of water, and what there is is often badly contaminated. Between 1981 and 1991 the UN declared an International Drinking Water Supply and Sanitation Decade to try to improve health in developing countries – with limited success, partly due to population growth raising demand. International Conferences on Water and the Environment were held in Dublin in 1992 and 1999; there have been more recent gatherings, and these have helped focus attention on water resource issues. The International Water Management Institute (IWMI) works to improve management of freshwater resources, supporting studies and development of irrigation, river management and so on. The World Water Council has helped organise a number of World Water Forum meetings, the first in Marrakesh in 1997, which brought together NGOs, UN agencies and aid donors. In 1997 the UN published a *Comprehensive Assessment of Freshwater Resources of the World*. Biswas and Tortajada (2001) warned there was a need to increase concern for water, especially to break away from focusing on the present to look more to the future. The International Year of Freshwater in 2003 emphasised supply and access, during which UNESCO published *The UN World Water Development Report 2003* http://www.unesco.org/water/wwap or http://www.berghahnbooks.com (both accessed October 2003). This Report presents key issues and provides a global overview of the state of the world's freshwater resources.

Fresh water is obtained from:

- short-lived flows following precipitation (overland flow or temporary streams);
- stable springs and stream flow. The latter may still fluctuate a great deal and prompt developers to construct dams;
- groundwater (underground supplies) held in water-storing rocks (aquifers); renewed by precipitation and surfacewater seeping into the ground, or by artificial recharge – with good management sustainable indefinitely;
- desalination, but hopes that it may provide a widely available alternative water supply are unfulfilled because current technology is costly;
- trapping mist, fog and dew, in a few cases, for crops and drinking-water.

Some groundwater accumulated in the past when the climate was wetter or has collected very slowly – so essentially it is unsustainable. Groundwater may be contaminated with natural pollutants, agrochemicals, contaminants leached from landfill sites, or have escaped from damaged storage tanks, pipelines or sewage. Other common contaminants are salts – seeping from saline groundwater, irrigation wastewater or seawater, or concentrated due to capillary rise and evaporation. Decontamination may be very difficult and slow, or even impossible. This is because groundwater often has long residence times and is difficult to flush clean. There is less exposure to air, ultraviolet light (sunlight) and fewer micro-organisms to attack chemical or biological contaminants. If overused, potentially sustainable groundwater resources may also be damaged or permanently destroyed because the aquifer collapses when pumped dry and will not then store water.

Water resources in rich and poor countries are falling into the control of large companies. Some worry that this is expropriation of communal resources, while others welcome the ability of commerce to invest in improving supply quality and quantity (Shiva, 2002). There have been forecasts that privatised water promises to be as profitable in the twenty-first century as oil is now. What was once a free or a state-owned resource is becoming privatised; that may not be all bad, because business may be better able to fund development and conservation, and is more motivated to recover costs (Barlow and Clarke, 2002; Holland, 2005). However, attention is needed if cartels and exploitative control by business are to be avoided. As the world's cities expand they compete with others for water, especially for agriculture. The likelihood is that expanding cities will adopt flush sewerage systems and the problems of sewage disposal will increase. Urban people tend to have a stronger voice than rural, and this could also distort access to water. Some current estimates suggest that by 2025 around two billion people will live in areas where water is scarce. Postel (1992: 28) defined 'water scarce' as anywhere with less than 1,000 m^3 per person per day. Whatever the human demand, whether urban or for irrigation, it means less natural water for river, wetland and lake ecosystems. Currently, water managers tend to see nature and human usage as competing and at best seek a simple division of supplies between the two. This will not continue to work – human demands and nature have to be carefully integrated (Hunt, 2004). Integrated watershed or river basin management offers possible ways forward – using biogeophysical units, with adequate overall control and powers to effectively co-ordinate the management of water, environment and human activities (Falkenmark and Rockstrom, 2004). Control of an entire drainage system gives administrators the opportunity to deal effectively with soil erosion, pollution, flooding and so on. Unfortunately, these approaches have frequently failed to yield their promise.

Water shortage and drought are not precise terms; consequently comparison of data is often difficult. Recently there have been proposals for a Water Poverty Index to give a practical and integrated assessment of scarcity (Sullivan, 2002). Societies respond to insufficient water in different ways: some innovate and reduce demand or find alternative supplies, others may abandon settlements or cease activities. To date, the reduction of contamination of water supplies by chemicals and pathogens has been relatively neglected, especially in developing countries.

Water resources management should not be confused with environmental management; it is focused on engineering, economic and perhaps social goals, but not so much on environmental issues. Water management is commonly poorly integrated with other activities, and those involved tend to have a short-term outlook and a project focus. Water managers may also see themselves as serving shareholders, citizens of a city. During the past thirty years or so bodies such as the IWMI and the US Corps of Engineers have been working hard to improve water management and broaden its outlook. Environmental managers can assist at a strategic level, advising on river systems management needs and ensuring that issues are not overlooked, integrating and co-ordinating developments, and encouraging developers to seek optimum development and environmental conditions.

At the 2002 Johannesburg Summit on Sustainable Development a target was set of halving the number without access to satisfactory domestic supplies from around 2.2 billion in 2002 to 550 million by 2015. It will be interesting to see whether such a goal attracts support – poor people are not likely to offer much profit for investors. Israel has responded to shortages by making its agriculture less demanding of water and by reusing wastewater. However, efficient irrigation systems are currently relatively costly to install and maintain, and are thus more suitable for commercial crops than food production in poorer nations. Postel (1992: 104) estimated that less than 0.5 per cent of the world's irrigation relied on low-waste techniques – so there is potential for water savings – if costs fall and maintenance becomes easier. There is also potential for soil and water conservation techniques such as mulching, stone-lines, terraces and micro-catchments, especially if these are based on low-cost, locally available materials and labour. These reduce soil degradation and improve groundwater recharge.

Sewage pollution is a widespread problem; also livestock waste, storm drainage from urban areas, agrochemical run-off from farmland, and effluent from mining and minerals processing, sugar, oil palm, rubber, aquaculture, alcohol production, and other agro-industrial activities. Sewage and livestock waste could be used for methane production, sent to composting plants, or dried and burned as fuel for district heat and electricity generators. But the waste must be of a suitable quality, the supply adequate, and the investment available to install biogas digesters, composting facilities or furnaces and generators. Peri-urban areas of developing countries can be used for agriculture, ideally with careful controls. Sewage could be chlorinated cheaply and harmful wastewater from storm drains or industrial areas diverted. The resulting effluent could be used for trees, fodder or amenity area irrigation, rather than food crops, so the disease, heavy metals and other contaminant risks would be less important. The Hyderabad Declaration on Wastewater Use in Agriculture 2002 acknowledges that wastewater is widely used and seeks to reduce the risks.

Some organisms spread via water channels, and markedly affect water resources and human health; for example, burrowing crab species have spread to new locations, and cause serious damage to native species and riverbanks. The unpleasant human parasitic disease schistosomiasis can easily be introduced, and may be difficult to eradicate.

Some countries have responded to water supply problems by importing cheap grain, and neglecting their agriculture (Yang and Zehnder, 2002). Altered tastes, especially those of numerous city people, reinforce this. Demand is shifting from traditional crops to imported wheat and maize. This will affect global food reserves and more people will depend on fewer producers thereby increasing risk. In the grain-importing countries rural employment will deteriorate, reinforcing the drift of people to cities and overseas. The rural depopulation can cause environmental degradation as traditional land uses are neglected.

Developing rivers

Rivers are sensitive and valuable ecosystems that have been widely damaged world-wide (see Chapter 13). The usual response to water shortage has been a 'technical fix' – large-scale engineering aimed at storing river flow or transferring flows from water-rich to water-poor areas. Barrages divert some flow from a river, so there is no large reservoir; however, they do not store enough to meet demand in dry periods, and support limited electricity generation. Where water storage or power generation are wanted from a river with fluctuating flows, a dam is the usual solution.

Large dams have been constructed on most major rivers since the 1920s. Between 1950 and 2004 over 35,000 were constructed, more than 19,000 in China alone. Globally, about 400,000 km^2 have been inundated by reservoirs, and it has been claimed that between forty and eighty million people have been displaced since 1960 (http://www.gfbv.de/gfbv_e/uno/geneva03/item_7_en.htm – accessed February 2004). After the oil price rise of 1974 there was incentive for countries to cut petroleum imports, and hydro-electricity seemed to offer an alternative and give other benefits such as irrigation supply. There were signs of renewed interest at the 2003 World Water Forum where calls were made for more dams (Khagram, 2005; Pearce, 2005; Trottier and Slack, 2005). There are currently plans for huge hydroelectric schemes (e.g. on the River Congo) to serve a proposed pan-Africa electricity grid.

Dams, and to a lesser degree barrages, cause difficulties by altering the flow of water and sediment and restricting the movements of fish and other organisms. At first those involved in dam construction had to make do with inadequate knowledge about tropical riverine ecosystems and the impacts of regulation and impoundment; as time passed, experience should have helped reduce unwanted problems. However, mistakes are still repeated, and it is striking how slowly engineering and dam management has become more appropriate and flexible. River development is overshadowed by inflexible and insensitive 'solutions', which for political, economic and cultural reasons continue to be implemented (Adams, 1992; Scudder, 2005). Some of these problems result from insufficient attention to environmental and social management; others are inherent in large-scale modification of river flows. Some seem to result from what Newson (1992: xx) calls a 'point-problem' outlook; i.e. the developers do not want to look upstream or downstream at the whole issue.

The World Commission on Large Dams (WCD) was established in 1998 to debate and review developments, and it formed an independent body to set guidelines and assess alternatives. The WCD terminated in 2005 after publishing its final report (World Commission on Large Dams, 2000, 2004). A body still interested in improving large dam planning and management is the UN Environment Programme Dams and Development Project. One dam problem does seem to be getting more attention – releasing artificial floods to try to maintain downstream environmental conditions. There is reluctance to do so because managers want to retain water for power generation or irrigation – a water management, rather than wider environmental management, focus. A number of countries have spent a great deal on flood protection levees – raised banks alongside a river. This seldom gives adequate flood protection, can shift flooding to less protected areas, and prevents flood silt from fertilising riverside lands. Worst of all, if a levee is breached, floodwater drainage is hindered and people may have a false sense of security and be little prepared.

Riverine ecosystems are probably the most affected by humans. Many people live close to rivers, dispose of their waste in them and rely on them for drinking-water, irrigation, transport, power and so on. Worldwide, the range of pollutants entering river systems and groundwaters has greatly expanded. Some can suppress the immune systems of wildlife and humans or cause reproductive problems and cancer, even at low levels of contamination (Malmqvist and Rundle, 2002). Many rivers have been damaged by pollution, flow regulation, dredging and channelisation; hopefully, some at least can be rehabilitated (see Wohl (2005) for a discussion of the situation in the USA).

Numerous grand-scale interbasin transfers have been proposed and some implemented. Interest in large integrated basin development schemes and grand interbasin transfers is currently growing. The latter are planned for Latin America (Grand Hidrovia canal system), and China is pressing ahead with a huge (over 1200 km) south–north water transfer scheduled for completion in 2007. This will channel flows equivalent to about one-tenth of that of the Mississippi from the Yangtze River to Beijing. India has proposed the Indira Gandhi Canal and a national water grid to redistribute river flows on a continental scale. Large-scale transfers may help reduce shortages in the Middle East if conflicts are resolved and terrorism becomes less of a threat. Before transfer schemes are embarked on, it should be asked whether a better approach would be to control water wastage and discourage growth in the water-scarce areas.

Most environmental scientists now accept that global warming is likely. Even limited change could markedly alter precipitation patterns. Already there has been considerable shrinkage of glaciers in many countries, threatening a number of important rivers in

Asia, the Indian Subcontinent and the Americas. Many of the world's larger rivers are shared by more than one country. If tension grows over a shared river there are two possible ways forward: (1) consultation, agreement and co-operation; (2) power politics ('hydropolitics') and possibly conflict (Ohlsson, 1995; Klare, 2001; Wolf, 2003). A somewhat sensationalist literature has been warning of the risk of 'water wars' since the 1970s. However, so far, inter-state disputes have almost never led to conflict. As demand for water increases and more supplies are contaminated, the risk of conflict may rise (Ashton and Ashton, 2002; Kalapakian, 2004; Westcoat and White, 2004). Swain (2001) noted that agreements on sharing rivers may be possible while, say, 80 per cent of total flows run to the sea, but feared that if future demands rise, and if environmental change and pollution cut supplies, agreements may not hold and new ones would be less likely.

The goals of shared river management should be: peaceful agreement, sustainable management, social equability, and opportunities for integrated and comprehensive development and maintenance of environmental quality. Most of the negotiations so far have focused only on quantity of flows and less on quality, yet the problem of contamination is growing – and few address the other goals listed above (Beach et al., 2005). A Convention on the Protection of Transboundary Watercourses and International Lakes was signed in 1996, but is only a start towards resolving water quality problems.

Laws relating to the sharing of river flows and groundwater are not very well developed. The UN Convention on the Law of the Non-Navigational Use of International Waters (adopted by the UN General Assembly in 1997) provides a framework for negotiations between countries. This defines the obligation not to cause harm to another, and the right to reasonable and equitable use by riparian nations – but it is not law. Spain and Portugal were beset by their worst drought for over fifty years in 2005, and, in spite of long-standing agreements, fell into dispute over sharing the Tagus and Douro (*The Times*, 4 July 2005, p. 53). Drought had hit Spain so badly by late 2005 that there were fears of a shortage of olive oil (much of which comes from Spain).

Lakes and ponds

Lakes and other water bodies have suffered around the world as a consequence of pollution, silting-up, the introduction of alien species, diversion or flow regime disruption of inflowing rivers, disturbance related to tourism and so on (see Chapter 13). Some tropical lakes are of great age and escaped the worst impacts of Quaternary environmental changes; they contain rare endemic species which are now vulnerable (Beeton, 2002). Aquaculture can cause serious impacts through effluent, deliberate or accidental introduction of competitive species, introduced diseases, and use of chemicals which may affect lake organisms. African lakes, including Chad, Victoria, Nakuru, Tana, Songor, Tanganyika, Malawi, Tonga, Djoudji and others, have been shrinking according to 2005 UNEP reports. Lake management needs better overall co-ordination, which integrates all the relevant fields and stakeholders, and which is preferably administered through a single lake development authority with adequate enforcement powers (this may need to be multinational). In Russia there are fears that Lake Baikal, which has already suffered some industrial pollution, may be endangered by a planned oil pipeline conducting oil eastward.

Smaller ponds and tanks in most countries are suffering from silt deposition, industrial and agrochemicals pollution, and probably from acid deposition. Some South Asian and North African tanks have been part of sustainable farming for centuries, but have recently fallen into neglect as a consequence of recession in crop prices, unrest and migration of labour (which has prevented adequate regular maintenance).

Irrigation, run-off collection and rain-fed agriculture

Huge efforts and expenditure have been directed at adapting land to fit crops, and much less has been spent on adapting crops to fit the environment. Social scientists would also add that the norm has been to make people fit innovations, rather than innovate in ways that suit their needs and offer sustainable results. The usual strategy has been to develop large-scale commercial irrigation which is wasteful of water, difficult to sustain, and liable to raise groundwater levels and contaminate rivers with agrochemicals and salt. However, there are alternatives which demand less water: run-off cultivation and improved rain-fed agriculture. These alternatives may be the only practical solution for many regions. The distinction between irrigated, run-off and rain-fed agriculture is not clear-cut; also, improved moisture use, sustainable agriculture and soil conservation are generally closely interrelated. Wasted water tends to erode soil and remove nutrient-rich debris.

There are many ways of improving rain-fed agriculture (i.e. that reliant upon precipitation and not irrigation): developing crops that have shorter growing seasons, the introduction of tractor ploughing, drought-resistant crops, fallowing, careful use of fertiliser and green manure, stall-feeding livestock, and much more. These do not demand much water and still improve security of harvest, boost yields, and through soil conservation allow sustainable production. There are also opportunities for developing crops which can make use of saline water or salty soil, or which fix atmospheric nitrogen and effectively provide fertiliser for themselves.

Large-scale gravity-fed irrigation schemes are increasingly difficult to implement, because most suitable sites have already been developed. In addition, developing and running such schemes is becoming more expensive. Sustaining large-scale irrigation is also a challenge, and it is not unusual for a project to fail to repay its investment costs before falling into disrepair. The area of irrigated land per person (worldwide) has shifted from marked increases during the 1960 to 1990s to present-day decline (Postel, 1992: 50–51). Large-scale irrigation feeds many people, but it has also received much of what has been invested in agriculture.

Large-scale irrigation wastes water, damages the environment, and can increase the transmission of malaria and schistosomiasis. About two-thirds of the world's irrigation is in need of repair, but it is still common for schemes to be abandoned and new ones started elsewhere. Huge sums of money have been spent on large-scale commercial irrigation and some of the problems encountered are similar to those of large dams: inflexibility, insensitive engineering 'solutions', wishful thinking by outsiders, reluctance to work with local people, and a failure to learn lessons in spite of considerable hindsight experience. More integrative and environmentally aware management would resolve some of these problems.

People can grow crops and water their livestock by means of a range of rainfall harvesting techniques – run-off agriculture. These are part run-off control and part soil conservation methods. Archaeology has helped prompt interest; studies in the Negev Desert of Israel, North Africa, the Andes, southwestern USA and Latin America has yielded promising strategies. In the past, peoples in those countries used techniques which modern agriculture would be hard put to match, and fed thousands in harsh environments. Some of these strategies could be appropriate for modern small farmers; being inexpensive, they do not condemn users to dependency on outsiders; techniques can often be improved by modern inputs (Pearce, 2005). Run-off farming demands little that is not available locally at minimal cost. One strategy is to use run-off collection catchments larger than the cropped plots, another is to dig microcatchments which concentrate moisture and soil around a single tree or patch of crops. The latter may be used on virtually any gradient if soil and other conditions are suitable (Evanari *et al.*, 1982; Reij *et al.*, 1996). These techniques magnify rainfall for cropped plots or can feed water to storage cisterns.

Air

Most take air for granted, yet as a global commons it is vulnerable and cannot be adequately monitored and protected without international agreement. In the past a few cities have legislated against smoke and smells, and the rich usually settled upwind from pollution sources. Europe and eastern USA experienced serious smog and acid deposition problems as early as the late nineteenth century, and London fogs killed thousands in the 1950s. Legislation and a shift from coal-burning to less polluting fuels helped resolve the problem. In the 1970s concern about nuclear weapons testing led to a test ban and subsequent explosions have mainly been below ground. The main air pollution problems today are vehicle related, sulphur emissions, acid deposition, tropospheric ozone pollution and stratospheric ozone scavenging (the main impact of which is to reduce ultraviolet radiation shielding). Airborne dust generated by ploughing, land degradation and bushfires has caused problems and there can be damaging levels of pesticides and nitrous oxide caused by nitrogenous fertiliser use. Atmospheric pollution with greenhouse gases (carbon dioxide, methane, and several other gases) has attracted a huge amount of attention in the past decade or so. Other atmospheric pollutants have attracted less attention. California has led the way in controlling partially burnt hydrocarbons from vehicles, and in richer nations catalysers are now common and there is pressure to shift to non-polluting motor vehicles. Acid deposition has been the subject of transboundary disputes and agreements in Europe for more than twenty years. With India and China and a number of other countries industrialising and with growing use of fertilisers, acid deposition will become a widespread problem. There is usually a lag between a region suffering acid deposition of dust fall-out and the admission and control by the sources.

Regional air pollution problems have helped establish the need for transboundary and global air pollution agreements. For example, the burning of oilfields in the Gulf War; Mount St Helens (a volcanic eruption which illustrated how widely air pollution can spread); the Chernobyl disaster; and smoke pollution in Southeast Asia, Central and South America when settlers burned forest and scrub to clear farmland. Agreement on control of ozone scavenging CFCs has been relatively successful, similarly on control of sulphur emissions.

Energy

Energy, it may be argued, is a key resource because, given adequate supplies, almost anything can be done: seawater desalinated, fertiliser produced, pollutants treated, and much more. Global warming has so far been caused mainly by the use of hydrocarbon fuel (vegetation clearance and agricultural modernisation leading to methane emissions are increasing and may become dominant). The 1974 Organisation of Petroleum Exporting Countries (OPEC) oil price rise triggered a brief energy scare, and fears about oil dependence have been growing in the past couple of years (partly as petroleum companies have admitted a shortfall in new discoveries of reserves), and because of the carbon pollution associated with its use. Sustainable replacement(s) for oil and coal are not established, although it seems likely that the shift will be to a mix of non-polluting energy sources, including hydrogen, solar, biofuels, wave and wind. *En route* to the adoption of such alternatives, the next few decades are likely to see a massive increase in coal and oil usage in India and China, and possibly natural gas elsewhere (and possibly friction and competition for the resources).

Throughout history a number of energy sources have been exploited:

● Fuelwood is probably the oldest source of energy and in a number of countries is still important. The modern, sustainable zero carbon emissions version is the bioenergy plantation. Various societies used their tree cover faster than it could regenerate, and in most cases land degradation resulted; a number of countries are in the throes of such a woodfuel 'crisis' today (such as Haiti).

● Throughout history human slaves have provided many cultures with much of their energy. Reduction of slavery in late Roman times led to widespread abandonment of cropping on more marginal land.

● Flowing water has long been used to power milling, metalworking, pumping water and manufacturing cloth, and from the early twentieth century it has been used for generating electricity. A recent development is mini-hydro plants for small communities.

● Whale, seal, seabird and fish oils – mainly for lighting.

● Wind has met similar demands to water, plus driving ships. It is increasingly being developed as a non-carbon-emitting energy source.

● Draught animals and riding animals have provided transport, power for threshing and lifting water.

● From about AD 1752, coal usage was partly prompted in the UK by deforestation; coal use in Europe and the USA peaked between the 1830s and 1940s. Still important for electricity generation, usage is likely to increase markedly in China and India.

● Petroleum took over from coal between the 1930s and the present. Many feel that the world is currently at 'peak oil' and there is a pressing need to find alternatives.

● Nuclear power has been used since the 1950s. So far this has been restricted to expensive and dangerous fission reactors, which some countries have abandoned. Effective fusion reactors would be a huge breakthrough – the ultimate energy source if they could be developed.

● Methanol and ethanol have been used in several countries in times of petrol shortage, and extensively in Brazil since the 1970s. The Brazilian 'Alcool' programme has substituted alcohol for more than 25 per cent of petrol usage.

● Solar energy, tapped by parabolic reflectors, more passive forms of heat collection and photovoltaic cells, is already locally important and may be used more widely in the future.

● Hydrogen, methane, geothermal sources, and sea waves or sea currents are used in a limited way at present.

Energy sources should be non-polluting, sustainable and affordable, but dominant sources generally do not currently meet those requirements. It is also desirable that energy is accessible – otherwise there could be international pressures and even warfare. Access to oil resources was probably a factor in a number of twentieth-century conflicts and the world seems to have passed 'peak oil' availability – without an easy transition to alternatives there could be strife as well as energy shortages. Most eco-luddites who call for a return to pre-industry and oil lifestyles forget that most food is won with the input of petroleum.

The so-called 'peak oil' seems to have been passed, although there is some debate. There are some indications that finds of petroleum and natural gas reserves have not been keeping up with demand. Some oil companies appear to have exaggerated their reserves, and exploration costs are rising. There were plenty of Internet sources in 2005 warning of a global oil crisis and of a need to shift fast to non-petroleum/natural gas energy. Biomass energy may be one possible route. However, while this may cut carbon emissions, it could also encourage a shift from food crops and cause shortage or even famine. Some biomass production plantations can offer biodiversity conservation and help improve the environment in other ways. In the UK in 2005 there were trials with 'elephant grass' (*Miscanthus giganteus*), a tall grass which can be burnt in district generation/heating stations. Capable of yielding the equivalent of thirty-five barrels of oil a year in temperate climates, it offers cover for wildlife. Other biofuels include oil crops such as soya, rapeseed and sunflower (yielding biodiesel), soya, cassava and sugar cane (feedstock for alcohol production), and woodchip crops for fuel or chipboard manufacture (willow, eucalyptus and so on). Not all of these fuels burn cleanly and refining some may threaten to pollute rivers.

Wind power has proved more problematic than many had hoped and alone it is unlikely to compensate for declining oil supplies. Sweden and Denmark have made some progress with using sewage and refuse to fuel cars, buses and trains, but it has required heavy subsidies leading to high taxation. Energy conservation can help, but there is probably going to have to be some use of nuclear fission power for a few decades to come.

Preventing greenhouse forcing is a priority for twenty-first-century environmental managers, and with China and India about to expand, coal usage alternatives and pollution reduction technology have to be promoted. Some think hydrogen will replace petroleum; however, it would have to be produced in a cost-effective and non-polluting manner – and that is yet to be perfected. Chernobyl frightened many planners away from nuclear fission. Even if nuclear fusion reactors can be built, the technology is unlikely to be ready for at least forty years. For the next half-century there may have to be less than ideal, and possibly unsustainable, energy solutions – and nuclear fission may have to be adopted. Wind, wave, geothermal, biogas and solar power have been adopted in a limited way, but expansion awaits technological improvements.

Energy supply companies have huge funds at their disposal, so are both formidable opponents to, and potential catalysts of, change. New energy technology is likely to be part developed and promoted by petroleum or coal companies – the pumps dispensing petrol today may be providing hydrogen or electricity in the future but will have familiar logos on them. There are opportunities for generating methane from refuse and sewage in urban areas and in some rural settlements, if the latter have the right mix of waste. Vegetable 'diesel fuels', such as rapeseed oil and cultured algae, may have potential as sustainable fuel sources.

Land and soil

Knowledge about soil qualities and distribution, let alone current fertility, degradation status, or vulnerability to various threats such as global warming or acid deposition is inadequate. UNESCO and the FAO published a *Soil Map of the World* in 1974, and in 1990 conducted a Global Assessment of Soil Degradation, which provided an estimation of the extent and severity of soil erosion and decline in fertility – although at a coarse scale. The UNEP, International Soil Reference and Information Centre (ISRIC), FAO, and other bodies conducted a country-by-country world soil degradation assessment, published in 1991 as the *Global Assessment of the Status of Human-Induced Soil Degradation* (GLASOD). An assessment of soil degradation in Southeast Asia may be found at: http://www.iiasa.acResearch/LUC/GIS/soil-deg.htm (accessed August 2005), and for Central and Eastern Europe, the EU is currently conducting a Pan-European Soil Erosion Risk Assessment. Africa is widely held to have a soil degradation crisis. Serious soil degradation may be found worldwide – even in rich nations; for example, the US National Resources Conservation Service recently estimated US soil erosion costs at US$44 billion (for land degradation data by continent see http://www. soils.usda.gov/use/worldsoils/papers/land-degradation-overview.html – accessed June 2005). In 2003 the European Soil Forum was established to explore threats to soil, and to set up monitoring and research (http://www.forum.europa.eu.int/Public/irc/env/ soil/home – accessed July 2005). Soil Conservation and Protection Strategies for Europe (SCAPE) were being established at the time of writing (www.scape.org – accessed August 2005). Environmental managers need better soils data than is often available; unfortunately, soils specialists are in short supply.

Unless an adequate portion of the profits or sufficient labour is reinvested in land husbandry, yields will fall and soil will degrade. Failure to take care of the soil may result from poverty, population growth and insecurity of some kind, insufficient income to pay for soil upkeep, greed, or possibly ignorance. Lack of secure tenure may be a consequence of landownership patterns, tradition or civil unrest. Many land users have no adequate documentary proof of land rights, leaving them vulnerable to eviction. Consequently they are reluctant to invest in sustaining production. In many regions land reform is as important as agricultural innovation.

Researchers and agricultural extension services have developed guidelines for better soil and water conservation (Lal, 1995), but the challenge is to disseminate them and to win genuine adoption and support. In some countries novel participatory methods are being tried, and also attractive, farmer-friendly information – delivered as comic books or embedded in popular TV and radio programmes. Unfortunately, soil conservation services in many countries are seriously underfunded.

Well-managed soil can be a sustainable resource, and one that may even be improved; however, when misused the risk is that it will be degraded or even wholly lost. It can be difficult, costly, and even impossible to rehabilitate degraded soil. Soil degradation may be rapid and obvious – gullying, severe sheet erosion, crusts or hard layers – or gradual and insidious. The latter is worrying because it can go largely unnoticed until a threshold is reached: the minimum depth to sustain crops or natural vegetation followed by a sudden disaster. Soils are damaged through natural and human causes. Eroded material can be removed by flowing water, wind, or both. Wind erosion commonly follows droughts, ploughing, fires, or other ground disturbance; the sediment can end up in streams, or it can also be blown great distances before it settles. Wind-blown dust can have a considerable environmental impact where it settles, some being fertile and some harmful. Volcanic eruptions are a natural source of ash and dust (often very abrasive and acidic); sometimes enough ash and aerosols get into the atmosphere to harm huge regions and even affect world climate. Soil degradation can damage agriculture, alter water availability, and by releasing soil organic carbon into the atmosphere, seed global warming.

Soils can be degraded through human and natural causes – manifesting this in one or more ways, including: decline in fertility, structural damage, loss of organic matter (soil organic carbon), and sometimes chemical changes such as acid sulphate damage or acidification. There should be plenty of knowledge following the USA Dust Bowl disaster of the early to mid 1930s, and problems in many other parts of the world. Nevertheless, soil management has not had anywhere near adequate attention or funding.

With perhaps 45 per cent or more of the world's land surface significantly affected by soil degradation, funding for soil studies and remedial activities is inadequate. Some governments have cut back on their soil survey, monitoring, and management in recent decades.

Soils may lose their fertility through poor farming, overgrazing or pollution, they may become salinised (salt or alkali contaminated), compacted, develop impermeable layers, get waterlogged, their carbon content may be oxidised away, they can suffer acidification, or be covered by buildings. Even slight alteration of vegetation cover or drainage can trigger soil changes, possibly leading to permanent damage.

Pollution (such as acid deposition), global environmental change, the arrival of refugees and other exogenous factors can affect soils. Eroded soil can blow or wash far away and cause problems where it settles.

'Desertification' is a term now widely accepted to mean a process resulting in land degradation, which may be difficult to reverse. Although widely used, it is imprecise and emotive. Nowadays, desertification is usually seen as the product of mismanagement of vulnerable environments, manifest as the loss of plant cover followed by soil degradation. Common causes are: overgrazing, woodfuel collection, bushfires, salinisation, pollution, the introduction of new species and so on. It may sometimes be partly or wholly due to natural causes (http://www.fao.org/desertification/intro.asp and http://www.ciesin.org/docs/002–193.html – both accessed December 2003). When faced by drought or desertification, the first questions which should be asked are: 'Is it natural?' 'Is it exaggerated by humans?' 'Is it wholly due to humans?' If nature is to blame, attempts to control the problem will probably be a waste of resources; however, if humans are partly or wholly a cause, control should be easier (Kassas, 1999).

Where there is seasonal or periodic shortage of precipitation, high rates of evapotranspiration, extreme temperatures and freely draining soils, stress on the vegetation can lead to desertification. Worries about desert spreading in Africa were voiced in the 1930s (Stebbing, 1938; Aubreville, 1949), and grew after the 1960s to 1970s droughts in the Sahel. There is a tendency for land use to expand and intensify during a good rainfall period and fail to adjust when precipitation declines, leading to desertification. Desertification is more likely in drylands and uplands where vegetation is under stress and soils, once exposed, are quickly damaged. While the claim is often made that deserts spread, most desertification occurs *in situ*. Desertification happens in rich and poor countries, including Europe and the USA; it also occurs in humid environments where soils drain freely – even Amazonia and Iceland. In a number of countries regions with seasonal rainfall shortage have rising livestock and human populations for a number of socio-economic reasons, including improved healthcare and the process of marginalisation. However, desertification also happens where human populations are very low (e.g. Australia). It can result from the ill-advised provision of wells, which prevent livestock from moving to avoid overgrazing, or if veterinary care allows herders to accumulate too many livestock. Environmental managers must be cautious in ensuring that they correctly identify causes and do not waste resources treating symptoms or wrongly diagnosed problems.

Following severe drought and desertification in the African Sahel between 1969 and 1973 the UN called a UN Conference on Desertification (UNCOD) in 1977, which drew up a Plan of Action to Combat Desertification. Countries were then urged by UNCOD to make their own plans, but a quarter-century on there has been poor progress. Various agencies and NGOs continue to make extravagant claims that as much as one-fifth of the world's land area is 'desertified'. In 1994 a UN Treaty on Desertification was signed by a number of countries, and the UN currently has a Secretariat supporting its Convention to Combat Desertification which claims that over one billion people were affected (http://www.unccd.int/main.php – accessed February 2004). While there is a serious problem, the data to back up some assertions are questionable. Much of the anti-desertification effort has been 'whistle-blowing' and advocacy. Insufficient effort has been expended on reliable research, monitoring and the development of desertification countermeasures. There has also been a failure of signatories to UNCOD to honour pledges made – between 1978 and 1991 only around 10 per cent was actually paid up.

Drought is often a recurrent problem, and one which can spiral into greater seriousness when there is poor land management. Drought management is undertaken in developed and developing countries, and is a complex field, partly because politics is often involved: authorities tend not to want to admit degradation or failure to act, and sometimes ignore a 'backward' or 'wayward' region's plight. The need for drought forecasting and management is likely to increase as populations in vulnerable areas grow, as marginalisation takes place, and as climate changes. International oversight of drought management is desirable to counter the aforementioned shortcomings.

Response to drought can include de-stocking grazing areas, improving water supplies, food aid and so on. Some of these measures must be applied with caution to avoid dependency or causing people to relocate. Quite a few 'anti-desertification' or 'anti-soil degradation' efforts have had little effect on the environment or locals' well-being. It is not unusual in a region of overgrazing and impoverished agriculture to install a costly, 'high-tech' irrigation scheme producing export crops, which do little to improve local food supplies, and provide limited employment. In addition, such measures can displace locals into more marginal environments where they degrade the land. Some governments see marginal lands and their peoples as subversive or backward – fear of desertification is used an excuse for increased controls. Reij *et al.* (1996: 3) observed that a crisis narrative suits some agencies and politicians, because it enables them to claim that their intervention is needed. Desertification remedies have tended to be technical treatments of physical manifestations (symptoms), when what is needed are employment, subsidies to counter unfavourable market prices and so forth.

The idea that there is a crisis of desertification must be treated with caution, since there are a number of ways in which environmental conditions and human welfare can be misread, including:

- misunderstanding physical, biological and human social conditions through inadequate knowledge or cultural blindness;
- adopting a 'snapshot view' – observations too constrained in time or space (or both). It is easy to over-generalise about desertification or land degradation, and short-term trends may be interpreted as indicating longer term patterns;
- interpreting limited 'hotspots' as proof of a widespread problem;
- using indicators that mislead (e.g. one study maps vegetation at the peak of a wet season or rainy climatic phase, another at the end of a dry season or during a dry phase);
- mistaking symptoms for causes;
- mistakenly seeking 'average' conditions in harsh environments where conditions fluctuate a lot, leading to mismanagement;
- an apparently resilient but vulnerable environment can suddenly be altered. This could happen when annual vegetation flowers or has set seed – fire or a brief spell of grazing might break the cycle and initiate degradation;
- the distortion of conditions by periodic or random climate fluctuations – for example, an El Niño event might trigger a drought;

- the initiation of feedback by a small change: there are cases where vegetation clearance altered the albedo, made it hotter, and started a trend of drying that was difficult to reverse;
- accepting received wisdom without checking its source.

A variety of people are involved with land degradation and desertification, and some have their own agendas: aid acquisition, local empowerment, regional vote catching and so forth. One is well advised to heed Warren's (2002) warning that land degradation should be judged in a spatial, temporal and cultural context. Too much of what is published has not been based on reliable facts. By the early 1990s a number of researchers were questioning the received wisdom that some regions had serious environmental degradation and desertification (Thompson et al., 1986; Ives, 1987; Ives and Messerli, 1989a, 1989b; Thomas and Middleton, 1994; Fairhead and Leach, 1996; Leach and Mearns, 1996; Lomborg, 2001). In 2002 there were reports that conditions in the Sahel had markedly improved, apparently due to more precipitation, and because farmers have been adopting soil and water conservation.

Some attempts to improved vegetation cover to aid soil, water and biodiversity conservation have the opposite effect. Plantations of eucalyptus trees may transpire more water than the scrub replaced. The eucalyptus intercept fine precipitation and then shed it as large and erosive drops from as much as twenty metres above ground.

Beneath the trees little wildlife may survive and few jobs are generated. Efforts have often been wasted on insensitive schemes, which seek to impose terrace or check-dam construction to counter sheet erosion and gullying. Farmers build a few so long as there is a grant or food aid to do so and, if not convinced of the value and that maintenance effort is worthwhile, they will abandon them. Remedies must be sustainable, address causes, not just symptoms, and locals have to believe in their worth. Thus sheet or gully erosion results from rural depopulation, structural adjustment, market forces and so on; building check-dams is just treating the symptoms. The real causes might be better treated through economic policy changes, employment schemes or subsidies.

The keys to countering land degradation and desertification are broadly: to improve vegetation cover and soil fertility, encourage soil and water conservation to support a shift to alternative land uses, which cause less damage, and try to establish alternative livelihoods. The ideal is to seek long-term, rather than short-term improvements, and to stimulate land users to initiate changes, wherever possible using their own labour and funds and building on established (traditional) methods to make improvements. However, soil and water conservation efforts may not seem especially attractive to poor farmers – the return may not be quickly obvious, and it may appear to benefit others rather than those paying for it. Development of new approaches to assessing the needs, attitudes and capabilities of local people, notably rapid rural appraisal and participatory rural appraisal, should assist efforts to counter soil degradation and desertification.

Food

Food may be seen as an indicator of developed status. For Western nations the lack of food shortages since 1945 has led to some degree of complacency. In recent EU negotiations supporting and stimulating agricultural production appear to have had a remarkably low profile. World food reserves – supplies to feed people in the event of one or more widespread failed harvests – are limited. Were situations like the poor weather in 1815 and 1816 to recur today there would be shortages. That disaster was probably due to a relatively moderate volcanic eruption – another could easily happen.

Too much food comes from too few areas and too few types of crops. It would be wise to encourage the world's food producers to diversify crops and generate and store greater reserves. World trade and economics do not support such goals. The Indian Ocean tsunami in 2005 encouraged awareness of environmental disasters; probably a major difficulty with world food supplies will be needed to provoke change. In 1972 and in 1975 the USSR had crop failures and bought grain, sending global food prices soaring. With commercial pressures discouraging food reserves storage, a similar situation may well bite harder; if farmers have shifted from corn to growing industrial crops there will be less to fall back on. Sustainable development strategies should not lose sight of a need for safe food reserves.

For the past fifty years or more agriculture has been relatively successful. Some of that achievement has been cancelled out by population increase and there are growing fears that a decline may set in. Much of the improvement has come from irrigation and green revolution crops dependent on inputs of agrochemicals. Irrigation expansion in the future faces challenges: current schemes often degrade rapidly, and agriculture-related pollution is a problem. The hope is for technological and crop-breeding breakthroughs which reduce the need for water and agrochemicals, but progress is seldom a stable rising curve and population still climbs. Global environmental change could alter environmental conditions enough to disrupt farming. There may also be a hidden crisis, whereby recent innovations have maintained crop yields yet promoted soil degradation. Worldwide urban growth is destroying considerable areas of productive farmland, some of it the best available. The GLASOD programme suggested that between 1945 and the early 1990s, about 23 per cent of the world's productive land had degraded to the point of uselessness. Environmental management must encourage objective review of strategies to seek vulnerability reduction through promoting resilience and adaptability. In addition, efforts are needed to encourage production and storage of adequate emergency reserves of food and diversification of production methods, inputs and so on to make food supply less dependent on a few areas and strategies.

While food may be seen by many as the key resource, it is to a large extent dependent upon others, including water, soil, energy and biodiversity (Pierce, 1990). As with other resources it is the poor who first feel and who most suffer any shortage. Rising populations and demands for better standards of living put agriculture and fisheries under growing pressure. Strategies used at present place the emphasis on yield increases, and much less on sustainability, security of harvest, reduction of environmental damage, equitable access to produce and employment generation. The promotion of environmentally sensitive and sustainable strategies has been termed the *doubly green revolution* and some see this as the only effective way forward. Much of today's agricultural research and investment is directed at commercial producers in favourable locations. There is less interest in the world's small farmers and herders who often live in relatively harsh and remote environments. In many regions small farmers and herders are suffering a breakdown in traditional livelihood patterns as a consequence of various development pressures. Agricultural improvements are needed to counter such problems and to try to control the resulting poverty, land degradation, urban migration and emigration.

Globally, food production has risen to an all-time high (World Bank, 2003: 84). In 1997 world food production, if divided on a per capita basis, should have given everyone around 2,700 calories per day – adequate for most (Conway, 1997: 1). Yet, in early 2001, food emergency situations arose in thirty-three countries and affected more than sixty million people. In Africa alone, eighteen million needed food aid in 2001 to 2002. Recent estimates suggest that, worldwide, around 830 million lack adequate access to food. About 210 million of these are in sub-Saharan Africa; 258 million in East Asia, 254 million in South Asia, with Latin America plus parts of the Caribbean and North Africa having about 8 per cent, 5 per cent of the total. Today people are hungry, not because there is insufficient food per capita, but because it is unavailable in their region or they are too poor to afford it. In the future it may be less easy to meet per capita needs.

What can environmental management contribute to improving food supplies? A useful input is to offer an overview sufficiently removed from actual production to be dispassionate. Archaeology shows that some pre-agricultural societies had more abundant and better balanced diets than many enjoy today, sometimes for much less labour input; and, because they could move around easily, they were less vulnerable (Fagan, 2004). Agriculture, when it functions well, feeds many more than hunter-gathering can, but it has drawbacks. Estimates suggest that by AD 2020 the world will have 2.5 billion more mouths to feed.

Hunger is an ongoing problem of too little food being available; famine is the catastrophic impact of hunger. Malnutrition is caused by diet that is inadequate in some way: possibly insufficient quantity, or poor quality, or even an excess of some foods. Famine can result from socio-economic or environmental causes. Often multiple factors trigger it and these may vary from event to event. Sometimes the problem is actual food availability decline – a shortage; alternatively, there may be supplies, but people cannot get access – because of poor transport; or they earn too little to pay the going price; or they are denied food for socio-political reasons. The economist Amartya Sen studied how insufficient income can reduce people's entitlement (access) to food, even though supplies are present. He noted how in Bengal (India) in 1943 rural labourers received lower wages than urban workers, so could not afford food at a time when prices were rising, and starved. Getting food may be as much related to obtaining adequately paid employment or land reform as it is to improving harvests. There is also the challenge of providing better nutrition – especially improving supplies of protein for developing countries. When food supplies falter, aid can consist of emergency rations, assistance with research, extension and training, help with infrastructure, or development projects; ideally whatever is needed to restore adequate long-term sustainable production.

Environmental managers, briefed by environmental historians and other specialists, should warn present-day food supply bodies of challenges to stable food production and suggest ways of reducing risk. They could also encourage the consideration of alternative means of food production which powerful agribusiness is unlikely to initiate.

The reasons a number of countries have witnessed a decrease in hunger since the 1940s include improved transport and storage; agricultural development (especially better seeds, chemical fertilisers, irrigation and pesticides) and socio-political developments (Action Against Hunger, 2001). However, many of the world's farmers are still poor, and many live in harsh environments with bad communications and are in poor health through diseases like malaria. Their livelihood is subsistence agriculture or similar, so they have little cash surplus to save against failed harvests or invest to sustain and improve production. Because such agriculturists have little money and grow low-value crops, agricultural industries tend not to invest in improving their lot because there is little profit in it. Consequently, commercial agriculture may be flourishing, while close by subsistence agriculture supporting considerable numbers of people is weak.

There is a growing trend towards monocropping; genetically identical crops are generally more vulnerable than traditional diverse crops, their mechanised production means less employment, and they might cause environmental and consumer health problems – a 'high price for cheap food'. Modern agricultural developments are likely to give agroindustry growing control over inputs and marketing (Lang and Heasman, 2004).

This could drive many small farmers from the land and have adverse environmental impacts. Environmental managers at national and international level need to be vigilant for these sorts of threats. Much of the soya exported from developing countries goes to developed countries to feed livestock, especially the EU, which has shifted from purchasing from the USA because it is likely to be genetically modified. In the EU livestock and cropping are increasingly diversified: livestock manure is a serious problem and is generally not used as fertiliser; crops therefore get chemical fertilisers, and animal feed is imported. In the past most farms had cropping and livestock which meant less need to import feed or fertiliser (and manure disposal was not a problem).

Nations such as the UK had famines in the past with much lower populations than now (e.g. in AD 1294 and repeatedly up to the late 1840s). The causes include one or more of the following: inclement weather, crop, livestock or human disease – the latter hit labour input; warfare, and unfavourable economic conditions. Over the past fifty years a number of former food-exporting nations have become food importers (including China). Globally, food production per capita is declining, and so are seafood landings.

Improving food resources

Once, all humans were hunter-gatherers; now most get their food from agriculture, although fishing is still largely high-tech hunting. Aquaculture may one day replace fishing, but so far development has focused on luxury products, and is seldom sustainable and environmentally sound. Many of the world's ocean and freshwater fisheries are in a sorry state.

Agriculture is a complex process, or more accurately an ecosystem and socio-economic system, pursued via many different strategies. To succeed, each strategy has to effectively manage a complex mix of environmental, natural-resources-exploitation, cultural, political, social, technological, institutional, economic and legal issues. Those discussing agricultural development often start by attempting to draw up a typology (classification) which helps deal with the complexity. Without going into detail, it is possible to divide agriculture into sedentary and non-sedentary. Most of the world's agriculture is now sedentary, and relies upon the ongoing use of a given land area. There are still large numbers of non-sedentary shifting agriculturists in some countries, who increasingly fail to sustain adequate production and must either find alternative livelihoods or make a transition to sedentary agriculture.

The world's agriculture may also be crudely divided into family and non-family, or into subsistence and commercial. But a more relevant division might be into those willing and able to invest in sustaining production and innovation, and those unable or unwilling to do so. Before the 1930s cereal yields in developed and what are now developing countries were roughly comparable. By the 1940s a number of richer countries had vastly boosted their harvests through improved seeds, fertiliser applications and better techniques (including pesticide and herbicide use). This 'intensification' through improved crop varieties has depended greatly on irrigation and agrochemical use. Since the mid 1960s the hope has been that, with the right aid, large parts of the world can also intensify; this strategy has been called the Green Revolution.

The alternative to intensification is 'expansion' – extending food and commodity production into new areas. In 1990 about 11 per cent of the Earth's land surface was cropped (used to grow food or commodities – this excludes grazing land), most lying within 35° of the Equator (Pierce, 1990: 35). Easily usable land is occupied – further expansion will require breakthroughs in crop breeding and techniques that will overcome lack of moisture, saline soils, infertile land and pests. It will also probably require social changes and community developments. It is reasonable to say that the world currently relies for most of its food on Green Revolution agroecosystems producing wheat, rice, maize and barley. Those agroecosystems may not be sustainable under current practices. There has been a trend towards more imports of developed-country-produced wheat into developing countries. This is sometimes a cause and sometimes a consequence of falling cultivation of traditional crops. There is also growing demand for wheat prompted by Westernisation and urbanisation.

Food security may be defined in many ways: one is a mechanism which ensures that people would not starve if faced by one or more year of scarcity. That means adequate surplus stored in more than one place, sufficiently dispersed to prevent total loss in a major disaster. History has shown that people can be hit by a number of successive poor harvests. Over 60 per cent of cereals produced by developing countries in 2002 was consumed by livestock, a large proportion of which is in developed countries. Thus finding surplus to store may be difficult.

In 1996 the World Food Summit in Rome pledged to seek food security for all and a halving of world undernourishment by 2015. The task is not getting easier with growing populations, global environmental change, pollution, land degradation and difficulties in obtaining satisfactory irrigation water (Cohen, 1996; World Bank, 2000). Areas of East Asia, sub-Saharan Africa, Central America and the Caribbean face serious hunger. HIV/AIDS will soon infect over thirty-six million people worldwide which will reduce labour availability and seriously disrupt family livelihoods, especially in sub-Saharan Africa, where some countries have infection rates of over 30 per cent (FAO, 2001: vi). In 1999 world agricultural output increased by about 2.3 per cent (with good harvests of cereal in the Sahelian countries), but growth slowed to around 1.0 per cent by 2001 with cereal demand outpacing production between 1999 and 2001 (FAO, 2001a: 3, 17).

Sub-Saharan Africa is where food production is lagging farthest behind population increase; between the 1960s and 1999 the continent's per capita food production fell by about 20 per cent and food imports increased (Goodman and Redclift, 1991: 155; Pretty, 1999). While grain prices tend to be rising, other food or commodity prices may be behaving differently – for example, coffee fell markedly in 2000 to about one-third of what it was in 1993. Where agriculturists produce such a commodity they must sell much more to obtain vital farming inputs and buy food. In such circumstances farmers and herders tend to neglect land management; migrate to find employment, leaving less able people to manage; or undertake exploitative production of livestock, narcotics and so forth, which further damages the environment (Robbins, 2004). Similar impacts may be caused by withdrawal of subsidies, grants and low-cost loans – often prompted by structural-adjustment programmes or by World Trade Organisation impacts.

Food and agricultural commodity prices are affected by a diversity of global, as well as local and national factors – e.g. oil price rises are likely to boost cotton and natural rubber prices because competing synthetics are made more expensive. The Uruguay Round of the multilateral trade negotiations meeting at Marrakesh in 1994 to discuss a General Agreement on Tariffs established the World Trade Organisation (WTO). WTO agreements have affected how countries can subsidise inputs and apply tariffs and other controls to support domestic production and discourage foreign competition; hopes for free trade and 'open' markets may have unwanted impacts on agriculturists and their environment.

Agriculture and fishing are conducted in the face of environmental change, extreme weather events, climatic fluctuations, tsunamis, volcanic eruptions, movement and evolution of pests and diseases, and many other variables. A significant portion of past famines can be blamed, at least in part, on natural causes (Davis, 2001). In the past pests such as locust and livestock disease such as rinderpest could spell disaster.

Effective pesticides, vaccination and other controls used between the 1940s and 1980s have caused a false sense of security. Recent warfare and austerity measures in many parts of the world have hindered the monitoring and control of locust, and the cost of pesticides and stricter environmental pollution prevention measures have also made resurgence more likely. The outbreaks of foot-and-mouth and BSE in the UK in the late twentieth century offer a warning of the need for ongoing vigilance, research, and expenditure and proactive policies (for statistics and maps showing pest threats see FAO, 2001b: 204–213).

Apparently sustainable production of natural products can be problematic: crops such as rubber, oil palm, timber or wood-pulp trees are often grown on common land, which results in biodiversity reduction and dislocation of people who may then clear and destroy flora and fauna elsewhere. There is relatively limited experience of plantation sustainability – few pre-date the 1950s, so it is uncertain how long they will be productive and how often trees can be renewed. There has also been little attention directed towards finding ways to make plantations more able to support wildlife (Berkes, 1989). Soya is in demand for humans and livestock in rich nations, and production is expanding into Amazonian forests, having already helped reduce wildlife across large swathes of savannah. The spread has been prompted by the development of varieties which can withstand humid climate and unfavourable soils and the buoyant market for 'non-GMO' soya produced in Brazil. Often, export crop processing is decentralised to ensure the produce does not deteriorate before it reaches the factory; consequently, if effluent controls are not effective, pollution is likely to be widespread.

Aquaculture is often hailed as a route to improved and sustainable food production; in practice much of the huge expansion is not sustainable and is luxury food for export. Several Latin American, Asian and Southeast Asian countries have serious problems as a result of developing lucrative tiger prawn and fish aquaculture export production. Mangrove areas are destroyed to construct ponds, and water resources are exploited and returned to the environment contaminated with waste and chemical pollutants. In order to feed the prawns, inshore waters may be netted for juvenile fish and plankton, which destroys wildlife, local fisheries and spawning areas. Less damaging aquaculture should be possible but requires willingness to invest. Currently, developers make big profits and have little incentive to improve things. The environmental manager faces a challenge – to break established malpractices and prompt a change to non-polluting intensive 'closed-cycle' aquaculture and sustainable sources of stock and feed.

Where natural products are being over-exploited the problem can sometimes be effectively addressed by imposing controls – extraction quotas, wardens, export restrictions, exploitation seasons, limited numbers of extraction licences, compulsory methods and so on. Profits from the licences may be ploughed into environmental management, but often get used for other things. It is also possible to establish protected areas, where some biodiversity is retained but extraction is also allowed, or to develop intensive production of commodities or foods or substitute with an alternative. There has been considerable interest in using bioregional planning to support conservation. A bioregional approach could be a practical way of dealing with conservation and other environmental activities at a manageable scale (Sale, 1985; Stolton and Dudley, 1999: 215–223).

Better storage of food

Much of the world's food is lost between harvest and consumption to rot and pests – loss reduction methods are needed. Those who store food on a large scale are developed country or multinational commercial organisations seeking to maximise profit.

It would be wise to encourage more local food storage, regional food reserves, secure and sustainable livelihoods, and so on. Another possibility is to somehow encourage non-commercial storage. In recent years grain surpluses have been falling, storage has been cut back, there is more demand for grain for livestock feed in developed countries, and growing interest in agricultural feedstocks for energy or industry. These trends discourage large- and small-scale production and storage of food.

Concern over the use of fungicides and pesticides has made it more costly and difficult to reduce losses. Better, non-polluting, low-cost means for improving storage are needed at local and larger scale: solar drying, pest-proof storage and so on.

Timber

Even thousands of years ago various cultures around the Mediterranean and elsewhere had destroyed vast amounts of timber. Fuelwood, charcoal, land clearance, grazing, building and ship construction have taken their toll. By the eighteenth century some countries were attempting conservation and replanting. Unfortunately, replanted trees seldom restore the diversity of natural forest; nowadays replanting is often in the form of exotic conifer or eucalyptus plantations, which cause even more disruption. The demand for timber, woodchips and veneer has provoked the deforestation of much of the world's lowland tropical forest and a good portion of temperate old forests, with what was the former USSR currently being stripped. This exploitation often yields a one-off profit in return for the destruction of a potentially sustainable harvest and some biodiversity conservation if there were adequate forest management. Frequently the removal of forest results in land degradation and little use for the land after timber removal. There can be further problems, even altered water resources and climatic change.

Monitoring deforestation has not been very accurate: causes are varied and commonly complex, and controls have generally been ineffective. Books on the problem and possible ways to resolve it have proliferated (Grainger, 1993; Rietbergen, 1993; Dudley et al., 1995). However, destruction continues at a worrying pace in many countries.

Biodiversity

Biodiversity is the diversity of different species together with genetic variation within each species – Myers (1985) called biodiversity the 'primary source'. It is material vital for sustainable development because:

- Crops, livestock and pharmaceutical products are constantly challenged by a inconstant environment and evolving pests and diseases, and must satisfy new demands and fashions; biodiversity is needed to breed new solutions to these challenges.
- Without the 'raw' genetic material there will be great difficulties in developing new crops, pharmaceuticals, fish for aquaculture, improved and novel livestock, bacteria and yeasts, and many other innovations.
- Biodiversity has philosophical and aesthetic value and may inspire new ideas and scientific advances (Posey, 1999).
- A case can be made that humans have a moral obligation to maintain biodiversity in trust for the future.
- Sustainable development dictates that humans should pass on to successive generations the same amount of biodiversity that they enjoy.

• Biodiversity-poor environments may be less resilient and less able to recover if disturbed. However, there is considerable debate on this issue (Tisdell, 1999: 38–40; Adams, 2004).

Nobody knows what may be needed to ensure a stable, secure and sustainable future environment. Biodiversity conservation keeps open future options; organisms cannot be re-created once lost. For some time biodiversity losses have exceeded the natural rates, and it is getting worse – to the extent that many argue that the world's greatest mass extinction is under way and humans are responsible.

Improved agricultural productivity has often led to reduced biodiversity through replacement of natural vegetation with areas dominated by very few crop species. Agro-industrial corporations promote fewer crop varieties, and these are often bred or genetically engineered so that there is no chance of farmers saving viable seed. Growers become more dependent on business, and this is probably reducing food security. For most of the time that humans have practised agriculture, domesticated varieties have grown in fields with wild ancestor species nearby, so that there has been ongoing cross-breeding which generates useful new varieties and maintains genetic diversity in the crops. New farmers are compelled by commercial forces and legal controls to plant a limited number of varieties, so that older types are lost while environmental degrada-tion depletes the surrounding pool of wild species. Morals aside, companies which develop new crops and biotechnology seek to recoup their investment in research by patents and similar measures aimed at protecting intellectual property rights and by restricting access to their gene banks (Khor, 2004). Often the progenitors of domesti-cated plants, livestock and fish, and material for biotechnology and pharmaceutical use have restricted distributions in the wild, and can easily become extinct as land is disturbed, climate changes, there is acidification, or water bodies are polluted or drained.

Many food crops have already been contaminated by DNA from genetically modi-fied organisms (GMOs). There is a risk that this could take place worldwide. It could also affect material in gene banks because viability declines and stored material is planted and seed re-harvested at intervals, at which point there could be contamination by pollen on the wind or carried by insects. There has been some response to GMO threats – in 2003 the Cartagena Protocol on Biosafety came into force; this is a legally binding international agreement which governs the transboundary movement of living GMOs (http://www.biodiv.org/default.aspx – accessed March 2004).

What is needed is more biodiversity conservation and open access to conserved bio-diversity. Unfortunately, even important conservation bodies such as Kew struggle for resources. What is needed is international agreement and funding to create a perma-nent, legally binding biodiversity conservation system, which involves all countries. The outcome of failure to do so could be either: increased plunder for private collections by businesses trying to take control of genetic resources; or each country demanding sovereign rights over biodiversity, which it can then sell to the highest bidder. Recently there has been lobbying to provide some rights for the source areas and peoples who traditionally used 'wild' genetic material which may be developed by business (Posey, 1990).

The Convention on Biological Diversity was negotiated just before the 1992 Rio 'Earth Summit', and came into force in 1993. It accepts that those countries with rich biodiversity would better conserve their resources if they could make money from them. As well as encouraging conservation the Convention supports sustainable use of biodi-versity and equitable sharing of benefits (McConnel, 1996; Convention on Biological Diversity: http://www.nhm.ac.uk/science/biodiversity/cbd.html and http://www.biodiv. org/doc/publications/guide.asp – both accessed January 2004). In some cases controlled trade in valuable species might be used to pay for conservation; the terms of the Convention on Trade in Endangered Species need reforming to support such ends (Hutton and Dickson, 2000). In spite of sluggish support from the USA, Canada, Australia and New Zealand, there have been efforts to agree that in return for open access to biodiversity, gene banks would ensure that their collections were available.

How can remaining biodiversity be best protected? The main ways are to promote conservation, and to discourage selfish property rights, unwise trade practices, damaging fashions and investment which destroys biodiversity (Frankel et al., 1995; Dobson, 1996). Efforts should be made to promote awareness of environmental changes which might affect conservation areas and to counter any pollution which threatens biodiversity. People are subjective in selecting what to protect. An organism may be saved because it is perceived to have potential economic value, or it seems deserving (Shiva et al., 1991; Perrings et al., 1995), or conservation occurs because a site or the biota are sacred (Berkes, 1999). Tisdell (1999: 27) noted that the loss of the dodo is often mourned, while the loss of numerous other species has attracted minimal attention. Where biodiversity is seen to have potential to generate profits, commercial interests may undertake conservation. For other species, various forms of aid will play a crucial role in conservation. Large animals like elephants, dangerous predators and those which migrate cause problems due to their mobility and because they pose a threat. Marine mammals and migratory fish also pose conservation problems because they do not stay in territorial waters.

Biodiversity stock-taking is an important first step (Groombridge and Jenkins, 2002). An assessment should then be made to determine what is vulnerable (Heyward, 1995). Next, efforts should be made to identify the root causes of loss; these are often indirect, cumulative and complex, and so difficult to trace (Wood et al., 2000). Once all this has been done it should be possible to develop a biodiversity management strategy (O'Riordan and Stoll-Kleemann, 2001). There are now many specialist NGOs which focus on specific ecosystems or species, as well as those with more general biodiversity conservation interests.

Often conservation and livelihoods appear to conflict: efforts to protect forests may be greeted by locals involved in the timber industry as a threat; attempts to restrict fishing methods or access to areas that have been traditionally used can result in opposition (Figure 10.2). Many conservation authorities argue that there is a need to involve local people and gain their support and, wherever possible, to offer them livelihood opportunities associated with conservation (Munasinghe and McNeely, 1994; Borrini-Feyerabend, 1996: Lewis, 1996; Jeffery and Vira, 2001; Koziel and Saunders, 2001). If not involved, locals may poach or resist conservation in other ways. Locals are increasingly involved in policing, administering and servicing conservation areas, can bring to bear considerable traditional knowledge and are adapted to local conditions. Community participation (co-management by the authorities and locals) has been married to adaptive management (and adaptive environmental management), to develop adaptive co-management strategies.

Community-based conservation has been spreading since the mid 1980s; for example, the Communal Areas Management Programme for Indigenous Reserves (CAMPFIRE) in Zimbabwe (Hasler, 1999). Typically efforts are made to involve and empower locals, tap their knowledge and skills, and ensure institution building provides stable foundations for supporting conservation (Pimbert and Pretty, 1995; Leach, et al., 1997). There are critics of the fashion to integrate conservation with development and the encouragement of local participation who ask: 'What should take priority: poor people or irreplaceable biodiversity?' In a number of places, including West Africa, participatory approaches may not be as successful as proponents claim. Oates (1995, 1999) argued

Figure 10.2 Species-rich tropical montane rainforest clothing the Cameron Highlands (Malaysia) *c.* 2,000 metres above sea-level. A new highway cuts across the slope and has caused severe landslides. The highway allows access to once remote areas and may aid deforestation, prompt further unsustainable farming and cause exhaust pollution. However, it could bring in visitors for eco-tourism and help support a change to sustainable organic agriculture.

that conservation should be for the intrinsic value of the biodiversity, not a subordinate part of socio-economic development. Participatory conservation initiatives need to be carefully planned and monitored, and not simply established and left unsupervised (Child, 2004).

There is potential to combine biodiversity conservation with sustainable livelihoods if it is done carefully; for example:

- *Tolerant forest management* – the extraction of products and some cropping, while striving to maintain as much of the original forest or other vegetation cover as possible.
- *Extractive reserves* – conservation areas where local people or other approved groups can remove products in ways that do minimal damage. The extraction may help pay for the conservation. Brazil first created extractive reserves in 1990.
- *Green tourism* – efforts are made to reduce environmental impacts and possibly use some of the profits for environmental management; tourism is often dependent on natural features or wildlife (see Chapter 14). Green tourism includes: golf courses, which restrict pesticides and herbicides and seek to encourage wildlife; watersport resorts which try to protect reefs and marine life; archaeological sites which pay for roads and generate funds for local communities.

- *Ecotourism* – this is a stronger form of green tourism, which contributes a significant portion of profits to environmental management (see Chapter 14). Typically, local people act as guides, and staff the accommodation and other services; the attraction is usually wildlife, scenic beauty, archaeology, or tightly controlled hunting and fishing or photographic tours. An attempt is made to keep the impact of tourists and support facilities to a minimum, and to educate the tourist to respect nature. Sometimes tourists even pay to work on environmental projects. Frequently there are excessive claims for the benefits of poorly managed ecotourism. There may be a problem of seasonality, so that ecotourism only provides a satisfactory livelihood for part of the year. As with other forms of tourism, there is a certain amount of vulnerability to sudden changes in fashion or scares about travel caused by distant crises, which can suddenly hit profits and any environmental management which has come to depend upon it (Tribe, 2000).

Conservation can be linked to various benefits; for example, a forest reserve may also lock up atmospheric carbon, protect a slope or catchment, or support tourism. A conservation project should wherever possible be integrated with other beneficial activities; an example is the Great Barrier Reef World Heritage Area Strategic Development Project, which links conservation with over sixty organisations in reef-related food production, job creation, recreation, cultural heritage protection and so on (Raymond, 1996).

Developed countries benefit from funding biodiversity conservation in developing countries. Nowadays, people are using more products derived from areas well away from where they live, so they have an impact. They should be made aware of this and of their obligations to pay something towards resolving any problems that their consumer habits cause. Developed countries can fund advertising to counter fashions which endanger species. Treaties and controls can help reduce the trade in endangered species. Sales controls and taxation can reduce trade in endangered species and fund conservation activities, as can tourism taxes, airport taxes and levies. The processes of globalisation and free trade agreements need to be closely monitored to ensure that these do not trigger biodiversity losses.

One way of funding biodiversity conservation and other environmental management has been the, sometimes controversial, debt-for-nature swap (Tisdell, 1999: 63). Many conservation activities are expensive to establish, demanding land purchase, infrastructure, vehicles and so on; but they cost much less for ongoing management, so foreign aid and measures such as debt-for-nature swaps can provide crucial start-up money.

Some reserves are too small to be viable long term, or are in localities that are prone to environmental change. If possible, reserves should be duplicated. Linking a number of reserves with corridors can enable species to disperse and adapt to change. Another beneficial measure is to establish buffer zones around reserves to reduce impacts on the actual conservation area (Oldfield, 1988). Although a reserve may be large enough to maintain biodiversity indefinitely and have sufficient resilience against expected climatic change, there are still disasters which can strike: bushfires, the arrival of pest species, pollution from distant activities, warfare, encroachment by squatters, fuelwood collectors, illicit miners, poachers and so on. In New Zealand recently one 'eco-terrorist' introduced predatory opossums to bird reserves because he felt disgruntled.

Some countries and a number of funding bodies now insist on biodiversity impact assessments before approving projects or policies. In addition, a number of environmental management systems have recently integrated biodiversity concern into their standards and prompt developers to publish biodiversity action plans outlining what they intend to do to reduce impacts (Porter and Brownlie, 1990; Barrington, 2001).

Minerals

Mining impacts include noise, pollution of streams and air pollution. Mining activities can be divided into large and small scale; both cause problems, but the former are easier to police. Large mining operations and small bands of miners in various parts of the world spread diseases, hunt out game animals, and pollute the environment with compounds such as mercury which kill fish and other organisms even at low concentrations (Cleary, 1990; Warhurst, 1999). Papua New Guinea has serious difficulties with the waste from large copper mines polluting rivers. Many of the world's large dams service mineral extraction and processing, and also cause serious environmental and socioeconomic problems (Barretovianna, 1992).

Ideally, a mining scheme should be required to store topsoil for rehabilitation and bank funds for restoration work, pollution control and compensation in the event of unwanted impacts. In time, abandoned mining land can be restored to something like its former state or can be of considerable conservation value even without much restoration, especially if it is managed to encourage biodiversity.

Mining may be for energy resources, for manufacturing, agricultural inputs (nitrates and phosphates), building materials, and wholly or partly for aesthetic value (gemstones and precious metals). Mining impacts upon the immediate vicinity, affecting workers and near bystanders with noise, blasting, airborne and waterborne pollution and subsidence. Those further afield are affected by transportation of ore, processing, hydroelectric generation for refining and workers' settlements. Stream and groundwater may carry pollution great distances, and some mining-related pollutants are very harmful and persist for a long time, even after mining has ceased. Most mining is unsustainable, although salt evaporation and geothermal emissions can be managed indefinitely. Abandonment may leave an expensive clean-up for people in the future, or in certain cases it offers sites for biodiversity conservation, fishponds and recreation. In some countries mining is one the few employers, and workers migrate to it, causing labour shortage in the areas they desert, which can lead to environmental degradation. Deep ocean mining has yet to develop, but may in the future challenge environmental managers – petroleum drilling is easier to inspect and monitor than, say, dredging phosphate nodules from the seabed. Gravel, tin, gold and diamond dredging are established in some shallow seas.

Environmental management tends to be dealt with by mining sector specialists. There has been considerable interest in the threats posed by mineral exploitation in vulnerable environments such as the Arctic and Antarctica. Diamond (2005) examined mining in Montana and the Far East, finding very different environmental management practices; in some cases mining companies were as good as anyone could reasonably expect. Large mining companies have huge resources and some embrace environmental management, while others can be cavalier.

Wetlands

Wetlands are very diverse; some are permanently waterlogged and others only during periodic flooding (see also Chapter 13). Some are literally at sea-level and some at high altitude. The nutrient regime varies from nutrient rich to poor and from acidic to alkaline. Some of the world's most productive and stable ecosystems are wetlands, and a number of sustainable agriculture strategies are types of wetland. Swamps, peatlands, floodlands, coastal marshlands, mangrove forests and so on are being lost at a worrying rate and are valuable environmental assets (Roggeri, 1995). Some are rich in biodiversity and many act as vital 'stepping stones' for migratory species. Coastal wetlands and mangrove swamps are important breeding grounds for marine organisms, provide coastal protection, and are valuable sources of game and wood for many people. The removal of coastal wetlands can also expose offshore coral to silt, which might otherwise have been trapped. The main initiative seeking to protect wetlands is the 'Ramsar Convention', the Convention on Wetlands of International Importance Especially as Waterfowl Habitat. Many countries signed this in 1971, and the Convention maintains a permanent Bureau which works closely with the IUCN to monitor and support wetland conservation.

Wetlands are often drained for irrigation schemes, aquaculture, docks, malaria control, real-estate development and so on. Larger wetlands such as the Sûdd in the Sudan and the extensive marshes of Iraq have suffered through agricultural drainage and warfare. In South America the huge Pantanal wetlands could be damaged by proposed navigation and land development projects. In Amazonia areas seasonally flooded by rivers carrying fertile silt – *várzeas* – are being developed for rice and ranching. Tropical forest conservation has attracted the attention of aid donors and charities, tropical wetlands and grassland. There is considerable literature on wetland environmental management in Europe and North America (e.g. Turner and Jones, 1991; Mitsch and Gosselink, 1993; Turner *et al.*, 2003), but tropical wetland environmental management is less developed.

Mangrove forests have suffered worldwide (Barrow, 1991: 121–123). Seagrass beds have also suffered in many countries. Peatbeds have been destroyed in temperate, cool and tropical environments by agriculture, fire during dry periods, and by digging for use as horticultural mulch or fuel. As peatlands are destroyed or dry out, the carbon they have locked up is released to the atmosphere, accelerating global warming. Furthermore, where peat is frozen there may be quantities of methane locked in it – another strong greenhouse gas.

Marine natural resources

Some years ago a TV series on marine nature opened with the adage: 'Two-thirds of the earth are covered by sea. Most of this is unexplored.' Whaling has largely been discontinued, world fisheries are under intense pressure and a number of marine fish stocks have collapsed. Apart from oil, sand and gravel from relatively shallow waters, mineral exploitation is limited. However, exploration is improving and technology capable of recovering resources from deep waters is being developed. Already, recently discovered hydrothermal vents have yielded useful material for biotechnology.

For many years several nations have disposed of hazardous waste in ocean deeps. Control of pollution and better management of fish, shellfish and sea cucumber are priority needs. Some have expressed concern at the decline in sharks as a consequence of demand for shark fins.

There is a growing awareness of the need to monitor oceans for potential hazard – tsunami and shifts of ocean–atmospheric circulation (which could seriously affect climate on land). Considerable attention is being directed at the latter, especially the threat of diminished flows in the North Atlantic, which position the Gulf Stream. The impacts of ENSO events have also been attracting attention in recent decades. Acidification of oceans has attracted less attention, but with global warming, seas might become acid enough to alter plankton populations, with potentially huge impacts.

Coral reefs

Once flourishing, coral reefs are now often dead and dying, due to one or a combination of causes, including: climate change; fishing gear damage; cyanide poisoning; pollution by silt, sewage, industry and agrochemicals; raised UV levels; starfish; and disease. The most widespread damage – often resulting in coral bleaching – seems to be bacterial disease, perhaps following weakening by other factors, and temperature rise (as little as 1°C is enough). Sea-temperature increase and rapid sea-level changes seem possible as a consequence of global warming, and are both likely to inflict damage even in little-polluted areas. Reef death means loss of biodiversity, damage to potential tourism, and possibly coastal erosion.

Environmental managers must seek to cut background levels of pollution in the oceans, and control pollution by silt, sewage, agriculture and industry. There must also be efforts to limit rapid global warming. There may be opportunities to establish new reefs at slightly shallower depth using materials such as scrap metal and old tyres, and in California a company now offers people reef-forming burial sites.

Indigenous peoples and natural resources

Since the 1970s indigenous peoples around the world have been establishing and exercising their rights over natural resources, notably in the USA, Canada, New Zealand, Brazil, Mexico, Greenland and Australia. The idea that indigenous peoples are always sympathetic to their environment, and so will not exploit it, is wishful thinking. There are cases of indigenous peoples causing environmental degradation, and sometimes central governments have found it difficult to intervene because rights had been granted. Usually locals know the land and its biota, and can effectively police and manage conservation and development. Some indigenous peoples have successfully resisted large mining companies and other potentially exploitative developers. A recent development – the extractive reserve – combines conservation and sustainable exploitation with local livelihood provision (Stolton and Dudley, 1999: 215–223).

Summary

- Modern development demands a diversity of natural resources; shortage of one or more could cause serious problems.
- Demand for water is increasing, so supplies must be better managed and valued more highly. Generally the best supplies have been developed and the quality of what is available is likely to be less than optimal.
- Water resources have mainly been developed by water managers with a relatively narrow outlook – usually concerned with economic and engineering goals. Environmental management applied to water resources promises to better integrate the needs of all stakeholders and to give a more comprehensive overview.
- Hunger is a serious problem, which could still menace even developed countries. Food production and food security need more attention and investment.
- Agricultural improvements since the 1950s have been based mainly on irrigation, which is often unsustainable. Rather than alter the environment to suit crops it would be better to adapt crops and techniques to suit the environment.
- Agricultural improvements since the 1950s have been masking worsening soil degradation and are causing agrochemical pollution. In the future agricultural

improvement will need to take the form of a doubly green revolution which boosts yields, sustains production, causes much less environmental damage, and copes with global environmental change and pollution.
- One route to resolving some of the current challenges faced by agriculture is genetic engineering; this could allow humans to respond more effectively and rapidly to problems and opportunities. While genetically modified organisms (GMOs) may help provide sustainable and less environmentally damaging food production, they could run out of control and pose serious threats.
- Wherever possible conservation should spread risks, duplicate reserves and *ex-situ* genetic collections.

Further reading

Adams, W.M. (1992) *Wasting the Rain: rivers, people and planning in Africa*. Earthscan, London.
   Water development in the real world. Excellent critique of inappropriate approaches.

Adams, W.M. (2003) *Future Nature: a vision of conservation* (revised edn). Earthscan, London.
   Excellent introduction to conservation and biodiversity issues.

Barrow, C.J. (2005) *Environmental Management and Development*. Routledge, London.
   Chapters 3–8 explore key resources.

Conway, G. (1997) *The Doubly Green Revolution: food for all in the 21st century*. Penguin, London.
   Argues that improvement of food production must show concern for the environment.

Hunt, C.E. (2004) *Thirsty Planet: strategies for sustainable water management*. Zed Books, London.
   Readable and radical coverage of freshwater ecosystems which focuses on human usage and management. Presents alternative approaches, which should protect nature as well as supply people with water.

McCully, P. (2001) *Silenced Rivers: the ecology and politics of large dams* (updated edn). Zed Books, London.
   An excellent text on dams and large development schemes.

Madeley, J. (2002) *Food for All: the need for a new agriculture*. Zed Books, London.
   Lively and readable coverage of food supply and agriculture.

Rees, J. (1995) *Natural Resources: allocation, economics and policy* (2nd edn). Routledge, London.
   Updated edition of a widely used introductory text on renewable and non-renewable resources.

Shiva, V. (2000) *Tomorrow's Biodiversity*. Thames & Hudson, London.
   Shiva champions the rights of peasants and plants against commerce and neo-colonialism; readable and thought-provoking, but offers little clear alternative.

www sites

Clean water for poor people in developing countries from One World Action (London) http://www.oneworldaction.org (accessed October 2002).

Consultative Group on International Agriculture (CGIAR) – this body supports fifteen centres around the world which have promoted the Green Revolution and continue to work to improve agriculture – http://www.cgiar.org/ (accessed February 2004).

Coverage of large dams UNESCO http://www.unesco.org/courier/200 World Commission on Large Dams http://www.dams.org (both accessed January 2004).

Food and Agriculture Organisation (FAO) http://www.fao.org (accessed March 2004).

International Rivers Network encourages equitable and sustainable methods of drinking water and energy supply, and flood management. Via http://www.mylinkspage.com/earthsummit.html (Earth Summit Info. Website Section 46 provides this and other water websites) (accessed April 2004).

International Soils Reference and Information Centre (ISRIC) – world soils and sustainable land-use information http://www.isric.org/index.cfm (accessed June 2005).

International Water Management Institute – improving water and land resources management for food, livelihoods and nature http://www.iwmi.cgiar.org/ (accessed April 2004).

IUCN *Red List* – regularly updated list of endangered species – http://www.redlist.org/&y= 02F3FC (accessed April 2004).

Peak oil crisis – numerous sites, many of which are biased; one reasonably objective site is http://www.oilcrisis.com (accessed October 2005).

Soil and Water Conservation Society http://www.swcs.org/t_top.htm (accessed February 2004).

UNEP World Conservation Monitoring Centre (WCMC) maintains a database with details of the world's protected areas, budgets, information sheets, and much more – http://www.unep-wcmc.org/right.htm (accessed March 2004).

World Conservation Union (formerly International Union for Conservation of Nature and Natural Resources) (IUCN) http://www.iucn.org/ (accessed March 2004).

11 Global challenges

- Identifying the challenges
- Transboundary issues
- Future priorities
- Summary
- Further reading

Before modern food production people were aware that they were vulnerable and depended on nature (God or gods); nowadays many have a false sense of security, assuming that technology and governance will support them. There may have been some awareness of limits by the 1830s, but as populations grew there was still space for settlement. By the early twentieth century German political thinkers and geographers and their equivalent in the UK (notably Sir Halford Makinder) were aware of crowding and competition for living space. In America by this point it was clear the 'frontier was closing' and that space for expansion was not infinite. Steamships, trains and the telegraph have also helped to shrink the globe. By the 1860s Britain ruled or influenced perhaps a quarter of the globe and commonly dealt with global political and economic challenges.

Two world wars and serious economic depression had a global impact which, with lack of knowledge about the structure and function of the global ecosystem, ensured environmental management issues were of little concern before the 1960s. The International Geophysical Year of 1957 to 1958 helped encourage sharing of data, standardisation of units of measurement, and environmental and biological research, all of which assisted in the recognition of global environmental management challenges. Ocean fisheries and whale stocks had recovered during wartime, helping prompt efforts to control whaling and fishing from the 1950s. Wildlife conservation and landscape conservation grew in importance after 1945. Sovereignty claims were largely put aside in 1959 when the Antarctic Treaty was signed by twelve major nations, recognising that 'it is in the interest of all mankind that Antarctica shall continue forever to be used for peaceful purposes'.

Fears as well as altruism were brought to bear; concern over nuclear weapons proliferation and testing led to the 1968 Nuclear Non-Proliferation Treaty. Unfortunately it has not led to significant disarmament of existing atomic weapons possessors, nor prevented India, Israel, Pakistan, North Korea, and possibly other nations, from obtaining them. The Test Ban Treaties have helped discourage atmospheric weapons testing, although the wish to conceal preparations has also helped. Fall-out from nuclear tests clearly had global effects by the 1960s – even well away from test sites levels of radioactive isotopes were elevated (so much so that radiocarbon dating could no longer operate beyond 1950). In 1962 Rachel Carson published *Silent Spring*, warning that levels of pesticides such as DDT were worrying worldwide. Further impetus was given in the early 1970s by the views of the Earth taken from space which made clear that it was finite and vulnerable (see Chapter 1). From about 1970 a number of ecology books and warnings of planetary damage appeared (Ward and Dubos, 1972), and in 1972 the first major environmental conference was organised in Stockholm (UN Conference on the Human Environment), which included calls for action on global environmental issues such as acid deposition.

Three crucial global challenges have been accepted in the past fifty years: the need for peace; reduction of poverty; and better environmental management (in chronological order). After 1970 Western citizens became aware of the words 'ecology' and 'environment' and, in academic circles, with the concept of sustainable development. Awareness was raised by the 1972 publication, by Meadows *et al.*, of *The Limits to Growth*, the foundation in the early 1970s of the United Nations Environment Programme, and a number of environmental 'scares' – marine oil spills, the Seveso chemical leak in Italy, the Love Canal contamination in the USA, and others. In the USA the National Environmental Policy Act 1969 (NEPA) came into force in early 1970 and required proactive environmental assessment. Texts on global environmental issues and ecopolitics began to appear (Pirages, 1978; UNEP, 1981), although much of the writing was Malthusian advocacy. There had been success in addressing the global challenge of smallpox, and high hopes in the 1960s and 1970s that malaria could be controlled. There was still reluctance to address global environmental problems, which in the 1970s were sometimes seen as a luxury or even 'green imperialism' – a conspiracy by rich nations to control the poorer using environmentalism. Gradually change took place towards awareness, and some willingness to negotiate and invest in combating global environmental problems. The change was helped by the appearance of *Our Common Future* in 1987, by the Rio 'Earth Summit' in 1992 and by growing information from palaeoecology and environmental history that the world environment is changeable and vulnerable. It is also likely that the Chernobyl disaster of the mid 1980s helped, as did recognition in 1987 that stratospheric ozone scavenging was a global threat. By the 1990s there was widespread support for environmental challenges (Mannion, 1991; Middleton, 1995; Graves and Reavey, 1996). In the last couple of decades films and bestsellers on environmental disasters (e.g. the 2004 climate shift film *The Day After Tomorrow*) have generated awareness. The Indian Ocean tsunami in late 2004 made fears more real, and triggered probably the greatest ever humanitarian response to misfortune beyond national borders. Over the same period statesmen and many scientists have focused on negotiations to reduce or halt global warming and to counter poverty, especially in Africa.

Identifying the challenges

Late twentieth- and early twenty-first-century media, NGOs, governments and citizens are aware of broad challenges (peace, poverty and environment), but these are too broad to really get to grips with. There needs to be systematic review of global challenges to identify priorities. Problems and opportunities are largely dealt with in an *ad hoc* manner when they are already happening. One of the first comprehensive assessments of global challenges was presented in 1980 (published two years later): *The Global 2000 Report to the President* (Council on Environmental Quality and Department of State, 1982); although widely read, this was largely ignored by policy makers. Another assessment of global challenges was *The World Development Report* (World Bank, 1992). In the late 1990s the approaching millennium provoked a number of assessments of challenges.

Sometimes government agencies or intelligence services prepare forecasts which may later be published. For example, the *CIA World Fact Book* or NASA's *Global Change Master Directory*: http:// www.gcmd.gsfc.nasa.gov/ (accessed July 2005). Some institutes, universities and NGOs also publish annual or occasional assessments (e.g. the Worldwatch Institute annual *Worldwatch Reports*). The end of 2000 stimulated forecasting and goal setting, one example being the *Millennium Development Goals*. Another is a guide to world challenges – the *One Planet. Many People: atlas of our changing environment* (UNEP, NASA, USGS and University of Maryland, 2005; or online at http://www.na.unep.net). Africa has attracted attention because it is seen to present particular challenges – see the *African Environment Outlook* – http://www.grida.no/aeo (accessed July 2005). Remote sensing and GIS have made assessment of challenges easier; recently, the UNEP (and five other agencies) have been mapping the Earth's land cover at high resolution (Globcover Project – due to be published in 2005).

Aid agencies, NGOs, media, pop stars, research institutes, international bodies, businesses and so on may identify and lobby for responses to global challenges. The Limits to Growth debates of the 1970s were triggered by efforts by a non-aligned group of concerned individuals (the Club of Rome). Groupings of government representatives such as the G7 or G8 are involved in pressing for or steering action on global issues. Thus environmental management has to work with a diversity of actors and stakeholders, master media relations, global politics and governance issues (Levy and Newell, 2004).

Chapter 10 examines key resources, sustainable supply or use of some of which are clearly problematic. Hydrocarbon fuel use contributes to global warming, and water supplies are widely agreed to be under growing stress. Phosphate fertiliser is in limited supply and other widely used agrochemicals cause pollution. Finding energy and growing food for the world's expanding population without excessive pollution will be a challenge.

Transboundary issues

Now, at the start of the twenty-first century, environmental management faces a range of global challenges. When people, states or businesses cause difficulties which cross borders they are likely to be held liable and will be required to pay for solutions. New persistent pollutants and the globalisation of trade have also increased transboundary effects. Technology and environmental awareness mean that problems are identified. Even though a single body, state or individual may not have jurisdiction, issues have to be pursued. Since the second half of the twentieth century the management of transboundary issues has been raised in various assemblies to prompt resolution, and legislation is being gradually developed to address issues (Kiy and Wirth, 1998; Linnerooth-Bayer *et al.*, 2005).

Global challenges fall into two broad categories which often overlap: human-caused and physical, and demand co-ordinated efforts to cover costs of avoidance or mitigation, establish early warning, monitoring and so on. Human-caused and physical challenges often involve politics and international competition, and demand development of impartial monitoring, policing and legal negotiation. Not all of those involved are keen to admit that a problem is real, let alone co-operate. For natural global challenges the political and legal issues and policing are often less problematic. Co-operation between groups of people is becoming more necessary as human populations expand and technological impacts increase. Global challenges involve temporal as well as spatial scales, because to pursue sustainable development there must be intergenerational sharing – current expenditure may be needed to ensure that something is passed on to the future. Altruism, which benefits people at a distance spatially and in time, is quite a new demand; most governance, legal systems and human behaviour are not geared towards it. However, there are promising signs: countries have been co-operating to deal with some issues and people are sometimes willing to pay tax or make donations of aid for groups they are not close to.

Perception of threat can be fickle and, even when people are aware, they may be slow to take it seriously. Often disaster is needed to focus attention: the threat of tsunami held little attention before late 2004, but now it has greater weight. Currently, governments and influential scientists focus on the threat of global warming – while other issues may not be getting enough attention. With limited resources and logistical support, global issues have to be subjected to some sort of triage so that the important ones are concentrated on. It is not clear who should choose and whether a body would be sufficiently trusted and empowered to do so. Some threats may be rare but cause terrible damage, so avoidance or vulnerability reduction must be proactive. Some recent studies by environmental historians have helped educate the public and administrators about how past disasters have had tremendous impacts upon people, and might recur (Fagan, 2004; Diamond, 2005). Technology and governance may not protect people as effectively as they hope – environmental managers should promote general vulnerability reduction and make people adaptable and resilient against anything, even unforeseen and sudden problems.

Systems, actors and rules change (Pirages, 1978: 32): there has been a shift from livelihoods based on expansion to intensification; colonial domination by a few powers has given way to negotiations and a wider group of powers, cartels and businesses. The 1945 to 1990s Cold War has broken down and new power relations are developing – countries such as China and India will exercise great influence within decades. Nations once craved coal, then petroleum, and fifty years hence, food or water may be the key resources. Feudalism gave way to *laissez-faire* development in large areas of the world during the past century or so, but proactive and strategically co-ordinated management is supplanting it. Currently, Western liberal democracies hold sway, but the nuclear balance is no longer stable – about ten nations have atomic weapons and terrorism has developed in ways which past generations would not have predicted. Some argue that a 'global village' will develop, based on English-language use, the Internet, and the evolution of present systems. Western labour is becoming expensive, so patterns of manufacturing and investment are shifting. Some would put anti-poverty before environmental care; others are more concerned with religious than secular issues. Respect for authority and learning has broken down in some societies, coupled with demands for softer lifestyles and more consumption. The world is unstable, environmentally, politically, socially and economically. To peace, poverty and environment may be added a further two challenges: human population growth and consumerism. Consumerism is the drive to acquire material possessions and to discard and update them regularly; currently most items consumed are manufactured goods, which cause environmental damage. Growing population plus increasing demand for consumer goods places severe stress on the environment.

It is far from clear what pattern of relations, willingness to co-operate and shared vision will develop in the future. Global challenges will have to be negotiated and solutions found in the face of uncertainty. The grand goal and challenge is sustainable development.

Transboundary issues caused or affected by human activity

Humans live in a single shared global environment; all depend upon and can affect critical life-support systems. Transboundary impacts have become apparent partly because population has increased. In the 1960s and 1970s 'Malthusian' environmentalists tended to see population growth as *the* root cause of most major problems. The linkages are not that simple; in some regions environmental degradation may result from population decrease and there are plenty of situations where problems are in no way related to large population. Nevertheless, the world is finite and human numbers mean increasing demands, and make it less easy to respond to challenges. So far there has been little resort to draconian population control measures, although a few countries penalise large families through taxation and by withholding state welfare. Poverty is also widely identified as a cause of environmental and social problems; however, again the linkages can be complex. There are countries with relatively low per capita incomes which have made progress with social development and environmental care, and there are rich nations with serious environmental degradation. In the 1970s poverty was seen to have priority over the 'luxury' of environmental management; today environmental care is less often seen to be second priority or unrelated to combating poverty. Poverty reduction and environmental care are now commonly seen to be interwoven. In the 1970s the mismanagement of technology was blamed by some environmentalists as a cause of environmental problems – part of the Western democratic (Judaeo-Christian) worldview, which led to mal-development. Environmental problems and other ills of the world are still seen by some to be due to faulty Western ethics that have spread like a cancer to affect other countries – 'a global terminal illness'. The cure, it is argued, is to cast off such ways; alas, with the world much altered by human activities and with large human populations, a return to some pre-modern idyll is not possible. To seek that would condemn millions to starvation and other dangers, and would not restore damaged ecosystems through neglect. A human-altered globe must now be steered by careful managers, but in a sensitive and informed way that has not been done in the past. The cost of failure is disaster. Technology is not something to be discarded; it is the key to sustainable development and a better future – but it must be better designed and managed.

Market forces have prompted problems and have not controlled them as economists may have hoped. Profit drives much of what modern humans do, and it is needed to achieve results. Thus economic reform will play a vital part in resolving major environmental issues. Already, a significant amount of the negotiations on recognised global environmental issues has focused on how to pay and who pays. Some promising agreements have foundered when funding pledges have failed to materialise. Allied to the creation of transboundary issues is the improved ability to make contact across borders and around the world. Better transport, media and telecommunications make it possible to address global issues. The weakening of the Cold War and opening of many once-closed borders helps countries share ideas and information. After 1945 the establishment of international bodies such as the UN, FAO, UNEP, and international courts and laws has supported those addressing global challenges.

Various groups of people, businesses and countries are more interdependent than in the past. For example, an earthquake in, say, Kobe could wreak havoc worldwide because it would interrupt electronic chip supplies. In addition, much of the world's grain comes from a few vulnerable regions; petroleum production is concentrated in a limited number of places; and producers in one place are affected by markets elsewhere (trade and environment are discussed in Chapter 4). International linkages increasingly bind trade, livelihoods, cultures and much more – globalisation. Globalisation is a growing influence on the environment (see also Chapter 4). It might be argued that globalisation is inevitable and a good thing if 'Spaceship Earth' is to be effectively managed; however, many are bitter opponents. Those objecting to the spread of globalisation have been very apparent at international meetings for several years (Kieley and

Marfleet, 1998; Lucas and Woodin, 2004). In recent years environmental concern and a desire to establish sustainable development has spread globally. One key challenge has to be the promotion of sound environmental goals when globalisation is encouraged.

Human activity can directly cause serious problems, act jointly with natural problems, or trigger a natural cause. Human activity acts at a range of scales: that of the individual consumer or small farmer; that of groups, companies, sectors; nationally and globally. The sum effect of individuals or groups can have worldwide impact. The consequences of actions are often not apparent to those involved, and unless pointed out by environmental managers and discouraged by controls such as taxation or restrictions, will continue. Sometimes the activity is prompted by desperation to earn a livelihood or by greed, so controls are difficult to effect. When activity is the result of ignorance or a fashion, change may be quite easy to initiate. Resources may be in common ownership, owned by none, under state control, or may be individually owned. Whichever, there is a possibility of environmental damage. No society has avoided environmental problems, whether it be Western free enterprise, socialist state control or tribal and so on.

For the past fifty years environmentalists have attempted to explain why environmental problems arise, citing the following explanations:

- *Malthusian and neo-Malthusian* – population increase exceeds environmental (and possibly social) limits leading to problems;
- *Poverty and marginalisation* – people are forced to misuse resources;
- *Ignorance* – the implications of the activity are unknown, perhaps because knowledge about the structure and function of the environment is poor;
- *Greed* – selfishness, and more recently consumerism, advertising and fashion prompt exploitation;
- *Warfare* – damage is incurred as a side-effect of strife (collateral damage, displacement of people) or is undertaken deliberately to hurt others (scorched earth, environmental warfare);
- *Common resource ownership* – individuals seek to maximise their use but there is no effective control;
- *Expropriation* – colonial power, occupying force, powerful company or state take control of resources and are not moderated;
- *Dependency* – weak groups are compelled and encouraged to cause damage by more powerful/controlling groups;
- *Faulty ethics* – humans dominate nature because religion or worldview encourages it. Humans see themselves as above nature or as having a right to control and exploit it;
- *Misvaluation* – the value of something is not understood and, in a world driven and moderated by economics, has not been valued in a way that encourages market control over use;
- *Policing problems* – users cannot be monitored or easily controlled (e.g. small farmers in remote areas, users of shallow groundwater, resource use where law and order is weak).

Humans do not behave in rational ways; they can expend huge resources on something of little value to their survival and welfare or benefit to the environment, and then haggle and fight to avoid spending far less in tax to achieve real benefits. In negotiations on global environmental issues some nations and business groups are perfectly willing to allow others to pay to resolve things, while others may be altruistic. Most wait to be convinced by proof and then spend if assured that costs will be reasonably borne by all, and that benefits justify expenditure. In a world of rich and poor nations a case may be made for the rich paying more for addressing global issues and for aid to poor nations. However, some poor nations have growing populations, and are poised to consume energy and resources and to pollute hugely. Blaming rich nations for 200 years of pollution is common, but most of those blaming have been happy to use the fruits of that industrialisation and development – better transport, modern medicine and so on. In negotiations there is a need for reliable and unbiased data and some adjudicator with 'teeth' who is respected by all. The UN currently falls short of this need.

Some countries have or will have an elderly population compared with others, so seeking agreements that impose a per capita payments solution can have a very different impact on these than it would on those with a young population. Ageing populations is one global issue which needs to be addressed; in some cases it may offer advantages, in others it poses special problems of vulnerability, drain on expenditure and reluctance to adapt. Some countries are run by small elite groups who may not have general welfare or world conditions as one of their priorities. Some countries have a harsh environment and poor resources endowment compared with others; the foundation from which to address global challenges is not a level one. People react differently to challenges and opportunities; a similar situation may result in a totally different response in different countries or even the same country at different points in history. Box 11.1 lists transboundary issues caused or affected by human activity.

Global warming

Global warming is now accepted by most of the scientific community and many policy makers (there are dissenters – see: *New Scientist* 2005 vol. 185 (2486): 39–40 for a brief review). Global warming fears dominate international environmental discussions at present (2005); perhaps there is too much focus on seeking carbon emission controls, resulting in the neglect of other threats. Change is likely anyway due to pollution already released – it may be argued that some of that spent fighting carbon emissions would be better spent on reducing vulnerability, and increasing adaptability to warming (and other disasters). During the 2005 Gleneagles G8 Summit the scientific academies of all eight nations challenged their governments to act to cut greenhouse gases emissions. The Kyoto Protocol is an international and legally binding agreement to reduce greenhouse gases emissions; signed by 141 nations in 1997, it came into force in early 2005 (it builds on the UN Framework Convention on Climate Change agreed at Rio in 1992). The USA contributes about 25 per cent of the world's total anthropogenic greenhouse gas emissions, and refused to ratify the Protocol in 2001 (Australia also). The terms of the Protocol were subsequently weakened to try and get agreement – so far not achieved. By late 2004 Russia had signed, but the USA, Australia, India and China had not endorsed the Protocol – which required *developed country* signatories to cut greenhouse gases emissions back from their 1990 level by at least 5.2 per cent by 2010. The Protocol allows the trading of emissions quotas between countries which have surplus and those likely to exceed allowances.

Unfortunately, by late 2005 hopes that the Protocol would return greenhouse gas emissions to something resembling 1990 levels by 2010 were looking unpromising, and agreement of all nations was still elusive. The EU has established a Carbon Emissions Treaty Scheme to try and cut greenhouse gases. This allows participants to import carbon dioxide credits to add to national allocations – it is a tradable emissions quotas-type scheme (and is due to start its second phase in 2008). Companies or other bodies likely to emit carbon or use energy enough to pose a problem will have to become involved in tradable emissions – 'carbon trading' – and this could affect even quite

Box 11.1

Transboundary issues caused or affected by human activity

- *Food production* – The achievements of the Green Revolution may not be ongoing unless major changes are initiated. There is also environmental degradation and the risk of global environmental change to cope with. More food should be stored in case of failed harvests, and food production should be diversified and made less vulnerable.

- *Loss of biodiversity* – Demands for land, minerals, water, timber, food, recreation, building land, and other causes are destroying species (there is effectively an ongoing mega-extinction event).

- *Water resource problems* – Water supplies are often inadequate in both quantity and quality. Poor management and faulty technological fixes have resulted in environmental damage and loss of biodiversity: shrinking lakes, disruption of floodland, wetland and riverine ecosystems; collapse of aquifers and subsidence; conflict related to competition for water, and many more difficulties. There are signs of renewed interest in building large dams and large-scale inter-basin transfer, some of which could cause inter-state conflict.

- *Soil degradation* – Impact upon food production, biodiversity, and release of soil organic carbon which may accelerate global warming.

- *Mineral problems* – Depletion of resources; conflict over access; mining/drilling-related impacts.

- *Manufacturing* – Air and water pollution; solid waste from manufacture, packaging and final disposal. Loss of land to factory sites. Global background pollution of air, land and oceans by PCBs, heavy metals and so on. Damage to stratospheric ozone layer through use of CFCs – progress in control and international agreement.

- *Hydrocarbon electricity generation impacts* – Greenhouse gas emissions; sulphur emissions; particulate emissions; acid deposition. Impacts may be felt hundreds of kilometres away even in the high Arctic.

- *Hydrocarbon vehicle fuel use* – Oil spillage at sea and pollution from drilling and refining; exhaust gas pollution. Major cause of air pollution in many countries. Some countries add lead to combat precombustion, and this impacts upon wildlife and humans downwind of highways.

- *Hydroelectric generation impacts* – Disruption of river flows and movement of organisms; impoundment of large reservoirs/relocation of people. Risk of transboundary conflict.

- *Nuclear energy* – Risk of fission reactor accidents, disposal of waste/reprocessing; danger of radioactive material reaching terrorist hands or aiding nuclear weapons access by further countries.

- *Agriculture/food* – Huge impact upon global ecosystem as a result of efforts to grow or catch food. Biodiversity impacts, soil degradation, emission of greenhouse gases, dust which blows great distances, waste agrochemicals and domesticated species (pesticides, herbicides, fertiliser, feral crops/livestock escaped into the environment, GMO crops).

- *GMO issues* – Risk of uncontrollable escape of a modified organism. Differential access to GMO skills and materials causes disadvantage and dependency.

- *Transmission of pests/diseases by improved transport* – Fears about avian influenza since the 1997 Hong Kong fatalities. Diseases such as flu, viral haemorrhagic fever, multidrug-resistant TB, or bubonic plague could spread fast and far before it is recognised; if controls or a cure are ineffective there is potential for a rapid global disaster. WHO is developing a Global Surveillance and Response System to improve disease control (partly operative in 2005).

- *Warfare* – Major cause of relocation of people and marginalisation. Direct impacts such as burning oilfields. Indirect impacts such as spread of diseases by dislocated people and troops. Waste of resources and hindrance of environmental care. Potential for major world environmental damage if nuclear, chemical or biological weapons are used. Weapons testing has so far been controlled and environmental contamination has been limited since the 1970s; this may break down.

- *HIV/AIDS (SIDA)* – Many countries are already severely affected and there is likely to be considerable further deterioration. Already some countries have more than 25 per cent infection rates. There are serious impacts upon livelihoods, land use and national economies in addition to human misery from death and disability.

- *Malaria* – There are hopes that new drugs may enable control in coming decades.

---

small businesses or institutions such as office blocks or colleges. Some fear the scheme will increase EU energy costs without cutting emissions by much, and might make European manufacturing uncompetitive with other nations. Between 1990 and 2000 emissions from the then industrialised countries fell by about 3 per cent, although richer nations increased by approximately 8 per cent. The collapse of the Soviet economy and decline of heavy industry and coal usage in the UK helped. While the USA (Federal/National government) has not signed the Protocol, a growing number of its states have signed agreements to meet Kyoto standards.

The USA, Australia, Japan, South Korea, India and China currently consume around 45 per cent of the world's energy and cause approximately 50 per cent of the global carbon emissions. In late 2005 these countries agreed to try and develop cleaner and more efficient technology as a way to reduce emissions without damaging economic growth – the Asia–Pacific Partnership on Clean Development and Climate (see *New Scientist* vol. 187 (2515) of 3 September 2005 – special report). This might prove a welcome addition to Kyoto, although some regard it as an attempt to undermine the Protocol. Supporters of the partnership claim it is easier to develop technology and promote its use than to alter economics and people's energy use habits on a wide scale. Whether technological solutions such as safe fission power and practical sequestration of carbon dioxide can be developed fast enough remains to be seen. What is likely is that China and India will greatly increase combustion of coal – with improved combustion technology the impacts might be reduced. The pressures to find new energy sources after the recent oil peak could fit in with the partnership technology improvement strategy. To some the technology route seems more promising.

Recently an Oxford academic argued that, if carbon emissions can be proved to cause threats to humans, then it becomes a liability issue. Companies proven to emit could be sued for damages. Fear of such litigation and the funds generated from it could help control emissions better than any Protocol – the challenge is to establish a legal precedence (Allen, 2005). NGOs, agencies and businesses have also been trying to control greenhouse gases emissions (Berkhout *et al.*, 2003). For example, in the UK the Carbon Trust works with companies to try and cut carbon dioxide emissions.

The Intergovernmental Panel on Climate Change (IPCC) predicts that current trends will lead to a global temperature increase of between 1.4°C and 5.8°C by 2100, and a sea-level rise of between 0.1 m and 0.9 m by 2100. Even without reaching the higher values there would be major impacts; for example, a 0.5-m sea-level rise would displace over six million people from Bangladesh alone. It is also possible that successful control of sulphur, dust and smoke pollution or other positive feedbacks could further increase global warming. In the past there have been repeated sudden drastic regional climate changes; one of these affected Western Europe and North America as a consequence of the weakening of the Gulf Stream. The cause is believed to be related to post-glacial warming which reduced Atlantic salinity by dilution with meltwater. Human-caused global warming might trigger similar cooling, with disastrous results. There is no shortage of other global warming impact assessments, focusing on sea-level rise, health, agriculture, severe weather events and so on (McMichael *et al.*, 1996; Martens, 1998; Lynas, 2004) (for information on potential global warming impacts see: http://www. geoconnections.org/ccportal/ – accessed July 2005). For a review of the global carbon and climate change debate and IPCC assessments of likely change impacts, including the implications for sustainable development, see Munasinghe and Swart (2005).

The impacts of global warming may be differential, with some countries benefiting and others suffering problems. Knowing in advance what the impacts may be could give nations and businesses opportunities for profiting as well as making genuine preparations to mitigate or avoid difficulties. As well as temperature change, the increasing level of carbon dioxide in the atmosphere may well affect plant photosynthesis and other organic and inorganic processes – one result may be that they become more acidic. There could be impacts on oceanic plankton productivity – which would have serious consequences. There could be changes in agricultural productivity; if poor (mainly tropical) countries lose out and richer temperate countries benefit, the gap between North and South could widen. Forecasting what will happen is nowhere near precise. Large projects, infrastructure, crops, agricultural practices and sustainable development strategies should be designed to deal with uncertainty – to be as flexible as possible. Currently, that is often not the case; the design is to meet economic and engineering criteria and the results are inflexible.

Hazardous waste

Hazardous waste has become a global challenge not simply because population growth and increasing industrial manufacture cause more pollution, but because technological progress has led to compounds being more toxic or disruptive than are found naturally. Some of these compounds are also very persistent. Particular difficulties are caused by PCBs, CFCs, radioactive compounds, those which disrupt wildlife and human reproduction, heavy metals, and many other pollutants (see Chapter 12). Most pollutants are difficult to deal with once released into the environment; some become a global problem as they spread through the ocean or atmosphere, and cause difficulties at a distance from the point of release. Problems may arise when exported materials, especially food, are contaminated and affect consumer nations.

There are difficulties with waste disposal: unscrupulous states or companies may seek to dump material in the ocean or send it for landfill in a developing country where controls are relatively lax. Pollution of a shared river becomes a transboundary problem. International agreements are starting to be negotiated, such as the Basle Convention on the Control of Transboundary Movements of Hazardous Waste and their Disposal (which came into force in 1992 having been signed by 149 nations) but there is a long way to go before controls are adequate.

Many countries have a growing problem of nuclear waste disposal, which is potentially a threat for all nations. There have been regional disasters already – one reason is that high-level nuclear waste generates heat and can become unstable (it may even support a fission reaction). Without careful disposal and ongoing management to contain and cool the waste, an explosion and scattering of dangerous material is possible. That management may need to continue for thousands of years. The waste may be attractive to terrorists so there is a serious security challenge. For material that demands storage for centuries it is difficult to guarantee no leakage into groundwater and escape into the environment. Thus developing repositories which are safe is costly and people seldom want such a site anywhere near them. A few countries may have remote areas and be willing to develop international facilities (charging for disposal) – Krasnoyarsk in Russia may be such a site.

Transboundary issues caused by natural processes

There has been a tendency to overlook the fact that natural environmental change has had a considerable effect on human fortunes until quite recently (see also Chapter 3). As mentioned above, this may be in part because of distaste for what was seen as crude environmental determinism, which held sway from the 1940s to the 1980s. Climate change attracted attention before the 1980s, in the 1960s and 1970s fears of a return to ice age conditions tended to hold sway, and since the late 1980s global warming has dominated. One thing is certain – climate is changeable. The impact of climate change depends on the degree of shift and on the speed of change. People can adapt if change happens slowly and they are aware of it, but sudden change or gradual unrecognised alterations can spell disaster. Food and water supplies depend a great deal on climate and there is now widespread concern about possible global warming impact. Sea-level rise is associated with warming and this too is generating debate and calls for agreements, contingency planning and so on. Soil degradation is another global challenge but it has attracted less, especially popular, attention. Biodiversity loss is widely perceived but controls and expenditure on it have been limited.

Environmental threats may have a long recurrence interval, so humans have little or no recollection of them. Some natural disasters recur so infrequently that there may be a temptation to write them off as too rare to be bothered with. Unfortunately, some infrequent events have a huge impact on well-being and survival, so ignoring the threat is unwise. Since the mid nineteenth century natural sciences have tended towards uniformitarianism and catastrophic events have not attracted enough attention. However, it has been realised that while much of the time change and evolution is slow, this is occasionally punctuated with sudden gross change; Ager (1993) compared Earth history to warfare: long bouts of boredom and brief spells of terror. Box 11.2 lists a number

Box 11.2

Transboundary issues caused by natural processes

The tendency has been to focus on human-caused environmental change and less on natural threats; many assume technology and modern governance have reduced vulnerability – population growth and complex interdependent lifestyles may have had the opposite effect:

- *Climate change* – Periodic and quasi-periodic (ENSO, Gulf Stream, and others), which might be forecast; random and unexpected; or major global climate trends that may be recognised as they develop in time to allow useful responses. The past 18,000 years have been relatively stable and benign but this is unlikely to last. Human impacts have altered natural patterns, making the future less certain. Even minor change could upset food production; shifts similar to some evident during the past 20,000 years could lead to millions of eco-refugees and serious problems/conflict over relocation.

- *Stratospheric ozone damage* – Could be caused by volcanic eruption or other natural events (as well as by human pollution).

- *Major volcanic eruption or fissure outpouring* – This might cause climate change; also ash deposition over large areas; possibly major changes to atmosphere enough to damage plant production and possibly affect respiration of organisms. One threat, as yet unproved, is the 'vernshot', a huge blast of hot carbon dioxide, other gases and molten rock which is projected into the upper atmosphere in greater quantities and for longer than normal volcanoes (Ravilious, 2005). The cause is a hot plume rising in the Earth's magma, which breaks through the crust (possibly in areas where vulcanicity is unexpected). If this were to happen there would be little that could be done to mitigate or avoid the impacts, and some geologists suggest this has been the cause of past mass extinctions – including the K/T extinction.

- *Lesser volcanic eruption* – More frequent than the last threat above, it could cause a loss of crops in many countries for one or a few years. Possibly damage to the stratospheric ozone layer and impacts associated with this.

- *Evolution of epidemic diseases* – Awareness of the threat from influenza; other potential problems are attracting limited attention.

- *Tsunami* – Normal and mega. The Lisbon disaster of AD 1755 shook Western views that major natural change was unlikely. The late 2004 Indian Ocean tsunami generated a new awareness of the threat of natural disasters and some improved willingness to invest in early warning and mitigation. The threat of mega-tsunami is largely ignored, although it presents a serious challenge.

- *Ocean/atmospheric circulation shifts* – Serious concern is being shown at present about the threat of a weakened Gulf Stream (possibly due to global warming). Other ocean circulation phenomena may also be a threat.

- *Asteroid, meteor, comet strike* – The past five years have seen a considerable increase in awareness and signs of willingness to make some investment in early-warning and mitigation.

- *Less well perceived threats:*
  - ocean outgassing of methane or sulphides – leading to sudden global warming;
  - major solar flare or cosmic ray bombardment;
  - ocean turnover problems – sudden upwelling of deoxygenated and nutrient-poor water which damages plankton and other organisms;
  - acidification of oceans which damages plankton productivity;
  - alteration of global cloud cover;
  - geomagnetic variation (weakening and then field reversal) during which more radiation penetrates to the Earth's surface.

of global issues caused by natural processes which may be described as catastrophic. Some of these threats can be assessed and plans made to deal with them and to watch for early warning signs. However, it is unlikely that all problems or opportunities will be foreseen and some cannot be avoided; it is important to ensure human vulnerability is reduced and adaptability is enhanced (Posner, 2005). Recovery rather than total disaster is made more likely by that route.

Recently there has been more willingness to take the threat of large-scale natural disasters seriously, partly because of recent tsunami, hurricane and earthquake damage. Not long ago any proposals to consider comet/meteorite early warning would have been dismissed as unrealistic; now there is at least a little interest. A clear challenge is to reduce human vulnerability and improve adaptability. Too many decision makers and citizens have the idea that modern civilisation is safe and more secure than were past civilisations. However, there is a thin veneer of science/technology and governance. People in general are less able to cope with disruption than in the past, and rich nations have specialised and invested in such measures as computer systems, which are not at all resilient in times of disaster. Food reserves are not really sufficient to protect against one or two widespread harvest failures, too few key facilities are sited and designed to give protection against disaster, and much more could be duplicated for security. Too often the attitude is 'the past is history' – irrelevant, rather than offering important lessons. Sustainable development will be more likely if we learn from past civilisations.

Some challenges are difficult to spot because they develop gradually or in an insidious way. The challenge is first to actually identify them and then to convince people that slow change is worth worrying about. Gradual sheetwash removal of soil is one such problem. Other gradual problems are slow biodiversity loss, acidification of an environment, decline of fisheries and so on. Tracing the chain of causation may be difficult when these are indirect and cumulative. Recognising opportunities deserves more attention; the focus tends to be on threat recognition. Improved exchange of research information is one way to stimulate recognition of threats and opportunities.

Future priorities

In broad terms the way forward should be through:

- recognising threats, opportunities and limits;
- developing ways to reduce threats, maximise use of opportunities and avoid exceeding limits (and if they are overshot to restore systems as fast as possible);

- making critical human needs less vulnerable;
- finding ways of improving human resilience.

Considerable progress has been made in the recognition of threats, opportunities and limits. The three other goals listed above are a challenge. The ability to agree solutions, pay for measures and police activities is limited. Yet there has been progress in spreading environmental awareness, and countries are coming together to discus issues – and even to agree and fund workable solutions. Environmental law applicable to international situations is developing. Economics is starting to become environmentally aware. Environmental management tools and strategies are improving. There is some integration between environmental science, planning, social studies, economics and political studies. Will enough be done, well enough, soon enough, to avert environmental disaster?

When a threat is recognised it may or may not be perceived to be important; for example, a few scientists in the 1970s opposed powerful chemicals, domestic products and cosmetic companies to lobby against the use of ozone-damaging propellants in aerosol cans. They succeeded, and within a few years in the Americas there had been a shift to ozone-friendly propellants and alternative packaging methods. They had presented a scientific case strongly enough to convince, they had won over public opinion, and were fortunate in that their problem was luxury products, which people can live without, and which manufacturers could 'turn green' and capture sales. When the thinning of stratospheric ozone was spotted in 1987 the scientists involved had first to widen acceptance that there was a real problem. They then had to prompt other researchers to make the link with what were believed to be 'harmless' CFCs, and from then on they faced a battle with industry and consumers. CFCs were an established and cost-effective way to cool buildings and cars, run refrigeration, degrease components in manufacturing, and discourage pests in stored food. The battle was not against a small number of luxury goods manufacturers, and the solutions were less easy and more costly. They won the battle and achieved controls (through the Montreal Protocol and its revisions), yet there are still people and businesses willing to flaunt agreements and damage the environment for profit – smuggling of CFCs takes place and some countries are reluctant to ratify the Protocol, even with the support offered for adopting safer technology.

Seldom do all parties involved in a challenge immediately see eye-to-eye; when acid deposition problems were first raised in an international forum, at the 1972 Stockholm Conference, the countries and bodies blamed went into denial. For years British electricity-generating authorities lobbied for further research and claimed no clear evidence was available. When the weight of evidence was undeniable, negotiations started and polluters had to find ways of paying for controls and, if necessary, passing on costs to customers. Luck was with the polluters in that a shift was already taking place to less dirty fuels such as natural gas, and economic and political development had shut down some of the worst emitters (e.g. UK heavy industry). Current negotiations seeking atmospheric carbon controls are beset with disputes between those backed by hydrocarbon industries and those opposed to pollution; conflict between those demanding controls and those fearful of the impact on employment and national manufacturing costs; and arguments between developed and developing countries over apportionment of blame and share of control costs. There are some environmental scientists who feel global warming fears are mistaken and some who argue that expenditure on it is unjustified, given the range of other challenges which need to be addressed. Stott (*The Times*, 4 September 2004) stressed that natural climate change is the norm, and that global warming was a politico- (pseudo)-scientific construct, and that hopes to manage climate change by adjusting just one factor like carbon fuel use were mistaken.

Since 1972 there has been the development of regular global summits and conferences addressing challenges. This has helped focus debate and those involved to network in ways unimaginable fifty years ago. Proposals were put forward at the 2005 G8 Summit at Gleneagles for a global early-warning system for natural disasters based on the established WMO hurricane prediction and warning system. The Internet has also had a beneficial impact, especially when warnings about a challenge are first being broadcast. Various commissions have been established, but what is often needed is a powerful body which all trust, and which has teeth. The UN is the closest there is, and bodies such as the International Atomic Energy Commission do have considerable powers. However, national sovereignty can easily oppose concern for a global challenge: for example, prior to the Iraq War inspectors were resisted or decoyed. Paying to address global challenges is itself a major issue. Agreements are often signed and then within a year or two some signatories fail to come up with the funding – some pledges made early in 2005 to address the Indian Ocean tsunami have not been honoured (and some defaulters are rich Western nations). Some have called for a form of global taxation to generate funding to address transboundary issues; various mechanisms have been suggested, or even imposed:

- a tax on geostationary satellites;
- a tax on armament export sales;
- an air travel tax (some people already pay a voluntary carbon emissions levy – British Airways asked passengers for such payment (e.g. £25 for London–Sydney) but only about one in 200 agreed to it in 2005);
- a tax on transfer of gold bullion.

Already there are funds to support poor countries with environmental management (see Chapter 4 – Global Environmental Facility), but these are limited and mistrusted by some (Young, 2002). Insurance companies have long dealt with some environmental challenges which threaten individuals, communities or businesses. If the risk is not too widespread the impacts can be spread so that many pay a manageable annual sum, and if a disaster happens those affected can receive a reasonable pay-out. Unfortunately, some global environmental challenges threaten to impact upon so many (or are unpredictable 'acts of God') that it is not really possible to spread the risk sufficiently (Kasperson *et al.*, 2003). There are already fears that, as global warming affects weather patterns, established insurance in various countries will face such an increase in claims that it will break down.

Can currently recognised global challenges be sorted into some order of priority? Threats which are under way or definitely recur, and which seriously endanger human survival, deserve a high ranking. In very rough order of importance these are:

- *Global warming* – widely accepted. Potentially seriously threatening. Nations are negotiating and some are willing to alter practices and pay. The costs involved in mitigation and improving adaptability are huge. Too much focus on reducing emissions and not enough on reduction of vulnerability. There is a risk that so much attention paid to global warming is diverting attention and resources from other threats.
- *Serious epidemic* – recently this has received more attention (fears over influenza), but in general too little is being done and too few disease threats are taken seriously enough. Vulnerability reduction should be reasonably affordable.

- *Food shortage* – too little awareness and limited response to the threat (in spite of bodies such as the FAO established half a century ago). Improved agriculture since the 1960s has given a false sense of security. Soil and water problems contribute to food shortage.
- *Soil degradation* – environmental scientists recognise the challenge, but citizens and governments do not give it enough attention.
- *Inadequate knowledge of the state of carbon sinks* (oceanic, terrestrial vegetation, and soil) – sudden releases could drastically speed up global warming.
- *Water resources* – currently awareness is growing. Generally people undervalue water.
- *Biodiversity loss* – serious damage already (and no way to restore lost species). Some citizens are aware and governments provide some support. Too little is being done and too little is being spent.
- *Loss of the Gulf Stream* – causing chill weather conditions in Europe and eastern USA. Threat taken seriously enough to trigger recent studies. Little or no contingency planning so far.
- *Increasing pollution of the global environment with compounds* – subtle effects on the ecosystem and human health mean more expenditure on monitoring is needed.
- *Major volcanic eruption* – a real threat, which is inadequately perceived. Potential to suddenly cripple world food supplies.
- *Asteroid/comet strike* – a real threat, which has attracted some attention recently. Monitoring for early warning started in a very limited way. Studies of avoidance techniques have begun. More should be spent but risk will remain because some bodies will approach suddenly.
- *Tsunami* – normal tsunami risk (i.e. up to 15-m waves) realised since 2004 and some expenditure on early warning – but not for Atlantic or North Sea. Risk of mega-tsunami (possibly > 100-m waves) has been flagged but so far there has been little serious debate or practical preparation.
- *Ozone layer damage* – threats accepted and reasonable success in agreements and threat reduction.
- *Seismic threats* – citizens often well briefed, some mitigation measures and contingency planning. All that can be done without better early warning has probably been done.
- *Eco-refugees* – displaced by any of the previous problems. There are predictions of huge numbers in the future if global warming, soil degradation, warfare and sea-level rise are not dealt with. Displaced people can in turn damage the place to which they relocate and trigger conflict.

Reduce vulnerability

One of the reasons for the success of humans has been their adaptability. Increasingly people are less mobile, populations have grown, environments and resources are under stress and in some cases degrading. Livelihoods are commonly specialised and depend on a chain of factors, any of which could be disrupted by natural or socio-economic change. Poverty and urbanisation have reduced the options for adaptation for many. Few people today are in possession of survival skills and specialists are needed to maintain current lifestyles. In spite of modern technology many are less resistant to disaster and have reduced resilience (chance of recovery) than people were in the past (Barrow, 2003: 100–122). Administrators and citizens often have a false sense of security. In richer nations in particular people are less adaptable then ever before and if governance broke down would soon suffer. In addition, as pointed out earlier, the environment in recent centuries has been unusually stable – that cannot be expected to last. Efforts have to be made to assess and reduce vulnerability.

Environmental management can recognise and assess threats, help develop monitoring and early warning, and co-ordinate the diverse specialists involved (UN, 2002). Civil defence arrangements have been reduced or abandoned in many countries as the Cold War threat has been seen to diminish – ignoring natural threats and human problems other than conflict. Around the world traditional risk avoidance or disaster response strategies evolved and served people for generations. The pressures of development have in many cases eroded or destroyed such strategies. Thus it is not only the rich who are more vulnerable than before; the poor, especially rural, folk are also at increased risk (Adger *et al.*, 2004). Slightly over half of the world population is concentrated in only about fifty large cities – dispersal tends to reduce vulnerability and crowding the opposite.

Assessing the risk that a threat will happen allows some prioritisation of efforts to react; however, there are too many to prepare for specifically, and knowing the risk of occurrence does not say when it will happen. Some problems may be averted or mitigated with good environmental management, but the world will still suffer surprise events. It is better to focus on predicting scenarios such as global warming and then preparing in better time – rather than expending time and vast sums of money negotiating emission controls, which may not be adequately enforced. Attention must be given to other issues, especially those which have a predictable threshold, or that offer early warning signs. However, much more effort should be directed at making food, water, biodiversity resources, stores of knowledge and hazardous waste repositories less vulnerable. Diversification and duplication in separate sites would be wise. At present some key crop collections are in single facilities where a disaster, even something like a power failure that stops refrigeration, could destroy material. Gene banks and data stores/libraries could be sited in hardened and safe localities to reduce risks from tsunamis, storms and so on. Vulnerability reduction and improvement of resilience is the only wise response to the threat of unexpected problems.

Not all threats are natural: some result from human error – it was feared that the Y2K 'bug', a design shortcoming, would cause chaos (and potentially it could have done). Evolving computer technology has made disks from before the 1990s largely inaccessible. Much on the Internet is discarded and not stored, and electronic data systems are vulnerable to computer viruses and power failures. Present disk storage may be unreadable in fifty years' time because the software and drives have been thrown away and radiation or fungal rot has corrupted CD/ROM surfaces. Secure archives both in terms of the media used and the storage locations are vital.

Decision makers in companies, institutions and governments now plan for the short term; long-term development is someone else's problem. Attention is directed towards what can be done and seen to be done in a term of office. While environmental issues often demand planning and management encompassing centuries, the real world rarely looks beyond ten years ahead. In a democracy people tend to support measures which give short-term gain. Research concentrates more and more on what can be seen to yield useful results (and 'useful' may be interpreted with reference to political correctness and what non-scientist managers deem valuable). In time this is likely to discourage original, speculative study, which in the past has given so much to human development. Research funding and institutes of learning are also often geared towards discouraging research into new 'unknown' fields; and there is also an establishment resistance to new ideas. A global challenge is to ensure that there is encouragement to study subjects of relatively unknown value.

In famine conditions or major epidemics, sudden climate shifts or whatever, people become desperate, many become refugees, and normal ethics break down. Conflict is then much more likely (Woodbridge, 2004). There needs to be threat recognition and contingency planning in advance; ideally, food stores should be increased, and future developments should seek to reduce vulnerability.

Sustainable development

Achieving sustainable development has become established as a goal. Continued interest in sustainable development is arguably the most important paradigm shift of the late twentieth/early twenty-first centuries. However, many pay it only lip-service, and making it work is a huge challenge. Sustainable development strategies must consider livelihoods, supplies of essential resources, waste disposal, biodiversity conservation, managing institutions in a workable fashion, and much more. The strategies must also be adaptable, robust and resilient (Markandya and Halsnaes, 1992; Turner, 1995). Ideally, initiatives should be diverse, duplicated and widely spaced. Then, if one element/strategy is damaged or destroyed there may be inputs and skills somewhere safe, which can be used to recover. When there is a need to set up a new element/strategy practice, inputs from elsewhere may be imported and adapted, and if faulty substituted. Some elements/strategies will dovetail in symbiosis with others, exchanging wastes or offering some service. It may well be that a given locality has several sustainable development elements/strategies close together or overlapping; effective overall co-ordination (and ultimately global overview) is vital. Co-ordination will be as challenging as 3-D chess. The UN has declared 2004 to 2014 the decade of Education for Sustainable Development, and in 2002 the World Summit on Sustainable Development was held in Johannesburg (Dodds, 2001; Middleton and O'Keefe, 2003).

Cut poverty

Poverty, environmental degradation and sustainable development are linked, so it makes sense to address them in a co-ordinated way. When people suffer poverty they are unlikely to give much attention to environmental issues. Sustainable development demands investment of some current resources into maintaining things in the future, and for poor people with little to spare this can be a dilemma (and one they will need aid to deal with). Before 1992 a number of countries saw environmental management as a luxury or even a conspiracy to hold back their development. There is less suspicion now but lack of funding often slows progress. Today the expectation is for poor people to get something out of environmental care, which counters their poverty. Poverty reduction and environmental management/sustainable development should, whenever possible, be dovetailed to be mutually supportive. However, there may also be occasions where the two demands interfere. It is difficult to oppose poverty and social development demands – citizens can exert pressure. However, some environmental degradation cannot be undone and some losses have huge long-term implications for human well-being on a wider scale. Care must be taken in such cases that local poverty reduction does not have widespread negative results which outweigh it. There must be careful, just, dispassionate assessment.

Summary

- Recognising important global issues does not necessarily mean that they can be reliably forecast so, for some, preparation may have to be limited to mitigation and recovery rather than avoidance.
- Modern peoples are still vulnerable to natural and human-caused disasters, probably more so than in the past.
- Identifying a threat is no guarantee: people or governments will prepare for it or avoid doing so, especially if the latter is costly.
- Much effort and money are being spent negotiating global warming controls. It may be better to plan for adaptation to warming, and to try to reduce vulnerability and improve resilience – to this and other threats known and unexpected.
- Concern at the passing of 'peak oil' has been increasingly expressed since mid 2005.

Further reading

Cartwright, F.F. and Biddis, M. (2001) *Diseases and History: the influence of disease in studying the great events of history.* Sutton, London.
Explores the impact disease has had in the past.

Chasek, P.S. (ed.) (2005) *The Global Environment in the Twenty-First Century: prospects for international cooperation.* United Nations University Press, Tokyo.
A collection of papers which explore the prospects for and problems associated with international environmental co-operation.

Diamond, J. (2005) *Collapse: how societies choose to fail or survive.* Allen Lane, London.
Neo-environmental determinism/environmental history – a fascinating book that should be read by all those interested in sustainable development, which reviews the past for warning and guidance.

Fagan, B. (2004) *The Long Summer: how climate changed civilisation.* Basic Books, New York.
Administrators should not assume that the environment is stable and benign – modern civilisation is like a large oil tanker: slow to alter course, and surprisingly easy to overwhelm; past societies were like small responsive boats – how can we better steer things in the future?.

Houghton, J. (2004) *Global Warming: the complete briefing* (3rd edn). Cambridge University Press, Cambridge.
Comprehensive introduction.

Kemp, D. (1990) *Global Environmental Issues: a climatological approach.* Routledge, London.
Clear review with many illustrations focusing on atmospheric issues.

Middleton, N. (1995) *The Global Casino: an introduction to environmental issues.* Arnold, London.
Widely used introduction.

Maslin, M. (2004) *Global Warming: a very short introduction.* Oxford University Press, Oxford.
Concise, yet thorough and readable coverage.

Officer, C. and Page, J. (1998) *Tales of the Earth: paroxysms and perturbations of the blue planet.* Oxford University Press, New York.
Readable review of ongoing natural threats.

Oldfield, F. (2005) *Environmental Change: key issues and alternative approaches.* Cambridge University Press, Cambridge.
Past and current environmental change focusing on key questions.

www sites

FoodFirst site on the green revolution/food in the future – Institute for Food and Development Policy http://www.foodfirst.org/media/opeds/2000/4-greenrev.html (accessed July 2005).

Intergovernmental Panel on Climate Change http://www.ipcc.ch (accessed June 2005).

Worldwatch Institute http://www.worldwatch.org (accessed July 2005).

Full text of the Kyoto Protocol http://www.unfcc.int/resource/docs/convenkp/kpeng.html (accessed July 2005).

For a listing of www sites and other sources on global issues see http://www.nottingham.ac.uk/sbe/planbiblios/bibs/Greenis/A/06.html (accessed July 2005).

12 Pollution and waste management

- A brief history of pollution and waste problems
- Pollution and waste associated with urbanisation and industry
- Radioactive waste and pollution
- Electromagnetic radiation (non-ionising)
- Treating pollutants and waste
- Agricultural problems
- Recycling and reuse of waste
- Summary
- Further reading

Sustainable development demands the ongoing management of *outputs* (waste and pollution) as well as *inputs* (food, water, living space, and other resources). Historically, humans have spent much more effort attending to inputs than to outputs. However, growing populations and the invention of potent compounds have made it vital to attend to outputs. It is doubtful that many before the 1960s would have taken it seriously that pollution could threaten the global environment, and some even welcomed pollution as a sign of progress (a side-effect of earning money). Often outputs can be recovered to become useful inputs to help sustain development.

Pollution may be defined as the introduction by humans, deliberately or inadvertently, of substances or energy (heat, radiation, noise) into the environment – resulting in a deleterious effect (O'Riordan, 1995). Contamination is the presence of elevated concentrations of substances in the environment, food and so on. Contamination is not necessarily harmful or a nuisance. Pollution involves contamination, but contamination need not constitute pollution. Nature as well as humans can generate toxic or nuisance compounds (e.g. volcanic ash). Waste may be defined as movable material that is perceived, often erroneously, to be of no further value. Once discarded, it may not be a problem, a nuisance or a hazard, but it can turn out to be a valuable resource (Hill, 1998). As waste may give rise to pollution, it is necessary to view both together. Pollution and waste management can focus on (1) prevention and avoidance, or following escape or release: (2) collection and disposal, or (3) reclamation/treatment/mitigation (which may sometimes be difficult and costly or impossible). Prevention involves catching waste or pollution before release, while avoidance seeks development without waste or pollution generation (Young, 1990; Bradshaw *et al.*, 1992). A hierarchy of desirability can be agreed. The following is that favoured in the USA (Middleton, 1995: 217):

- reuse;
- waste reduction;

- recycling;
- resource recovery;
- incineration;
- landfill.

Wherever possible, waste and pollution disposal should be integrated with employment generation, food and commodity production and appropriate industrial development. There are examples of huge expenditure on waste disposal based on developed country experience that are costly for poorer nations to maintain. Various groups are affected differently by pollution and waste, especially in a multicultural society, where dissimilar lifestyles means varied exposure and different concerns. For example, there have been studies in the USA which show that the poor accumulate much higher levels of pesticides in their bodies, because their diet is more likely to be contaminated and they are exposed during menial work and through poor housing (Hurley, 1995). In the UK it is quite noticeable that poor people tend to live in the east of towns and cities – downwind of most pollution sources. Children and women may be differently exposed and physically more vulnerable than the male working population for a variety of reasons. Community-based approaches are currently seen to have promise for urban improvement, which means environmental managers must have familiarity with social issues and with participatory approaches (Hasan and Khan, 1999).

Pollution can be 'primary', having an effect immediately on release to the environment, or indirect ('secondary'), the product of interaction after release with moisture, other pollutants and sunlight. The effects may be direct, indirect or cumulative, felt intermittently or constantly, immediately or after a delay; affecting the atmosphere, soil, oceans, water bodies, groundwater, or be restricted to certain organisms, produce or localities. The effects of pollution may be short term or longer term; pose a hazard or a nuisance; be toxic or non-toxic; take the form of a chemical, biological, radiation, heat, light, noise, dust, or smell problem. Pollution may have local, regional, transboundary or global impact. The environment may render pollution and waste harmless until a threshold (the absorptive capacity) is exceeded, after which, if there is no effective control, there will be gradual or sudden problems. Loss of absorptive capacity may be very difficult to recover from, so it is important that environmental managers model and monitor to avoid exceeding thresholds.

The risks from pollution and wastes are far from fully understood, and available standards and monitoring techniques need constant improvement. What was considered safe twenty years ago is often no longer accepted and what is acceptable today may not be in the future; also new compounds are released. A 'safe' background level of a pollutant may become dangerous to organisms near the top of a food web as food organisms concentrate it by feeding or absorption (bioaccumulation or biomagnification). Some pollutants become concentrated in certain tissues of higher organisms (e.g. DDT and polychlorinated biphenyls (PCBs) in the fat; radioisotopes like strontium-90 in the bone; and radioiodine in the thyroid). This can damage the affected and surrounding tissues (Odum, 1975: 103). Wastes and pollutants can also be concentrated by tidal action, sudden rain-out by storms, chemical bonding to certain soil compounds, localised interception of contaminated rainfall and so on to form hotspots. Some pollutants change little after release, some decay or disperse and become harmless, while others are unstable and may be converted into harmful compounds.

Pollutants or wastes initially discharged into water or the atmosphere may exchange between these two systems; for example, dust may settle on water and sink, or polluted water may form aerosols or contaminate groundwater which surfaces at a different time and place. Pollution sources may be: point (e.g. an explosion), linear (e.g. a road) or extensive (e.g. dust from a desert). Releases can be continuous, single, repeated, brief, extended, random events, or periodic. The distance pollution and waste disperses depends on its qualities and how it is released. Gases or dust are affected by the height of the release, their temperature relative to the air, weather conditions (especially wind-speed), and their density or particle size, presence of inversion layers, whether any obstacle is encountered and the texture of that obstacle, and many other factors. Dispersal in water is affected by an equally diverse set of factors. Even large particles may be widely scattered by natural or artificial explosions, especially if material is projected into a jet stream, storm or ocean current. It is not uncommon for a temperature inversion in the atmosphere effectively to put a lid over an area, trapping pollution. Water bodies (a sea or lake), and the atmosphere, may become stratified, so only limited levels mix and disperse compounds; consequently conditions become stagnant and pollutants become concentrated.

A brief history of pollution and waste problems

Even wandering hunter-gatherers or those living in scattered hamlets or small towns contaminated water supplies and faced health risks as a consequence of careless waste disposal and living in smoke-filled dwellings. However, real problems followed urban development, population concentration, industrial activity and technological development. In London the smell from tanning, the operation of lime-kilns, and the shift from wood to coal burning polluted the air enough for Edward I to legislate in AD 1306 (Brimblecombe, 1987). There is evidence of widespread smoke and heavy metal (especially lead) pollution in Roman times (Wellburn, 1994). Chemists have developed man-made compounds since the 1920s which have serious effects on the environment, such as pesticides, radioactive isotopes, PCBs and chlorofluorocarbons (CFCs). These can be toxic, carcinogenic, mutagenic, or harmful in a variety of other ways, even at very low concentrations, and some persist, posing a danger for organisms or biogeophysical cycles for a long time. Losses of marine mammals in the Baltic and elsewhere may be due to viral infections triggered as background levels of compounds like PCBs reduce immunity.

Before the 1960s it was common for industry and government bodies to ignore or hide harmful on-site or off-site impacts. There were wide gaps in health and safety laws and it was often difficult for workers, consumers or bystanders to seek damages. When action was taken it was usually cleaning up rather than preventive. Consequently, the burden of pollution and wastes was generally not borne by those who pollute, and sometimes people far removed and unrewarded suffered. The impacts were often indirect in time, with effects felt and clean-up costs suffered perhaps generations after pollution occurred or waste was discarded.

Pollution and waste occur worldwide: both socialist modernism and Western capitalism have neglected environmental management (Feshbach and Friendly, 1992; Mnaksakanian, 1992). Governments, international agencies and NGOs have increased monitoring and control of pollution and waste, and there is a shift towards making the polluter-pay and encouraging prevention. This shift is far from complete. Changing technology, expanding industrialisation and growing urban populations mean that provision for pollution control and waste disposal is continually challenged. Many sites where pollution and waste have accumulated have not been recorded yet pose a hazard. Dangerous material in containers could escape when seals deteriorate or if acid deposition makes compounds in soil more mobile (chemical time bombs). Pollution and waste are increasingly transboundary problems. Before the 1970s such transboundary threats were largely unrecognised; now the environmental manager must seek pollution controls that can be applied to more than one country or the global environment.

Over the past few decades pollution and waste management has been aided by:

- growing adoption of the polluter-pays principle (Box 12.1);
- a trend towards more proactive planning and management (e.g. use of EIA);
- increased release of information on pollution and waste risks and nuisance – as a result of government, NGO, media and international agency activity;
- improved legislation to define, monitor and control pollution and waste;
- spread of better environmental quality standards more widely applied – which aids monitoring and exchange of information, and provides yardsticks for legislators and enforcement;
- development of better policing, and of self-regulation and joint agreements between regulators and potential polluters;
- better methods and equipment for monitoring and assessment;
- some shift to treating pollution before discharge;
- the end of the Cold War and improved international co-operation and exchange of information (Young and Osherenko, 1993).

Regulatory authorities, industry and so on have to balance costs of pollution control and waste management against the value of environmental quality or human well-being. Too strict a control, and business suffers and may relocate or go bankrupt; too lax, and people and environment suffer. The goal is the best possible environmental option

---

Box 12.1

The polluter-pays principle, NIMBY and NIABY

Polluter-pays principle

This principle aids pollution and waste control by ensuring that manufacturers, agriculture and public realise and pay the full costs for goods and services (i.e. incorporating pollution damage and pollution control costs into prices). However, even within one country people hold different values so its adoption involves complex sociopolitical and economic interactions, and implementing it on a wider front, like the EU or internationally, is a challenge (O'Connor, 1997).

The agricultural sector is a major source of pollution in many countries but has lagged behind in adopting the principle (Baldock, 1992; Seymour *et al.*, 1992; Tobey and Smets, 1996).

NIMBY and NIABY

The public are becoming better informed on environmental issues, and may object to real or imagined pollution and waste hazard or nuisance. Special-interest groups or the public can adopt a not-in-my-backyard (NIMBY) attitude. NIMBY is often invoked when authorities are proposing to site a hazardous or 'nuisance' waste or pollution treatment plant (Hunter and Leyden, 1995). The reaction may not just reflect local views: some proposals are opposed by environmental activists and NGOs *wherever* they are to be sited (for example, deep underground storage of high-level nuclear waste) – a not-in-anybody's-backyard (NIABY) response.

(BPEO) (O'Riordan, 1995). Pollution and waste control are now a significant force in politics (Weale, 1993), and ideally should be integrated with economics, social and political policies (Haigh and Irwin, 1987).

The identification of the loss of stratospheric ozone and the relative success in reaching agreement for international controls on CFCs (see also Chapter 10) may be seen as an example of successful environmental management. However, the discovery was largely accidental: a British Antarctic Survey researcher, who found the 'hole', was researching something else and earlier concerns in the 1970s had been largely marginalised. This is hardly the precautionary principle in action, more serendipity. Similar comments may be made about the linking of global carbon dioxide levels and climate change – long-term recording of CO_2 on Mt Mauna Loa, Hawaii was not undertaken to identify environmental threats. There is a double problem here: the lack of environmental management 'trouble-shooting', and a growing tendency around the world to fund only sharply defined and applied research. Neither is conducive to identifying 'unknown' threats and opportunities.

Pollution and waste associated with urbanisation and industry

The term 'urban' broadly means 'the concentration of people in cities and towns'. Some urbanised areas are the most altered, unhealthy and contaminated of the Earth's environments. Cities affect other, often-distant environments, and rapid urban expansion, especially in poor countries, can stress services and infrastructure to the point of breakdown. Many cities, even in developed nations, have dangerous levels of atmospheric pollution, and sewage and waste pose problems.

Built-up land has a different albedo, heat storage characteristics and roughness than non-urban. There may also be considerable waste heat from homes and industry. These combine to cause a heat-island effect – city areas are warmer than their surroundings, causing local airflows that may recycle pollution. Spreading cities destroy farmland, biodiversity and amenity areas. The FAO estimated that worldwide between 1980 and 2000 about 1,400 million ha of arable land would be lost through urban sprawl (WRI, IIED and UNEP, 1988: 42). Most cities have considerable areas of dereliction, which could be used to reduce demand for new land, but to decontaminate and rehabilitate it for settlement can be costly, although it can be converted relatively cheaply into amenity areas by tree planting and landscaping (which reduces the heat-island effect).

Drains and sewers speed up run-off of sewage, and silt-contaminated storm water from cities. Leaking sewerage contaminates groundwater beneath – London has such a problem – which means alternative supplies must be taken from surrounding areas, reducing streamflow ('sewerage' refers to the infrastructure/pipework concerned with conveying sewage). Before the late nineteenth century sewage was only a problem in urban areas, which depended on cesspits, latrines and street collection by nightsoil carts, or open-channel sewerage systems that emptied without treatment. Cities in Europe and North America began to install water-borne sewage disposal systems in the 1850s, following rising incidence of faecal–oral diseases and smell.

Urbanisation shifted sewage management from reclamation to disposal, from resource use to resource waste. Many of the world's sewerage systems are becoming overtaxed, and waste disposal consumes large amounts of water that might be used for other purposes. Poor districts of developing country cities are likely to discharge sewage contaminated by small-scale industrial activity – especially oil and solvents.

Sewers in cities which grew before the 1930s are often crumbling and need costly refurbishment. Modern sewerage design can reduce silting up, for example, by installing stepped or ovoid cross-section pipes, but there are still many problems associated with water-borne sewage disposal:

● the cost of installing, extending and maintaining sewerage;
● failure to separate storm water, sewage and industrial waste, which makes treatment and disposal more difficult;
● waste of often-scarce water;
● treatment of sewage before discharge, which is seldom satisfactory.

Many of the world's rivers, lakes and seas are so polluted that they pose a health risk and adversely affect tourism and wildlife. Where long sea-outfalls were once satisfactory, population increases and anti-discharge regulations mean treatment costs will rise. More appropriate alternatives to established water-transported sewage disposal should receive more attention (e.g. composting latrines). Most sewage treatment generates sludge contaminated with pathogens and toxic heavy metals, and effluent rich in phosphates, nitrates, and many other dissolved pollutants. In the past sludge was sent to landfill sites, spread on agricultural land, or dumped at sea. The last option has been outlawed in the USA since 1993, and was illegal in the EU from late 1998. In Europe and the Americas sludge is increasingly pumped on to farmland but that option is inadequate. In the USA and Europe treatment is widely used to reduce the pathogen content, but in the UK only about 25 per cent was so treated in 1998. Much of the UK's sewage has been disposed of as untreated sludge – on to farmland (and this is unlikely to be reduced much in the immediate future). Disposal of raw sewage sludge on to farmland risks contaminating agricultural produce, groundwater and streams and ultimately, perhaps, domestic water supplies with problem micro-organisms such as *Cryptosporidium* or harmful strains of *Escherichia coli*. Disposal options are increasingly outlawed, which leaves high-temperature incineration. Europe and the USA will probably increasingly de-water sewage and incinerate the solids. Village- or household-scale biogas production can offer safe, cost-effective sewage disposal and supply gas for heating, cooking and lighting. Waste from the biogas digestion may be used as relatively safe compost. Unfortunately, even cheap systems may be too costly for very poor communities, and there needs to be an optimum mix of sewage and farm or household waste, which is not always to be found.

There are a number of water-conserving 'waterless' sewage disposal systems: low-volume water flush to septic tank, earth-fill latrines, household or village composting toilets, electric incineration and chemical digestion toilets (from which safe waste can be periodically removed and spread on the land or safely disposed of in some other way). Septic tanks (limited sewerage systems) are widely used and are effective, *provided* the soil and groundwater conditions are suitable and the operation and regular solids removal and treatment are well supervised. At the village, large farm or small town scale, composting sewage mixed with agricultural waste such as straw may prove an effective method of disposal, yielding a safe, useful end-product.

Refuse (trash or garbage) disposal can be based on self-help, which cuts the cost and may benefit poor people. Such an approach includes street corner skips, which are collected and taken to the tip or recycling plant by a rota of local people. An alternative is to offer an incentive to people to collect waste and bring it to a recycling plant, composter or incinerator. Recycling waste can provide employment, and local authorities or aid agencies could assist by providing protective clothing, supervision, healthcare and refuse handling and processing equipment (where informal and less healthy garbage picking would probably develop anyway).

Cities generate large quantities of refuse: in developed countries this can be 500 to 800 tonnes per day per million people. This may be landfilled, incinerated, dumped at sea, recycled or composted. Domestic waste is a mixture: in the UK it is typically (approximately): 7 per cent plastics; 8 per cent metals; 10 per cent glass; 10 per cent fines (dust); 12 per cent miscellaneous textiles; 20 per cent waste food and other easily decomposed material; and 33 per cent paper products (*The Times*, 14 June 1993: 32). Poor countries tend to produce waste which is less rich in plastic and paper packaging (which makes landfill less problematic); however, the trend is towards an increase. There has been a world decline in the use of reusable glass bottles since the 1960s, matched by an increase in plastic bottles. This mixture can make sorting and treatment a problem. Tin and steel can be recovered with magnets, and aluminium cans are fairly easy to recycle, but plastics are more of a challenge. A number of developing country cities have begun to aid the poor to sort and recycle garbage by providing washing facilities, supplies of boots, gloves, tetanus injections, and health treatment (Perera and Amin, 1996).

Over 90 per cent of UK refuse was disposed of by landfill in the mid 1990s (North, 1995: 164); at the same time in the USA over 70 per cent was similarly disposed of. Landfill sites should be located to avoid nuisance and the risk of contaminating streams or groundwater: a minimum of a layer of clay should be put in place before tipping, and after completion used to cap and seal the tip (Figure 12.1). Many redundant tips and some of those currently in use meet none of those standards and present a serious future hazard because they are poorly sealed. Sometimes domestic refuse is mixed with industrial wastes (co-disposal), power-station flue-ash or sewage for convenience, or in the hope that it will assist breakdown of harmful materials. Whether this is wise is debatable.

It is important to locate landfill sites to avoid nuisance to surrounding areas, to reduce the risk of ground and surface water contamination, to minimise damage from escaping methane or underground fires, and to ensure that vermin are not a problem. Houseflies and rodents can cause difficulties some distance from landfill sites if they are allowed to breed, and scavenging birds may pose a serious threat to airports within a few miles (Clark *et al.*, 1992; North, 1995: 164–186). Provided that decomposition of organic matter is vigorous, a refuse tip should generate enough heat to kill most harmful organisms. Chemical contaminants are a different matter: 'tip archaeology' in the USA suggests that paper products may be a major source of contamination of groundwater, streams and surrounding environments due to the printing inks, waxes and sealants used on them. Methane, heat from spontaneous combustion, toxic compounds and subsidence limit the future land use of tip sites, although in some cases gas can be collected and used for power generation or heating.

Plastic waste is a problem in landfill, as litter on land and adrift in the oceans. Some plastics degrade slowly and can cause considerable harm to wildlife (for example, marine turtles are injured by eating floating polythene bags) and equipment such as pumps. Phantom fishing – lost or discarded plastic nets and long-lines – does tremendous damage to wildlife. Both problems might be reduced if biodegradability could be built in. Unfortunately, fishermen do not want equipment to deteriorate before it is lost and packaging has to last a reasonable time. Laws in the USA, Sweden and Italy insist on biodegradable plastics for certain types of packaging. Germany requires *manufacturers* to arrange for proper disposal of the packaging materials around their products, and The Netherlands has a covenant with manufacturers which aims to reduce and simplify packaging to aid recycling. About 60 per cent of the world's shipping *should* be covered by a 1989 treaty requiring no dumping of plastics at sea (Annex V of the International Convention for the Prevention of Pollution from Ships). However, a glance at

Figure 12.1 Landfill refuse disposal site in Belgium. This is a state-of-the-art facility with geosynthetic clay and high-density polyethylene lining to prevent leaching. Each pit can hold about 250,000 tonnes of waste. The gas emitted during decomposition is captured for electricity generation.

Source: Press release photo from Bitumar N.V. Belgium (1998)

any strandline confirms that compliance is lax, and much plastic also gets into the sea from rivers and sewers.

Lead pollution of drinking-water and food peaked in the nineteenth century, after which lead pipes and soldered food cans were gradually discontinued. Paints still pose a threat of lead contamination, and redecoration may release dust from material applied decades previously. About 90 per cent of the atmospheric lead today probably derives from leaded petrol. Ice cores from Greenland clearly show the pattern of pollution rising after the 1750s and accelerating from 1925. Lead reduces birth weights and children are vulnerable as they accumulate the metal and may suffer retarded mental development, especially if exposed in the early years of life. The poor are likely to suffer greater contamination. Some countries, starting in 1972, have insisted on the use of non-leaded petrol. Such controls are not universal or always enforced. Nevertheless, atmospheric lead in the UK has fallen from a peak in 1974 (and similar patterns may be seen in countries that have banned leaded fuel).

Between the 1860s and 1950s many UK and European cities had sulphur dioxide-rich winter smog (smoke and fog combined) problems caused by domestic coal fires. That problem has been much reduced; however, there has been a worldwide increase in warm weather nitrogen dioxide-rich (photochemical) smog, mainly caused by petrol-engine vehicles. The 1956 Clean Air Act, the 1968 Clean Air Act, the 1974 Control of Pollution Act, and a consumer shift to less polluting natural North Sea gas cured UK smogs (British Medical Association, 1991: 11). Photochemical smogs are likely where there are sunny and still weather conditions and traffic pollution, especially at high altitudes where there is less shielding from UV. Other emissions are caused by vehicles in urban areas – partially burnt hydrocarbons including the dangerous volatile organic compounds (VOCs) (which include toluene, ethylene, propylene and benzines), toxic dust, heavy metals and noise. VOCs are formed mainly from diesel exhausts, cause respiratory diseases and may be carcinogenic; they also play a part in tropospheric ozone production and acid deposition. Pollution continues downwind of cities or busy road systems. A common problem is tropospheric (the lower few kilometres of the atmosphere) ozone formed from partially burnt hydrocarbons in vehicle exhaust or power-station flue gases. The WHO considers 60 ppbv of ozone to be dangerous to humans (the UN suggests 25 ppbv as a 'safe limit') – in 1992 Mexico City exceeded 398 ppm on more than one occasion. Many other cities exceed safe ozone levels in hot, still weather. Crop damage from tropospheric ozone throughout the USA may amount to a 5 to 10 per cent depression of harvest. A car manufacturer has recently introduced a radiator catalyser which it hopes reduces ozone in the air passing the car, utilising waste heat to do so.

Air pollution from combustion-engine vehicles can be reduced by exhaust catalysers, cleaner fuel (especially reduced sulphur and lead content), lean-burn and direct-injection petrol engines, capture of evaporated fuel, non-polluting vehicles (driven by electricity, natural gas, hydrogen or fuel cells), restrictions on use of polluting private cars, such as road tolls for city driving or high parking charges and other vehicle use restrictions, and improved public transport (Association of County Councils, 1991). (Note that exhaust catalysers do not reduce carbon dioxide emissions.) In some countries (including the UK) total mileage travelled by car has increased and markedly reduces the effect of better emissions controls. Some cities have tried restricting car use. Paris and some other French cities recently banned car use in certain areas on specific days to encourage alternative transport use, and Singapore and London monitor each car and charge for the distance covered. It is also important to control disposal of lubricants, hydraulic fluid and other harmful transport-related pollutants. What is needed in many countries, including the UK, is more foresight and an integrated approach to planning and managing land use, transport and manufacture.

Pesticides are widely used by city health authorities to control mosquitoes; it is not only malaria which is carried, there is also yellow fever, Nile fever, and in 2005 many Southeast Asian cities were fighting dengue. There are safer alternatives that might be used more widely to reduce pesticide applications: net screens on windows, application of oil or kerosene to standing water to prevent mosquitoes reaching maturity, stocking water bodies with mosquito- or snail-eating fish, trapping rodents, and enforcing laws that prevent pest-breeding sites.

Pollution and wastes are produced during extraction and processing of raw materials, transportation, manufacture, product use and disposal. The pollution or waste may be gaseous, particulate, liquid, debris, radiation, heat, light or noise (see Figure 12.2). Fine particles and gas, radiation and noise are difficult to counter once released and dispersed (and are better managed before release); water-borne waste and solid waste (large particles) are more easily intercepted and managed. Effluent may be moved via pipes and sewers, by road, rail tanker or ship to chemical treatment plant, recycling plant, incinerator, landfill site, deep underground repository, injection borehole, or the ocean.

Management should be an ongoing process: landfill may contaminate groundwater or streams years after burial, through poor site choice, inadequate sealing, bad management, disturbance by burrowing animals, erosion, earth movements, acid deposition, or human interference. Pollutants are often disposed of in the sea, in the hope that dilution will neutralise them. Unfortunately, in shallow estuaries, enclosed seas and continental shelf shallows, mixing and dispersal may not happen. These waters also receive pollution from rivers and atmospheric fall-out (e.g. dust, acidic aerosols). Waste sealed into containers and dumped in ocean deeps may escape through corrosion, the activities of

Figure 12.2 Sungai Besi tin mine, south of Kuala Lumpur, Malaysia. This is one of the world's largest holes in the ground (a large lorry mid-photo gives an idea of scale). Once worked out, the pit will flood. Spoil (sandy tin tailings) from mines in this region has caused considerable nuisance, choking streams and lying as an infertile layer over large areas.

marine organisms, trawlers, anchors or undersea landslides. Once it has been dumped into the deep ocean, inspection and remedial action is difficult (Clark, 1989).

Waste motor tyres are a fire and pollution risk, and are best treated by: remoulding (only possible for some worn tyres); combustion in district combined heat/power plants and cement kilns, or they may be dumped at sea. Tyre 'reefs' reportedly attract fish, become encrusted with marine organisms, do not leak pollutants, and offer storm protection and sites for conservation (Mason, 1993: 5). Another possibility is to shred waste tyres for road-surfacing material or construction materials. A lot of land contaminated by industry is abandoned or resold (Syms, 1997). Known contaminated land presents a rehabilitation challenge. However, in many countries record keeping has been poor and contaminated land has been unknowingly built upon. Infamous cases occurred in the 1970s, notably Love Canal (USA) and Lekkerkirk (The Netherlands). The solution is legislation, adequate record keeping, rehabilitation and land-use restrictions.

Pollution from thermal power-stations and industry can affect the environment at considerable distances off-site, one impact being acidification. Acid deposition may be in the form of contaminated snow, mist or cloud droplets (wet deposition), or as dust, aerosols or gases, especially sulphur dioxide (SO_2) (dry deposition). Uncontaminated precipitation is usually slightly acid (pH above about 5.6). Acid deposition is generally recognised when pH falls below 5.1 (Elsworth, 1984: 5). (Note that pH 4 is 10,000 times more acid than pH 8.)

Acid deposition may:

● damage plants and animals directly;
● alter soil chemistry or structure;
● alter plant metabolism;
● alter metabolism or species diversity of soil micro-organisms, leading to change in fertility or soil chemistry;
● damage man-made and natural structures;
● mobilise compounds in soils, waste dumps and water (notably phosphates, heavy metals, aluminium).

The impact varies: some localities may be exposed to prevailing winds, others get occasional storms of acid rain, and some receive sudden snowmelt carrying the accumulated deposition of a whole winter. Some soils or water bodies can withstand more acidification than others (certain soils may become more fertile): they are said to 'buffer' the pollution through alkaline material within or reaching them from underlying basic (alkaline) rocks. In temperate and colder environments soils over slow-weathering, non-alkaline bedrock are more likely to be affected; in warmer climates already acidic, aluminium-rich soils are vulnerable. Soils which receive a dressing of ammonium-rich fertiliser may suffer acidification whether or not there is acid deposition, and agricultural modernisation often uses such agrochemicals (Eriksson, 1989; Kennedy, 1992). Volcanic eruptions, sea spray, weathering of gypsum and gas emissions from forests, grasslands and marine plankton can lead to natural acid deposition. In the Pennine uplands of the UK, blanket peat's acidification since the 1750s damaged plants vital for continued peat formation. Land has eroded and the land is now dominated by only two species (Usher and Thompson, 1988).

During the 1960s acidification of Scandinavian water bodies was linked to air pollution from Europe and the UK, and in Germany dieback of conifers was noted by the 1960s and attributed to the same causes. At the 1972 UN Conference on the Human Environment in Stockholm, concern was voiced but was met with scepticism. By the mid 1980s precipitation of pH 3.0 was not uncommon in Central Europe. Five years later Western Europe, parts of North America and several other countries were suffering serious damage to acid-sensitive plants and animals, and increased maintenance costs for infrastructure. Gradually the problem was accepted as real, as were its causes. Acidification may cause aquatic systems to suffer mercury methylation (release of harmful levels of mercury from sediment or bedrock).

Most years volcanoes vent less sulphur dioxide (SO_2) than the UK's power-stations did in 1987, but some eruptions release huge amounts, and have been known to affect winter temperatures for a few years. Human SO_2 emissions have more significance in terms of acid deposition than climatic cooling. Elsworth (1984: 6) suggested that about 70 per cent of acid deposition was due to SO_2 pollution (largely produced by combustion of coal), and about 30 per cent due to nitrogen compounds – nitrogen dioxide (NO_2) and nitric oxide (N_2O) mainly. Greenland ice cores show a two- to threefold increase in sulphate and nitrate deposition during the last c. 100 years, mainly attributable to acid deposition. By 1988 about half of the sulphur in the Earth's atmosphere could be attributed to human activity. The distribution is not uniform: over Europe the anthropogenic component would probably have been about 85 per cent and over the USA about 90 per cent (Rodhe and Herrera, 1988: 11). If SO_2 pollution is cut there are fears that global warming might increase.

Even in regions that generate little pollution, wildlife, agriculture and buildings can suffer damage – an infringement of the polluter-pays principle. Until recently, acid deposition was a problem for Europe and northeastern America. It is now spreading due to increasing combustion of coal by industries in developing countries (Park, 1987: xii; Rodhe *et al.*, 1992). Northern polar regions receive acid deposition mainly in the spring (visible as atmospheric 'Arctic haze' and as soot particles in the snow), along with aerosols, dust, pesticides, heavy metals and radioactivity (Heintzenberg, 1989). The sources are Eurasia, Europe and North America. There is concern that this haze will trap solar radiation and warm the Arctic enough to cause problems. Another difficulty is that slow-growing Arctic lichen and mosses may accumulate pollution and die, or grazing animals may suffer heavy doses of pollutants (Soroos, 1993). Other tundra vegetation appears to be vulnerable to acid deposition damage.

It is possible to map areas of vulnerable soil, vegetation and water bodies, and to superimpose forecasts of future acid deposition. Large areas of Southeast Asia, Asia, Africa and Latin America have soils that are already acidic and with high concentrations of aluminium and other heavy metals which mobilise to damage plants if the soil is further acidified. Upland cloud forests which intercept precipitation are vulnerable to acidification, as are epiphytic plants and acidic tropical rivers. By the time there are obvious signs of acid deposition there will have been serious damage to sensitive ecosystems. How acid deposition damages vegetation can be difficult to unravel and impacts vary, even from plant to plant of the same species. Plants may not be damaged directly: it may be that symbiotic bacteria or fungi and other soil micro-organisms are affected and a plant then has less support in its quest for nutrients or resistance to disease and pests. Vulnerability of plants is affected by position, altitude, soil, moisture availability, and various other factors – so it is difficult to forecast (Park, 1987: 110). Plants that grow in exposed positions can trap pollution and, as they are in a harsh environment, they are already under stress and vulnerable. Broad-leaved trees appear less susceptible, although in Europe and North America they are increasingly showing dieback. The process can be slow, taking up to forty years, so is probably under way in many areas without having become manifest. In 1988 the cost of acid deposition to Scottish foresters was estimated to be about £25 million (Milne, 1988: 56). By the mid 1980s probably over half of Germany's coniferous forests were showing signs of dieback, and about 560,000 ha were 'devastated' (Elsworth, 1984: 18). In addition to serious forest damage and loss of fisheries, if acid precipitation is not checked, soil suffers and so damages agriculture and wildlife. Wellburn (1988: 52) reported that cereals and grasses might benefit from *slight* acid deposition (but once levels rose above 60 ppmv SO_2 productivity fell). Some wild and crop species suffer, including rye, salad vegetables, barley, oats, wheat, tomatoes, apples and pears. McCormick (1988: 5) estimated the value of crop losses to acid deposition in Europe at US$500 million a year. With acidification environmental managers must deal with a threat that is often episodic, complex and insidious. There is a need for sensitive monitoring.

Radioactive waste and pollution

Natural radon emissions damage human health; production of uranium, plutonium and other (unnatural) radioactive materials affects miners, enrichment plant workers and the global environment, especially through atomic weapons testing, military and civil nuclear power plant accidents, and contamination from industrial and medical isotope sources. Between 1945 and 2005 there were about 2,000 nuclear test explosions (*c.* 130 above ground in the atmosphere, the rest mainly underground). The 1963 Limited (or Partial) Test Ban Treaty sought to end test explosions in the atmosphere, under water and in space. However, tests continued below ground and a number of countries are not signatories. The 1967 Nuclear Weapons Test Ban ('Test Ban Treaty') further reduced above-ground testing although some non-signatory nations still do it.

Some underground test sites are failing to offer containment of radioactivity, which seeps into groundwater and thence to rivers or the sea. There are weapons test-contaminated areas: in the deserts of southern USA, in the Soviet Arctic around Nova Zemlya, in what was Soviet Central Asia, the Gobi Desert (PRC), near Muroroa Atoll (French Pacific), Montebello Island (Indian Ocean), Maralinga (South Australia), in Pakistan and India. Accidents have led to the loss of several nuclear submarine reactors at sea, and there have been at least fifty-four re-entries of nuclear isotope-powered satellites, some of which scattered radioactive debris above or upon the Earth's surface. Nuclear weapons have been lost at sea, and a few have broken open on land. Gourlay (1992: 62–64) estimated that there have been at least fifty such accidents.

Radioactive wastes can be highly hazardous and very long lived (with half-lives of thousands of years). Stored high-level wastes generate heat and gas pressure and radiation damages containers. There must be adequate radiation shielding and protection against hazards such as earthquakes and some repositories need cooling equipment. Radioactive materials are also attractive to terrorists. Thus management is expensive, difficult and a very long-term demand. Nuclear waste may be stored in shallow or deep repositories, discharged into rivers or the sea, pumped down deep boreholes, dumped in containers in deep ocean, or reprocessed (Berkhout, 1991). Most of these options are imperfect, and some are now felt to be inadvisable or illegal. Low-level waste is generally disposed of by shallow landfill. Worldwide nuclear installations hold huge quantities of often high-level waste in temporary storage awaiting long-term disposal. For over fifteen years there has been a moratorium on dumping nuclear waste at sea (Japan, the UK, the USSR, France and some other nations failed to observe this fully up to 1993), so managers of high-level waste are tending to focus on underground storage or reprocessing. Underground storage demands thick, non-fissured, impermeable rocks, which are not prone to earthquakes, and where sea-level rise and tsunami are not a threat – even where the geology is suitable there is likely to be NIMBY resistance, and the costs are high.

There is still a lot to be learned about safe levels of exposure to gamma, beta and alpha radiation. Concern has been voiced about alpha-particle-emitting tritium (often released from atomic power-stations and other nuclear installations in the belief that it poses little hazard), which easily leaks from containment (Fairlie, 1992). Low-level waste containing caesium, strontium and plutonium has in the past been pumped into the sea, but now some countries have hotspots or contaminated wildlife. Obsolete or failed nuclear power-stations, weapons production plants and military reactors pose problems. Disassembly of nuclear facilities using robot equipment will be needed, and when completed much of the waste still has to be put in a repository. Like Chernobyl, which suffered a meltdown of one of its reactors and release of radioactivity in 1986, it seems likely that many contaminated sites will be buried under a mound of concrete, clay or pumped sand to save money. Whether such containment is effective for long enough remains to be seen; Chernobyl's concrete sarcophagus is already disintegrating. Decommissioning the UK's obsolete nuclear stations may cost £30 billion (Pasqualetti, 1990; *The Times*, 3 May 1993: 5) plus disposal of ten nuclear submarines. In late 1989 there were at least 356 nuclear power-generation reactors in thirty-one countries, either operating or under construction. By late 1988 worldwide 239 units had been shut down and 100 were being decommissioned. Japan, South Korea, the PRC and the USA were still building atomic power stations in 1993. Sweden has ceased to use atomic power, no easy decision, given that 52 per cent of Swedish electricity came from nuclear generation in 1992. Other countries will continue to depend a great deal upon it for decades: in 1989 the former USSR received about 14 per cent of its electricity from nuclear reactors, France 73 per cent, Japan 27 per cent, Belgium 59 per cent, the UK 23 per cent, Germany 28 per cent, Switzerland 40 per cent and Spain 40 per cent – overall, about 17 per cent of the world's electricity is generated by nuclear reactors (Gourlay, 1992: 59).

There are dangers in nuclear generation, but burning natural gas, oil and coal is a waste of valuable industrial feedstock and a source of greenhouse gas emissions – until good alternatives are developed the true costs of various energy sources need to be weighed before nuclear power is blindly opposed (North, 1995).

Electromagnetic radiation (non-ionising)

Electromagnetic force (EMF) emissions are produced by microwave ovens, radar transmitters, power cables, transformers, radio and TV broadcasting, telecommunications equipment (including mobile phones), computers and high-voltage transmission lines. Stray EMF can cause difficulties with legitimate radio and TV broadcasting, hospital equipment, research activities, control systems in cars, aircraft, weapons and so on, and measures are taken to shield against it and to legislate to control sources. Epidemiological studies in the USA and by the Swedish National Board for Industrial and Technological Development suggest high-voltage power cables might cause childhood leukaemia, cancer and brain tumours; worries about portable telephone receiver and transmitter station emissions are unproved. So far there is no convincing proof that EMF of less than 100,000 hertz is dangerous to humans (Hester, 1992). However, until proven completely safe, EMF should be treated seriously. It may prove necessary to shield equipment much more carefully and to zone land use to keep transmission lines and housing apart.

Treating pollutants and waste

Pollution and wastes are hazardous if they threaten human health or environment by virtue of their toxicity, ability to cause cancer, reproductive or genetic disorders, or because they transmit disease or pest organisms. Less hazardous material may be a nuisance or unsightly. Hazardous pollution and wastes may be grouped as chemical hazards, biohazards and radiation hazards. In addition, emergency services and health and safety planners usually recognise explosives and fire hazards. Chemical hazards include organochlorine compounds and PCBs, which, once released, pose a long-term threat even at low concentrations. At very low concentrations PCBs can mimic hormones: one effect of this is to cause cancer, another is to disrupt reproduction in fish, reptiles, birds and mammals. Nicknamed 'gender benders', these and some other pollutants have already disrupted fish, bird and reptile stocks in various countries, and there are fears that they may be reducing human sperm counts. Unfortunately, these compounds are utilised in the manufacture of plastics and other widely used materials.

Hazardous materials must be effectively labelled, carefully handled, stored and used. They must either be securely isolated from the environment (sealed containment) or treated chemically or biologically, or incinerated to render them safe. Pumping materials into rivers, the sea or the ground is widespread but unsafe. Containment is storing material, and hoping time will reduce the danger or it is someone else's problem. Treatments seek to neutralise a material chemically or biologically or to bind it to something (e.g. vitrification) or destroy it by heat. To avoid emission of dangerous fumes or dust, incinerators must achieve complete combustion at high temperature – to treat PCBs effectively requires over 1,200°C for at least sixty seconds (British Medical Association, 1991). Even with back-up filtration of flue gases and oxygen injection things can go wrong, so it is wise to site hazardous waste incinerators in remote areas or on board ships that can move to a suitable place. However, there has been criticism of shipboard hazardous waste incineration (it may be difficult to oversee, and an accident means widespread and untreatable contamination); in EU and North Sea waters a moratorium on their use is in force. America has companies which offer mobile (trailer-mounted) incinerators, which can be taken to where decontamination is needed. In the future particularly dangerous compounds may be treated in incinerators at over 9,000°C using solar power or plasma-centrifugal furnaces. Current treatments can be expensive – PCBs, incineration or bioremediation (treating with micro-organisms) cost US$2,000 to 9,000 per tonne of soil/waste treated (in 1995). Many countries export hazardous waste for such treatment, either because they do not have the facilities or as a way in which commerce can avoid tight environmental controls at home. At present there is no cheap, effective way to decontaminate fissured rocks or clays that have been deeply infiltrated by materials such as PCBs or dioxins. Topsoil can be ploughed up and formed into banks, treated with bacteria and left for bioremediation or can be transported for treatment at a decontamination facility. Fermentation and oxidation may be sufficient to treat pollutants. Bacteria and yeasts are being bred to neutralise hazardous compounds (including toxic chlorinated hydrocarbons and waste oil) in bioreactors. For organic wastes, composting or fermentation can be suitable, yielding useful compost and methane. There has been considerable interest in some of the bacteria found deep underground or around deep ocean hydrothermal vents, in the hope that they may be used to effectively convert heavy metal pollution into recoverable sulphates.

Chemical treatment of wastes ranges from simple disinfection (e.g. maceration and chlorination or ozone treatment) to complex detoxification plants that chemically convert materials such as nerve gases. Asbestos, widely used for construction, insulation, fireproofing, and in vehicle brake and clutch linings, poses health problems during manufacture through dust it liberates when disturbed. Blue and white asbestos present the greatest threat; brown asbestos is less of a hazard. Inhalation or ingestion, particularly of white or blue asbestos, causes asbestosis, a chronic, debilitating, often fatal respiratory disease that can manifest itself decades after exposure. The dust can be carried on the wind and workers using the material may contaminate people downwind, and their families and friends through dust on clothes. In developed countries controls have been tightened in recent years but in many developing countries they are still woefully inadequate.

Illegal waste disposal – fly-tipping – poses health threats, and damages the landscape and wildlife. It is one of the most widespread means of side-stepping the polluter-pays principle. Fly-tipping may be by householders, traders or manufacturers, or by a dishonest contractor a client has paid for proper disposal. The solution is surveillance and checking waste for clues to its origin, then enforcement of severe penalties. Transporting waste or a pollutant does not solve the problem of disposal, it merely shifts it. As pollution controls are tightened in developed countries there is a temptation to export hazardous substances to where regulations, labour costs and public resistance are more favourable. There are two ways of doing this: (1) a factory can be relocated in a developing country, or a subsidiary company can be established; (2) waste or pollutants can be shipped for 'disposal'. If hazardous processes are transferred to a less developed country, employees and local people may not appreciate the risks, or may be forced by circumstances to accept them in return for employment. Companies may make inadequate declarations about the materials they are using for fear of regulations or loss of trade secrets (Ives, 1985: 76). There is a need for better labelling of materials and inspection of sites and carriers, so that all involved know what is present, whether there is risk, and what safety measures are needed.

Efforts have been made to improve controls on the export of hazardous waste. The EEC introduced regulations in 1988 which, like similar legislation in the USA, aimed at improving access to information so that monitoring cargoes would be easier for governments and NGOs. The Basle Convention, which came into force in 1993 (amended 1995), is intended to regulate international trade in hazardous waste and especially to ensure that hazard is not exported to developing countries. Unfortunately, although 105 countries signed, it has gaps, and a number of nations did not ratify the Convention.

A lot of 'less hazardous' waste is sent to shallow landfill, often poorly sited and managed. Choice and supervision of these sites has become stricter in recent years, at least in developed countries. Suitable landfill sites are getting difficult to find in many countries and their selection must be integrated into land-use planning (Clark *et al.*, 1992). With disposal at sea increasingly outlawed, the alternatives are composting, recycling and incineration. Landfill is widely used to cope with sewage sludge, domestic refuse and agricultural wastes. Switzerland and Denmark incinerate a large proportion of their wastes; many other countries still rely on landfill. A problem (discussed above) is that 'less hazardous' waste may develop into something threatening after burial, as packaging, inks and so on decompose, mix, react, and leach to the groundwater or streams. There is essentially a choice: well-engineered, carefully sited and managed landfill with the risk of leaching; or incineration with the risk of air emissions.

Agricultural problems

Intensification and specialisation of agriculture may lead to pollution and waste from agrochemicals, livestock manure, livestock feedstuffs, crop residue and crop processing. Those responsible for pollution and waste may not perceive it or bear the costs and this can hinder remedial measures.

Chemical fertilisers

Between 1952 and 1972 UK agricultural output rose by about 60 per cent, thanks largely to artificial fertilisers, although changes in the crops grown make it difficult to assess how much. On a world scale, fertilisers, particularly N-fertilisers, have played a key role in increasing crop production (Pinstrup-Andersen, 1982: 148). In 1950 the world used about fourteen million tonnes of N-fertiliser; by 1985 this had risen to about 125 million tonnes (Saull, 1990). In the late 1970s on average the developing countries used 28 kg ha^{-1} and the developed countries 107 kg ha^{-1}. Most of the fertiliser used in developing countries is for large-scale grain and export crop production. The world's food and commodity production is clearly dependent on chemical fertilisers, and their use is likely to increase. There are uncertainties about the long-term impact of chemical fertilisers. There is some indication that where use of fertiliser has replaced crop rotation and livestock manure, problems arise, in particular a net loss of organic matter from the soil, and in some areas acidification and zinc or sulphur deficiency. Artificial fertilisers offer the following advantages over organic fertilisers:

● They can be easier to store, handle, apply and transport than most natural fertilisers currently in use.

- There is less smell, and lower risk of pathogenic contamination (although well-composted organic material is virtually pasteurised).
- Land spread with manure cannot be properly grazed for some time because livestock object and there is disease risk.

Arable and livestock farming are often no longer integrated – the former must rely on chemicals and the latter has an animal waste disposal problem. The costs of disposing of agriculture waste may one day bring agriculture full circle, to recycling livestock manure and crop residue, possibly together with domestic refuse and human sewage. But to do so will require composting facilities. An alternative is to incinerate these wastes and recover electricity and district heating (as in Denmark). If they are not applied with caution, artificial fertilisers can cause eutrophication of water bodies and increased nitrates in groundwater. Phosphates have been accumulating in soils, river and lake sediments for decades, as a consequence of the use of fertilisers, spreading of livestock manure, disposal of sewage and leaching of poorly sealed landfill sites. This poses a threat, particularly in Europe and North America. Studies in Europe suggest that, even if the application of phosphates is controlled, steady leaching and possibly more rapid mobilisation if there is soil acidification or global warming will lead to a six- to tenfold increase in river and groundwater contamination. Such levels would raise problems for domestic water supply, and for the ecology of rivers, lakes and shallow seas (Behrendt and Boekhold, 1993).

In parts of the USA irrigation using N-fertilisers seems to be a major cause of groundwater nitrates. In the UK, borehole studies suggest that conversion of pasture to arable with N-fertiliser use causes high groundwater nitrate levels (Conway and Pretty, 1991: 186). Fertilisers are either mainly nitrogenous or contain proportions of phosphate and/or potash. Phosphates are mined in only a few countries and supplies are dwindling (they are also vital to many industrial processes); hopes of mining phosphate nodules from the ocean floor have not been realised. It would make sense to cut back on the waste of phosphate fertilisers (which is common), to conserve supplies and to reduce groundwater and stream pollution.

There are ways of controlling nitrate fertiliser use: reduction of price supports for crops; regulation of crops grown; quotas or permits which seek to limit expansion; set-aside – the withdrawal of land from production; taxation of nitrate fertilisers (Clunies-Ross, 1993). Even if such controls were adopted, improvement would come slowly because nitrates may take up to fifty years to reach groundwater, depending on the geology (Hornsby, 1989). Conversion of farmland to some other use could, because agricultural liming ceases, lead to increasing soil pH and greater releases of nitrates, phosphates and heavy metals. Costly slow-release liming treatment may be needed. In temperate environments, planting winter wheat with white clover might help to reduce nitrate leaching, and would cut costs of fertiliser inputs and discourage pests. Authorities will be forced to treat domestic water to remove nitrates, blend contaminated and pure supplies or store water in surface reservoirs for long enough to reduce nitrate content.

Pesticides

Pesticides are compounds used to kill, deter or disable pests for one or more of the following purposes:

- to maximise crop or livestock yields;
- to reduce post-harvest losses to rodents, fungus and so on;

- to improve the appearance of crops or livestock;
- for disease control (human health and veterinary use);
- for preservation and maintenance of buildings, clothing and boats;
- to control weeds which hinder transport and access;
- for aesthetic or leisure reasons, lawn care, garden flowers, golf-courses.

Some natural pesticides are available ('organics'), but are not necessarily harmless alternatives to synthetics: some are very toxic or carcinogenic. DDT was one of the first synthetic compounds (mainly organochlorines or chlorinated hydrocarbons), initially synthesised in 1874, rediscovered in 1939 and adopted for louse and mosquito control during the Second World War, and from the 1950s for agricultural use. The second main group of synthetic pesticides, the organophosphates, was discovered in the 1930s. After the 1940s other synthetic organic compounds were developed and widely used for agriculture and public health measures. There has been a trend to replace many of these pesticides with 'safer' organophosphate and pyrethroid insecticides and fungicides. Organophosphates can be more toxic than organochlorines but are less persistent (Conway and Pretty, 1991).

Successful pest control commonly reduces field and storage losses by 20 per cent or more. But it is difficult to quantify the benefits and the risks of pesticide use – in developing countries a large proportion of what is used is applied to luxury export crops, not staples; there may also be off-site pollution that is difficult to trace back. Pests may flourish if predators are poisoned and they survive. Pesticides are also used because consumers demand blemish-free produce, and growers seek to ensure ripening to assist gathering and processing. There have been suggestions that, in spite of pesticides, crop losses have increased in the past few decades – but would things have been worse without pesticides?

Recognition that 'safe' pesticides caused environmental problems came by the early 1960s, the public having being alerted by Carson (1962). DDT was found to concentrate in the fat of higher organisms through 'biological magnification'. In 1972 the use of DDT was banned in the USA (but not its manufacture and export). Weir and Shapiro (1981: 4) publicised how the export of pesticides banned in the USA still had an impact there through contaminated food imports. The problems associated with pesticide use may be summarised as follows:

- poor selectivity of many pesticide compounds (not narrow-spectrum, i.e. specific in terms of what is killed or injured);
- overuse;
- toxicity and slow breakdown;
- tendency to be concentrated by the food web;
- misuse or unsafe methods of application;
- the effects of long-term use of pesticides on soil fertility is little known;
- the impact of cumulative effects on the global environment is not known.

Ideally, a pesticide should kill, disable or deter a specific pest and affect nothing else. Unfortunately, most compounds are far from specific: non-pest organisms may be directly or indirectly affected. There are other possible impacts: on-farm (contamination of workers, livestock, crops, soil, wildlife and groundwater); off-farm (contamination of nearby woods, hedges, housing, streams); and global contamination. The impacts may be short term or long term, are often indirect, and may have cumulative (synergistic) effects. Tracing impacts (and proving liability) from pesticide use back to the point of application can be difficult. Much pesticide is used pre-emptively and may not be necessary. Use increased rapidly from about 1950, partly reflecting the Green Revolution and the spread of modern crop varieties. About 50 per cent of all pesticides are applied to wheat, maize, cotton, rice and soya; much of the rest goes on to plantation crops such as cocoa and oil palm. About half of known pesticide poisonings and at least 80 per cent of fatalities have occurred in developing countries, yet these use only 15 to 20 per cent of the world's pesticides (Pimbert, 1991: 3).

Pesticides are costly to develop; even those acting with the best of motives and care may not be able to test them fully, and there could be reluctance to withdraw a compound if there is some fault. Manufacturers may resist developing specific pesticides because it restricts sales. Side-effects may only become apparent after extensive use, and pesticide developers may neglect pests if there are limited profits to be made from their control. Pesticide problems can be reduced by the following:

- banning dangerous compounds;
- developing alternatives such as biological control or integrated pest management;
- restricting trade of pesticide-contaminated produce;
- controlling pesticide use by monitoring, inspection and licensing to ensure sensible procedures;
- developing less dangerous pesticides;
- controlling prices of pesticides to discourage excessive use;
- education to discourage unsound strategies;
- rotation of crops to upset pest breeding and access to food;
- hand- or non-chemical weeding;
- encouraging agencies to cut funds for pesticides;
- treating drinking-water to remove pesticides.

Most countries have established departments responsible for reviewing pesticide use which have powers to initiate controls, but there are still problems in disseminating information about pesticides and their effects, in monitoring, and with political and economic aspects of control (Ghatak and Turner, 1978; Boardman, 1986). In 1986 the FAO issued an International Code of Conduct on the Distribution and Use of Pesticides and in 1990 got 100 countries to sign a code of conduct on pesticides. The UNEP, WHO, OECD, ILO, EC, the Pesticides Action Network (PAN) and other international bodies and NGOs make efforts to improve pesticide use and controls, but there is a long way to go before controls are satisfactory. Various databases and networks are now established to assist with monitoring and control. The FAO and WHO have set up the Codex Alimentarius Commission ('Codex System') to establish food standards. One of its tasks is to check on pesticide residues in produce (and each year to publish information to assist in this). Under GATT agreements the Codex has increased influence over the way countries set their food and agriculture standards (Avery et al., 1993).

In addition to compounds intended to destroy or control insects – pesticides – other materials designed to control certain animal and plant pests include the following:

- molluscicides – snail and slug control;
- anti-fouling compounds – for treating ship hulls and pipes to prevent weed and mollusc growths;
- fungicides – to control mould and fungi;
- herbicides (weedkillers) – to control unwanted plants;
- growth 'hormones' (to stimulate growth, flowering, ripening);
- anti-helminths – to destroy worms in livestock and humans;
- rodenticides – to control rats and mice and similar pests.

Some of these have non-agricultural uses for medical and veterinary purposes to control disease vectors, fungal infections and so on. Some are used in food storage, construction, households, anti-narcotic or warfare activities (defoliants). Anti-fouling compounds have caused serious damage to wildlife in various marine and freshwater aquatic environments.

Integrated pest management (IPM) should reduce the use of pesticides and make pest control more focused. IPM involves study of the pest(s) and the context to diagnose the best mix of crop and pest control techniques to use. IPM must be co-ordinated with conservation, land and water management, social and economic development, and so on, and uses pesticides only as a last resort in a judicious manner. As with chemical pesticides, there is a need for caution over biological controls. History has taught that control organisms may become a problem. Genetic engineering may also be a double-edged sword. It offers alternatives to pesticides but also threatens serious problems if dangerous traits were to be passed to another species, or a modified organism runs out of control.

Agricultural waste

This includes animal excreta; silage effluent; cereal straw and other crop residue. Agricultural waste problems can be countered by the following:

- waste treatment and incineration;
- quotas – limits on quantities a farm may produce;
- composting (discussed earlier) or use as a raw material (e.g. strawboard/cardboard);
- set-aside – withdrawal of land from production A risk is that remaining agricultural land will be more intensively used.

Manure, once a resource that sustained cropping and pasture, is now often a problem. A UK farm of 40 ha with just fifty cows and fifty pigs can present waste disposal problems equivalent to a town of nearly 1,000 people. UK livestock produce 2.5 times the total human sewage (Conway and Pretty, 1991: 276). In 1993 The Netherlands had a *c*. forty million tonne manure mountain – more than twice that which could be safely disposed of on all the farmland. Consequently, The Netherlands (and Denmark) has established waste-processing plants (some contributing to district heating or greenhouse heating) and some manure is composted. Stored in pits or lagoons, livestock waste generates methane, ammonia and hydrogen sulphide, which cause nuisance smells, damage vegetation downwind (because of the ammonia) and act as greenhouse gases. If such slurry escapes it can cause serious stream, lake or groundwater pollution (through chemical oxygen demand – COD, biological oxygen demand – BOD, harmful bacteria and parasites, excreted antibiotics or growth-promoting hormones, steroids and sometimes heavy metals – especially copper and zinc added as growth accelerators to pig feed). Disposal by spraying on to farmland is impractical as it may transmit diseases, the heavy metals from feedstuffs can concentrate in the soil, and nitrates and phosphates leach to contaminate surface and groundwaters – de-watering and composting or incineration is needed.

Silage has become popular in Europe over about the past twenty-five years as livestock feed. When it is made, moisture is released; it is common for 330 litres of effluent to be formed for each tonne of silage made. This means that farmers must store and dispose of large quantities of acidic (often pH 3.4) effluent. Escapes damage soil, aquatic life and groundwater. Stored in lagoons or pits, the effluent gives off ammonia and hydrogen sulphide, both active as greenhouse gases. The solution may have to be de-watering and incinerating or composting.

Cereal straw burning has been a problem in Europe for some years until recently when legislation began to curb it. In Brazil, Mauritius, parts of Australia and the Caribbean, sugar fields are burnt before harvesting. Crop residue burning helps control weeds and pests but also destroys harmless or useful wildlife, damages soil, and gener-ates soot and greenhouse gases. Modern cereal straw is not as strong as in the past, and thus less useful for thatching, but there is still potential for manufacturing strawboard, paper, cardboard, or for on-farm or district heating. Clearing land for agriculture is often done with fire, which generates soot and greenhouse gases, damages soil and kills wildlife. The problem has recently got seriously out of control in Indonesia, Amazonia, Venezuela and Mexico. In Europe, Australia and the USA fires are more likely to be accidental or set by arsonists, rather than be intended for agricultural clearance. Simply banning burning is no solution, because regular fire may be vital to prevent occasional serious fires.

Processing rubber, sugar, meat, fish, and many other products generates effluent. In West Malaysia and Sabah palm-oil processing takes place in local factories to ensure that treatment is not delayed to get a high-quality product. Consequently few major streams have escaped pollution in spite of legislation since the late 1970s. In Brazil the processing of sugar, cassava and yams to produce alcohol for automobiles results in about thirteen litres of high-BOD effluent for each litre of fuel; rivers, especially in the northeast, have suffered. Crop processing often demands fuelwood, and large areas may be deforested for tobacco curing, tea drying, and preparing many other crops. Leather tanning with oak bark, wattle bark or other natural compounds produces acidic, high-BOD effluent, smell and nuisance from flies. Toxic chemicals are increasingly adopted for tanning: some contain chromium or mercury, and may damage aquatic ecology and contaminate groundwater.

Recycling and reuse of waste

Waste recovery and waste recycling are terms that can lead to misunderstandings: a country might recover 80 per cent of its waste paper but recycle little, instead using it for district heating; another may recover 10 per cent but recycle/reuse most of it. There is usually a need to sort, transport and treat recovered waste. Sorting can be done by the state, companies, householders, individual 'scavengers', and the waste producers. In the USA 'reverse vending' has been tried – a waste skip credits a company for return of cans, bottles or whatever. Plastics and non-ferrous metals are at present difficult to recover: estimates suggested about 5 per cent of Japanese, 15 per cent of European and 10 per cent of USA plastics were recycled in the late 1990s. Even if plastics or metals of the same general type can be recognised, pieces vary in subtle ways and may be attached to other materials, or have an unwanted coating or contamination that is diffi-cult to remove. Some plastics absorb chemicals, reducing their value for recycling. Crude sorting is sufficient if the aim is simply to recover a limited range of materials such as aluminium, glass, low-grade plastics, iron and combustible material for fuel.

Developed countries can learn about refuse recycling and reuse from the informal sector of developing countries (Bouverie, 1991). Alternatively, they might export waste (mixtures of plastics packaging) to where cheap labour can sort it. Recovered material is often bulky for a given weight, making transport and storage costly, and may be of low value (Engstrom, 1992; Fairlie, 1992).

Recycling may not be as environmentally desirable as it first seems. Virtanen and Nilsson (1993) suggested that waste paper processing might generate more pollution than burning it for electricity generation and district heating (Kurth, 1992; Pearce, 1998). Controlling packaging materials and ensuring they are labelled should reduce the cost of recycling (Johnson, 1990; Gourlay, 1992: 185). Glass can be recycled indefinitely and each time saves on energy compared with production of new material (but, at the time of writing, Europe recycled only about 49 per cent of its glass, and some individual countries far less). Reuse of soft drink or milk bottles requires a decentralised network of manufacture. Centralised supermarket retailing in the UK, USA and Europe is unlikely to encourage a return to reusable bottles which are heavier, and so cost more to transport than plastic. Reusable bottles also get damaged and many are not returned, which raises the cost of recycling (North, 1995). Sometimes, if a firm arranges to recycle its products, it may be able to restrict sales of salvaged second-hand parts, and so profit.

Steel and aluminium recovery can be worth while: the latter saves *c*. 95 per cent of the electricity used in making fresh aluminium. Paper can be recycled up to four times before the fibres are damaged too much (*The Times*, 14 June 1993: 33). The increasing use of disposable nappies (diapers) may pose health hazards for those working in refuse disposal, and might be countered by establishing laundry delivery and collection services – but for this to work consumers need to be assured of very high standards of hygiene.

A field which seems likely to develop in coming years is that of urban agriculture or horticulture: the growing of crops to employ and feed the poor, and if possible to use refuse and wastewater. It is better for authorities to assist and monitor such activities than to pretend they do not happen. Better to pay to chlorinate sewage outflows than have it applied untreated to vegetables sold in city markets. A flourishing urban and peri-urban agriculture sector should improve food supplies, help dispose of wastes, and generate livelihoods for people otherwise unlikely to find useful employment (Honghai Deng, 1992). Since the 1980s an alternative technology development group has been researching the *chinampas* farming system used by the Aztecs. This consists of narrow raised fields constantly mulched with mud from narrow canals surrounding them. The canals received sewage and effectively treated it, producing crops, fish and poultry in the process – sustainable and very productive agriculture. Fortunately, an area still survives near Mexico City and has enabled researchers to assess how the strategy works; one discovery has been the identification of a particularly effective bacterium which promises to be valuable in sewage treatment systems appropriate for developing countries (Ayres, 2004).

Summary

- Sustainable development, whether in rural or urban environments, demands effective handling of outputs – waste and pollution – as well as adequate inputs – water, food, energy.
- Pollution management demands the establishment of ethics as well as regulations, monitoring and enforcement. There is an ongoing shift to a proactive approach and adoption of the polluter-pays principle.
- Many pollution and waste management issues are transboundary, and some are global. Solutions and controls demand co-operation and funding from both developed and developing countries.
- There is a move towards the polluter-pays principle. In the past polluters seldom shouldered the costs; increasingly they are being made to act responsibly or to do so.

Further reading

The field of pollution and waste management is very broad, which makes it difficult to recommend a specific introductory text. Coverage of more specific topics such as pesticides, radioactive waste, refuse, sewage is easily found on the Internet or in library catalogues.

Clayton, A., Spinardi, G. and Williams, R. (1999) *Policies for Cleaner Technology: a new agenda for government and industry.* Earthscan, London.
Suggests how to clean up industrial pollution and move towards sustainable development. Problems presented sector by sector – UK focus.

Farmer, A. (1998) *Managing Environmental Pollution.* Routledge, London.
A comprehensive introduction to the nature of pollution and its impacts, and solutions.

Jacobson, M.Z. (2002) *Atmospheric Pollution: history, science and regulation.* Cambridge University Press, Cambridge.
Good introductory text which includes global warming coverage.

Newson, M. (ed.) (1992) *Managing the Human Impact on the Natural Environment: patterns and processes.* Belhaven, London.
Like Park (1997), this covers pollution and waste as part of a more general text.

Park, C. (1997) *The Environment: principles and applications.* Routledge, London.
A general text, but with a simple and broad introduction to pollution and waste issues.

13 Environmental management in sensitive, vulnerable and difficult situations

- Are there areas of the world particularly prone to environmental problems?
- How sensitive and vulnerable are ecosystems?
- Environmental problems and developing countries
- Environmental problems and transitional countries
- Environments which challenge environmental management
- Lessons the environmental manager can learn from study of sensitive environments
- Summary
- Further reading

There are many reasons why an environment may be sensitive or difficult to manage: harsh climate; remoteness; the impact of natural disasters; as a consequence of easily damaged vegetation or soil; because of insularity; or as a result of excessive human demands. Excessive demands may be made in rich and poor countries, causing environmental degradation sometimes at low human population densities (as in parts of Australia, Amazonia or Siberia). Certain parts of the world are more likely to suffer natural disasters: seismically and volcanically active areas, those subject to hurricane or tornado, tsunami, avalanche, landslide, or sudden frost. These risks can be mapped by hazard and risk assessment, for insurance companies, civil defence and so on. Groups of people also vary in their vulnerability at any given moment and this may alter over time. Even within one social group different sexes and ages often differ considerably in vulnerability and resilience as a consequence of factors such as physiology and lifestyle. It is possible to map events and people that are vulnerable to disruption by, say, storm, pollution or erosion. Hazardous sites, such as chemical plants and potential military targets, can also be mapped. Various assessments and maps can be overlaid or combined in a GIS-type system to assess combined risks. Unfortunately, there are often no efforts to make such assessments.

The Millennium Ecosystem Assessment, launched in 2001 by the UN Secretary General Kofi Annan, seeks to gather information on ecosystem change, its consequences, and options for responding. The Assessment is also intended to provide a benchmark for future studies, and it is hoped that it will develop environmental management tools and help build institutional capacity. By 2003 the Assessment was yielding reports, and by mid 2005 it was clear that over 60 per cent of the Earth's ecosystems were degraded and matters were getting worse. Degradation was likely to hinder achievement of the UN Millennium Development Goals agreed in 2000 (http://www.millenniumassessment. org.//en.About.Overview.aspx – accessed July 2005).

Are there areas of the world particularly prone to environmental problems?

Given long enough, even the safest area could be subjected to natural or anthropogenic problems. Environmental managers must be alert to situations where risk and likely severity of problems may be altered by human activity – for example, land may be sensitised to drought or soil erosion by misuse. The effects of storms can be magnified if people remove vegetation cover from watersheds or coastal land. Development may drive people, or they may be attracted into areas where they trigger natural processes to cause environmental problems. People often evolve survival strategies to suit their environment – nomadic grazing, shifting cultivation or whatever. These seek to minimise risks to their well-being, often seek to be economical in terms of labour input, and enhance recovery if there is a disaster (coping strategies). Unfortunately, recently many of these strategies have broken down for various reasons – social, economic, environmental, political. The breakdown may also lead to environmental degradation. Environmental management should monitor to try and predict breakdowns and help develop cures or alternatives. However, there is a limit to resources, so such wide-ranging study is seldom possible. The solution may be for NGOs or international agencies to do such monitoring.

Human vulnerability

The breakdown of established livelihood strategies and (disaster) coping strategies has increased the vulnerability of many groups. Particularly in developed countries, people are becoming more dependent on technology and inputs that are not locally available, which increases their vulnerability to: (1) hazards posed by malfunction; (2) hazards posed by system breakdown, through mismanagement, accident, obsolescence, civil unrest or war; (3) dependency and the risk of interrupted supplies, and external political and economic pressures. The way to reduce this vulnerability is to have better regulations and controls to ensure safe equipment and fail-safe systems, and to seek local or national inputs. Insurance companies and public awareness may provide some safeguards, although the poor cannot afford the former. There are a number of aspects of current globalisation which could cause increased vulnerability and more likelihood of environmental stress (Shiva, 2004). Generally the rich can buy their way out of risky situations and have resources to weather difficulties. In 2005 it was mainly the poor who lingered in flooded New Orleans. Poor people are often fully aware of threats yet have no alternative but to face them.

Rich countries which have complex, interlocking technology and economics can be vulnerable. Complex systems are easily disrupted by nature and socio-economic problems, and need skilled specialists to repair and maintain them. They are difficult to run in times of stress and may be a challenge to salvage. Natural disasters or terrorism could have marked effects on complex computer systems needed by government and banking; a poor country may have few difficulties moving or salvaging robust card-indexes and decentralised services. There are currently fears that some computer virus could wreak havoc on electronic systems. If so, it may prompt governments not to rely so much on vulnerable and centralised electronic systems. Certain parts of the world are more likely to suffer natural disasters; there are also areas more likely to be affected by warfare and social difficulties. Global environmental change will alter the current threat situation; for example, sea-level rise will make more places vulnerable to storm surges, and warming climate may make storms more intense and alter their distribution. Activities such as mangrove clearance and overgrazing of slopes also alter the vulnerability of land.

Vulnerability assessment is growing; a scan of the Internet reveals numerous consultants and agencies active in the field. Some carry out stock-taking of existing situations and some specialise in predictive assessments, which like EIA can be inaccurate and lead to a false sense of security. Human response to threat is a field still far from adequately researched. Even the same group of people can react differently to successive similar impacts – vulnerability and resilience are quite variable. Regional vulnerability assessments can be very useful for civil defence and local disaster preparedness; most focus on specific threats rather than a full spectrum. For example, much of California has some degree of earthquake awareness, but much less knowledge of volcanic, tsunami, bushfire or storm threat. In the USA the Environmental Protection Agency has established a Regional Vulnerability Assessment Program (ReVa) which co-ordinates and supports various disciplines to assess threats (http://www.epa.gov/reva/ – accessed June 2005).

How sensitive and vulnerable are ecosystems?

The concept of ecosystem stability was discussed in Chapter 3 (Hill, 1987; Stone et al., 1996). Mitchell (1997: 51) felt that basic concepts of ecosystem diversity and stability were too simplistic adequately to describe reality; so, as ecosystems were inherently complex, environmental managers would need to accept that often they could not manage ecosystems, although they might manage human interactions with them.

On the whole, biogeochemical and biogeophysical processes tend to resist change and are self-regulating within limits, so one may expect a sort of dynamic equilibrium. However, some global cycles, environments and organisms (and groups of people) are more sensitive to change than others. Stability is in large part a function of resistance to change and resilience following it. Resistance (or sensitivity) may be defined as the degree to which a given ecosystem undergoes change as a consequence of natural or human actions or a combination of both. Resilience may be used to refer to the way in which an ecosystem can withstand change. It is widely held that ecosystem stability is to a significant degree related to biological diversity: the greater the variety of organisms there are in an ecosystem, the less likely is there to be instability (Pimm, 1984). However, it is quite possible that a change in some parameter could have an effect on all organisms regardless of diversity: thus diversity may help ensure stability but does not guarantee it. Resilience is often used as a measure of the speed of recovery of a disturbed ecosystem but can refer to how many times a recovery will occur if disturbance is repeated (Holling, 1973).

Worldwide, a growing problem is environmental degradation – a rough definition of this is: the loss of utility or potential utility, or the reduction, loss or change of features or organisms which may be difficult, costly or impossible to replace. Recognising degradation can be difficult: it may be slow and gradual, or it may take place long after disturbance. People may fail to notice change, which may sometimes be too imperceptible for a single generation to spot. Nowadays it is rare for an environment to be 'natural'. The chances are that there has been degradation by humans (e.g. development in southern France may degrade *maquis* scrubland, but it is already much degraded compared with the prehistoric forest cover it replaced long ago). The current condition of an ecosystem may not indicate what has been lost, or show whether there has been improvement. An ecosystem may not be stabilised when disturbed: it could already be much degraded or improved compared with its natural state, or it could be undergoing cyclic, more or less constant or erratic change (Kershaw, 1973: 65–84). Return to a pre-disturbance state when disturbance ceases is by no means certain; nor is the pattern of recovery always predictable (Burton et al., 1977; Blaikie and Brookfield, 1987; Goldsmith, 1990). For example, grazing can lead to increased scrub cover; a reduction in grazing might be expected to lead to a reduction of the scrub – but that sometimes causes a thickening of woody vegetation. Many poor people are wholly dependent on the land and have nothing to fall back upon if it degrades or if there is a disaster. Land degradation must be monitored and combated when it occurs. (For a practical handbook of indicators for assessing land degradation see Stocking and Murnaghan, 2001.)

Some threats are continent-wide or even global in impact such as a large asteroid strike or a huge volcanic eruption, no locality is wholly free of vulnerability, and local threat mapping usually does not include such infrequent but large-scale problems.

Environmental problems and developing countries

Marginalisation

People who are marginalised – forced or attracted on to poor-quality, perhaps easily degraded land, or in some other way prompted to live 'close to the edge' – usually become progressively more disadvantaged and vulnerable (a vicious spiral). The reasons for marginalisation are diverse and include: loss of common resources; efforts to escape unrest; the hope of employment or access to farmland; eviction from conservation areas or from the estates of large land users; altered trade opportunities; economic impacts of structural adjustment and national debt; changes in labour costs and availability; widowhood; reservoir flooding, and much more. Environmental or socio-economic change or technological innovation can cause people to become marginalised (or demarginalised) *in situ*; for example, drought; disease or pests; pollution; decline in demand for produce due to change in fashion, economic slump or substitution; warfare; rising labour costs; changes in communications; altered land-user attitudes; introduction of new crops; labour migration and so on.

There is considerable support for the view that growing environmental problems in developing countries are often caused by the disempowerment of local people, i.e. locals can no longer participate in resource management and are losing access to resources (Ghai and Vivian, 1992: 72; Harrison, 1992: 126; The Ecologist, 1993; Bromley, 1994). A widespread cause of loss of access to common resources is the penetration of capitalism (Tornell and Velasco, 1992). In India, Thailand or Brazil it may be companies seeking land to grow eucalyptus, or large landowners growing soya for export, that acquire common land; elsewhere it may be ranchers looking for more grazing land for export-orientated beef production or land and tax speculation.

What tends to happen through marginalisation is that the marginalised over-stress the resources they still have access to, and with nowhere to move to, or no means of moving, they become unwilling agents of damage and their own ultimate demise. Marginal land is likely to demand inputs and is less forgiving to users but is least likely to receive such investment. Sustained resource exploitation strategies may be disrupted; those disturbed may adapt their activities, cause environmental problems and suffer hardship. People forced to move are also likely to be disorientated, and may have left livestock and tools behind. Such people often suffer hardship and violence. Together with many of those who have willingly relocated, they probably lack the necessary local experience and resources to establish sustainable livelihoods. Many of those who practise degenerate shifting cultivation are 'shifted cultivators', people who have relocated.

Marginalisation can be caused by quite minor changes, for example, in attitudes, trade, weather and so on, particularly if the existing livelihood or land use was poised on a knife-edge. The terms of trade can be a root cause of poverty, which may drive people to damage the environment, and which also starves governments of funds to counter problems. Studies of poverty and environmental degradation suggest that three factors often combine to cause marginalisation: (1) rapid population growth; (2) land consolidation and agricultural modernisation in fertile agricultural areas; (3) prevailing inequalities in land tenure (Leonard et al., 1989: 5).

In a number of countries, particularly those with debt problems by the 1980s, economic austerity measures (especially structural adjustment programmes) have reduced spending on environmental management and may have increased the marginalisation of rural peoples. In addition, the spread of HIV/AIDS, the diversion of funding to priority social and economic development challenges, and natural disasters have hit environmental management and caused marginalisation in some countries. One example is the recent increase in locust activity in many poor African countries, with the risk of its spreading to much wider areas. Between the late 1940s and 1990s pesticides were used by locust control bodies to manage the problem; the situation deteriorated due to rising costs of aerial spraying, fears of pollution, wetter conditions in some countries, and civil unrest which has prevented access.

Population growth and environmental problems

Many developing countries struggling to maintain living standards in the face of growing poverty have little to spend on countering environmental problems. The growing populations of developing countries currently consume far less per capita of the world's resources and cause less pollution than do the populations of rich nations. Nevertheless, demographic increase puts some regions under stress and ultimately could exceed global carrying capacity. Population growth does not automatically mean environmental degradation: it can sometimes stimulate agricultural production and improvement of technology (e.g. population growth in Europe probably drove farmers to farm fertile but difficult clay-lands and shift from long fallow to annual cropping). It is simplistic Malthusian or neo-Malthusian determinism to say that population growth inevitably leads to problems; environmental impact is a function of population and standard of living, the technology practised and attitudes. Devastation may occur at low population levels and it is probably fair to say that, up to a point, population increase becomes a socio-economic problem only if food production technology fails to keep up. Caution is necessary when examining population–environment relationships – for example, there has been little research to check a common assumption that the presence of poor people correlates with environmental degradation (Kates and Haarman, 1992) (for an introduction to Malthusian and Boserüpian views see Harrison, 1992: 11–19).

An African crisis?

Africa is frequently singled out as having or being close to an environmental crisis, a development crisis or both (Commins et al., 1986; Ravenhill, 1986; Watts, 1989; Davidson et al., 1992), although there are some who feel this is exaggerated (Blaikie and Unwin, 1988). Things look grim for a large portion of sub-Saharan Africa. Harrison (1987: 17–26) concluded that there was a crisis, particularly an environmental crisis, and that, unlike the rest of the world, most of the African continent was not developing but was regressing, because:

1   there was a food supply crisis, manifest as a decline in per capita food production;
2   poverty was increasing;
3   there was a debt crisis which grew worse as Africa's exports fell in value and imports rose in cost;

4   there was an environmental crisis, growing worse as vegetation and soils become degraded.

To these difficulties may be added unrest (sometimes an important factor): Africa, with less than 10 per cent of the world's population, had almost 50 per cent of the world's refugees in the late 1980s (Harrison, 1987: 52). Another problem, HIV/AIDS, has taken a particular hold in parts of sub-Saharan Africa, and contributes increasingly to the continent's problems by depleting agricultural labour.

The trends for food production are worrying: in the 1950s most of Africa was self-sufficient, but now many states import, and it is the only continent to show a per capita decline in food production. By 2006 at least sixty-five of the world's 117 nations were suffering serious undernutrition, more than thirty of those in sub-Saharan Africa. Food security for Africa would probably demand a 4 per cent per annum increase in the continent's food production, plus a similar increase in export crops to provide foreign exchange for inputs. Over the past thirty years the average growth rate for food production has been *c.* 2 per cent and export crop production has shown a decline (added to which the prices for these crops have fallen, making it difficult to buy inputs and to encourage producers). There is widespread acceptance that Africa has a food problem but little agreement on why (Rau, 1991), although suspicions have been voiced that it reflects widespread traditional communal or state landownership, and that a move towards individual landownership or improved communalism might help.

At the 1972 Stockholm Conference on the Human Environment, problems of the environment were widely regarded in poor countries as matters for developed nations, since they were too poor to afford the luxury of worrying, and anyway 'the rich were to blame'. Until about 1985 it was common for developing countries to suspect calls to protect the environment of being a ploy to hold back their development and continue their dependency or to withhold aid. Virtually all developing countries would now accept that there are environmental problems which require attention (Schramm and Warford, 1990). The nature and causes of the problems and the cure for them are often less than clear.

Developing countries have tremendous diversity of environment, style of government, administration, historical background, degree of poverty and so on. Many have handicaps associated with being tropical: no season cold enough to kill pest organisms; soils that are often infertile and difficult to manage without causing environmental degradation; intense storms and so on (Huntington, 1915; Ooi, 1983: 2; Kates and Haarman, 1992). Adams (1990: 6–8) suggested that developing countries faced a double threat: a crisis of development and a crisis of environment. The first of these involves debt, falling commodity prices and poverty. The second is the result of global environmental change, the impacts of resource exploitation.

There has been growing interest in applying environmental management to developing countries, partly stimulated by large funding agencies which since the late 1970s have established environment departments and adopted policies of environmental assessment and EIA. By the 1990s most aid agencies and funding bodies had policies seeking to improve environmental management, and many published guidelines which have influenced a wider spectrum of companies and consultants.

From the late 1970s developing countries have established environmental agencies, so that by 1992 virtually all had such bodies (although their powers vary enormously). In parallel with these developments international and in-country NGOs were spreading. Nowadays NGOs maintain a network of contacts and command considerable funds and power. There are still risks that expatriate and even indigenous experts will misinterpret threats, opportunities and people's capabilities. Environmental management must work effectively with government agencies, international agencies and NGOs. If an agency has not done so, environmental managers will need to identify target groups and other 'players', establish what is needed and how to work with the various groups (or even control them). In developing countries a major task is often one of co-ordination, hindered by lack of funds and trained manpower, diverse goals, poor infrastructure, sometimes difficult environmental conditions, inefficiency and corruption. (For an introduction to developing country environmental management see Little and Horowitz, 1987; Koninklijk Instituut voor de Tropen, 1990; Montgomery, 1990b; Eröcal, 1991). Strategies and techniques formulated in developing countries may have potential for developed countries.

Misinterpretation of developing country situations is often compounded by a poor database. Influential stakeholders commonly (if not usually) have 'polarised perceptions' – they have a biased viewpoint. Commonly, data gathering has been conducted close to roads and mainly in the dry season; reasonably affluent city-dwelling researchers may not really understand poverty and local conditions. Scientists and social studies specialists too often accept received wisdom without sufficient checks (Fairhead and Leach, 1996; Leach and Mearns, 1996). Fashion may dictate data collection; currently an emphasis on participation and community development may clash with dispassionate and problem-focused study. Some would argue that if a goal is conservation, seeking to combine this with employment creation or community development results in a partial resolution of each. For example, a state may establish extractive reserves, make much noise about participation, and then essentially impose plans 'top-down'. The results are likely to be limited because the participation is not real and there is no local enthusiasm. Environmental managers must be as aware as possible of social issues and work closely with able social studies experts.

Environmental problems and transitional countries

Since the mid 1980s large portions of the world have shifted from centrally planned socialist government to more 'free enterprise' and 'Western' systems. These 'transitional economies' include the former USSR, China, and a number of the satellite states of each. Although the transitional states had been radically different from Western models of development there had still been massive environmental degradation and disasters (Pryde, 1991; Peterson, 1995; Bridges and Bridges, 1996; OECD, 2005). Today these nations are struggling to adapt; while there are new freedoms to monitor and lobby against environmental problems, there are also hindrances unknown before the 1980s – inflation; limited funding; powerful special-interest groups; rapid industrialisation and allocation of funding to economic development which may once have been free for other things. There is often a legacy of environmental neglect, rapid development and scarce funds. Freed from some of the pressures caused by the Cold War and with access to Western techniques and approaches (e.g. EIA), there has been some development of environmental management; however, techniques and strategies developed in affluent, Western free-market countries may be unsuitable or need considerable adaptation. The USSR and China play key roles in global warming and energy supply, and both have land degradation problems, which not only affect their territory but cause windblown dust problems globally. It is in the interests of all nations to aid the transitional countries to improve environmental management. Like developing countries, transitional nations have skills, approaches and expertise that can aid other countries with environmental management.

Here and there a few countries have yet to become transitional – Cuba's President Fidel Castro has called for a halt to the transfer of consumer society lifestyles and aspirations to developing countries, where he argues it fuels a growing environmental crisis (Castro, 1993). He also calls for sustainable development, a message similar to that of deep-green environmentalists.

Environments which challenge environmental management

Some environments require especially careful management because they are easily damaged, are prone to environmental threats or excessive human demands (often both), or demand a specialist approach (the following section is only a brief overview).

Environments used as common resources

The problem of over-exploitation of common resources and the 'tragedy of the commons' model have generated considerable debate (Berkes, 1989; Bromley and Cernea, 1989). Interest has recently been renewed (Feeney et al., 1990; Vandermeer, 1996; Elliott, 1997). An environment subjected to use as a common resource need not be particularly sensitive to suffer damage: an inappropriate resource development approach can easily cause stress. Where common resources have sustained livelihoods for a long time, thanks to local people developing social controls, exploitation may break down as a consequence of changing attitudes. For example, in certain traditional fisheries in Amazonia, fishermen left areas undisturbed, or only occasionally fished, allowing stocks to recover. The enforcement was through superstition and tradition, but this has broken down as outsiders have been seen to break the rules without mishap. Common land is also easy to expropriate; the users often have weak title to the use of the land, if any; so, a government or minister can easily sell licences or ownership to companies or individuals.

Oceans beyond territorial limits are subject to multiple use and are not nationally controlled. In addition, huge distances are involved, and deep oceans require sophisticated research and monitoring equipment such as submersibles. Remote sensing, GIS and unmanned monitoring devices have helped with stock-taking, but policing presents problems. A number of ocean fisheries are in danger, but conservation measures are hindered because exploiters can resort to flags of convenience, claim to catch forbidden stock as an accidental side-effect of seeking something else, or just rely on not getting caught. Agreeing international quotas for catches of endangered fish and shellfish is not enough. Even remote areas of the seas have obvious pollution with plastic waste, tar and so on, and global ocean background levels of some pollutants are worrying (Payoyo, 1994; Borgese, 1998). Forms of governance (management that is enforceable) have to be developed to manage common resources effectively, particularly when a resource or environment is under multiple use by different stakeholders (Marshall, 2005).

Urban environments

Cities have grown rapidly in the past half-century and continue to do so at an increasing rate, and relatively recently humankind has become more than 50 per cent urbanised. Urban growth in developed and developing countries leads to environmental problems: pollution of air and water, refuse disposal, loss of farmland and natural areas worldwide. In developing countries there are the added demands of urban areas' demand for fuelwood, large poverty-stricken slums, and little money to maintain and improve services such as waste disposal (Barrett, 1994). Rapid urban growth coupled with lack of funding is a particular challenge. In the twenty-first century urban environmental management is going to become much more important as those problems develop. A century ago few cities exceeded one million; today many are much larger, a growing number of these in poor countries. Size alone makes environmental management difficult, and it makes it especially problematic in disaster situations when millions have to be controlled, sheltered, and provided with food and water.

Some hold that it is possible to develop sustainable urban ecosystems and even to de-link the urban areas from rural, depopulate the former and remove pressure from the latter, allowing better biodiversity conservation and reducing environmental damage. Whether that is realistic remains to be seen; cities have huge eco-footprints.

Islands

A recurrent problem has been the decimation of endemic island flora and fauna, by accidental or deliberate introduction of alien organisms, for example, the decimation of native songbirds of the main Hawaiian islands, partly due to the introduction of disease-carrying mosquitoes (Elton, 1958). Woodcutting, overgrazing, building and more frequent fires have also caused damage. Island biogeography can assist those developing sustainable management strategies for islands (Mueller-Dombois, 1975; Gorman, 1979; Troumbis, 1987; Beller *et al.*, 1990; D'Ayala *et al.*, 1990). Managers of island environments must consider dispersal of biota as well as on-island conservation; for example, disturbance of migrant birds on a particular island may have a much wider impact if they are denied that island stepping-stone. Study is vital to uncover situations where the degradation of an island environment may have causes elsewhere; Margaris (1987) reported the breakdown of terrace agriculture in the Aegean Islands as a consequence of the falling demand for dried fruit and olives caused by European women entering full-time employment and changing to ready-prepared or frozen food, and to large-scale mainland production which floods the market with cheaper produce.

Mountains and high latitudes

In both the Old and New Worlds there has been considerable development of high-altitude environments (Price *et al.*, 2004). Worldwide, mountain (alpine) environments have attracted tourism. High-altitude and high-latitude ecosystems are subject to extreme conditions, not just low temperature but also high winds. Mountains experience marked diurnal temperature fluctuations, high levels of UV radiation, wind exposure and drought; high latitudes suffer all three, plus they may have permafrost soils which impede drainage of summer melt and are prone to cryoturbation (frost movements), and day length varies by season, with prolonged winter darkness (Bliss *et al.*, 1981). These areas cover a considerable portion of the Earth's land surface, and their vegetation and soils are sensitive to disturbance, are likely to be slow to recover, and may have a relatively low species diversity. With vegetation and soils under stress, mountain and high-latitude areas are vulnerable to transboundary pollution, especially acid deposition and damage by tourism (Figure 13.1).

A number of mountain environments have experienced considerable population increase in recent decades, leading to forest, pasture and soil degradation. In Europe and the Rockies of the USA, cross-mountain highways spread vehicle exhaust emissions into high passes where they may directly cause pollution or produce (tropospheric) ozone which damages vegetation – some countries have tried to force traffic off mountain roads in an attempt to halt damage to mountain forests. Disturbance of mountain

Figure 13.1 Alpine resort of Cervinia (Matterhorn mid-background). The slopes are subject to pressures from winter ski activity, car traffic reaches as high as 2,300 m above sea-level (foreground), and there is considerable hotel and chalet construction.

ecosystems may impact upon lower altitudes through avalanches, landslides and altered streamflow (Figure 13.2). In mountain ecosystems managers tend to adopt a valley, watershed or micro-watershed approach to ensure an integrated view of higher and lower slopes (livelihood strategies often operate at several altitudes).

There are growing cities and townships in many high-latitude regions, such as Canada, Alaska, Norway, Finland and Russia, and military and resource development activity. Any traffic across permafrost during the summer is likely to result in damaged vegetation and soil that will be slow to mend. In winter, mountain and tundra regions are less likely to suffer soil compaction and ground vegetation damage. However, skiers and skidoos easily damage plants emerging from the snow, such as young trees, and snow compaction may delay spring thaw and cause problems for wildlife. At high latitudes pollution may break down slowly, so accidents such as oil spills are a problem. Radioactive fallout can become concentrated in lichens and bryophytes, affecting grazing animals. Alaska, Canada, Scandinavia, Greenland, Spitzbergen, Iceland and Russia have oil and gas, and other mineral development in tundra regions (e.g. the Trans-Alaska Pipeline) and so far have managed environmental impacts quite well (Williams, 1979; Copithorne, 1991).

Northern high latitudes have experienced considerable social and technological change: the hunting and transport practices of indigenous people have altered, notably with the adoption of motor boats, skidoos and firearms. There has also been pressure from some NGOs for change in traditional hunting practices (Berg, 1969). The end of the Cold War and better scientific equipment and vehicles have facilitated study, monitoring and exchange of data, assisting northern high-latitude management (Perkins, 1995). Disposal of waste presents problems at high latitude and in mountain areas. In Antarctica regulations now strictly control waste disposal and most is returned to lower

Figure 13.2 Overgrazed land and poorly maintained terraces, High Atlas Mountains, Morocco. As vegetation is degraded at higher altitude, landslides, silted streams and erratic streamflow disrupt farmland and irrigation at lower altitudes.

latitudes for disposal (Harris and Meadows, 1992). Some mountain areas have growing waste problems, mainly associated with tourism and climbing activity, notably the Himalayas. Polar seas are vulnerable to pollution due to the slow growth of organisms and slow decay of pollutants and also as a consequence of ice cover, which can restrict mixing of water masses and trap pollutants; the risk of spills is increased by the movement of icebergs, which can damage oil extraction and other infrastructure. There have been a number of marine oil spills near Antarctica, in the Arctic, and off Alaska (e.g. the *Exxon Valdez* disaster). The CIS has a problem with radioactive waste dumping in the Barents Sea and other Arctic waters, and with radioactive contamination from military facilities along the northern seaboard. These are likely to be expensive to manage and are strategically sensitive.

There are virtually no tundras in the high southern latitudes, but there are extensive peat bogs and swamps in southern South America, the Falkland Islands and on some sub-Antarctic islands. These peatlands and the southernmost forests of Tierra del Fuego share many of the vulnerabilities of northern tundra areas. Sub-Arctic and sub-Antarctic islands have some of the problems of both high-latitude environments and island isolation.

Drylands

Like mountain and high-latitude regions, dryland (i.e. seasonally dry as opposed to extreme arid environments) vegetation, soils and fauna are under stress, and so are easily damaged, may be slow to recover and difficult to rehabilitate. Drylands are areas where agricultural productivity is limited by periodic shortage of moisture and where fire damage may be common. During wet seasons and wetter than average years people tend to over-exploit, and during drier times land degradation sets in, leading to permanently more arid conditions (Geist, 2004). When conditions fluctuate widely from year to year reliance on the 'average' results in over- or under-exploitation. Traditionally people recognised this and adapted to cope with boom-and-bust fluctuations. About 20 per cent of the world's people live in drylands and many of these are suffering as a consequence of the breakdown of traditional livelihood strategies and land degradation. Some drylands have had marked human and livestock population increases. The reasons for this are diverse, and include provision of medical and veterinary services, and improved water supplies.

Drylands degradation ('desertification') has attracted much attention and has generated many misinterpretations (Mainguet, 1994; Thomas and Middleton, 1994). Since the 1960s problems with drought and desertification have prompted interest in environmental management (Beaumont, 1989; Dixon et al., 1989; Stiles, 1995), improved rangeland management, savanna management (Mott and Tothill, 1985; Werner, 1991; Young and Solbrig, 1992), rehabilitation of degraded drylands, sustainable development of drylands, and coping with fire. There has been limited success in countering dryland environmental problems. This may partly be because these areas have experienced considerable unrest and warfare, but also because governments have neglected these areas or intervene in a heavy-handed way with inappropriate strategies. Peoples in drylands and mountain areas can be fiercely independent, which may offend national governments. In drylands care must be taken to be sure of the cause of problems and of the environmental and socio-economic parameters.

Areas with sensitive and vulnerable soils

There are areas with sensitive and vulnerable soil in many different environments (Figure 13.3). Some soils dry out quickly, some lose their organic matter easily through oxidation if disturbed (a problem in drylands) or drained (a problem with peatlands). Certain soils shrink, crack, and develop crusts or concretionary layers. Infertile, acid-sulphate soils may develop on drainage, and there are areas where aluminium and boron deficiencies pose an immediate or potential threat (especially if there is acid deposition). Loess soils and similar fine-grained loams wash and blow away easily if disturbed and require skilled land husbandry if production is to be sustained. Soils which are fine-textured allow salt-carrying groundwater to rise and evaporate leaving a saline crust unless there is enough rain or irrigation to leach salts away. Vulnerable soils may coincide with dryland environments and human over-exploitation – as in parts of China at present and Midwestern USA in the Dust Bowl era (c. 1933–1936).

China currently has serious dryland soil degradation, especially in Inner Mongolia and northwestern China. Dust Bowl conditions are widespread and Chinese dust is blowing into other parts of the country causing health and environmental problems. Dust is also causing problems much further afield. China has always had dust storms but land degradation seems to be worsening, with desertification affecting around one-third of the country. The government has introduced a programme of windbreak planting, such as the Green Great Wall Project, and is encouraging the improvement of vegetation cover (Williams, 2005). Various international agencies, such as the Asian Development Bank, the UN and the Global Environment Facility, are helping to fund Chinese anti-desertification. Unfortunately, efforts are beset with problems: different agencies often pursue conflicting policies; there may be a natural cycle of periodic drying; the measures adopted do not necessarily work in practice; there is probably global warming; and rising population demands more food so that land remains under pressure.

Figure 13.3 Vegetation damage leading to soil degradation. Deforested landscape in the High Atlas Mountains, Morocco. Note the pollarded trees (a) (goat grazing and fuelwood collection); gullying and sparsity of groundcover vegetation (b).

Environmental managers need to understand the soils they deal with and to promote appropriate soil and water management and re-vegetation – put simply, to ensure good land husbandry (Hudson, 1992). To some extent global warming and other fashionable development issues have sidelined concern for soil degradation, which is unwise, for without good soil, food production cannot be sustained. Soil and water conservation and better land husbandry demand commitment and reinvestment of profits and labour. This means trading off present benefits against sustainability. People may not always be able to do that unassisted, and frequently fail to see the benefit to themselves. Aid and education are likely to be needed.

Many agencies warn that soil degradation is severe in developed and developing countries and it rates as one of the world's major environmental threats. Nevertheless, soil conservation and land husbandry are not attracting priority attention and funding. Many countries are spending too little to counter soil degradation, and some have even cut back on their efforts. The growth of interest in environmental management in the UK, Europe and the USA has not been matched by stronger support for soils research and extension services. Indeed, in the UK quite a few geography or earth sciences courses at universities now have little or no coverage of soils. A historian looking back on the late twentieth and early twenty-first centuries may well be puzzled by this weakness.

Coral reefs

Throughout the world coral reefs have suffered as a consequence of collection for building and cement manufacture, the souvenir trade, from damage by anchors, and the use of dynamite for fishing. Pollution, and perhaps disease related to it, is now taking a toll (Wells, 1992; Gray, 1993). The loss of sediment-filtering mangroves, plus more turbid river flow caused by land development may be to blame for some reef damage. There has been suspicion that anti-fouling paint may be causing coral damage. A number of reefs have been damaged by the spread of the crown-of-thorns starfish. There seems to be a correlation between agrochemical use on northern Australian sugar plantations and damage to the Great Barrier Reef. Fears are voiced that background pollution of the world's oceans, UV damage from stratospheric ozone thinning, and possibly the effects of global warming, are damaging coral (Pernetta, 1993; Wilkinson and Buddemeier, 1994).

Damaged reefs means loss of biodiversity, nursery and feeding areas for fish and other commercially important species, and reduced storm protection for low-lying islands and coastlands. One proposal is to establish artificial reefs, perhaps with scrap cars or old tyres.

Forests

Forests are being degraded and lost worldwide. In the humid tropics there has been tremendous loss of lowland rainforest. Forests in the seasonally dry tropics have also suffered, as has tree cover in drylands and in temperate and cold environments. There has been some recovery of forest area in North America and the UK since 1900, although species diversity has been reduced; in Scandinavia, Western and Central Europe and some other areas acid deposition and other pollution has started seriously to damage conifer and, more recently, broad-leaved forests. Within the past few years large-scale logging has become a serious threat to the boreal forests of the CIS.

Although there has been considerable international concern, there is little sign of any slowing down of deforestation in many of the problem regions. Between 2002 and 2004 Brazilian Amazonia lost about 24,000 km² of forest a year, and current rates are unlikely to be less (Laurance, 2005). The destruction in Amazonia is driven by a complex of causes, including expanding soya production, logging, clearance by marginalised squatter settlers, land speculation/ranching, drought and bushfires.

The cause of forest damage and loss varies from area to area, although there may be shared factors. The causes are often difficult to identify precisely and may be multiple: sometimes logging is to blame, sometimes land clearance by small farmers or governments, pollution may play a role, or ranchers may be responsible. Clearing is usually facilitated by road building or the opening up of trails for power cables. The former may in part be for strategic reasons or to facilitate mineral prospecting and development. Areas of biodiversity-rich natural forest are sometimes cleared for monoculture plantation cropping, usually of eucalyptus or fast-growing pine species.

There have been efforts to improve environmental management of forest ecosystems, often linked with local people's participation, agroforestry or 'tolerant forest management' (the extraction of products, leaving as much of the natural forest as possible intact) (Anderson, 1990). Sustainable logging has been more difficult and is less common than some foresters care to admit, and few manufacturers claim 'product of sustainable forestry' on their labels. It is more likely to be 'product of a managed forest' – what 'managed' means is often not clarified.

Wetlands

Wetlands comprise a wide range of ecosystems (e.g. bogs, marshes, floodlands, swamps, mangrove forests), the functioning of which depends on water (see also Chapter 10). They include marshes, fens, bogs, peatlands, swamps, river-margin floodlands, delta areas, mangrove forests, floodlands, coastal marshlands and man-made wetlands – irrigated land, reservoir drawdown areas. About 6 per cent of the Earth's land surface may be classed as wetland (Maltby, 1986: 41). Some of the world's most productive habitats are wetlands: they may be breeding and feeding areas for fish and other fauna, they may contain rich biodiversity, some act as crucial flood regulators on rivers, and mangrove swamps offer valuable coastal protection. Some wetlands are potentially very sustainable cropland if they are managed well, so there is both incentive to develop (and exploit until damaged) and opportunities to crop sustainably (Fraser and Keddy, 2005). Some wetlands are rich in biodiversity and merit better conservation. Wetlands often play a vital role in regulating streamflow and river flooding, and may help cleanse run-off of pollution and excess sediment. Wetlands may be privately owned or a commons resource, and attempts to alter them generally result in impacts over much wider areas and populations (Whitten and Bennett, 2005). Worldwide in rich and poor nations wetlands are being degraded at a worrying rate (Turner *et al.*, 2003).

People often depend on wetlands for food, fuel or building materials, and there is potential for domesticating wetland plant and animal species for aquaculture. Some of these areas are heavily populated, such as the deltalands of Bangladesh or Egypt. Unfortunately, there are many ways in which wetlands can be damaged: by drainage; by dam or barrage construction; by canal building or channel improvement; by pollution; by over-exploitation of plants and animals; through climate or sea-level change; and by reduction or diversion of inflow. One of the world's largest marshlands lies in southern Iraq (ancient Mesopotamia), watered by the Tigris and Euphrates. Between 1991 and 2000 these marshes were all but destroyed by deliberate drainage ordered by Saddam Hussein and by reduced river flow as a consequence of dams built in Turkey. About 37 per cent of the original area had been restored by 2005 (*The Times*, 25 August 2005: 37; http://www.gsd.harvard.edu/mesomarshes – accessed August 2005).

Mangrove swamps have suffered worldwide as a consequence of land development for real estate, aquaculture ponds, oil spills, logging and clearance for agriculture. By 1990 it is likely that the world's mangrove forests had decreased by about 79 per cent and the loss is accelerating (Kunstadter *et al.*, 1989: 8). Global warming may cause even greater losses, leaving tropical coastlands more exposed to storm damage, resulting in serious loss of biodiversity and of habitats where a wide variety of marine animals, including commercially valuable species, breed and feed. The costs of mangrove damage have been realised and there is some interest in conservation and reforestation, and in sustainable management (Kunstadter *et al.*, 1989).

Around the world, coastal wetlands, marshlands, peatlands and floodlands are being converted to agriculture or cleared for building at an alarming and accelerating rate. For example, in America there is a chance that the huge Pantanal wetlands could be damaged by river navigation and canal projects, and the future looks gloomy for the Mekong Delta and many other wetlands. Often the benefits of 'development' are short-lived and land is left degraded. Where peatbogs are drained the oxidation adds to global atmospheric carbon (Barrow, 1991: 117–128; Turner and Jones, 1991; Mitsch and Gosselink, 1993; Roggeri, 1995). Wetland sustainable agriculture and fisheries strategies merit study, and man-made and some natural wetlands could prove more sustainable and more productive than some terrestrial cropping, with fewer damaging impacts. Unfortunately, sustainable wetland agriculture/aquaculture has received far less investment than has unsustainable and damaging conversion to aquaculture or rice (the aquaculture mainly producing king prawns and similar shrimp types). Many countries now generate huge incomes from unsustainable shrimp/prawn ponds.

Lakes and smaller water bodies

Worldwide, water bodies have been degraded directly and indirectly by humans (see also Chapter 10). The damage is due to pollution (caused by industry, agriculture, human and livestock sewage aquaculture, and much more), reduction of inflow, contamination with irrigation return flows, the introduction of alien species, over-exploitation of fish and other organisms, disturbance by boats, heat emission from power stations, and so on. Often water bodies become eutrophic and many suffer salinity increase. Smaller water bodies are vulnerable to drainage or silting. In southern Asia water storage tanks were numerous and kept free of silt by cheap labour, but rising wages have meant that many are now neglected. Increasing acid deposition can damage ponds and pools even in remote and undisturbed areas. Lakes may be of great age, and in lower latitudes escaped some of the worst Quaternary climate changes; they therefore can have unique endemic species. Loss of that biodiversity is common.

Lake management can be difficult if several countries have jurisdiction over its waters and surrounding land (and the feeder streams), and in warmer environments stratification can restrict mixing so that waters may become relatively easily polluted or depleted of oxygen.

Estuaries and enclosed or shallow seas

Worldwide, marine environments with restricted circulation have suffered from pollution and over-fishing, including the Baltic, Caspian, North Sea, Aegean and Japanese Inland Sea. Effective commercial management demands control over huge catchment areas that contribute pollution and in many cases international co-operation. Shallow-water weed beds, including sea-grass, are important breeding sites for marine life. These weed beds have been damaged in many countries, directly by disturbances such as dredging, and indirectly by sewage or other pollution.

Rivers

River development problems are also discussed in Chapter 10. Failure to take proper care of the environmental management of river systems can have severe consequences for the riverine ecosystem, adjoining floodlands, estuaries and nearby seas. Enclosed seas and lakes are especially vulnerable to poor river management. The Aral Sea is a clear example of the environmental degradation and socio-economic misery which result from failure to co-ordinate and control developments within a river drainage basin (Kotlyakov, 1991). Humans tend to settle by rivers and riverine environments have been disturbed more than most.

The main issues of concern to the environmental manager dealing with rivers are silt from eroded soil, pollution and regulation of flow mainly by dams or barrages. Dams pose a greater threat than barrages because they alter downstream flow and water quality far more, and pose a greater barrier to the migration of fish and other organisms. Dams are also more likely to impound an extensive reservoir, which has significant environmental impacts on an area, and may force the relocation and disruption of livelihood for large numbers of people.

Environments subject to multiple pressures

Any of the environments listed may be subject to simultaneous and ongoing multiple pressures (and if there is common ownership, monitoring and policing may be problematic). Few environments are now untouched by humans and many are beset by complex demands and many threats. Environmental management therefore involves mediation, deciding priorities, co-ordination, delicate public relations work, and much more. Seldom can a strategy be developed and left unadjusted for any length of time; constant proactive assessment and monitoring must feed in. For example, in the UK national parks are not areas set aside for conservation, they have farming, pastoralism, tourism, settlement, industry; management must balance tradition and the status quo with necessary changes to ensure conservation and 'the common good'. Other countries, as they become more populous, are also having to manage mosaic landscapes of varying usage; in Brazil the extractive reserve has been developed to try and manage multiple demands such as: resource extraction, conservation, tourism, mining and so on. In the highlands of Malaysia farming, forest product extraction, property development, tourism, and other demands have to be orchestrated for best environmental management without hindering economic development and existing livelihoods.

Common ownership does not necessarily mean lack of governance and poor environmental management. Traditionally many peoples had tribal or village rules, practices and taboos which worked to sustain livelihoods and reduce environmental damage. Recently such controls have been eroded; people may lose respect for taboos and tradition and revert to selfish and unsustainable ways, or they have the resources expropriated from them. Governments may sell land or license resources use to a state or private company which then ignores or displaces traditional users. Colonial settlement did this on a huge scale. It is common for forested common land to be cleared for a company to grow plantations of trees for woodchips, or to plant soya or cassava for export. Often developers simply fail to see how many local people there are and do not appreciate their traditional rights (Berkes, 1989; The Ecologist, 1993). There is also the drive to earn foreign exchange (and possibly personal wealth) by supporting the loss of common resources in order to export crops or minerals.

Lessons the environmental manager can learn from study of sensitive environments

A number of common points can be recognised in the environments discussed above:

- Damage often progresses covertly to become serious before the problem is accepted and action is taken (sometimes too late – as for the rainforests of West Africa).
- Adopting a careless approach to researching problems, often exacerbated by inadequate data and time, can lead to misassessment. Consequently, symptoms rather than causes of problems are focused upon and treated. It is sometimes convenient for those in power to make such mistakes: better for them to blame nature or the peasantry than admit misguided, perhaps personally profitable, policy decisions.
- Local resource users tend to be overlooked in favour of national interests, large companies and their investors. Worse, local people may be marginalised – rural folk are less likely to riot or vote out a government than are their urban cousins.
- Long-term effects are overlooked as a consequence of pressure to maximise shorter term gains.
- Each situation is special. It is dangerous and often difficult to generalise.
- A problem may be realised, but a ministry or other responsible body may lack power, funds or trained personnel to make a satisfactory attack on it.
- Crucial issues, such as soil degradation, may fail to attract enough support.

- A number of the problems listed above, plus many others, are, at least in part, due to lack of adequate co-ordination and overview.
- Problems are increasingly transboundary, making it difficult for environmental managers to have jurisdiction or powers to enforce solutions (or even to assess the threat).

There are clearly measures which could be taken to reduce, avoid or mitigate damage to vulnerable environments:

1 As far as possible leave them alone and find less damaging ways of getting the same resources (or, at the very least, ensure that some examples of the ecosystem are conserved). Environmental managers might do more to prompt those considering development to look at technological solutions, or better use of areas already developed, or rehabilitation of degraded resources.
2 The environmental manager should pay attention to the local conditions, not generalise. A point stressed by Johnson and Lewis (1995: 303) is that it is important to build on local knowledge and local traditions and be aware of local constraints and opportunities. However, co-ordination is needed to ensure that each local activity does not cause wider difficulties.
3 Planning tools such as strategic environmental assessment could help to highlight risks where there are complex environmental and socio-economic linkages.
4 Impact assessment can encourage policy makers and planners to check what they propose more carefully, and should identify most risks, so that they may be avoided or the development be modified or abandoned.
5 Risk and hazard assessment can encourage the timely development of contingency plans.
6 In most countries building regulations often demand that structures meet 'average' conditions; when something important or dangerous is to be stored, structures should be much more robust.
7 Better monitoring of environments and of socio-economic conditions is important.
8 A problem is to achieve more willingness to consider long-term impacts, and to take preventive or remedial action. That is as much a problem for governments, NGOs, international agencies, the media and the public as for environmental managers, although the latter should be catalysts.
9 One of the key inputs from the environmental manager is to co-ordinate and to encourage and facilitate a thorough overview of proposed developments and monitoring of the state of various ecosystems, even if they are not obviously being altered.

Vulnerable environments (assuming they are recognised) deserve particular attention from monitoring bodies, more care from planners and greater vigilance from NGOs, media and international bodies. There are agencies or NGOs which focus on particular problem environments or threatened organisms. Unfortunately, many lack sufficient funds and other resources to intervene effectively, and may find it difficult to tackle transboundary problems. Biodiversity conservation has generated a lot of debate, but not everybody supports it: marginalised people may clear forest to survive; businessmen may develop areas of scientific interest for profit (and generate employment in doing so); a government may be forced to weigh aid for the poor against protecting the environment; deer may fare better when hunted with hounds but public opinion finds the practice abhorrent; the ethics of conservation can be far from straightforward (for a discussion of the ethics of biodiversity conservation see Blench, 1998).

Preservation of the environment is often not practicable, given commercial forces and a growing human population. Johnson and Lewis (1995: 228) make the important point that human use of the Earth has two faces: 'creative destruction, the process by which the natural world is modified and sustainable land-use systems are developed.' The second is 'destructive creation' characterised by 'a failure to achieve long-term sustainability and by the initiation of progressively more serious patterns of land degradation'. An environmental manager has to accept that there will usually be environmental changes (good husbandry involves making changes). It is crucial to decide when 'destructive creation' has begun or is likely, and to act to stop it or prevent it.

Integrated area approaches

Frequently environmental managers have to work for a specific region or area, and it is likely to be subject to multiple demands. Regional planners have dealt with similar issues since the 1930s, developing comprehensive and integrated area approaches. One of the first steps towards such approaches was the Tennessee Valley Authority (TVA) in the Midwest of the USA. Although the real success as a comprehensive authority lasted for only a few years in the 1930s it was a seminal exercise, which helped prompt the development of integrated river basin planning. Integrated river basin planning has had patchy results but is currently attracting renewed interest as an approach which can use water as a tool for development and to integrate and help various stakeholders and specialists co-operate. In addition, effective environmental management of a lower or mid-river basin demands control of the whole basin. There are ways of working in manageable biogeophysical and socio-economic units adopting an integrative approach that would suit environmental management:

- Integrated river basin planning (water acting as an integrative theme).
- Watershed management (again water, but more specifically run-off control and soil and water conservation, is the integrative theme).
- Coastal zone management.
- Urban areas.
- Using ecotourism in a particular area to integrate development and help encourage co-operation.
- Island units.
- Mountain units.
- Bioregions/eco-neighbourhoods.

The key requirements are that:

- Area units are discrete, not ephemeral, have a manageable scale, do not overlap others and leave minimal 'gaps'.
- The unit facilitates study, monitoring and management of biological, physical, social and economic issues.
- There is a theme to integrate and encourage interdisciplinary approaches: river development; ecotourism; sustainable development; land husbandry/control of degradation and so on.
- If possible it has a 'sense of place' for local people.

Some of these issues have been addressed by those promoting bioregionalism (Sale, 1991; Aberley, 1993; http://www.greatriv.org/bioreg.htm – accessed February 2005), bioregionalism being rooted in a blending of biogeography and a search for alternative development paths – an 'organising of societies by commonality of place adopting a framework for sustainable living'. Eco-neighbourhoods have been proposed as practical units for the pursuit of sustainable development (Barton, 1999). In the USA especially, eco-neighbourhoods have generated a lot of Internet activity among environmentalist groups. Watershed and river basin management has a stronger profile, and is supported by many governments and international bodies. The management of watersheds (catchments) has been supported by soil and water conservation bodies, pollution control agencies, and by agencies working on healthcare, social development and anti-poverty initiatives. River basin development has attracted regional development planners and those concerned with use of shared rivers, including multi-state sharing. One issue which is relatively little developed is the potential for downstream groups to be taxed to pay upstream land users for making environmental management efforts; similarly, the taxation of upstream users for impacts suffered downstream is still far from adequately developed.

Adaptive and flexible approaches

There are a number of reasons why environmental management strategies should be adaptive and flexible. All environments can alter, and the change may be sudden and unexpected even with good baseline studies and ongoing monitoring. Some environments are particularly changeable, including many of those which are most challenging. Human behaviour is far from constant and predictable – fashion shifts alter demands, technology developments lead to new resource demands, conflict places areas and sources of raw material out of bounds and can generate refugees, and so on. Human development is anything but adaptable and flexible. Funding may take the form of large and rigidly controlled loans; it is often impossible to modify engineering (but frequently a little thought could have made it flexible). The approach to a challenge is often to assemble a team which is constrained by budget and time limits (there may even be early completion bonuses); soon after the infrastructure is completed specialists move on, and then problems appear. Developers commonly obtain enough money to construct and start a project, but not to maintain, repair and modify it. Added to these problems, Western science and administrators have tended towards a very short-term outlook since the 1940s. Victorian developers often planned for centuries and over-engineered things so that structures have coped with unforeseen problems and changed demands. Some of the problem of Western limited planning horizons may stem from the nature of democracy – results must be achieved within government terms of office, current taxpayers have to be satisfied, accountants work to a limited time span and so on. Many modern structures have quite short design lives: whereas brick and wrought-iron used before the 1930s will last, few modern plastics and concrete survive for more than thirty years.

Environmental management has to look at a wider and longer term picture and must try and encourage more adaptive, appropriate and flexible approaches. Adaptive environmental assessment and management (AEAM) and adaptive resource management have developed to address some of these problems (Holling, 1978; Walters, 1986; Gunderson et al., 1995; for a bibliography of adaptive environmental management see http://www.for.gov.bc.ca/hfp/annobib/ambib.htm – accessed September 2005). AEAM involves learning by experiment from complex systems and is an inductive approach which usually has a monitoring component and a response component – essentially, 'management is a continuous learning process'.

Summary

- This chapter examines situations that particularly challenge environmental management through physical and/or human causes. Often such situations can be usefully mapped.
- In a given situation, even the same group of people may react differently on successive occasions to the same challenge. Various groups are likely to vary in reaction, vulnerability and resilience; in addition, even within one social group different sexes and age groups can differ in these respects.
- Environments are usually subject to multiple threats and pressures. In addition, environments and human demands are changeable; environmental management must cope with complex and changeable systems.
- Increasingly environments are becoming degraded and this degradation may affect wider areas. Some challenging and vulnerable environments currently have increasing human (and sometimes livestock) populations.
- Generalisation is unsafe; every situation should be carefully assessed with care to avoid accepting received wisdom without question.
- Environmental management commonly has to adopt an integrative and co-ordinating role.
- Sometimes problems are caused mainly or partly by poverty and marginalisation.

Further reading

Adger, W.N., Kelly, P.M. and Ninh, N.H. (eds) (2001) *Living with Environmental Change: social adaptation and resilience in Vietnam.* Routledge, London.
Vulnerability, adaptation and resilience.

Blaikie, P.M. and Brookfield, H. (eds) (1987) *Land Degradation and Society.* Methuen, London.
Explores why management often fails to prevent environmental degradation.

Blaikie, P.M., Cannon, T. , Davis, I. and Wisner, B. (eds) (1994) *At Risk: natural hazard, people's vulnerability and disasters.* Routledge, London.
Disaster reduction and the promotion of a safer environment.

Glantz, M.H. (ed.) (1994) *Drought Follows the Plough: cultivating marginal areas.* Cambridge University Press, Cambridge.
Examines the links between society, climate shifts and land degradation.

Stocking, M. and Murnaghan, M. (2001) *Handbook for the Field Assessment of Land Degradation.* Earthscan, London.
Simple, practical ways to assess land degradation and farmers' interaction with the land.

www sites

UN Secretariat on the Convention to Combat Desertification (UNCCD) http://www.unccd.int/main.php. – accessed February 2005.

14 Tourism and environmental management

- Green tourism
- Ecotourism
- Ecotourism and sustainable development
- Tourism and environmental management in practice
- Summary
- Further reading

Tourism is a major contributor to the economies of many countries; however, the impacts sometimes overshadow the benefits and seldom is development as environmentally sound as it could be. Environmentally sound tourism can offer some areas otherwise unavailable funds for improving environmental management, it has the potential to offer sustainable livelihoods, and it can act as an integrative 'core' to bring together various stakeholders, who otherwise would probably not work together. The latter quality may be used as a way to establish sustainable development strategies – for example, using tourism profit to help establish sustainable agriculture or conservation, and then developing those as tourist attractions. When badly managed tourism degrades the environment, local culture visitors tend to pay less or to go elsewhere. Environmentally sound tourism can be more lucrative as well as being sustainable. Where low-quality tourism has degraded a locality and driven down profits, green tourism may be a route to restoration of environment and economy (see Figure 14.1).

There has long been what may be described as environment-based tourism; i.e. some aspect of nature attracts the visitors – walking, climbing, skiing, viewing scenery, sport-fishing, hunting. The idea that tourism could help pay to care for the environment and a growing awareness of the desirability of minimising the physical and social impacts of tourism have developed since about the mid 1980s, and this may be described as the 'greening' of tourism. Since the 1970s, but especially after the mid 1980s, there has been increasing interest in ensuring that tourism development is sustainable. Over the same period there has been a growing shift from advocacy and theoretical discussion to developing tools and strategies which seek to assess and minimise impacts, identify threats, and generally 'green' tourism wherever developers are willing to do so. Some of the impetus has come from planners and governments, or it is prompted by tourists and tourism companies becoming more environmentally aware. Obvious environmental and socio-economic problems have been caused by mass tourism in many areas, which helps encourage change.

The approach to tourism development has altered over the past few decades: twenty years ago, the main planning input would typically have come from economists; today the likelihood is that it will be multidisciplinary. There has also been a shift from a 'develop-now-cope-with-problems-as-they-appear' approach towards much more

Figure 14.1 The development of apartments and hotels in the Cameron Highlands (Malaysia). Building on steep slopes threatens the valley with landslides and if great care is not taken silt and sewage will pollute streams. Along with unsustainable farming, this building destroys montane rainforest and occupants add to traffic and other pollution. Ideally this sort of development should be sited on already degraded land near the valley floor where it would do less harm to the biota and intrude less on the scenery.

proactive planning, aware of the precautionary principle (i.e. seeking to err on the side of caution to avoid problems). There has also been a growing tendency to at least inform, and ideally encourage some degree of participation from, people affected by tourism (co-management). Planners and managers are also increasingly held responsible for their actions, and the media and the Internet make it likely that any problems are apparent to a worldwide audience.

Green tourism

Tourism is a huge money earner in rich and poor nations, and some countries are very dependent on it. Tourism currently provides more than one in fifteen of all jobs world-wide, it is expanding, and green tourism is one of the fastest growing and evolving sectors (Weaver, 2002). Tourists are influential, sometimes in a negative way, but they can also bring innovations, and sometimes return home with new ideas. Local people may acquire much of their view of the outside world and aspirations from contact with tourists; wasteful and damaging habits are easily learned and sound traditions cast aside. There is a huge diversity of green tourism initiatives, much little more than opportunism and 'hype', giving little if any of its profits to help the environment or local people.

Sometimes the only real basis for claiming to be environmentally friendly is that the tour company donates a tiny fraction of its profits to environmental charities and writes that off against tax (WWF, 1995). There is a complex green spectrum, ranging from deep (dark) green to shallow (light) green. The following categories are generalisations, but serve to illustrate the diversity of green tourism customers and the 'raw material':

- *Hard-core ecotourism* – minimum negative impact/maximum benefit to environment and host population. Tourist interest is genuine and deep, and tourists learn and have their attitudes affected. These tourists are willing to endure indifferent accommodation and catering to enjoy pristine sites and reduce impacts but they tend to object to abundant visitors.
- *Dedicated tourists* – limited negative impacts and reasonable benefits for the local economy and some contributions towards environmental care. These tourists do learn something and they are not just present because they are bored by mass tourism or keen to return with tales and photos to impress others. They tend to be keen on an active pastime such as bird watching, diving, hill walking and climbing, rather than general environmental or cultural interest. Standards of accommodation and availability of alternative activities and attractions are important.
- *Marginal* – little benefit to the locality and some negative environmental and socio-economic impacts. These tourists probably learn a little, but do not shed naive attitudes. Possibly they have become bored with mass 'sun and sand' tourism and are seeking 'an experience' to use to impress others. Comfort is important, their attention span is often limited, and attractions probably have to be 'enhanced' ('swimming with dolphins', crocodile farms, four-wheel-drive safari trips, and similar) (Figure 14.2). They are more tolerant of crowding, but unlikely to accept non-air-conditioned accommodation and basic food and beverages.
- *Casual* – on the whole a negative impact. They have little interest beyond an entertainment visit or two, make a minimal contribution to local economy and environment, learn little and retain original attitudes. Many will stay within the boundaries of hotel or beach resort, which has a considerable eco-footprint.

A mismatch is possible between green tourism initiatives and the aforementioned categories of tourists; planners must research the customers, ensure developments are co-ordinated and do not clash, and try to propagandise them before they arrive. *Casual* tourists are less likely to fit in with ecotourism. It is of course possible to reduce the socio-economic and environmental impacts of all categories of tourists through appropriate building, energy and water conservation, and zoning. The Mayan Riviera (Caribbean, Yucatán coast), Mexico, effectively caters for mass tourism in Cancún and for a few hundred kilometres beyond supports greener tourism. In recent years tourism has encroached more and more in remote and unspoilt environments. Better ski, diving, mountaineering and walking equipment allow access to situations that could only be reached by specialist expeditions thirty years ago. Improved air travel means more people can afford to travel to more remote sites. Places such as mountain tops, oceanic islands and isolated coral reefs are increasingly visited and can be very vulnerable. Trampling, wildlife disturbance and oil spill from cruise ships is a concern in Antarctica and the sub-Antarctic, and some of the highest peaks of the Himalayas are now littered with discarded gear, refuse and bodies; in such environments decomposition is very slow so it accumulates. In the tropics, beach developments, littering and tourist activity have seriously damaged a number of turtle-breeding beaches.

There are a number of situations where tourism and conservation areas have been established without sufficiently involving local people. The locals then feel alienated

Figure 14.2 Crocodile farm by Lake Tonle Sap (Cambodia). A tourist attraction which, by yielding skins, might take some hunting pressure off wild reptiles. If no attempt is made to license or tax owners no funds will be collected for environmental management.

and exploited and have often lost access to traditional resources. Protected species may leave reserves and kill livestock or people and damage crops; tourist behaviour may offend, and mainly outsiders meet employment and supply needs. Conservation and environmental care can beneficially tap local traditional knowledge and support if it wins people's confidence and offers benefits such as jobs. Controlled resource extraction can be compatible with tourism, and it may usefully provide livelihood opportunities if business is seasonal, erratic or insufficient for family support (Pasoff, 1991; Foucat, 2002). Tourism is often seasonal and can be easily damaged by shifts in fashion, bad publicity, disasters and unrest. Sole reliance on tourism is therefore not a good idea, and whenever possible authorities should seek a mix of mutually beneficial 'dovetailed' activities – agriculture or fishing and tourism, plus conservation and craftwork. Routes to sustainable environmentally friendly tourism generally include 'dovetailing' mutually supportive activities, using local resources, and involving local people. Duffy (2002: 98–126) outlined how tourism and indigenous peoples interact, drawing on Central American experience.

Some countries seek to control tourism impacts by restricting numbers of tourists, charging those who are admitted large fees (e.g. Bhutan); but not all localities will have that option. If tourism is made 'greener' through use of local building materials and styles (to reduce imports and save on air-conditioning), use of bathwater for irrigating gardens, and by consumption of locally produced food and beverages, local production of souvenir items, and so on, the locality tends to gain character and prove more attractive to higher value tourism. A relatively recent development is compulsory green taxation or voluntary contributions from tourists; the problem is ensuring the revenue generated is spent on environmental needs or socio-economic development. Tourism is part of globalisation, and tourists help to spread Western values; one could therefore say that ecotourism is a form of green imperialism – the exploitation of developing country resources by mainly developed country tourists.

Ecotourism

Even before the 1920s tourism was seen as a way of paying for conservation areas (e.g. the Yellowstone Park in the USA), and as a means of educating tourists to better themselves through contact with nature. This, however, was by no means mass tourism. As numbers of tourists grew in the 1980s, thanks partly to wide-bodied jets, the literature on tourism and recreation expanded, and included attempts to make planners more aware of environmental and social side-effects of tourism and the problems and opportunities posed by nature and local people for tourism (Edington and Edington, 1986). Tourists are increasingly seeking more differentiated and interesting attractions, and host countries like the promise of non-destructive natural resource exploitation offered by green tourism. Natural history programmes have become more frequent on TV in richer nations and this is helping encourage ecotourism.

Modern forms of ecotourism (or eco-tourism) appeared during the 1980s and spread rapidly – the First Asia Ecotourism Conference was held in 1995, and 2002 was the UN International Year of Ecotourism (http://www.uneptie.org/pc/tourism/ecotourism/iye.htm – accessed October 2005). Developed and developing countries have invested in ecotourism; the latter include: Galapagos Islands (Ecuador), Tanzania, South Africa, Belize, Rwanda, Costa Rica, Cuba, Yucatán (Mexico), Zimbabwe, and many others (Reid, 1999). Some of these depend a great deal upon it for conservation funding and it can be a major foreign exchange earner. Ecotourism is difficult to define precisely; the Ecotourism Society has made one of the many attempts to do so: 'responsible travel to natural areas which conserves the environment and improves the well-being of local people' (http://www.ecotourism.org/ – accessed September 2005). Ecotourism, sustainable tourism and sustainable development are all interrelated and overlap and share techniques, but there is variation in approach and interpretation. However, all are multidisciplinary activities with an integrative approach, dedicated to environmental protection and the improvement of human well-being (Pforr, 2001). Ecotourism should be a symbiotic relationship, whereby the environment attracts tourists, and tourists pay a significant amount for environmental management; hopefully the process is one of sustainable development (Lindberg and Hawkins, 1993; Fennell, 1999; Wearing and Neil, 1999; Page and Dowling, 2001; Weaver, 2001). Some ecotourism supporters argue that it should not only fund environmental management and contribute to local livelihoods, but that it should also educate the tourists towards more environmental awareness and responsible behaviour (Orams, 1995).

Ecotourism is essentially a set of principles which may be allied to any nature-related tourism. Some forms are more 'passive' – visitors just look; others are more active – such as paying volunteers working on conservation, land rehabilitation, environmental research and so on. Even destructive tourism activities can sometimes contribute to environmental management, and can possibly be sustained – but do not deserve to be called ecotourism, which should put back more than it degrades. The basic way to pursue ecotourism is to:

- assess the current situation and trends;
- identify suitable ecotourism solutions;

- plan a strategy that keeps tourist impacts to a minimum and ensures the behaviour of the host population does not deter tourists and that some benefits reach them;
- seek a strategy, which is appropriate, sustainable and flexible;
- establish adequate ongoing monitoring;
- ensure there will be recurrent funding to maintain and modify ecotourism.

To manage and steer ecotourism and to assess sustainability performance it is important to develop effective indicators and monitoring methods. The trend has been away from narrow focus indicators to those derived from a number of multidisciplinary inputs; various institutions have developed such sustainability indices or benchmarks which can be adapted. The following indicators and assessment tools have been applied to sustainable tourism and ecotourism:

- EIA – a structured assessment of existing and potential threats, problems and opportunities.
- Carrying capacity – an assessment of what can be tolerated. This can be misleading; for example, exploitation well within carrying capacity may falter if conditions alter and reduce it. There are a multitude of carrying capacities to be assessed – physical/environmental; social; cultural; aesthetic and so on.
- Limits of acceptable change system.
- Eco-footprinting.

EIA can be a valuable input to tourism planning, but it is not by any means 100 per cent effective, so planners and managers should not let its application give them any false sense of security. EIA has been used during coastal resort development in Peninsular Malaysia (Vun and Latiff, 1999). Eco-footprinting has been used to measure progress towards sustainable development by cities, businesses and service providers, and Gösling et al. (2002) explored it as a tool to assess tourism in the Seychelles. Eco-footprinting uses space equivalents to express appropriation of environmental resources by individuals, groups, companies and so on. An eco-footprint is an expression of the area required to support a lifestyle or activity, compared with the available area. The assessment tries to include all inputs and outputs, and even the environmental impacts of travel to the destination from country of origin.

Does ecotourism need to be small-scale, high-priced and elitist? Sometimes it is, but it need not be. Ecotourism must fit the carrying capacities (environmental, cultural, economic, or whatever) in the affected area. One strategy is to zone areas according to their sensitivity, so as to give maximum protection to pristine and vulnerable localities; buffer areas around these help to protect them from more intensive exploitation in outer zones. Less sensitive outer zones could be used for mass tourism, and zones near and possibly including buffer areas could support smaller scale ecotourism and occasional day trips by those based mainly in mass tourism zones. It is important that the pristine core areas are grouped close together if possible, ideally linked with wildlife migration corridors; and that sufficient regard is given to the possibility of climate and other environmental change, increasing pollution and natural disasters. Too rigid and unimaginative zoning with insufficient adaptability will not sustain flora and fauna. Even apparently 'safe' tourist development may damage the environment – various cave sites have suffered because visitors bring in fungus organisms and the lighting supports mould and moss growth – proactive problem spotting should help prevent problems, but there will be surprises, and any development should, above all, be flexible and adaptive. While mass 'sun, sea and sand' tourism tends to demand large and intrusive hotel complexes, ecotourism can make use of much smaller facilities, such as redundant mansions, old plantation buildings and so on (Figure 14.3).

Figure 14.3 Yucatán (Mexico): (a) Mayan pyramid – Chichen-Itzá. This and other archaeological sites in the region bring in many tourists who shop and stay in local towns. Beyond the excavated and restored sites there is little tourism impact and some profits can be spent on environmental management. (b) Some hotels in the region (Yucatán) are low-rise, adobe-walled and thatched. Traditional construction cuts out the need for imported materials and cement manufacture. The thatch and open-plan designs reduce or even eliminate the need for air-conditioning. Some revenue is generated for environmental management by tourist taxation.

The attraction of ecotourism for many administrators is that it can yield foreign exchange, has the potential to be sustainable and 'green', and can be established with reasonable investment and limited socio-economic change. The approach adopted depends on the attitude of administrators; most have an anthropocentric outlook, and some development agency and funding bodies place reduction of human poverty above environmental concern. Deep-greens value environment above improving human well-being, and generally argue that ethics, economics and human outlook have to alter (Duffy, 2002). Environmental managers must steer ecotourism for optimum results in each situation.

Ecotourism and sustainable development

Sustainable development was promoted in the early 1970s, but only gained widespread media and academic attention after about 1987, as a consequence of the Brundtland Report. There is no single universally acceptable definition of sustainable development; however, all recognise that there are environmental limits to development; that environmental protection and development are interrelated and mutually dependent; that together with environmental care there must be concern for the poor; and that there needs to be intergenerational equity (i.e. this generation passes on to future generations at least as good potential and range of options as we have enjoyed). Sustainable development not only demands an environmentally sound strategy, it also depends on establishing (if none are already present) and maintaining appropriate social institutions to support management and help adapt to challenges (World Bank, 2003). It is important to ensure that those involved in sustainable development initiatives can call upon adequate social capital – the arrangements, traditions and obligations which back up individuals, families and groups to help them survive or innovate. Social capital can decline, which may mean problems in maintaining environmental quality and the socio-economic conditions which favour tourism. Villages and regional officials may not recognise the often insidious loss of social capital and the damaging chain of events it can trigger, and instead blame difficulties on non-existent environmental change. Tourist–host population relations need to be monitored because both sides in all countries may behave in ways that can destroy sustainable tourism. Locals may be welcoming but disorganised, and so deter visitors; they may be apathetic and ignore them, or openly hostile. Tourists can corrupt local culture, offend people, undermine social capital, spread diseases, make heavy use of facilities locals pay for, cause crowding, and much more. Local enthusiasm and appropriate tourist behaviour seem to be quite important ingredients in sustainable tourism (including ecotourism).

Sustainable development efforts must be well researched and adaptable to socio-economic change. They must also cope with environmental change, natural periodic shifts (e.g. ENSO events), random extreme events, and ongoing global warming-related shifts (e.g. rising sea-level, upward shift of vegetation zones, altered precipitation, new patterns of disease transmission). Human-induced environmental changes are not restricted to global warming; there is a need to monitor for such things as acid deposition, the introduction of exotic flora and fauna, or smoke and smog pollution.

The ecotourism and sustainable tourism literature is large and growing rapidly. It includes introductory guides and handbooks, until recently focused mainly on Europe, North America or Australia (Cater and Lowman, 1994; Coccossis and Nijkamp, 1995; Harris and Leiper, 1995; Hall and Lew, 1999; Tribe et al., 2000; Eagles, 2002), and many articles and books dealing with discrete sectors or regional experiences. The latter includes fragile environments (Price, 1996): river basins, coastal zones, small islands, forests and mountain environments. The green tourism and ecotourism sector is expanding in both developed and developing countries (Cater, 1993; Lipscombe and Thwaites, 2003). Ecotourism and activity tourism overlap a great deal: golf-courses, trekking and bird watching can all be made nature-friendly (Berry and Ladkin, 1997; Garrod and Feyall, 1998; Mowforth and Munt, 1998; Stabler, 1998; McCool, 2002). Well-planned and skilfully managed ecotourism can be a route to sustainable development, and one which can overlap other land uses (Ashton and Ashton, 2002).

Ecotourism attractions may be created from dereliction through rehabilitation. In a number of countries derelict environments have been converted into recreation and tourism attractions, which do little environmental damage, and in some cases offer environmental benefits. In the UK, Kew Gardens has evolved a recreational role in addition to its long-established function as a botanical garden and herbarium, which is compatible and even pays for some of the research. In Cornwall (western UK), the Eden Project was opened in 2001, in what had been a derelict china clay pit (mine). It is a development of large enclosed environments filled with exotic plants which provides an attraction for tourists, and (like Kew) conserves endangered species, seeks to educate the public and provide a base for overseas conservation aid (Smit, 2001). The Project has also created employment and funds for environmental improvement in a relatively remote and poor region. Kew, the Eden Project and a number of other tourist attractions/plant and animal collections also allow researchers to conserve biodiversity and explore possible future scenarios, including global warming. As threats such as acid deposition endanger plants in their natural habitiat (in situe conservation), conservation facilities like Kew and the Eden Project allow some duplication in enclosed environments and gene banks. There are problems with facilities like Kew or the Eden Project in the coexistence of visitor attractions and the botanical/conservation role. This can lead to disputes between the scientific staff and visitor promotions or management who are keen to earn money to pay back debts.

Ecotourism is widely seen as an accessible 'engine of economic growth' which without excessively heavy investment can help a country generate foreign exchange; added to this is the hope it will be 'green' and sustainable (Weaver, 1998; Honey, 1999; Gösling, 2000; Duffy, 2002; Weinberg et al., 2002). With good planning and management and adequate reinvestment of profits it can be. Public relations and propaganda can be important elements in converting existing tourism into green, hopefully sustainable, ecotourism or establishing it from scratch. Tourism authorities can distribute informative brochures via airline seat pockets, or display them in shops, public buildings, car hire offices and hotel rooms; near attractions, they can establish visitor centres with displays; informative videos can be offered with in-flight movies or on TV in tourist source countries; articles can be placed in specialist magazines and newspapers; dedicated ecotourism wardens or tourist police can patrol tourist areas; and websites can be established to attract the right sorts of tourists.

The needs and capabilities of the host population and the characteristics of the environment must be considered when ecotourism is being planned, and monitoring once it is established. Ecotourism efforts do fail, and may also cause environmental and socio-economic damage. Tourism benefits often leak away, so the local people bear the costs and gain minimal benefit. Sometimes local people get some benefits but these are not adequate and locals may also lack empowerment (Boo, 1990; Cater, 1995). Ecotourism generally seeks to counter those faults as well as use tourism to pay for better environmental management.

Many sustainable development ecotourism initiatives will be managed locally, but a strategic overview is crucial to ensure different projects do not conflict with each other, and if possible aid each other. A strategic, co-ordinating body can also prepare visitor itineraries so they dovetail one activity and site with another to compile a varied vacation which is more attractive to them. It can also be a way to take the pressure off some sites and generally guide tourists. Ecotourism should be part of an overall sustainable development strategy: an ailing plantation industry could be subsidised and helped to reinvest (as in the case of highland Sri Lanka); local agriculture should supply sustainable tourism, rather than relying on food and beverages air-freighted from overseas; this can improve local livelihoods and should give the tourism more character (Fennell *et al.*, 2003). Many countries have a plethora of ministries and other bodies, each jealously managing a sector; persuading them to cooperate to support ecotourism can be a challenge, but a multidisciplinary development agency may manage it.

Tourism and environmental management in practice

Already it has been stressed that, if it is to work, sustainable ecotourism demands a proactive, adaptive and multidisciplinary approach plus good ongoing monitoring. Even the best techniques and skilled planning will fail to predict some challenges, monitoring may not give adequate advanced warning of problems, and tourist behaviour can be fickle. Sustainable tourism must therefore maximise adaptability. Strategies which support local participation (even co-management) and improve adaptability have been developed in fields such as conservation, forestry, resource extraction, agricultural development and healthcare provision (Brandis, 1998; Jiggins and Rowling, 2000; De Boo and Wiersum, 2002; Berkes, 2004 – see also discussion of adaptive environmental assessment and management (AEAM) in Chapter 6, this volume). Most monitoring and predictive assessments, like EIA, give 'snapshot' results – a relatively simplified picture with limited coverage in space and time. Such snapshot vision is far from ideal for managing constantly changing and complex situations. Adaptive environmental assessment and management approaches are thus valuable; efforts must be made to spot the unexpected before it happens (Jones and Greig, 1985). AEMA studies of livelihoods–environment interactions were conducted in a mountain valley in Austria in the 1970s; one of the findings was that, as tourism developed, locals invested in hotels, and dairy farming was relatively neglected. The consequences were marked and unexpected. Because cattle were stall fed and not grazed on pastures as carefully as they had been, the mountainside grasses grew and when snow fell offered a poor anchorage so that avalanches were more of a threat and ski pistes were thinner. Reduction of grazing also damaged the display of spring and summer wild flowers and had a negative impact upon tourism (Barrow, 1997: 81–82). It should be relatively straightforward to adapt these co-management-supporting approaches to meet ecotourism needs.

Like sustainable development, one of the values of ecotourism is that it can act as a unifying and integrative catalyst with which various interests can identify (Cater, 2000). Ecotourism and sustainable development share a key core feature: they demand reinvestment of adequate surplus into maintaining, and if possible improving, local (and if resources allow, wider) environmental and social issues. There is no reason why ecotourism revenue from mountain tourism could not pay for social improvements or environmental management in, say, a valley or a coastal zone – *provided* enough is spent locally to prevent deterioration and to sustain the tourist attraction.

There is a need to encourage viable sustainable ecotourism. Guidelines which can be enforced, and independent certification, will be crucial. The latter is likely to be through a suitable environmental management system (EMS) (WWF, 2000; Honey, 2002). Codes, guidelines and ethical standards have been published (e.g. by the Ecotourism Society of Australia and in the USA by the Ecotourism Society). Using strategic environmental assessment and strategic environmental management approaches could help strengthen strategic overview. Popular interest in green issues is expanding, and some established destinations are dissatisfied with their present 'mass tourism' and seek to shift it to a higher grade tourism by adopting a green tourism focus; other developers will establish green tourism in new areas. Provided there is no economic depression or travel fears (such as those provoked by the Bali bombings) there should be a growing number of affluent and older tourists interested in flora and fauna, craftwork, archaeology, and so on (Lück, 2002). The future for green tourism looks reasonably promising. Ideally, green tourism should try to develop fully into sustainable ecotourism.

Making use of tourism to aid the poor – pro-poor tourism – fits in with the demands of sustainable development and could be part of a sustainable ecotourism strategy (Ashley *et al.*, 2001). Non-tourism agencies concerned with poverty reduction have been developing approaches to provide improved sustainable livelihoods for the rural poor; these could also support sustainable ecotourism, providing tools for assessing and monitoring people's needs and capacities, and for institution building (Carney, 2002). In some situations poverty reduction may well be allowed to dominate tourism, and in others environmental considerations will attract adequate attention but not the welfare of local people – the challenge is to get the balance right.

Summary

- Tourism is a huge and expanding field which can earn vast sums. It is vital that tourism impacts are reduced and it is harnessed wherever possible to improve environmental management.
- Ecotourism is widely seen as an accessible 'engine of economic growth' which without excessively heavy investment can generate money; added to this is the hope that it will be 'green' and sustainable.
- Much ecotourism and green tourism is 'hype', a veneer of 'greenwash' disguising activities that are not as environmentally sound as developers would like tourists and the public to believe.
- Ecotourism is a symbiotic relationship, whereby the environment attracts tourists, the tourists learn environmental lessons and pay a significant amount towards environmental management; hopefully the process is one of sustainable development.
- There is no reason why revenue from, say, mountain tourism could not pay for social improvements or environmental management in a different region such as a coastal zone – *provided* enough is spent locally to prevent deterioration and to sustain the tourist attraction.
- Environmental managers must encourage ecotourism initiatives to be adaptable and beneficial to local people (and/or people and environments in need in other areas).

Further reading

Fennell, D.A. (1999) *Ecotourism: an introduction.* Routledge, London.
  Good introduction with numerous case studies from around the world.
Honey, M. (1999) *Ecotourism and Sustainable Development: who owns paradise?* Island Press, Washington, DC.
  Overview of ecotourism with case studies.

Smit, T. (2001) *Eden*. Corgi Books, London.
    The story of the conception, struggle for funding, construction and future hopes for the Eden Project in the UK.

Weaver, D.B. (2002) *Ecotourism*. Wiley, Chichester.
    Covers developed and developing country experience.

www sources

Ecotourism portal – free resources and networking http://www.ecotourism.cc/ (accessed September 2005).

Ecotourism Society – founded in the USA in 1991. The Society holds regular workshops and publishes guidelines, offers library material and so on – e-mail: ecotsocy@igc.apc.org; website: http://www.ecotourism.org/ (accessed February 2004).

UNEP Ecotourism/International Year of Ecotourism http://www.uneptie.irg/pc/tourism/ecotourism/home.htm (accessed September 2005).

15   Urban environmental management

There is no clear, simple, universally accepted definition of 'city'; in some developing countries urban areas are distinctly different from those in richer nations, and within each country cities differ, sometimes markedly. There is usually considerable difference between districts of any city. Some cities have been planned, and in some cases built on undeveloped rural sites. Many cities grew up in a wholly haphazard way, and reflect economic, defensive, or colonial needs. There are cities established for centuries, and some built within the past few decades.

Single-storey shanty towns in developing country cities may not have especially dense populations, and often there are large amounts of unoccupied land and few of the services familiar in metropolitan areas of richer countries. Cities in developed countries tend to have greater population density, more commuters and cars, consume more energy and water, and generate more problematic waste. Some countries also have large refugee camps which display many of the characteristics of impoverished urban areas. Such camps form rapidly and relatively imposing demands upon water, food and fuel supplies in a region, sometimes for decades; they also lack taxpayers to support services. The people of such camps have often lost possessions, and are traumatised and desperate, presenting a challenge for those seeking to improve things. Cities differ greatly in their degree of industrialisation, and whether it is dispersed among dwellings or contained in separate industrial zones. Some cities are 'Westernised' (e.g. Singapore), and some have extensive scattered low-rise building with a great deal of urban and peri-urban agriculture (e.g. Bangkok).

For several years now the world has been more urban than rural, in terms of population numbers, and much of the future population growth in developing countries will be in cities. Often, but not in all cases, cities have grown rapidly in recent decades. Rapid, and largely unplanned, urban growth poses serious problems for environmental management (Devas and Rakodi, 1993). Here and there it is possible to find cities which have made progress, improving urban environmental management; but most of these are relatively affluent city states or have flourishing economies based on petroleum or tourism. Singapore is one such example, having made impressive progress since the 1970s in spite of limited space and industrial growth (Khoo, 1991).

Urban environments

There is a growing interest in urban ecology in developed and developing countries (Ward, 1990; Schell *et al.*, 1993; Sukopp *et al.*, 1995). The following points seek to objectively list urban environment characteristics:

- Cities should be able to provide healthy and stimulating environments for their dwellers, and perhaps even generate funds and offer services to aid rural areas.
- Rapid population growth is not always a cause of urban problems; there are cities which have undergone rapid demographic expansion with limited ill-effects – rapid growth coupled with poor governance is likely to be problematic.
- While there are many large cities and most of these have appeared quite rapidly, much of the urban environment in developing countries consists of far smaller settlements.
- Mega-cities are not necessarily prone to problems; but, if these appear, they may demand considerable resources to cure.
- There is a huge diversity of urban environments: cities may vary a lot and within each city there are often marked differences, notably between rich and poor districts.
- Statistics are incomplete and inaccurate; consequently there is too much generalisation and false assumptions about problems. Problems vary from city to city and within a single city.
- Ways in which developed countries responded to rapid population growth and urban environmental problems in the past may not be appropriate for the future or for developing countries now.

The environmental manager's goals are to ensure that the cities give their inhabitants a sustainable healthy environment and employment, and that doing this does not unduly impact on other, extra-urban, areas. There are advantages in a city situation; in particular, the concentration of people gives economies of scale for provision of services; it is also easier to monitor and enforce environmental standards when activities are less scattered, and there are likely to be opportunities to dovetail activities (e.g. by-products can be passed to subsidiary industries).

Linkages between urban and rural areas are often well developed and complex, and must be unravelled and understood if sustainable urban development is a goal (Lynch, 2004). Various studies have applied adapted forms of rapid rural appraisal or participatory rural appraisal to urban areas to try and assess key issues and identify stakeholders (see rapid urban environmental assessment http://www.gdrc.org/uem/uem-rapidassess.html – accessed October 2005). City development, it has been suggested, should be 'de-linked' as much as possible from the rest of the world. This would involve encouraging large cities rather than smaller settlements, pursuing high-tech food production, and allowing remote areas to return to nature. Drastic changes like de-linking are unlikely, and continuation of established patterns of city management are what most environmental managers will probably work with. The message, which comes across from various sources, is that successful cities depend upon good governance.

Planners have tried to foresee and control city problems, with mixed results; few would hail Brasilia, constructed from scratch since the 1960s with its activities rigorously zoned into specialist districts and a street plan drawn up with the car and modern lifestyles in mind. Yet another Brazilian city, one comparatively 'unplanned' before the past few decades – Curitiba – has made progress in coping with its environmental and socio-economic problems in spite of limited funding (Rabinovich, 1992). Curitiba has recently developed green strategies and environmentally sound public transport strategies which rich cities such as London could learn much from.

In general, the amount of pollution in cities in developing countries is less on a per capita basis than it is in most developed countries' cities. For example, Dhaka (Bangladesh) in 1992 emitted about twenty-five times less carbon per capita than Los Angeles or Chicago (Hardoy et al., 1992: 187). Poor people consume less and much of what they discard is organic material, which can be composted; in general their refuse is less toxic than in rich countries (this is starting to change). The poor also use far less heating and air-conditioning, although they often depend on woodfuel and charcoal, which may cause local air pollution problems when burned. Woodfuel and charcoal supply usually causes serious vegetation damage and soil degradation for hundreds of kilometres around cities – where gangs of collectors strip the countryside. Where dung is collected soils are denied the return of organic matter and are likely to decline in fertility and degrade. The solution to deforestation and soil damage is to plant woodlots or encourage a shift in energy consumption to alternative fuels. Making a change may be very difficult for poor people.

Worldwide, urban areas modify run-off markedly, making it more erratic, usually polluted, with more extreme flows, and with the discharge channelled. Cities pollute the air flowing past, and the albedo of the built-up area may cause a 'heat island effect', causing warmer cool season and hotter warm season temperatures within and around them (Douglas, 1983; Harris, 1990; McGranaham, 1991). The supply of inputs for cities – water, energy and food – affects a considerable area, and the impact can be measured – an ecological footprint or eco-footprint – and may even be felt far overseas (Rees, 1992).

Until quite recently, illegal settlement of urban and peri-urban land usually prompted authorities to respond with force, evicting people, bulldozing shanty settlements and harassing the squatters to drive them elsewhere. However, since the 1980s the sheer numbers involved, plus the growth of civil rights and media interest, have reduced such treatment. Nowadays, city authorities cannot ignore their slums and squatter settlements. Growing numbers of authorities have started to try to provide improved water supplies, housing, waste disposal and public transport. In urban environments poor newcomers are less likely to have established effective coping strategies than poor rural folk and are often isolated from families and communities, so have nobody to fall back upon in times of hardship. Longer established squatter settlements usually make some effort as communities to improve their services, standards of housing, and to reduce disaster vulnerability (Maskrey, 1989, Aldrich and Sandhu, 1995). Improvements are often hindered by the largely unplanned urban growth of such slums. Illegal settlers are often a large proportion of total city populations in developing countries; for example, in Recife (Brazil) they account for over half and they can be vociferous.

Good urban environmental management adopts a forward-looking approach: (1) developing adequate services; (2) passing and enforcing appropriate legislation; (3) implementing protective measures. There is also a need for city administrators and planners to look at real needs, rather than adopting approaches suited to rich communities, or what was appropriate twenty or fifty years ago. Quite often cities were established for reasons that are no longer valid; what made sense for colonial administrators or with respect to past rail or river transport may no longer hold. Building standards may be derived from Europe or the USA and are inappropriate. Environmental improvements are ideally part of urban rethinking which finds new revenue and generates jobs, and wherever possible recovers costs.

Developing country cities usually have storm drains sewers, generally open channels, which easily block and overflow. Ill-drained areas become contaminated, and support the breeding of mosquitoes and other unwanted pest organisms, wells become contaminated and dwellings damp. There is frequently overcrowding, poor housing, lack of street lighting, unsurfaced paths, uncleared refuse, and the use of dangerous cooking and lighting methods – particularly kerosene stoves – in flammable shacks. All this, plus poverty, combines to damage human well-being through:

- diarrhoeal diseases – a major killer, especially of children;
- malaria and other mosquito-transmitted diseases: notably dengue, hepatitis-A, Nile fever, yellow fever, various forms of encephalitis;
- diseases spread in crowded, poor-quality housing, and made more likely by poverty: TB, HIV/AIDS, parasitic worms;
- debilitation by parasites;
- respiratory problems due to damp, crowding, and air pollution from fuelwood and charcoal;
- unusually high rates of accidents – burns, scaldings, falls, which result from poor housing, alcoholism, and unsurfaced and unlit tracks.

Squatter settlements commonly make use of steep or swampy land which is rejected by others. In sloping areas landslides beneath or on to housing are a risk whenever rainfall is intense. Steep land is also prone to gullying, and typically has dangerous pathways. Squatter settlements on floodland are regularly inundated with sewage-contaminated water. The children of the poor seldom have safe play areas, something that is frequently overlooked by planners.

Urban environmental problems may be subdivided into those associated with the home, the workplace and the neighbourhood. For poor people the home is often also the workplace, which means exposure of the whole family to noise, fire risk and harmful materials. The employment of child labour is common, and children are especially vulnerable to industrial accidents and pollutants. Scattered small workshops and householders undertaking contract work are difficult to supervise and protect, and accidents can affect surrounding housing (Figure 15.1). Innovations in larger factories are more likely to be monitored than those adopted by scattered small workshops, and health and safety and environmental education are difficult to deliver to scattered workshops. Very few, if any, workers in those workshops have any medical or injury insurance.

The problems of poor areas may not be contained there and can spill out to affect richer areas, including overseas to developed countries – dangerous diseases, such as multi-drug-resistant TB, dengue and Nile fever have been spreading well beyond slum areas in developing countries and to developed countries.

Figure 15.1 Typical Southeast Asian shop-house/workshop. This family (children included) make furnishings such as lampshades. Note child's cot, TV, and many drums of adhesive, paints and solvents – a particular threat to children (Hue, Vietnam).

Urban sprawl

Buildings and infrastructure expansion are destroying biodiversity and soil worldwide. The loss significantly counteracts any gains in food production made through agricultural modernisation; often it is the best and most accessible agricultural land which is built upon. Human disturbance and pollution associated with urban growth impacts upon large areas. Urban growth removes soil and vegetation 'sinks' for carbon dioxide, and so adds to the problem of possible global environmental change. In developed countries suburban gardens and lawns may help compensate for lost carbon sinks, but in poorer and more arid countries this is less likely unless urban and peri-urban horticulture is encouraged.

Urban areas in poorer countries are not adequately mapped at present, making it difficult to estimate the extent of the direct impacts of building and problems such as troposphere ozone, smog and acid deposition (Anathaswamy, 2002).

Urban problem areas

Within each urban environment there are usually problem areas – generally the districts where the poor have settled, and the areas affected by industrial activities. Other problem areas are those with soils which make buildings vulnerable to subsidence or earthquake, localities at risk of flooding or landslides, and where polluted air accumulates.

At the start of the twenty-first century, 'poverty is suffered by a minority of urban dwellers in richer nations and by the majority in poorer nations' (Hardoy *et al.*, 1992: 195). The people who become the poor of cities often have countryside skills which are of little use. Those suffering hardships in urban areas may be more vulnerable than rural folk; theirs is a cash economy, and limited opportunities to forage, and without enough money, survival is a struggle. Rural poor have famine crops and fallback strategies which are unavailable to city folk. This may be partially compensated for by the relative ease of access that aid agencies have to cities, and because governments may help urban people who are the most likely to give them their votes and pose a closer threat if neglected.

Pollution and waste associated with urban growth

Cities are among the most altered, unhealthy and contaminated of the Earth's environments. People generate waste, which may be defined as movable material that is perceived, perhaps wrongly, to be of no further value; once discarded it can be a nuisance or a hazard (pollutant), but may turn out to be a valuable resource. There is a huge diversity of types of waste; some may break down with time, or react with the environment or other waste to release pollutants. For management purposes it is possible to subdivide waste into toxic and non-toxic; or flammable and non-flammable; or compostable and non-compostable; or solid and liquid; or useful and problematic. Cities often rely on wells or are supplied through low-pressure pipes that may become contaminated from the soil; leaking drains, chemical and oil escapes. Managing waste is therefore crucial.

Wherever possible, waste and pollution disposal should be integrated with employment generation, food and commodity production and appropriate industrial development. There are examples of huge expenditure on waste disposal that are costly to maintain and often shift problems to a new (end-of-pipe) location. For example, the Cairo Wastewater Scheme was funded by the World Bank and UK DFID in the early 1980s to provide improved sewerage, which would not become choked with refuse. Unfortunately, while the system worked well, it discharged into the Nile with little or no effluent treatment because the disposal was not included in the scheme's terms of reference.

Most developing countries have serious traffic pollution; in a few nations where vehicle ownership is limited, or where technological 'leap-frogging' has led to the construction of improved public transport, there may be fewer difficulties. In many countries public transport is poor and car numbers are growing fast; in many countries the standard of motor vehicle maintenance is poor; so city authorities struggle to find effective pollution controls. Not all countries insist on fuel being lead-free and demand catalysers; the latter are also expensive to maintain once cars age a little, so levels of lead, VOCs, troposphere ozone and other pollutants present serious and growing risks. Even where there are impressive metro or modern light railway systems (e.g. Mexico City, Singapore, Bangkok, Shanghai and Kuala Lumpur), there are still difficulties dealing with masses of private motorists, taxis and buses. Taxi and bus companies often have to manage on shoe-string budgets because customers cannot pay higher fares.

Sustainable urban development

Sustainable urban development and sustainable urban design are attracting a lot of attention nowadays, but there is more rhetoric than actual success (Button and Pearce, 1989;

Stren and White, 1992; White and Whitney, 1992; Satterthwaite, 1999). One problem is how to appraise the sustainability of a city, policy or programme (for a review and coverage of scenario accounting approaches see Ravetz, 2000b). Some obvious measures can be taken: recycle as much water as possible; plant trees and shrubs along streets; reduce waste of electricity; discourage pollution; use trash/refuse and sewage productively; establish energy-efficient public transport systems; collect rainwater; and generate solar power.

The pursuit of sustainable urban development may be broken down into: seeking urban policies and plans that support sustainable development; striving to reduce the impact of a city on the environment; monitoring progress and sustaining the function of the city. The challenge is to be met in rich and poor countries, and both should be able to develop appropriate solutions (Zetter and Watson, 2002). The 1992 Rio 'Earth Summit' encouraged 6,400 local authorities in 113 countries to produce Local Environmental Action Plans (*Local Agenda 21*s). These were intended to integrate environmental objectives into development plans, emphasising participation, accountability and sustainability, but progress since Rio has been limited. Some cities have implemented something slightly different – City Development Strategies which seek broadly what the Environmental Action Plans proposed, but are more pro-poor. Hopefully, these will stimulate a shift to genuine action (Singh, 2001).

Ways of addressing individual urban problems are already available – these will be components in the sustainable urban development strategies which planners and environmental managers are developing. It will be important to select the right components for given situations and to find ways of implementing strategies, paying for them, and ensuring they continue to function. Japan is one of the most urbanised countries and has begun to make some progress with sustainable urban development (Tamagawa, 2006). Sustainable strategies must also be adaptable. There does seem to be a growing awareness of the need for cities to adapt to global environmental change, especially those sited close to sea-level (Singh, 2001).

Urban sustainable development has not been achieved if the wider environment is damaged achieving it.

Improved urban water supplies

Contaminated water is a major cause of ill-health in many cities. Overcoming the problem depends on: finding adequate supplies; treating and conveying them safely to consumers; and carefully disposing of waste water. There are a number of cities which have to rely partly or wholly on seawater desalination. The private sector has taken over many city water supplies and even in poor settlements may supply much of what is used.

Sometimes this is simply sale from tanker vehicles, but adequate piped water supply systems and sewerage improvements for some poor people may be paid for by business, rather than by city authorities or foreign aid (Cairncross and Feachem, 1983: 89; Holland, 2005). Another route is community self-help schemes (more often for sewage disposal than for drinking-water supply); typically an NGO or city authority provides pipes and other hardware, and local people put in the labour to install and run them. Waste water is seldom recovered for reuse, and few urban areas collect rainwater (exceptions being townships on Bermuda and Gibraltar). If storm water is kept separate from that polluted with sewage or by industry it can be applied to amenity vegetation and treated more easily. However, the usual practice is to allow storm water and other waste to mix; this also means that when there is a heavy rainstorm the effluent treatment plants can be overwhelmed and release untreated effluent.

Improved waste disposal

In developed country cities waste collection is generally adequately managed for both sewage and refuse; the problem is treatment and disposal, which may be far from environmentally sound (and in some cases cannot be sustained in current forms). In many urban areas of developing countries excreta and refuse are left on vacant ground or at street sides – a most unsatisfactory situation. If there is any organised removal it is by cart or tractor and trailer; and removal seldom means safe and environmentally sound disposal. Some settlements do convey refuse and excreta ('nightsoil') to surrounding agricultural land. Safer options are composting, biogas production, or incineration serving district electricity and heat schemes. When well managed these can be affordable, relatively safe, and less damaging to the environment (Mara and Cairncross, 1990). Each city district generates different waste types and quantities and poses its own collection challenges; the solution must be tuned to these and somehow funded. Recycling materials such as metals or glass and/or sale of compost may help pay, but also environmental protection and public healthcare demand some expenditure.

Refuse and sewage may be contaminated with heavy metals, pharmaceuticals and other troublesome chemicals, limiting opportunities for use as compost or irrigation. Poor districts of developing country cities are also likely to discharge sewage contaminated by small-scale industrial activity – especially oil and solvents. Managing such scattered and insidious discharges is difficult. Basic treatment of sewage to reduce the content of pathogenic organisms is not too challenging (simple chlorination or oxidation in a lagoon for a while may be enough), but removal of solvents, heavy metals, pesticides and other chemicals with available technology can be too costly for even rich countries. Biogas production, using sewage and refuse or farm wastes, has been explored in Sweden and Denmark. In Sweden there has been some development of biogas-fuelled trains, buses and cars – but this has been based on considerable support from citizen taxation.

City sewerage (i.e. the infrastructure) facilities worldwide are generally Western-style water-based systems. Often such systems offer only rudimentary treatment, and streams, lakes, seas and groundwaters become contaminated. There are two problems: first, removing waste effectively (a pressing challenge in most poor areas); second, adequate disposal – a challenge for many cities both rich and poor. Increasingly, cities have been extending low-cost sewerage systems (sewage removal) to poor areas, literally to protect public health. A popular approach has been for governments and NGOs to support the installation of cheap pour-flush (bucket-flush) sewerage systems. By supplying moulds for making the toilet slabs/floor area and subsidising piping, they can keep costs down. However, slums spring up unconstrained by any planning, there are no regular roadways and housing is scattered up hill and down dale – consequently it can be very difficult and costly to provide services such as piped water, mains electricity and sewers.

Where sewers choke with sediment it is possible to install new ones, or re-line existing ones with an egg-shaped, narrow at the base, or stepped cross-section – to ensure that even when flows are weak the debris is swept along better than with a normal-shaped conduit, reducing the risk of blockages and standing water that might seep into the ground. Inadequate sewage disposal is a key factor in determining well-being in developing countries, and calls have been made for huge sums of aid to improve sanitation and to protect drinking-water supplies.

Provision of toilets and sewerage (pipes) is less of a problem than the safe disposal of sewage (effluent); many systems simply discharge into streams or the sea or are connected to septic tanks. While this can be satisfactory for smaller settlements and in certain geological conditions, and does give less opportunity for flies and vermin to breed and come into contact with people, it frequently results in severe groundwater and surface-water contamination. In addition, water-based sewerage systems are easily overwhelmed during heavy rainstorms, especially where there is poor provision for surface-water drainage. If possible, sewage-contaminated water should be kept separate from 'grey' water – washing water and storm run-off – because the latter is easier to process and can be allowed to escape when there is heavy rain. Where there are many settlements along a river and if sewage reaches shallow and confined seas or lakes, pollution is likely to be bad. The problem is that those suffering sewage pollution are often some distance from the population causing it – urban–rural linkages are often ignored.

There are alternatives to water-based sewerage. The problem is making available effective, affordable systems that will be used. People are familiar with flush-toilet, water-based systems; even in areas without sewerage they see them on the TV. More companies are currently geared up to manufacture, sell and install water-based sewerage systems. Water-less systems have to overcome some consumer resistance and their development has been relatively neglected. Water shortages and effluent treatment costs may help establish such systems in the future.

Urban and peri-urban agriculture

A field likely to develop in coming years is urban and peri-urban agriculture (the former within, and the latter within and around cities); this could employ and feed the poor, and possibly use refuse and waste water (Mougeot, 2005). This could help poor cities with growing squatter settlement unemployment and poverty – rural migrants are often unable to find adequate employment in the city and know how to practise horticulture. Richer urban areas can use effluent for amenity area irrigation. There are promising urban agriculture-waste/sewage disposal strategies; one has been developed near Mexico City. Since the 1980s a Mexican alternative technology development group has been researching the *chinampas* farming system used by the Aztecs before the sixteenth century. This consists of narrow raised fields constantly mulched with mud from narrow canals surrounding them. The canals receive and effectively treat sewage, and the resulting slurry is used to sustain highly productive agriculture. One discovery made during *chinampas* experiments has been the identification of a particularly effective bacterium which promises to be valuable in other sewage treatment systems (Ayres, 2004). If authorities do not effectively collect, treat and control waste, the risk is that it will be spontaneously used by farmers, and that could lead to environmental damage and health problems that are difficult to monitor and manage.

A number of developing country cities have begun to aid the poor to sort and recycle garbage (Perera and Amin, 1996). For example, concrete sorting platforms can greatly help; most important are to try and improve basic health and safety facilities by providing washing facilities, supplies of boots, gloves, tetanus injections and health treatment. One of the most basic improvements is to organise regular refuse collection and removal to a site away from habitation. This does not need sophisticated garbage trucks: tractors and trailers will often suffice (UNCHS, 1988). In Cairo, for areas with narrow streets the authorities supported traditional waste collection with improved donkey carts, and it proved highly effective.

In more affluent cities householders are often asked to sort refuse for recycling, or it is compressed and landfilled or incinerated. In poorer countries garbage is often sorted by the informal sector (opportunist poor) and much can be done to assist them and make it less hazardous. There have been suggestions that the export of waste for sorting in developing countries would improve employment and enable effective recycling; however, there are ethical issues involved – notably the export of hazard and of demeaning employment.

If waste and sewage are contaminated with chemicals, heavy metals and pathogens it is possible to irrigate non-food crops with it, provided that care is exercised. Farms and aquaculture may use sewage and food waste to feed poultry, fish, pigs and so on. In the early 1990s Shanghai (China) was supplied with over 80 per cent of its vegetables from urban and peri-urban farmers using nightsoil (Hardoy *et al.*, 1992: 139). Around Calcutta sewage is added to fish-ponds to help feed the fish and to grow water hyacinth as fodder for other livestock; similar activities take place close to many other developing country cities. It is better for authorities to assist and monitor such activities than pretend they do not happen. Better to pay to chlorinate sewage outflows, than have it applied untreated to vegetables sold in city markets.

Urban and peri-urban agriculture could improve food supplies, help dispose of wastes, and generate livelihoods for people otherwise unlikely to find useful employment (Honghai Deng, 1992). A sustainable strategy linking waste and agriculture must be monitored by people with a broad knowledge of pollution, healthcare and so forth. Activities such as chicken or pig rearing near cities can pose serious threats even if waste recycling is not involved – avian flu and SARS seem to originate from contact between fowl, livestock, insects and humans in such conditions. Urban environmental managers must keep a close eye on urban and peri-urban agricultural activities.

Urban rehabilitation and improvement

Degraded areas of cities which are ripe for redevelopment have come to be termed 'brownfield' areas in the developed West, and are often former industrial sites, gasworks or filling (gasoline) stations. Redeveloping such land may mean very expensive decontamination work to protect against lead, asbestos and other contaminants. Aid agencies often finance construction but offer little recurrent funding; consequently, a problem for urban environmental management is fundraising for ongoing repair and modifications. The difficulty is pronounced in poor areas because the locals are unlikely to be able to pay much tax. Community-based approaches are currently seen to have promise, which means environmental managers must be familiar with social issues and with participatory approaches (Hasan and Khan, 1999). Slum upgrading in developing countries is now often conducted by giving local people the materials for housing and road improvements, and providing pipes and skilled labour for sewerage.

Squatter housing (slums, shanties) offers little proof against vermin, damp, cold or severe storms, and is vulnerable to fire. Flimsily built slums are unlikely to kill many by collapse during earthquakes, but they suffer from fire if the quake happens when people are cooking; however, where people use heavier mud construction, collapse may pose a danger to inhabitants. Where funds are available, governments can make considerable improvements to people's well-being by supplying materials for housing improvement: building blocks and roofing sheets can cut fire risk and damp; cheap doors and windows can help curb malaria transmission and crime.

Urban transport improvements

Developing cities rely a great deal on bus and taxi transport; and partly thanks to leap-frogging some have excellent systems. Many cities, however, lack adequate public

Figure 15.2 Cycles, rickshaws and motorcycles still outnumber cars in some Vietnamese towns and cities (Hanoi)

transport and tend to have few pollution control measures, so traffic pollution is a problem. In South Asia and parts of Southeast Asia motorcycles and three-wheeled two-stroke auto-taxis are common (Figure 15.2). The latter offer cheap and effective transport and replace human-powered 'cyclos'/rickshaws, consume little fuel, and cause less congestion than full-size taxis. However, they are noisy and seldom equipped with pollution control devices. Some cities have adopted rapid transit systems: state-of-the-art trams, metro or rail systems.

Without effective transport there is the risk that workers will crowd around manufacturing areas, which can mean safety risks and inner-city districts may develop tenements. Charging and quota access initiatives have had limited success in excluding private motor cars.

Improving urban energy supplies

Access to electricity for any but the rich is patchy in the developing world. Some of the poor illegally connect to supply networks; many make do without, using fuelwood, charcoal, kerosene or bottled gas. Even if electricity is available, cooking is likely to be with the latter fuels in all but the most affluent cities. Accidents are common and the fuels cost a fair proportion of poor people's incomes. With urban smog and often overcast conditions cooking with solar power using parabolic reflectors is unlikely to be practical for many households.

Household photovoltaic (PV) supplies are small solar arrays providing limited amounts of electricity to households, schools or community services. These PV systems have been falling in cost, and are now being adopted by the 'more affluent' poor as a means of powering a TV and possibly some limited lighting – which allows extended hours of working and evening study for children, but will not power cooking stoves. PV is used in poor townships by local cafés or bars to offer TV and telecom. Local authorities and private companies are also starting to install PV public telephones. Cheaper robust systems could gain even wider use. In richer cities households may generate enough for household use, and even sell to the grid, using larger PV arrays or household heating and electricity generation units (gas or diesel). So far such developments have been largely experimental. Cities in both low and high latitudes could make much more use of improved building design to trap solar radiation for heating or to reduce the need for air-conditioning. As PV and TV spread there are opportunities to reach more people via the media; a priority should be to educate and mobilise local action to improve environmental management (Abbott, 1996; Toteng, 2001).

The poor frequently shift to kerosene, because bottled gas is less readily available and more costly. Another alternative to fuelwood or charcoal is biogas; however, this is difficult to produce for small urban households. It is more feasible at a neighbourhood scale, especially if linked to excreta and refuse treatment. Unfortunately, not all settlements will have suitable waste. So far such biogas adoption has been mainly in China and India.

For decades a growing problem in many rural and urban areas has been the supply and use of fuelwood or charcoal (Leach and Mearns, 1989). The ideal is to dovetail sewage and refuse disposal with energy supply using biogas or district generation furnaces. Sustainable fuelwood plantations can also help with fuel supply, and this could be low net carbon-emission energy. However, such plantations are often eucalyptus stands on what was once common land, which means people get relocated and such trees can generate a range of unwanted environmental and economic impacts. Sustainable plantations of fast-growing plants – willows, reeds or aquatic plants – possibly fertilised with sewage or composted refuse/sewage, can be converted to biogas, charcoal briquettes or woodchips to burn in district generation stations. The selection of land for such plantations must be done with care. Aquatic bio-fuel might be produced from lagoons or convoluted channel systems, which also serve to trap, silt and extract excess nutrients from contaminated streams or to treat city sewage. Such alternative technology sewage treatment systems promise to be quite sustainable and yield useful products (fuel, charcoal, thatch and raw material for chipboard or paper).

Richer districts of developing country cities and towns are likely to increase energy consumption and per capita refuse production in coming years. Control through taxation is problematic, because if energy or other basic commodities are targeted it hits the poor more because buying such commodities takes a greater portion of their income. The rich spend a lower proportion of their income on these commodities and may be more able to evade tax.

Urban–rural linkages

Cities do not exist in isolation; they exert influence, often out to great distances (eco-footprints and socio-economic influence). Urban wage, usually being higher than rural, can cause rural–urban migration, which robs the countryside of labour and may trigger practices that lead to land degradation. Sometimes it is the lure of better services and excitement which attracts rural folk into towns. Rural–urban migration leads to squatter settlements and slums and pressure on urban services; in addition to drawing labour from the countryside it can also remove those willing and able to innovate. Consequently, the affected countryside may resist change more than it would have done otherwise. Urban markets affect economics; governments often hold grain prices down for the city poor and this tends to depress rural agricultural produce prices. Importation of cheap foreign grain to feed city people (a shift to bread as a staple in some countries is encouraging this) can have a similar effect. Low prices for agricultural produce cause farmers to abandon land, or neglect soil and water conservation through lack of income. The result can be environmental degradation and falling national food production, leading to more dependence on a few world food surplus areas and so increased vulnerability.

Cities generally demand large amounts of water, and may compete for it with rural users – making irrigation problematic, and there is also a demand for electricity, which is often met by hydroelectric dam construction. Dams can have severe environmental and socio-economic impacts and may be sited far from the cities they serve (McGregor *et al.*, 2005).

Cities and global environmental change

Environmental change can rapidly disrupt food and water supplies to cities, and in the past there have been catastrophic results (McCulloch, 1990). In the next few decades the greatest impacts on urban areas may well be through rising sea-levels. Many cities are sited on coasts and are vulnerable; for example, Alexandria (Egypt), Dacca (Bangladesh), New Orleans (USA), Venice (Italy). In addition to flooding and storm damage, rising sea-levels cause penetration of salt-water into estuaries and the contamination and displacement of fresh groundwater. Already some cities have installed coastal protection bunds and floodgates on rivers; these may not be an adequate long-term solution, probably do not justify the huge cost and can give citizens a false sense of security. The alternative is for planners to discourage vulnerable land use in areas likely to flood, and to provide strong flood and storm refuges and early warning as a civil defence measure.

Global climate change is likely to impact upon urban water supply; the result may be shortages or more flood risk, and altered susceptibility to landslides. The changes may not necessarily be local – a number of cities depend on rivers fed by distant snow-fields, and warming may reduce snowcover. Storm drainage and water supply systems are likely to be built to tight budgets; and once constructed changes are difficult to make and there is little flexibility for adaptation. The troubles of New Orleans in 2005 emphasise that storm occurrence patterns and severity might alter. Many cities are now so huge that a disaster would affect large numbers of people and make aid, evacuation and recovery difficult (as was shown by the 2005 New Orleans storms). The disruption of manufacturing, loss of services, and demand for aid can have national and global effects (Mitchell, 2005).

Environmental change may well have urban health impacts; but warnings are difficult to assess because many complex factors are often involved. Predictions are too often based on guesswork: for example, global warming is widely assumed to be likely to cause a spread and increased incidence of malaria. Yet, in Europe before the 1870s, malaria was common, although conditions were cooler than they are now, and were so back to the 1450s (the 'Little Ice Age'). Factors other than climate must explain the reduced transmission in Europe in the later nineteenth century.

Other likely impacts of global warming are shifts in energy use and consumption of goods as rich and poor countries implement economic controls and introduce new technology to honour agreements over carbon emissions and adapt to environmental change. Planners, administrators and the media are rather fixated on the expectation that there will be relatively gradual global warming. There is too little acceptance that such change could trigger sudden cooling in some regions and may not be as gradual as hoped. It is also important not to focus only on global warming and forget that a diversity of natural disasters can suddenly cause marked environmental alterations. Expenditure on global warming may be better spent on general adaptability measures, including more emergency supplies (especially of food) and so on. If anything, most cities have reduced civil defence since the 1970s.

Summary

- Sustainable development, whether in rural or urban environments, demands effective handling of outputs – waste and pollution – as well as inputs. Sustainable urban development is emerging as a goal but actual achievements are still limited.
- Water supply, refuse and sewage disposal are pressing urban problems. Until these are improved, human well-being suffers. In the coming decade a lot of attention will focus on these challenges. Energy, transport and housing difficulties are the next priority.
- Cities will have to cope with rising populations and the possibility of global environmental change.
- Urban environments, especially in developing countries, are very diverse. During the past decade urban populations have expanded to include more than half of all people. Much of that expansion has been in developing countries, where some cities are now huge. In both developed and developing countries, but especially the latter, environmental and socio-economic problems blight urban areas.
- Urban problems tend to result from rapid population growth, poverty and poor governance. City impacts are felt way beyond urban areas. Some developing countries' urban problems, such as diseases, could spread to affect developed countries; it is therefore in the interests of citizens of the North to aid improvements in the South.

Further reading

Drakakis-Smith, D.W. (1987) *The Third World City*. Methuen, London (2nd edn entitled *Third World Cities* (2000) Routledge, London).
Good introduction to developing country cities and towns and urban policy – Chapter 4 reviews environmental issues.

Environment and Urbanisation (October 1992).
Journal special issue on sustainable cities.

Hardoy, J.E., Mitlin, D. and Satterthwaite, D. (1992) *Environmental Problems in Third World Cities*. Earthscan, London.
Review of urban environmental problems in developing countries.

Inoguchi, T., Newman, E. and Paoletto, G. (eds) (2005) *Cities and the Environment: new approaches for eco-societies*. United Nations University Press, Tokyo.
Explores routes to urban eco-societies.

Wheeler, S.M. and Beatley, T. (eds) (2005) *The Sustainable Urban Development Reader*. Routledge, London.
Selected papers on sustainable urban development.

World Resources Institute (1997) *World Resources 1997: the urban environment.* World Resources Institute, Washington, DC.
  Coverage of urban environmental issues by a leading environmental NGO – available on the Web; see below.

www sites

Centre for Urban Development and Environmental Management (Leeds Metropolitan University, UK) http://www.leedsmet.ac.uk/as/cudem/ (accessed September 2005).

EU European Sustainable Cities and Towns Campaign –seeks to promote urban sustainable development http://www.ourworld.compuserv,com/homepages/European_Sustainable_Cities/introduct (accessed September 2005).

Global Development Research Center virtual library – environmental management http://www.gdrc.org/uem/how-tos.html (accessed January 2004).

Urban Environmental Management (Asia/Japan) http://www.iges.or.jp/en/ue/report3.html (accessed October 2005).

Urban Environmental Management Project (University of Waterloo, Ontario) http://www.fes.uwaterloo.ca/research/civics/uem/uem.htm (accessed September 2005).

Urban environmental management sources http://www.uneptie.org/media/review/vol23no1–2.vol23no1–2.html and http://www.gdrc.org/uem/documents.html (both accessed October 2005).

Urban environmental management toolkit (World Bank) http://web.worldbank.org/WBSITE/EXTERNAL/TOPICS/ETURBANDEVEL (accessed October 2005).

Urban Studies (Journal: Carfax Publishing, Abingdon) http://www.journalsonline.tandf.co.uk/app/home/issue.asp?wasp (accessed March 2004).

World Resources Institute (Downloadable form of reference listed above) http://www.wri.org/wri/wr-96–97focful.html (accessed January 2004).

THE FUTURE

PART

III

THE FUTURE

16 The way ahead

- Key challenges and new supports
- Looking at the future
- The 1992 UN Conference on Environment and Development, Rio de Janeiro, *Agenda 21*, and follow-up meetings
- Post-Cold War environmental management
- Politics and ethics to support environmental management
- Closing note
- Summary
- Further reading

> We have to understand that we can never control Nature. Nature's systems are far, far too complex. But we *can* manage them.
>
> (Charles Secrett, Director, Friends of the Earth: *The Times*, 28 April 1998: 8)

Some of the problems faced by environmental managers are reasonably clear: population increase, pollution, urbanisation, rising consumption (consumerism) and globalisation (Kiely and Marfleet, 1998). Greenhouse warming is widely accepted as a threat, while worsening soil degradation and loss of biodiversity attract less attention. The impacts of Hurricane Katrina (2005) in southern USA have helped elevate world oil prices and focus attention on the danger of unsustainable energy policies. During the past few years it has become apparent that Antarctic ice is melting at a worrying rate; it is also apparent that glaciers are in retreat worldwide, and the Southern Hemisphere stratospheric ozone hole is getting bigger. For some of these problems remedial action has been much too limited, and necessary future responses may have to be 'quick and dirty' with no time to wait for adequate data, better technology, change of public opinion and effective legislation. It is fair to say that environmental management is increasingly running on the spot, trying to keep up with growing environmental problems, and some fear that cumulative/synergistic problems may be emerging to make it even more difficult. North (1995: 105) warned of 'a blizzard of cliché and prejudice' surrounding environmental challenges. However, he was able to present a rational, often optimistic and readable assessment of the challenges being faced by environmental management. North felt that, though the world may have to support ten billion people within a couple of generations, it might be possible to do so and still care for the natural world.

Key challenges and new supports

Not all challenges faced by environmental management are about human survival and the conservation of biodiversity. Many concern aesthetics. For example, wind farms make sense as a means of supplying clean sustainable electricity, but in countries such as the UK siting them is proving controversial. Often environmental management will be invoked to give scientific respectability to government or public preferences, rather than to ensure sound research and rational choice (North, 1995: 119). Growing fears of a petroleum and natural gas shortage in the medium term have led UK Prime Minister Blair to urge voters to think carefully about the likely need to invest in nuclear power.

Environmental management has largely evolved in Western modern liberal democracies. It is starting to adapt to other situations and needs, and it is likely that conditions and outlook in the West will change a lot in coming decades. Non-Western nations such as China and India are growing fast and developing their technology and economies; Islam is a growing influence – environmental ethics, environmental law and environmental management must evolve to serve diverse and changing socio-political systems. Western secular consumerism and economic and technological dominance may not last.

There is a need to better integrate physical and social sciences, and to find a more problem-oriented problem-solving form of environmental science (De Groot (1992) has discussed such moves). Some of these issues are familiar to geographers, and human geography and environmental management can be usefully linked. Both environmental management and geography stress the importance of multidisciplinary or interdisciplinary approaches, but there are difficulties in seeking this – Marion (1996) warned of 'infoglut', the flood of data that has to be constantly sifted and made sense of. To be effective, environmental management must have mastery of 'infoglut', and effectively develop a clear overview of development scenarios and recognise what should and should not be supported.

Environmental management can draw upon palaeoecologists and historians to 'back-cast' (i.e. from an understanding of past events obtain warning of possible future challenges and of how environments and people might respond to various changes in the light of past responses). Business management has also developed future scenario prediction and assessment techniques (visioning) to aid in the identification of the best strategies. Diamond (1998, 2005) provides interesting insight into how human fortunes might be affected by environmental factors and past history; and Kennedy (1993) tried to produce an objective assessment of likely future scenarios using the approach of a historian. Environmental managers must have a broad and long-term view into the future, which brings them into contact with futures study and 'futurists' (Coates and Jarrett, 1989). These may often (if not always) be speculative, but they provide ideas, warnings, and prompt contingency planning (e.g. futures debates in the 1960s and 1970s helped prompt concern for limits and the concept of sustainable development).

Environmental management must deal with a diversity of stakeholders – ministries, NGOs, various groups among the public, international agencies and so on. That demands an ability to cope with complexity and conflicting demands. There has been progress in understanding and monitoring the world's structure and function, the development of environmental management standards and systems, and accessible and powerful computing systems, tools such as remote sensing, automatic instrumentation and GIS which permit much better data gathering, storage, retrieval and processing. The tools used for risk, hazard and impact assessment have also improved greatly since the 1970s. The spread of telecommunications, especially the Internet, makes contact between the environmentally concerned easier and cheaper, and should help prevent planners, governments or special-interest groups from hindering dissemination of information to the public, NGOs and various other bodies. The Internet has made it easier for people to blow the whistle on environmentally ill-advised activities, share information and promote environmental issues.

One of the key goals for environmental management is to achieve sustainable development. Although the concept of sustainable development is increasingly dominating environmental management, clear goals and tried-and-tested practical strategies are scarce (Carley and Christie, 1992; Barrow, 1995b). The literature is choked with clichés and wishful thinking about paradigm shifts to a sustainable society, and the need for eco-ethics, but little of this is of practical value for the near future. To improve environmental care and make a transition to sustainable development more likely will demand a change from current reliance on GNP and GDP as measures of national performance, to something more green (Henderson, 1994). *Agenda 21* signatories committed themselves to improving national accounting to include environmental costs, benefits and values. Environmental management approaches need to be developed and tuned for practical use: more effort could be spent on this. One such study, by Auty (1995: 262–271), made a comparison between strong green and weak green approaches to environmental management.

Environmental management makes decisions which affect future generations as well as the present generation. There is a need for better rules and ethics to guide environmental managers: what trade-offs between present and future are desirable. Cooper and Palmer (1992: 135–146) have examined the ethics of likely future environmental challenges.

Most environmentalists, numerous NGOs, growing numbers of politicians, and many citizens are now convinced that there is threatening global environmental change under way. In 2004 the Indian Ocean tsunami, and in 2005 Hurricane Katrina and the East Asian earthquake helped to raise public and government awareness of natural disasters. A few individuals, companies and economies are cashing in on these fears but mostly people show concern and will even pay a little to support controls. Adequate responses will demand greater payments and changed habits – until disasters start, a 'business-as-usual' scenario is likely to prevail. Faced by any global threat the likelihood is that some groups and countries will, whether consciously or unwittingly, wait for others to act and invest. Global warming is now attracting a great deal of concern, perhaps too much, with the effect of diverting attention and expenditure away from other threats (e.g. soil degradation, disease epidemics, and ocean pollution and over-fishing).

With respect to biodiversity conservation the responsibility of present-day environmental management is relatively clear. Living species are being lost at an alarming rate; once extinct, they are not recoverable and their value in ensuring environmental stability and providing many crucial benefits for humans is largely undetermined. Even excluding philosophical and moral beliefs that causing extinction is wrong, it makes sense to conserve biodiversity to keep open future options. Without the bark of one tree (for quinine) much of the settlement, trade and progress of the past few centuries would have been impossible; without access to yam species in the 1950s, modern oral contraceptives would probably not have been discovered and synthesised. Investment in conservation benefits everyone, but costs are mainly borne by one group or nation – that is the case with much environmental management: supporters and beneficiaries are different. E.O. Wilson (1992: 335) suggested that, regardless of a person's beliefs, 'the ethical imperative should therefore be, first of all, *prudence*. We should judge every scrap of biodiversity as priceless' (italics added).

In biodiversity protection and, I would argue, most other aspects of environmental management, prudence (the precautionary principle) should underpin all decisions.

Biodiversity conservation is not just about establishing and managing reserves, gene banks, zoological and botanical collections – it requires environmental management to ensure that there are no transboundary or global threats that endanger such collections, and to try to ensure that there are safeguards (duplicated collections located well apart, secure as possible sites and so on).

Biotechnology is of great value for developing new crops, pest control methods which reduce reliance on chemical pesticides, bioremediation of pollution, pharmaceuticals, biological nitrogen fixation and so on (Hector, 1996). The value of biotechnology to improve food and commodity production, to offer alternatives to agrochemicals (fertilisers, pesticides and herbicides), aid healthcare and treat pollution must be carefully weighed against risks. The main threats are the escape of genetically engineered organisms carrying recombinant DNA material to cause a serious environmental problem, or the use of biotechnology for commodity substitution. There have already been cases of substitution which have had severe economic and social impacts; for example, the adoption of high fructose corn syrup by the food and drinks industries in developed countries hit some developing country sugar producers badly (between 1983 and 1984 America cut sugar imports by US$130 million). When export markets collapse, farmers may be forced to produce other crops or to abandon land, both of which can cause serious environmental degradation. Biotechnology might make it possible for large companies to produce things such as cocoa butter substitutes or naturally decaffeinated coffee, or even alternatives to palm oil, which would severely upset countries that rely on these exports.

There is little disagreement that biodiversity is a world resource from which all should benefit, but in practice, seed companies, biotechnology companies and other commercial interests seek to profit and recoup their research and development costs. The call for patent rights by developers of biodiversity is growing as biotechnology develops (global free trade allows holders of rights to control huge markets for their products). In response, developing countries, indigenous peoples and NGOs have started to campaign for free access to 'raw material' for biotechnology (i.e. access to biodiversity), and some reward for and control over products developed from their indigenous biodiversity. Some governments have accused MNCs and TNCs and developed countries of 'bio-piracy', taking genetic material from poor nations, producing something from it with biotechnology, and then selling it back at a huge profit (Fowler and Mooney, 1990; Shiva, 1993).

Looking at the future

Futures studies received a boost in 1972 with the publication of *The Limits to Growth*, further impetus was delivered by the 1992 Rio 'Earth Summit', and again in 2000 when there was millennium speculation and stock-taking. The problem with futures studies is for users to separate sound speculation from unreliable (none will be anywhere near perfect). Sometimes futures assessments change what might otherwise have happened – without *The Limits to Growth* stirring debate, the realisation that there were threats may not have dawned for years; recently, the surviving authors have reviewed their warnings and advocacy for solutions and updated them (Meadows *et al.*, 2004). A number of bodies do regular stock-taking and issue futures predictions – repeated annual assessments at least show trends and refresh baseline data (UNEP, 2002; Worldwatch Institute, 2005, 2006). One of the key resources which clearly needs watching and, if possible, the adoption of a forward-looking approach, is food supply. Since 1990 it is generally agreed that agricultural production has slowed, while population growth is still running at around seventy-six million a year (Brown, 2005).

Predictions can be biased towards pessimism or optimism, but even if virtually un-biased, predictions are fraught with uncertainties. There may be more chance of reliable predictions when stock-taking, assessment of trends and prediction of future scenarios are done on a regional, national or local basis. Global futurology faces more challenges. Recently there have been a number of attempts to take stock and forecast trends in Africa (e.g. the African Environment Outlook initiative: http://www.unep.urg.dewa/ Africa/publications/AEO–1/2662.htm – accessed October 2005). The USA and its agencies (e.g. the CIA) have conducted similar assessments of American and global trends.

The UN Millennium Project drew up a set of goals, some social, some environmental, and some a combination. The Millennium Development Goals (MDGs) were published together with suggestions of how to pursue them (Cheru and Bradford, 2005). Clearly these will greatly influence development efforts and environmental management for decades (Sachs, 2005). It is likely that the MDGs are too ambitious and will be only partially achieved. While sustainable development is included to a fair extent, environmental management has a much more limited place. The worry is that sustainable development is being used as a sort of attractive 'mantra' without really having accepted how much it will depend on better environmental management. Social development features a lot in the MDGs, and care may be needed to ensure that environmental issues do not become sidelined.

The 1992 UN Conference on Environment and Development, Rio de Janeiro, *Agenda 21*, and follow-up meetings

One commentator felt that the main achievement of the Rio Earth Summit was to 'put the world's nose against the window' (i.e. it made environmental issues matters of serious interest for administrators, commerce and the public) (Thomas, 1994). The Earth Summit produced a programme of action, *Agenda 21*, which published goals and targets enunciated in the Earth Charter. It has had some influence on policy making in Europe, North America, many other countries, and a number of international agencies (Young, 1994; Henry, 1996; Voisey *et al.*, 1996). The Earth Summit and *Agenda 21* have been an important and effective catalyst for environmental management and sustainable development. What the 1992 Earth Summit agreed was a Convention on Climate Change; a Statement of Principles on Forests; a Biodiversity Treaty (which the USA would not sign, largely because it threatened biodiversity patent rights); *Agenda 21*; the establishment of a UN Commission on Sustainable Development (possibly the most important achievement); and a Declaration on Environment and Development (Holmberg *et al.*, 1993; F. Dodds, 1997). Since Rio, the UN General Assembly has held (in 1997) a Special Session in New York, dubbed 'Earth Summit II' (or 'Earth Summit + 5'). This was intended to take stock of what progress had been made in meeting the commitments made at Rio (Osborn and Bigg, 1998). Further summits addressed sustainable development (Johannesburg 2002) and followed up Rio at five-year intervals.

Post-Cold War environmental management

Not only has the Iron Curtain fallen, making it easier to exchange information and to co-operate on environmental care, but also the Western free enterprise capitalist system seems to be spreading (what impact future oil shortages might have remains to be seen).

To have much effect, environmental management will probably have to work with and manipulate commercial interests. With growing populations and limited resources, developing countries should be targeted for environmental aid to spread better environmental management (Colby, 1990; Erocal, 1991; UNDP, 1992). In many countries over the past decade there has been a trend towards privatisation, decentralisation and economic restructuring. The private sector has taken on much of what the public sector once did, so that private companies, non-profit-making bodies and NGOs are playing and will in future probably play greater roles in areas such as environmental management and resource management (Carney and Farrington, 1998).

Sovereignty is a problem: states have the right to exploit their own resources, but this can affect other nations. Nation states are here for the foreseeable future and ruling elites will continue to influence their political decisions and development policies. There also appears to be a shift towards supranational controls (Stott, 1995: 12). Current Western thinking on the global environment is dominated by a faith in the essentially compatible nature of humanity, rationality and enlightened self-interest. Since the late 1940s unity and co-operation between the nations of Europe have grown. The European Economic Community (EEC) evolved into the European Union (EU) in 1957, and is still gaining members and developing links with non-EU countries. The EU consumes a significant part of the world's resources and plays a strong and increasing role in shaping the modern global economy, influencing the world's environmental agreements and providing aid to poor countries. As the EU expands and becomes more integrated it offers environmental policy makers opportunities for wider co-operation and enforcement (for an overview of EU environmental policy and country eco-profiles see Hewett, 1995).

The Cold War may have ended (at least for the near future between Western nations and the former Soviet Union), but conflicts with serious environmental impacts continue – the Iraq–Kuwait conflict caused serious pollution through burning oilfields. Nuclear, chemical and biological armaments are spreading, there are poorly guarded stockpiles of atomic weapons, plus fears of economic stagnation or slump, and a growing underclass of poor disenfranchised people – the Cold War may have ended, but there are still abundant threats. In addition to ongoing global environmental change, India and China are industrialising, undergoing economic growth and have expanding populations, all of which could lead to much increased environmental damage and global competition for resources. The balance of power, and controls exercised by NATO, the USA and the UN are starting to change; poorer nations see (real or imagined) Western lifestyles and either aspire to achieve, migrate to seek them, or resent them. The world is not a safe or predictable place.

How can nation states be prevailed upon to adopt good environmental management? Stott (1994) examined these issues and expressed hope that international bodies could be instrumental, arguing that the UN had already established a Commission on Sustainable Development (in 1992), and that the Intergovernmental Panel on Climate Change (established in 1988) had brought scientific legitimacy to global warming predictions. The change might also be prompted by environmentally aware businesses. There has also been some progress since the 1994 (Cairo) International Conference on Population and Development in dealing with human population growth, plus some signs that there may be the start of a fall in birth rates in some countries. Unfortunately, even if there is establishment of relevant international bodies, and assuming most nations sign agreements, there remains the problem of enforcement – signing a treaty, convention or agreement is no guarantee that the signatory will abide by it or contribute to funding. The process of lobbying for a convention and then progressing to a workable treaty may take decades – inadequate in the face of sudden and rapid environmental and socio-economic change. Yet adaptation to challenges is possible: Cuba has weathered severe hardship since the collapse of its sugar-based economy and cheap Soviet oil imports with little outside help; currently it has an adequate lifestyle drawing on agriculture closer to sustainable development than most other nations (Castro, 1993). Some hope the market will drive environmental management; others are looking beyond that (Redclift, 1992). Perhaps novel and relatively painless new taxation measures could be developed by the UN to help pay for environmental management: levies on use of geosynchronous orbits for satellites, or eco-tax on all air travellers or golfers? Compared with expenditure on armaments at the height of the Cold War, or even at present, the cost of curing the most pressing environmental problems is relatively affordable (see Figure 16.1). Environmental problems, resource shortages and disasters could either

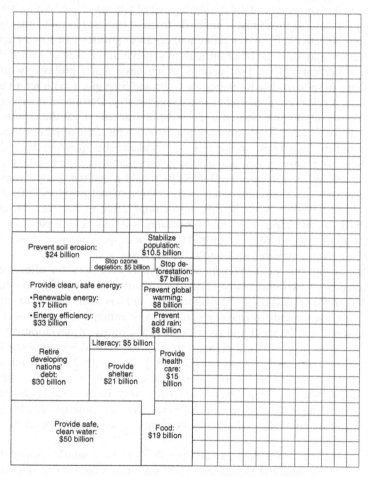

Total chart denotes total annual world military expenditure ($1 trillion); each square denotes 0.1% of annual world military expenditure ($1 billion)

Figure 16.1 What the world wants – and how to pay for it

Note: Figure shows annual costs of various global programmes. Each programme is estimated to be sufficient to accomplish its goal worldwide. The combined total cost of all these programmes is about 25 per cent of the world's total military expenditure in 1994 (in US$, US billion and trillion).

Source: Henderson (1994: 128, Fig. 2)

cause conflict or prompt closer international co-operation. There is a need to establish how the latter outcome can be encouraged. Perhaps developments such as better media and Internet availability will encourage global lobbying, referenda and ways of organising solutions.

Politics and ethics to support environmental management

Environmental management is a politicised process (Wilson and Bryant, 1997: 85). There has been considerable debate about the most supportive forms of politics for environmental management: is it better to seek centralised or decentralised, citizen-led or state-managed, liberal or authoritarian control? Goldsmith *et al.* (1992: 23) suggested there were two groups of future strategies (which may be of value for environmental management): (1) those that counter destructive trends; (2) those that help foster more positive objectives. However, more authoritarian politics can force change, which in a democracy may be delayed or difficult to achieve in the face of selfish, irrational, ignorant or indolent voters. Two nations which have made progress with environmental problems in spite of disadvantages are Cuba and Singapore, both relatively authoritarian regimes. A relatively poor and authoritarian country, Dominica, managed to enact effective conservation and anti-land degradation measures in the 1990s. The People's Republic of China recently announced a nation-wide shift of labour to reforestation and has toughened controls on logging – stimulated by growing land degradation and flooding. So, relatively non-democratic governments are able to take proactive environmental management measures. Democrats would argue that people must have the right to make mistakes and learn – however, environmental problems affect more than one nation and mistakes could be difficult or impossible to resolve. Current citizens hold the environment in trust for future generations – in business, trust funds usually demand a firm and impartial administrator. There is a slippery slope between achieving these ends in a democracy and shifting to 'eco-fascism'.

Many argue that democracy in some form is necessary for effective environmental management. Environmentalism and green politics developed in Western democracies and has so far tended not to be authoritarian and has been largely reactive to problems, whereas environmental management needs to be anticipatory. A democratic system may allow public involvement and some degree of scrutiny of development, but it may also slow decision making (for a discussion of ecology and democracy see the 1995 special issue of *Environmental Politics* 4(4): 1–321) (Figure 16.2). A portion of environmentalists and green politicians is to a large extent essentially anarchistic. Their decentralised, local-focus, often far from dynamic approach could be even slower coping with environmental challenges than present democracies. Worse, some environmentalists and green politicians are romantics who would be happy to see a rejection of many modern tools and strategies.

Popular concern for posterity and people in some countries is not strong and needs shaping. Environmental management will sometimes have to go beyond the will of the people, or continue in spite of loss of interest or fashion changes. How, then, can environmental management deal with popular self-interest, inertia or misguided hostility without resorting to authoritarian 'eco-fascist' powers? If a liberal democratic approach is favoured, should it be moral, popular or pragmatic in outlook? Appealing to a people's sense of decency or altruism is probably too much of a gamble, and anyway, liberalism tends to be anthropocentric. One way of countering the environmental inertia of democracy might be to adopt a World Charter on the Rights of Nature (efforts to do so at the 1992 Earth Summit failed).

Figure 16.2 Singapore, a city state which, in spite of a dense population and a challenging humid tropical environment, has made impressive progress with urban environmental management. In a number of fields the city is among world leaders, notably in efforts to control car traffic and provide adequate public transport. The approach adopted has been quite 'top-down'.

Already energy needs have led countries to conflict over resources and transboundary pollution. Environmental problems can lead to political conflicts and vice versa. To be anticipatory, environmental management will have to involve political analysts as well as ecologists, specialists in ethics and law, social scientists and economists. Science may be used by environmental management to reduce the polarisation and squabbling that can be generated during negotiations (Brenton, 1994).

There are situations which demand a large-scale approach to their solution; for example, investment in vital research, problem avoidance and mitigation may be beyond the level a single region or country could afford. Most sustainable development strategies will have to be tuned to local conditions, but need co-ordination at a higher level and possibly funding from national or international sources. EIA has moved towards a tiered approach, as have some environmental management systems (e.g. SEM). With such an approach, environmental management could be applied to local conditions – adjoining areas may have quite different approaches, and here and there (not necessarily adjacent) there may be similarities. Overall co-ordination, at regional, national and global levels, would look for conflicts, ideas that might be shared, and monitor resources (notably biodiversity, crop varieties and knowledge) that should be duplicated far enough apart for security, so that if one locality suffers a disaster there are possibilities for recovery. The overall pattern would be like mosaic tiles, a global picture with considerable local diversity, simplicity of organisation, and sometimes duplication of units at different locations to give security against loss of infrastructure, skills, biodiversity and so on in the event of a disaster.

Noting that environmental management is a multi-layered process, Wilson and Bryant (1997) suggested that international, state and non-state bodies could undertake it.

At present environmental management is under central, state control in some countries, while in others it is decentralised. There are also grass-roots environmental managers (e.g. peasants seeking to protect their forests), MNCs and TNCs which have global environmental policies, NGOs (often with a sectoral focus – e.g. active in protecting whales and dolphins), and individual activists/scholars (e.g. Vandana Shiva, Anil Agarwal, Ignacy Sachs). The future probably lies in ensuring that environmental management operates as a multidisciplinary multi-layered process dealing with human–environment interaction, as suggested by Wilson and Bryant (1997). In practice, some individuals and bodies are good at being watchdogs, others excel at campaigning, and others at checking and researching or developing strategies.

Economists and natural resource specialists are warning that the world has passed the 'oil peak' (i.e. affordable supplies of petroleum, on which modern lifestyle is very dependent, are probably in decline) and an oil crisis is developing. It is prudent to expect oil supplies to dwindle and become problematic within sixty years, possibly much less. Food production, energy supply, transportation, and much more must find alternatives and do so fast. So far that adaptation is not happening anywhere near fast enough. If the threat were real it would make sense to discourage current waste and even to introduce rationing. Hopefully, post Hurricane Katrina (which struck southern USA in 2005) oil price rises might help awaken concern. If a smooth transition to a post-petroleum future is not made in time, it is possible that there could be a decline to a new 'dark age' – much of the present supply of food depends on oil inputs (Figure 16.3). Since about 2000, numerous authors and research bodies have been warning of this threat, some linking it to other developments to predict a multi-faceted cumulative catastrophe

Figure 16.3 Mass rapid transport, Singapore's metro system which is helping reduce road traffic

within forty years or so (Kundstier, 2005). Petroleum companies are worried about rising costs of exploration and further significant finds have recently been limited. There have been recent scandals over oil executives exaggerating known reserves. The threat is taken seriously by the oil industry: one major company has adopted the slogan 'Beyond Petroleum'. If fears of an oil (or oil plus other problems) crisis are correct, societies and economies are going to have to change – it is unlikely that alternatives will simply be smoothly substituted in less than fifty years.

Manufacturing and service provision costs are low in India, China, and several other countries, which now have strong economic expansion. Developed countries will find it difficult to compete, and it is likely that there will be altered patterns of manufacturing and employment. This will impact upon environmental management and corporate social responsibility as companies in some countries cut back on costs and in the expanding economies where rapid expansion takes place and where, in the past, there has been limited environmental concern.

Closing note

In the 1950s few people expressed concern that there was an environmental problem. In the 1960s environmentalists began to publish 'messianic' warnings. In 1972 the UN Conference on the Human Environment (Stockholm), one of the first world gatherings on environmental issues, closed, with many of the delegates, especially those from developing countries, seeing environmental management as a luxury, and some even suspecting that environmental concern might be a new type of green imperialism. In 1972 only a handful of nations had environmental ministries, and generally media and the public showed limited interest. In 1992 the UN Conference on Environment and Development attracted a huge attendance, and there were few delegates willing to state publicly that environmental care was not a vital component of development. By 1992 virtually every country had an environmental ministry or agency; most newspapers and television channels had environmental correspondents; and the public were following events. Interest in sustainable development was well enough established by 2002 for a major international conference to be held by the UN. Global warming had become a real threat demanding responses by 2004, with some nations prepared to spend considerable sums responding to it.

There has been progress which few could have expected in the 1960s. The extensive achievement of sustainable development goals may look unlikely to current sceptics, but if a similar degree of progress takes place in the coming forty years then the prospects are reasonable. The rapid economic growth in China and India and the problem of declining oil reserves may prompt more rapid technological development and opportunities to design-in sustainable development.

Unthinking opposition to modern agriculture or commerce will not counter environmental challenges, may exacerbate some, and would probably cause widespread starvation. A doubly green revolution involves refocusing technology to avoid environmental damage, not the rejection of technology. That refocusing has to happen soon and ensure adaptation to oil shortage and dwindling supplies of phosphates, and to counter the worsening soil degradation and environmental pollution caused by agriculture. Competition for water supplies and possible climate change will make the task difficult.

There is a need for ways for the various layers involved in environmental management to communicate and arrive at decisions. Commerce and the appropriation and privatisation of resources cause some of the current environmental and socio-economic problems. Sometimes privatisation might be an improvement, but it should be monitored and, if necessary, controlled. So too should the process of globalisation. Hildyard (1992: 154) called for a 'Liberation Ecology' which would empower local people to counter pressures from MNCs and TNCs or governments and promote better environmental management and improved livelihoods. Practical, binding agreements and an adequate Earth Charter were not achieved at the 1992 Rio 'Earth Summit'. Nevertheless, there has been considerable progress in four decades – so there are grounds for optimism.

There is a need for environmental management to ensure that governments and research bodies do not neglect 'blue-sky topics' in favour of applied studies. There is also a need to address infrequent but potentially catastrophic risks, which state governments might argue they could not afford to waste funds upon. There was, for example, reluctance to fund the studies which in 1987 gave warning of a growing stratospheric ozone loss. In addition, funds for maintaining checks on atmospheric gas levels were hard to come by in the 1950s, yet without long-term monitoring carbon dioxide and methane changes would be difficult to understand and extrapolate. Luckily those areas of research found support.

Until recently the risk of comet or asteroid strike was given little serious thought. Then, in 1991, asteroid 1191B narrowly missed the Earth (and was big enough to have caused a catastrophe). Then there was the spectacular collision of fragments of comet Shoemaker Levy-9 with Jupiter (visible Earth-size explosions on impact), and the discovery of the 2-km-diameter asteroid 1997xf11, which it is hoped will narrowly miss the Earth in AD 2028. Such astronomical warnings, together with the debate about possible strikes on Earth in the past, show the need for measures to warn of and react to astronomical threats (Huggett, 1990; Ahrens and Thomas, 1992; Lewin, 1992; Chapman and Morrison, 1994; Steel, 1995; Gribbin and Gribbin, 1996). Some people feel that current technology could be reasonably easily developed to give some protection (for the first time in roughly 4,000 million years of the Earth's history). There are other slight but worrying threats: geologists suspect that massive outpourings of lava floods in the past may have had serious impacts on the atmosphere and climate, both nationally and regionally. The problem is that all these threats have not been taken seriously for decades, even when there have been sound data. Response is usually made after a disaster happens – only now is there expenditure on tsunami early-warning systems for the Indian Ocean (other oceans look likely to be missed). More has to be done to embed the precautionary outlook in the world's administration.

Space research at present seems to be purely academic to most of the public. However, it has yielded remote sensing and GIS, and is giving interesting insight into the Earth's processes, helping unravel issues such as global change, natural periodic climate change, vegetation changes, patterns of pollution, and much more. One day environmental managers as well as science fiction writers may have to worry about terraforming Mars, Europa or other planets (Robinson, 1992) (terraforming is the alteration of entire planetary conditions towards something more suited to human needs). Terraforming (i.e. interfering in the fate of worlds) prompts the question: What are environmental managers driven by? A referee who read the draft of this book commented that environmental managers could be in the position of selecting one of many possible alternative futures, and in so doing preventing other possibilities. The profession should bear in mind the old adage 'those whom the gods wish to cast down, they first inflate with pride'. The practice of environmental management – developing policies, collecting data, implementing developments, co-ordinating, trouble-shooting, or whatever – must all be guided by principles that are based on prudence and sound ethics and should be overseen by laws which serve rather than hinder environmental care.

Summary

- Some of the problems faced by environmental managers are reasonably clear: population increase, pollution, urbanisation, rising consumption (consumerism) and globalisation. However, many problems will be unexpected. Watchfulness, monitoring, and adaptive strategies are vital.
- Prudence and sound ethics should underpin environmental management.
- The future is not reliably predicted – unexpected challenges will have to be dealt with. Shortage of some key resources (notably oil) within a few decades and levels of environmental degradation seem to many to pose serious threats, threats which governments are not taking seriously enough and for which inadequate efforts are being made to adapt. There is too much sense of false security in current technology and governance.
- Environmental management is a political process – doing what is best for the environment and nature, possibly with little initial support and with a key element being the winning of votes/support.

Further reading

Adams, W.M. (1996) *Future Nature: a vision for conservation.* Earthscan, London.
Explores the way forward for conservation.

Brown, L.R. (2005) *Outgrowing the Earth: the food security challenge in an age of falling water tables and rising temperatures.* Earthscan, London.
Readable source on a critical resource.

Coates, J.F. and Jarrett, J. (1989) *What Futurists Believe.* Lomond Publications, Mt. Airy, MD, and The World Future Society, Bethesda, MD.
Widely used futures book.

Goudie, A. and Viles, H. (1997) *The Earth Transformed.* Blackwell, Oxford.
Environmental change overview.

UNEP (2002) *Global Environmental Outlook 3: past, present and future perspectives.* Earthscan, London.
Futures predictions by the UNEP.

Van Ginkel, H., Barrett, B., Court, J. and Velasquez, J. (eds) (2005) *Human Development and the Environment: challenges for the United Nations in the new millennium.* United Nations University Press, Tokyo.
Reviews major trends, problems and prospects.

Worldwatch Institute (2006) *State of the World 2006.* Earthscan, London.
An annual stock-taking and forecast.

www sites

Chronicle of the Future (scenarios to 2050) http://www.42explore2.com/future.htm (accessed October 2005).

International Journal of Futures Studies http://www.systems.org/HTML/fsj-room.htm (accessed October 2005).

World Futures Society http://www.wfss.org (accessed October 2005).

For information on futures http://www.ag.arizona.edu/futures/fut/datdf21.html (accessed September 2005).

Glossary

Adaptation – adjustments in ecological, social or economic systems in response to actual or expected (perceived) change.

Agenda 21 – a set of goals and proposals published following the 1992 Rio 'Earth Summit' which promotes sustainable development. All countries have been encouraged to adapt the proposals and incorporate them into governance and management.

Best available technique not entailing extra cost (BATNEEC).

Best possible environmental option (BPEO).

Biological oxygen demand (BOD) – water enriched with a pollutant(s) prompts excessive microbial activity (eutrophication) and algae growth, and becomes depleted of oxygen and enriched with ammonia, leading to extensive kills of higher aquatic life forms. A body of water can only handle so much pollution before BOD and/or *COD* (see below) overwhelms it. Once seriously damaged, the water body can render much less pollution, if any, 'safe'. Nitrogen in sewage and other organic pollutants and phosphates from fertilisers or detergents are usually responsible for BOD.

Bovine spongiform encephalopathy (BSE) – a cattle 'disease' (there is still some debate about causes) which can affect humans if they consume contaminated material, ultimately causing severe mental deterioration.

Cassandras – those catastrophist and pessimist environmentalists who preach environmental disaster through human mal-development. The opposite are the *cornucopians* (see below). A cruder shorthand for Cassandras is 'prophets of doom'.

Chemical oxygen demand (COD) – pollutants cause chemical reactions, which deplete water of oxygen and kill aquatic life.

Cold War – a period (about 1946 to 1986) of mutual distrust; power struggle, arms race and propaganda contest between the First and Second Worlds.

Commodities – non-food agricultural produce (e.g. cotton, coffee, tea, rubber, forest products (timber, etc.)).

Common resource – resource accessible to a number of users; over-exploitation prevented for centuries by disease, warfare and so forth, then population overcomes controls (e.g. better healthcare, law and order). The assumption has been that users maximise their exploitation leading to a crash. However, many argue that there is seldom 'open access' to common resources – tradition, taboos, rules prevent over-exploitation. Sometimes these controls may break down or be inadequate for modern demands. In recent years many common resources have been 'privatised' by commercial or state bodies, with little opposition because traditional users have no legal clout or documentation to resist. The dispossessed may move to degrade other marginal land or lose part of their subsistence.

Consumerism – hopes and demands for material possessions, typically luxury goods, which rapidly become 'obsolete' and are discarded. Part driven by fashion, advertising, and the globalisation of Western culture.

Cornucopian – applied to optimists who feel (the implication is uncritical) that an environmental crisis will be overcome without too much trouble. These optimists have also been called 'prophets of boom' (see: *Cassandras* above).

Corporate social responsibility – voluntary approach to environmental management by business (as opposed to command and control by government). Business takes on some of the responsibility previously held by governments.

Cultural ecology – study of the adaptation of human societies or populations to their environments, emphasising the management of technology, economy and social organisation, through which culture mediates.

Desertification – loss of vegetation/soil degradation tending towards drying conditions (desiccation). Degradation of plant cover and soil, likely to result in less precipitation infiltrating the soil, and possibly leading to regional climate change (due to changed albedo and more wind exposure as vegetation deteriorates). Today mostly attributed directly or indirectly to poor soil and water management rather than to natural causes.

Development – increasing the capacity to meet people's needs and improve the quality of human life. Conscious pursuit of better quality of life for people.

Drought – there are many definitions, none universal. The following is probably as good as any: a deficiency of precipitation from the expected norm which means there is too little moisture to sustain agriculture, vegetation cover, and often groundwater under existing demands. Drought is often seen to be caused by climatic fluctuations and to be of a temporary nature – that may not be the true situation. Drought is frequently (by no means the only) cause of hunger and famine.

Earth Summit – UN Conference on Environment and Development (UNCED) held at Rio de Janeiro in 1992. The second major UN environment and development conference (the first being at Stockholm in 1972).

Ecologism – deep-seated objections to the way the environment is affected by society. Adherents are not content with correcting a problem; they seek to examine why it was generated and how social change, new outlooks or ethics can avoid it.

Ecology – study of the relationships between living organisms and/or between living organisms and their environment. probably first used by the German naturalist Ernst Häckel in 1866 to describe the study of the 'web' of organisms and environment.

Ecotourism (eco-tourism) – environmentally sensitive tourism, which may or may not be based on nature, and which is characterised by investing significant amounts of the revenue it makes in environmental management.

El Niño–Southern Oscillation (ENSO) – a large-scale ocean climate phenomenon. Climatic conditions in the Pacific off eastern SE Asia alter ocean circulation, which in time weakens the upwelling of nutrient-rich cold deep water off South America. The impact upon climate and wildlife is marked. These El Niño events are recurrent but not quite predictable. The opposite of an El Niño (strongly 'normal' conditions off South America) is an El Niña. ENSO events have extended impact to a large part of the world for some time after starting to manifest off South America.

Empowerment – process by which individuals, typically including the poorest, are assisted to take more control over their lives so that they become agents of their own development.

Environmental determinism – concept that human fortunes are predominantly caused by physical conditions (and/or genetics) and not exercise of free will, learning and so forth. The crude and simplistic determinism of the 1870s to 1940s is now rejected; however, there has been renewed interest in more cautious neo-environmental determinism since the late 1980s. This suggests that nature does impact upon human development but does not determine the outcome.

Environment – the sum total of conditions within which organisms live; the result of interaction between non-living (abiotic) physical and chemical, and, where present, living (biotic) components.

Environmentalism – planned intervention to secure improvement in environmental quality. Huge diversity of types of environmentalism. While not a clear-cut division, one could separate 'green' activists from environmentalists, because the former are more politicised – both share a desire to protect and improve the environment.

Environmental possibilism – the idea that environment plays some role in human fortunes, but does not determine the outcome (i.e. it limits and prompts but does not control).

Ethics – standards of proper conduct.

European Union eco-management and audit system (EMAS) – not to be confused with the eco-management and audit regulation (EMA). Originally the intention was to make the scheme obligatory; however, lobbying led to its becoming a voluntary (but encouraged) measure – which came into force in 1995 (Council Regulation of the European Community – EEC No. 1836/93).

Expert system – a computer program which assembles the knowledge of experts in certain fields (and which can often learn and improve with use). It is used by relatively unskilled staff to make acceptable decisions, having first entered available information. For example, symptoms of a disease are fed in to prompt a diagnosis for a healthcare worker unable to consult a doctor.

Famine – not a precise term. Essentially a food shortage sufficient to cause disastrous levels of debilitation and death; fatalities are often through hunger weakening immunity to various diseases rather than starvation alone. Society and family behaviour is likely to break down (see *malnutrition*).

Gaia hypothesis – formulated by James Lovelock in 1979. This postulates that the biosphere operates as a single 'organism'. Complex checks and balances between the living and non-living maintain the Earth's environment (a homoeostatic system). Initially treated with scepticism, but there is now growing interest in the hypothesis. The implication for development, if it is correct, is that humans must work carefully with earth systems – 'fit in'.

Gender benders – slang term for compounds with an oestrogen-like effect (androgen disrupters). At low concentration PCBs (see below), dioxins, compounds used in plastics manufacture (including plasticiser compounds – especially phthalates), some pesticides, fungicides, and industrial wastes such as PCBs, sewage with contraceptive pill hormones and so on can result in a region's fish and reptiles becoming predominantly female or neuter. There are also fears that human sperm counts have been lowered (Lomborg (2001) offers a possible alternative behavioural cause, rather than pollution).

Global Environmental Facility – a fund set up in haste in 1991 by rich developed countries to support developing nations which undertake works that promise global environmental gains. Managed by the World Bank, it has been accused of secretiveness. In addition, it only funds activities promising global benefits. It is supposed to help developed countries deal with biodiversity, climate change, desertification and persistent pollutants.

Governance – the means by which policies are implemented and monitored through administration, policy making and the rule of law. In short: the manner of governing.

Greenwash – misinformation disseminated by an organisation so as to present an environmentally responsible public image.

Hazard assessment – this is often confused with risk assessment; to illustrate the difference: imagine a person wishing to cross the Atlantic – the hazard is drowning. They have a choice of making the journey by row-boat or ocean liner, each of which clearly presents different risks of drowning. Hazard assessment seeks to identify threats and their nature; risk assessment is also concerned with the probability, severity, frequency and timing of threats.

Heavy metals – often toxic at low concentrations, and may be concentrated by organisms near the base of the food web so that organisms near the top are seriously affected. These include: mercury, cadmium, zinc, lead, copper, chromium and silver. Domestic waste, human sewage and some livestock waste can be rich in heavy metals.

Industrial ecology – may be defined as a system that can support economic and social development by adjusting the flows of matter and energy (industrial metabolism) and abiding by natural limitations.

Laissez-faire – approach to development, trade and so on which advocates minimal interference by the state.

Limits to growth debate – many lines of evidence suggest that humans may be approaching crucial limits. Drawing upon systems modelling, the Club of Rome published a report which stimulated the 'limits to growth debate' from 1972 (see Meadows, *et al.*, 1972, 1992).

Little Ice Age – widely placed between about AD 1350 and 1850. A phase during which it appears that weather conditions (at least in the Northern Hemisphere) were cooler, wetter and more changeable than today.

Livelihood – the capabilities, assets (material and knowledge-based) and activities required for a means of living. What supports a person or family. To be sustainable a livelihood must cope with disruption and change, and not undermine the resources on which it is based.

Malnutrition – a deficient diet (the deficiency may be overall quantity, inadequate protein, lack of minerals or vitamins, or a diet somehow harmful) leading to poor development of children and illness and debilitation in adults. Some, especially in affluent societies, suffer through an excess of certain foods rather than an inability to afford a healthy diet. Malnutrition probably affects about two-thirds of all humans.

Malthusian – derived from the views of Thomas Malthus in the early nineteenth century. A viewpoint, widely held between the mid 1960s and late 1970s, which argued that human population growth would lead to environmental stress and ultimately disaster.

Multinational corporation (MNC) – a business which has subsidiaries and is active in a number of states.

Non-governmental organisation (NGO) – a body which is not controlled by a state. Mainly special-interest groups or charitable organisations. Some operate internationally and command enough support to exercise considerable power; others are smaller, often local 'grass roots' NGOs (e.g. Friends of the Earth).

'Not-in-my-backyard' (NIMBY) – often invoked when authorities propose to site a hazardous or nuisance facility, this attitude may not just be a local reaction. NGOs and activists oppose some proposals wherever they are to be sited (NIABY – not-in-anybody's-backyard). Some facilities are unavoidable; environmental managers must somehow find a place for sewage treatment, waste processing and so on; and in a populous area it is a challenge not to have it in someone's backyard.

Persistent organic pollutants (POPs) – some of the most dangerous compounds. May cause cancer, and nervous system, reproductive system and immune system damage even at very low concentrations. Because they are persistent organisms one can accumulate them, and they are difficult to remove from the environment. Some are derived from industrial activity; others are pesticides used in agriculture. POPs include dioxins, PCBs, furans, DDT and various organochlorines. The worst twelve of these POPs are the focus of the 204 Stockholm Convention on POPs. This Convention legally binds governments to cease production immediately as well as any release of existing stocks. At the time of writing over fifty countries had signed the Convention.

Political ecology – an approach to explaining human–environment interrelations which has in part evolved out of political economics. It explores how power relations affect environmental conditions.

Political economy – there is a diversity of political economy theories seeking to analyse political and economic structures and processes (e.g. Marxism). Environmental problems are entangled in a political-economic context.

Polychlorinated biphenyls (PCBs) – cause damage at very low concentrations and may persist without degrading for a long time. Already there is widespread contamination – and they are a transboundary as well as a local or regional problem. Treatment of PCBs before release into the environment is difficult and costly. Monitoring is important, and it is in the interests of all nations to support this. The levels in oceans is already a matter for concern, leading to contamination of some fish stocks, and may account for recent deaths of marine mammals and some coral reefs. The source is mainly industrial activity.

Scoping – initial stage in an environmental or social impact assessment, which seeks to identify the appropriate scope of study, terms of reference, approach and so on.

Social capital – ways in which people work together for common purpose in groups and organisations. It demands shared values and the subordination of individual interests. It can help people to adapt and develop.

Stakeholders – individuals or groups who have some sort of a claim on something.

Stewardship – management of something, such as an estate, forest, company or whatever which does not exploit for short-term gain. Rather, the intent is to sustain function and if possible improve it over time. Likely to require reinvestment and effort.

Structural adjustment – faced with growing debt in the early 1980s, many developing countries embraced, or were forced by funding bodies, to adopt austerity measures (structural adjustment). This often led to cuts in government spending and unemployment, which resulted in poverty and environmental damage and neglect of environmental management.

Subsistence agriculture – essentially land use which provides enough crops or livestock for a family to eat and sometimes obtain other things such as paying for weddings, funerals and other cultural obligations. Little or no money crosses the farm boundaries. This may be replaced by subsistence cashcropping (produce sold for money), whereby the farmer produces food or some other commodity which is sold to buy food and other very basic needs. In this case money does cross the farm boundaries, and the producer is likely to be vulnerable to market price shifts, gluts and problems selling produce, indebtedness, and so on. Some are wholly subsistence producers, some fully cashcroppers, and there are many gradations between the two.

Sustainable development – development that meets the needs of the present without compromising the ability of future generations to meet their own needs.

Volatile organic compounds (VOCs) – materials such as ethylene, benzene, acetone and others; these evaporate readily and contribute to air pollution directly or, through photochemical reactions, yield secondary pollutants like tropospheric ozone and various nitrates.

Woodlot – small tree plantation, usually on or close to agricultural land.

Y2K – the 'millennium bug'; early computer programmers were keen to reduce demands on the then limited computer memories, and so used dates without reference to the century. The problem was that at the end of 2000 systems might think it was AD 1900 and crash.

References

Abbey, E. (1975) *The Monkeywrench Gang.* Avon, New York.

Abbott, J. (1996) *Sharing the City: community participation in urban management.* Earthscan, London.

Aberley, D. (ed.) (1993) *Boundaries of the Home.* New Society Publishers, Gabriola Island (Canada).

Aberley, D. (ed.) (1994) *Futures by Design: the practice of ecological planning.* New Society Publishers, Philadelphia, PA.

Abu-Zeid, M. ((1998) Short and long term impacts of the River Nile projects. *Water Supply and Management* 3 (3), 275–283.

Acharya, R. (1991) Patenting of biotechnology: GATT and the erosion of the world's biodiversity. *Journal of World Trade* 25 (6), 71.

Action Against Hunger (2001) *The Geopolitics of Hunger 2000–2001: hunger and power.* Lynne Rienner, Boulder, CO.

Adams, P. (1991) *Odious Debt: loose lending, corruption and the Third World's environmental legacy.* Earthscan, London.

Adams, P. and Solomon, L. (1985) *In the Name of Progress: the underside of development aid.* Earthscan, London.

Adams, R., Carruthers, J. and Haml, S. (1991) *Changing Corporate Values.* Kogan Page, London.

Adams, W.M. (1990) *Green Development: environment and sustainability in the third world.* Routledge, London.

Adams, W.M. (1992) *Wasting the Rain: rivers, people and planning in Africa.* Earthscan, London.

Adams, W.M. (2001) *Green Development: environment and sustainability in the third world* (2nd edn). Routledge, London.

Adams, W.M. (2004) *Against Extinction: the past and future.* Earthscan, London.

Adger, W.N., Kelly, P.M. and Ninh, N.H. (eds) (2004) *Living with Environmental Change: social vulnerability, adaptation and resilience in Vietnam.* Routledge, London.

AEO (2005) *Africa Environment Outlook: past, present and future prospects.* Earthprint, London (available online from http://www.grida.no/aeo/).

Agarwal, A. (1992) Towards global environmental management. *Social Action* 42 (2), 111–119.

Agarwal, A. and Narain, S. (1991) *Global Warming in an Unequal World: a case of environmental colonialism.* Centre for Science and Environment, New Delhi.

Agarwal, B. (1992) The gender and environment debate: lessons from India. *Feminist Studies* 18 (1), 119–158.

Agarwal, B. (1997) Environmental action, gender equity and women's participation. *Development & Change* 28 (1), 2–44.

Ager, D. (1993) *The New Catastrophism: the importance of the rare event in geological history.* Cambridge University Press, Cambridge.

Ahmad, Y.-J., El Serafy, M. and Lutz, E. (eds) (1989) *Environmental Accounting for Sustainable Development.* World Bank, Washington, DC.

Ahrens, T.J. and Thomas, A.W. (1992) Deflection and fragmentation of near-Earth asteroids. *Nature* 360 (6403), 429–433.

Alberti, M., Caini, L., Calabrese, A., and Rossi, D. (2000) Evaluation of the costs and benefits of an environmental management system. *International Journal of Production Research* 3 (17), 4455–4466.

Albrecht, D.E. and Murdock, S.H. (1986) Natural resources availability and social change. *Sociological Inquiry* 56 (3), 380–400.

Aldrich, B.C. and Sandhu, R.S. (eds) (1995) *Housing the Urban Poor: policy and practice in developing countries.* Zed Books, London.

Alexander, D. (1990) Bioregionalism: science or sensibility? *Environmental Ethics* 12 (2), 191–173.

Alfsen, K.H. and Bye, T. (1990) Norwegian experiences in natural resource accounting. *Development* 3–4, 119–130.

Allan, N.J.R. (1987) Impact of Afghan refugees on the vegetation resources of Pakistan's Hindukush-Himalaya. *Mountain Research and Development* 7 (3), 200–204.

Allen, G.R. (1977) The world fertilizer situation. *World Development* 5 (5–7), 526–536.

Allen, J.P. (1991) *Biosphere 2: the human experiment.* Penguin, New York.

Allen, M. (2005) Climate and punishment. *New Scientist* 187 (2519), 422–423.

Allenby, B.R. 1998) *Industrial Ecology: policy framework and implementation.* Prentice-Hall, New York.

Allenby, B.R. and Richards. D.J. (eds) (1994) *The Greening of Industrial Ecosystems.* National Academy Press, Washington, DC.

Alpert, P. (1993) Conserving biodiversity in Cameroon. *Ambio* xxii (1), 44–49.

Alvarez, W. and Asaro, F. (1990) What caused the mass extinction? An extraterrestrial impact. *Scientific American* 263 (4), 76–84.

Amin, A. (ed.) (1994) *Post-Fordism: a reader.* Blackwell, Oxford.

Ananthaswamy, A. (2002) Cities eat away at Earth's best land. *New Scientist* 176 (2374/5), 9.

Anderson, A.B. (ed.) (1990) *Alternatives to Deforestation: steps toward sustainable use of Amazon rainforests.* Columbia University Press, New York.

Anderson, A.B. (1992) Land-use strategies for successful extractive economies in Amazonia. In D.C. Nepstad and S. Schwartzman (eds) *Non-Timber Products from Tropical Forests: evaluation of a conservation and development strategy* (Advances in Economic Botany No. 9). New York Botanic Gardens, New York, pp. 67–77.

Anderson, I. (1992) Can Nauru clean up after the colonialists? *New Scientist* 135 (1830), 12–13.

Anderson, K. and Blackhurst, R. (eds) (1992) *The Greening of World Trade.* Harvester Wheatsheaf, London.

Anderson, W.T. (1993) Is it really environmentalism versus biotechnology? *Bio-Technology* 11 (2), 236.

Anon. (1990) Draft convention on environmental impact assessment in a transboundary context. *Environmental Policy and Law* 20 (4–5), 181–186.

Anon. (1992) Where should GATT go next? *Nature* 360 (6401), 195–196.

Anon. (1993) Cakes and caviar? The Dunkel Draft and Third World agriculture. *The Ecologist* 23 (6), 219–222.

Anon. (1995) Environment information systems in sub-Saharan Africa: an Internet resource. *Bulletin of the American Society for Information Science* 21 (4), 24.

Anon. (1996) Environment – plastics waste trading on the Internet. *Chemical Weekly* 158 (19), 17.

Applegate, J.S. (2000) The precautionary principle: an American perspective on the precautionary principle. *Human and Ecological Risk Assessment* 6 (3), 413–443.

Armitage, D. (1995) An integrative methodological framework for sustainable environmental planning and management. *Environmental Management* 19 (4), 469–479.

Arnold, J.E.M. (1990) Social forestry and communal management in India (Overseas Development Institute, Social Forestry Network Paper No. 11b). ODI, Agricultural Administration Unit, London.

Aronsson, T., Johansson, P.-O. and Löfgren, K.-G. (1997) *Welfare Measurement, Sustainability and Green National Accounting: a growth theoretical approach.* Edward Elgar, London.

Asante-Duah, D.K. (1998) *Risk Assessment in Environmental Management: a guide for managing chemical contamination problems*. Wiley, Chichester.

Ashley, C., Goodwin, H. and Roe, D. (2001) *Pro-Poor Tourism: expanding opportunities for the poor* (Pro-Poor Tourism Policy Briefing No. 1 – Overseas Development Institute, IIED and the Centre for Responsible Tourism). University of Greenwich, London.

Ashton, P.S. and Ashton, R.E. (2002) *Ecotourism: sustainable nature and conservation based tourism*. Krieger Publishing, Melbourne, FL.

Ashworth, G. and Kivell, P. (eds) (1989) *Land, Water and Sky*. Geopers, Groningen.

Association of County Councils (1991) *Towards a Sustainable Transport Policy*. ACC, London.

Athanasiou, T. (1977) *Slow Reckoning: the ecology of a divided planet*. Secker & Warburg, London.

Atkinson, A. (1991a) *Green Utopias: the future of modern environmentalism*. Zed Press, London.

Atkinson, A. (1991b) *Principles of Political Ecology*. Belhaven Press, London.

Atkinson, G. (1997) *Measuring Sustainable Development: macroeconomics and the environment*. Edward Elgar, Cheltenham.

Aubreville, A.M. (1949) *Climats, Forêts et Désertification de l'Afrique Tropicale*. Société d'Editions Geographiques, Maritimes et Coloniales, Paris.

Auburn, F.M. (1982) *Antarctic Law and Politics*. Indiana University Press, Bloomington, IN; World Bank, Washington, DC.

Auty, R.M. (1995) *Patterns of Development: resources, policy and economic growth*. Arnold, London.

Avery, N., Drake, M. and Lang, T. (1993) Codex Alimentarius: who is allowed in? Who is left out? *The Ecologist* 23 (3), 110–112.

Ayres, A. (2004) Xochimilco's sunken treasure. *New Scientist* 182 (2442), 50–51

Ayres, R.U. and Ayres, L.W. (eds) (1996) *Industrial Ecology: towards closing the materials gap*. Edward Elgar, Cheltenham.

Ayres, R.U. and Ayres, L.W. (2002) *A Handbook of Industrial Ecology*. Edward Elgar, Cheltenham.

Ayres, R.U. and Rohagi, P.K. (1987) *Bhopal: lessons for technological decision-makers* (Research Report RR-87–10). International Institute for Applied Systems Analysis, Laxenberg, Austria.

Baarschers, W.H. (1996) *Eco-Facts and Eco-Fiction: understanding the environmental debate*. Routledge, London.

Bahro, R. (1982) *Socialism and Survival*. Heretic Books, London.

Bahro, R. (1984) *From Red to Green: interviews with 'New Left Review'*. Verso, London.

Bailey, R.G. (1986) Delineation of ecosystem regions. *Environmental Management* 7 (3), 365–373.

Bailey, R.G. (1993) *Eco-Scam: the false prophets of ecological apocalypse*. St Martin's Press, New York.

Bailey, R.G. (1996) *Ecosystem Geography*. Springer, New York.

Baldock, D. (1992) The polluter pays principle and its relevance to agricultural policy in European countries. *Sociologia Ruralis* 32 (1), 49–65.

Ball, S. and Bell, S. (1991) *Environmental Law*. Blackstone, London.

Barbier, E.B. (ed.) (1993) *Economics and Ecology: new frontiers in sustainable development*. Chapman and Hall, London.

Barde, J.-P. and Pearce, D.W. (1990) *Valuing the Environment*. Earthscan, London.

Barkey, M.B. (ed.) (2000) *Environmental Stewardship in the Judeo-Christian Tradition: Jewish, Catholic, and Protestant wisdom on the environment*. Acton Institute for the Study of Religion and Liberty, Grand Rapids, MI.

Barlow, M. and Clarke, T. (2002) *Blue Gold: the battle against corporate theft of the world's water*. Earthscan, London.

Barretovianna, M.D. (1992) Environmental management tools: experience of the aluminium industry in Brazil. *Journal of Clean Technology and Environmental Sciences* 2 (3–4), 187–203.

Barrett, B. (1994) Integrated environmental management: experience in Japan. *Journal of Environmental Management* 40 (1), 17–23.

Barrett, B. (1995) From environmental auditing to integrated environmental management: local government experience in the United Kingdom and Japan. *Journal of Environmental Planning and Management* 38 (3), 307–332.

Barrett, J. (2001) Component ecological footprint: developing sustainable scenarios. *Impact Assessment and Project Appraisal* 19 (2), 110–130.

Barrett, J. and Scott, A. (2001) The ecological footprint: a metric for corporate sustainability. *Corporate Environmental Strategy* 8 (4): 316–325.

Barrington, R. (2001) Biodiversity: new trends in environmental management. *Corporate Environmental Strategy* 8 (1), 39–47.

Barrow, C.J. (1981) Health and resettlement consequences and opportunities created as a result of river impoundment in developing countries. *Water Supply and Management* 5 (2), 135–150.

Barrow, C.J. (1991) *Land Degradation: development and breakdown of terrestrial environments.* Cambridge University Press, Cambridge.

Barrow, C.J. (1995a) Sustainable development: concept, value and practice. *Third World Planning Review* 17 (4), 369–387.

Barrow, C.J. (1995b) *Developing the Environment: problems and management,* Longman, Harlow.

Barrow, C.J. (1997) *Environmental and Social Impact Assessment: an introduction.* Arnold, London.

Barrow, C.J. (1998) River basin development planning and management: a critical review. *World Development* 26 (1), 171–186.

Barrow, C.J. (2003) *Environmental Change and Human Development: controlling nature?* Arnold, London.

Barrow, C.J. (2005) *Environmental Management and Development.* Routledge, London.

Barrows, H. (1926) Geography as human ecology. *Annals of the Association of American Geographers* 13 (1), 1–14.

Barry, J. (1999) *Environment and Social Theory.* Routledge, London.

Barton, H. (1999) *Sustainable Communities: the potential for eco-neighbourhoods.* Earthscan, London.

Barton, H. and Bruder, N. (1995) *A Guide to Local Environmental Auditing.* Earthscan, London.

Batchelor, M. and Brown, K. (1992) *Buddhism and Ecology.* Cassell Publishers, London.

Bate, J. (1991) *Romantic Ecology: Wordsworth and the environmental tradition.* Routledge, London.

Baumol, W. and Oates, W. (1988) *The Theory of Environmental Policy* (2nd edn). Cambridge University Press, Cambridge.

Baumwerd-Ahlmann, A., Jaschek, P., Kalinski, J. and Lehmkuhl, H. (1991) Using KADS for generating explanations in environmental impact assessment, in H.J. Bullinger (ed.) *Human Aspects in Computing* (Proceedings of the HCI International, 1–6 September 1991, Stuttgart). Elsevier, Amsterdam, pp. 866–876.

Baxter, M. and Bacon, R. (1996) Which EMS? – your questions answered. *Environmental Assessment* 4 (1): 5–6.

Beach, H.L., Hamner, J., Hewitt, J.J., Kaufman, E., Kueki, A., Oppenheimer, J.A. and Wolf, A.T. (2005) *Transboundary Freshwater Dispute Resolution: theory, practice and annotated references.* United Nations University Press, Tokyo.

Beardsley, D., Davies, T. and Hersh, R. (1997) Improving environmental management: what works, what doesn't. *Environment* 39 (7), 6–9, 28–38.

Beaumont, J.R. (1992) Managing the environment: business opportunity and responsibility. *Futures* 24 (2): 187–198.

Beaumont, J.R., Pedersen, L.M. and Whitaker, B.D. (1993) *Managing the Environment: business opportunity and responsibility.* Butterworth-Heinemann, Oxford.

Beaumont, P. (1989) *Environmental Management and Development in Drylands.* Routledge, London.

Beck, P. (1986) *The International Politics of Antarctica.* St Martin's Press, New York.

Becker, E. and John, T. (eds) (1999) *Sustainability and the Social Sciences: a cross-disciplinary approach to integrating environmental considerations into theoretical reorientation.* Zed Press, London.

Becker, H. and Vanclay, F. (2003) *International Handbook of Social Impact Assessments.* Edward Elgar, Cheltenham.

Beckerman, W. (1995) *Small is Stupid: blowing the whistle on the Greens.* Duckworth, London.

Beek, K.J. (1978) *Land Evaluation for Agricultural Development.* International Irrigation Research Institute (IIRI), Wageningen.

Beeton, A.M. (2002) Large freshwater lakes: present state, trends, and future. *Environmental Conservation* 29 (1), 21–38.

Begossi, A. (1993) Human ecology: an overview of the relationship man–environment. *Interciencia* 18 (3), 121–132.

Behrendt, H. and Boekhold, A. (1993) Phosphorus saturation of soils and groundwater. *Land Degradation & Rehabilitation* 4 (4), 233–243.

Bell, C.L.G and Hazel, P. (1980) Measuring the indirect effects of an agricultural investment project on its surrounding region. *Australian Journal of Agricultural Economics* 62 (1), 75–86.

Bell, C.L.G., Hazel, P. and Slade, C. (1982) *Project Evaluation in Regional Perspective: a study of an irrigation project in northwestern Malaysia.* Johns Hopkins University Press, Baltimore, MD.

Bell, D. (1975) *The Coming Post-Industrial Society.* Basic Books, New York.

Bell, D., Keil, R., Fawcett, L. and Penz, P. (eds) (1998) *Political Ecology: global and local.* Routledge, London.

Bell, S. (1997) *Ball and Bell on Environmental Law: the laws and policy relating to the protection of the environment* (4th edn). Blackstone Press, London.

Bell, S. and Morse, S. (1999) *Sustainability Indicators: measuring the immeasurable.* Earthscan, London.

Beller, W., D'Ayala, P. and Hein, P. (eds) (1990) *Sustainable Development and Environmental Management of Small Islands.* UNESCO, Paris.

Bello, W. and Cunningham, S. (1994) Dark victory: the global impact of structural adjustment. *The Ecologist* 24 (3), 87–93.

Bennett, G. (1991) The history of the Dutch National Environmental Policy Plan. *Environment* 33 (7), 6–9, 31–33.

Bennett, G. (1992) *Dilemmas: coping with environmental problems.* Earthscan, London.

Bennett, M. and Jones, P. (eds) (1999) *Sustainable Measures: evaluation and reporting of environmental and social performance.* Greenleaf Publishing, Sheffield.

Bennett, R.J. (1984) System analysis and environmental management: development and future trends. *Mitteilungen der Österreichischen Geographischen Gesellschaft* 126 (1), 29–49.

Benôit, R. (1995) Current and future directions for structured impact assessments. *Impact Assessment* 13 (4), 403–432.

Benôit, R. and Podesto, M. (1995) Environmental assessment: transition of a decision support system to the environmental management of projects. *Impact Assessment* 13 (2), 117–133.

Berg, G. (1969) *Circumpolar Problems: habitat, economy, and social relations in the Arctic: a symposium for anthropological research in the north, September, 1969.* Pergamon Press, Oxford.

Berg, P. (ed.) (1978) *Reinhabiting a Separate Country: a bioregional anthology of Northern California.* Planet Drum, San Francisco, CA.

Berger, P.L. (1987) *The Capitalist Revolution.* Wildwood House, Aldershot.

Berkes, F. (ed.) (1989) *Common Property Resources: ecology and community-based sustainable development.* Belhaven Press, London.

Berkes, F. (1999) *Sacred Ecology: traditional ecological knowledge and resource management.* Taylor & Francis, Ann Arbor, MD.

Berkes, F. (2004) Rethinking community-based conservation. *Conservation Biology* 18 (3), 621–630.

Berkhout, F. (1991) *Radioactive Waste: politics and technology.* Routledge, London.

Berkhout, F., Leach, M. and Scoones, I. (eds) (2003) *Negotiating Environmental Change: new perspectives from social science.* Edward Elgar, Cheltenham.

Berry, S. and Ladkin, A. (1997) Sustainable tourism: a regional perspective. *Tourism Management* 18 (7), 433–440.

Billings, W.D. (1978) *Plants and the Ecosystem* (3rd edn). Wadsworth Publishing, Belmont, CA.

Birley, M.H. (1995) *The Health Impact of Development Projects.* HMSO, London.

Birnie, P.W. and Boyle, A.E. (1992) *International Law and the Environment.* Clarendon Press, Oxford.

Bishop, R.C. and Romano, D. (1998) *Environmental Resources Evaluation.* Kluwer, Dordrecht.

Biswas, A.K. and Tortajada, C. (eds) (2001) *Integrated River Basin Management: the Latin American experience.* Oxford University Press (India), New Delhi.

Black, R. (1994) Forced migration and environmental change: the impact of refugees on host environments. *Journal of Environmental Management* 42 (3), 261–277.

Blackburn, T.C. and Anderson, K. (eds) (1995) *Before the Wilderness: environmental management by Native Californians* (Anthropological Papers No. 40). Ballena Press, Menlo Park, CA.

Blaikie, P.M. (1985) *The Political Economy of Soil Erosion in Developing Countries.* Longman, Harlow.

Blaikie, P.M. (1988) The explanation of land degradation in Nepal, in J. Ives and D.C. Pitt (eds) *Deforestation: social dynamics in watersheds and mountain ecosystems.* Routledge, London, pp. 132–158.

Blaikie, P.M. (1989) The use of natural resources in developing and developed countries, in R.J. Johnston. and P.J. Taylor (eds) *A World in Crisis? Geographical perspectives* (revised edn). Blackwell, Oxford, pp. 125–150.

Blaikie, P.M. and Brookfield, H. (1987) *Land Degradation and Society.* Methuen, London.

Blaikie, P.M. and Unwin, T. (eds) (1988) *Environmental Crisis in Developing Countries* (IBG Developing Areas Research Group, Monograph No. 5). Institute of British Geographers, London.

Blakeslee, H.W. and Grabowski, T.M. (1985) *A Practical Guide to Plant Environmental Audits.* Van Nostrand Reinhold, New York.

Blench, R. (1998) Biodiversity conservation and its opponents. *ODI Natural Resource Perspectives* 32, 4pp.

Bliss, L.C., Heal, O.W. and Moore, J.J. (eds) (1981) *Tundra Ecosystems: a comparative analysis.* Cambridge University Press, Cambridge.

Bloemhofruwaard, J.M., Vanbeek, P., Hordijk, L. and Van Wassenhove, L.N. (1995) Interactions between operational research and environmental management. *European Journal of Operational Research* 85 (2), 229–243.

Blowers, A. (ed.) (1993) *Planning for a Sustainable Environment.* Earthscan, London.

BMA (1991) *Hazardous Waste and Human Health.* Oxford University Press, Oxford.

BMA (1998) *Health and Environmental Impact Assessment: an integrated approach.* British Medical Association (BMA), London.

Boardman, R. (1986) *Pesticides in World Agriculture: the politics of international regulation.* Macmillan, London.

Bocking, S. (1994) Visions of nature and society: a history of the ecosystem concept. *Alternatives* 20 (3), 12–18.

Bodansky, D. (1991) Scientific uncertainty and the precautionary principle. *Environment* 33 (7), 4.

Boehmer-Christiansen, S. (1994) Politics and environmental management. *Journal of Environmental Planning and Management* 37 (1), 69–86.

Bohoris, G.A. and O'Mahoney, E. (1994) BS-7750, BS-5750 and the EC's Eco-Management and Audit Scheme. *Industrial Management & Data Systems* 94 (2), 3–6.

Boo, E. (1990) *Ecotourism Potential and Pitfalls* (2 vols).World Wildlife Fund, Washington, DC.

Bookchin, M. (1972) *Post-Scarcity Anarchism.* Black Rose Books, Montreal.

Bookchin, M. (1980) *Towards a Social Ecology.* Black Rose Books, Montreal.

Bookchin, M. (1982) *The Ecology of Freedom: the emergence and dissolution of hierarchy.* Cheshire Books, Palo Alto, CA.

Bookchin, M. (1986) *The Modern Crisis.* New Society Publishers, Philadelphia, PA.

Bookchin, M. (1990) *Remaking Society.* Black Rose Books, Montreal.

Boons, F.A.A. and Baas, L.W. (1997) Types of industrial ecology: the problem of co-ordination. *Journal of Cleaner Production* 5 (1–2), 79–86.

Booth, D. (1997) *The Environmental Consequences of Growth.* Routledge, London.

Borgese, E.M. (1998) *The Oceanic Circle: governing the seas as a global resource.* United Nations University Press, Tokyo.

Bormann, B.T., Brookes, M.H., Ford, E.D., Kiesler, A.R., Oliver, C.D. and Weigand, J.F. (1993) *A Broad Strategic Framework for Sustainable-Ecosystem Management* (Eastside Forest Ecosystem Health Assessment vol. V). USDA Forest Service, Washington, DC.

Born, S.M. and Sonzogni, W.C. (1995) Integrated environmental management: strengthening the conceptualization. *Environmental Management* 19 (2), 167–181.

Borri, F. and Boccaletti, G. (1995) From total quality management to total quality environmental management. *TQM Magazine* 7 (5), 38–42.

Borrini-Feyerabend, G. (1996) *Collaborative Management of Protected Areas: tailoring the approach to the context.* IUCN, Gland.

Boserüp, E. (1965) *The Conditions of Agricultural Growth: the economics of agrarian change under population pressure.* Allen & Unwin, London.

Boserüp, E. (1981) *Population and Technology.* Blackwell, Oxford.

Boserüp, E. (1990) *Economic and Demographic Relationships in Development* (essays selected and introduced by T.P. Schultz). Johns Hopkins University Press, Baltimore, MD.

Bouchet, J.-J. (1983) *Management of Upland Watersheds: participation of the mountain communities.* FAO, Rome.

Boulding, K.E. (1971) The economics of the coming Spaceship Earth. *Development Digest* (1), 12–15 (reprinted from 1966 in H. Jarrett (ed.) *Environmental Quality in a Growing Economy.* Johns Hopkins University Press, Baltimore MD).

Bouverie, J. (1991) Recycling in Cairo: a tale of rags to riches. *New Scientist* 130 (1775), 52–55.

Bowander, B. (1987) Integrating perspectives in environmental management. *Environmental Management* 11 (3), 305–315.

Bown, W. (1990) Trade deals a blow to the environment. *New Scientist* 128 (1742), 20–22.

Bowyer, J.L. (1994) Needed: a global environmental impact assessment. *Journal of Forestry* 92 (6), 6.

Boyce, M.S. (1997) *Ecosystem Management: applications for sustainable forest and wildlife management.* Yale University Press, New Haven, CN.

Boyle, A.E. (1992) Protecting the marine-environment: some problems and developments in the Law of the Sea. *Marine Policy* 16 (2), 79–85.

Bradshaw, A.D., Southwood, R. and Warner, F. (eds) (1992) *The Treatment and Handling of Wastes.* Chapman and Hall, London.

Braidotti, R., Charkiewicz, E., Häusler, S. and Wieringa, S. (1994) *Women, the Environment and Sustainable Development: towards a theoretical synthesis.* Zed Press, London.

Bramwell, A. (1994) *The Fading of the Greens: the decline of environmental politics in the West.* Yale University Press, New Haven, CN.

Brandis, P. (1998) *Science in and for Adaptive Management – a help or a hindrance?* Graduate School of Environment, Macquarie University, Sydney.

Bremen, A. (2004) *Social Research Methods* (2nd edn). Oxford University Press, Oxford.

Brent, R.J. (1997) *Applied Cost–Benefit Analysis.* Edward Elgar, London.

Brenton, T. (1994) *The Greening of Machiavelli: the evolution of international environmental politics.* Earthscan, London.

Briassoulis, H. (1986) Integrated economic-environmental policy modelling at the regional and multiregional level: methodological characteristics and issues. *Growth and Change* 17 (1), 22–34.

Briassoulis, H. (2001) Sustainable development and its indicators: through a (planner's) glass darkly. *Journal of Environmental Planning and Management* 44 (3), 409–427.

Brick, P. (1995) Determined opposition: the Wise-Use Movement challenges environmentalism. *Environment* 37 (8), 16.

Bridges, O. and Bridges, J. (1996) *Losing Hope: the environment and health in Russia.* Avebury, Aldershot.

Briggs, D.J. and Courtney, F.M. (1985) *Agriculture and Environment: the physical geography of temperate agricultural systems.* Longman, London.

Brimblecombe, P. (1987) *The Big Smoke: a history of pollution in London since medieval times.* Methuen, London.

British Standards Institution (1992) *BS7750 British Standard for Environmental Management.* British Standards Institution, Manchester.

British Standards Institution (1994a) Part I: Auditing of environmental management systems (guidelines for environmental auditing – audit procedures, ISO/CD14011/1, Document 94/400414DC). British Standards Institution, Manchester.

British Standards Institution (1994b) *Environmental Management Systems BS7750.* British Standards Institution, Manchester.

British Standards Institution (1994c) *Draft BS ISO14040: life cycle assessment: general principles and practices.* British Standards Institution, Manchester.

British Standards Institution (1996) *Environmental Management Systems: specifications with guidance for use.* British Standards Institution, Manchester.

Broecker, W.S. and Denton, G.H. (1990) What drives glacial cycles? *Scientific American* 262 (1), 42–50.

Brokensha, D.W. (1987) Development anthropology and natural resource management. *Uomo II* (2), 225–249.

Bromley, D.W. (1994) The enclosure movement revisited: the South African commons. *Journal of Economic Issues* 28 (2), 357–366.

Bromley, D.W. and Cernea, M.M. (1989) *The Management of Common Property Resources: some conceptual and operational fallacies* (World Bank Discussion Paper No. 57). World Bank, Washington, DC.

Brower, D.J., Schwab, A.K. and Beatley, T. (1994) *An Introduction to Coastal Zone Management.* Island Press, Washington, DC.

Brown, D.J.A. (1995) EU Eco-Management and Audit Scheme (EMAS). *Journal of the Institution of Environmental Sciences* 4 (3), 4–7.

Brown, J.R. and MacLeod, N.D. (1996) Integrating ecology into natural resource management policy. *Environmental Management* 20 (3), 289–296.

Brown, L.R. (2005) *Outgrowing the Earth: the food security challenge in an age of falling water tables and rising temperatures.* Earthscan, London.

Brown, L.R., Flavin, C. and Postel, S. (1992) *Saving the Planet: how to shape an environmentally sustainable global economy.* Earthscan, London.

Brunner, R.D. and Clark, T.W. (1997) A practice-based approach to ecosystem management. *Conservation Biology* 11 (1), 48–58.

Brush, S.B. (1986) Farming systems research. *Human Organization* 45 (3), 220–238.

Bryant, R.L. and Bailey, S. (1997) *Third World Political Ecology.* Routledge, London.

Bryant, R.L. and Wilson, G.A. (1998) Rethinking environmental management. *Progress in Human Geography* 22 (3), 321–343.

Buarque, C. (1993) *The End of Economics? Ethics and the disorder of progress.* Zed Press, London.

Buchholz, R.A. (1998) *Principles of Environmental Management: the greening of business.* Pearson-Prentice Hall, London.

Buckley, R.C. (1993) International trade, investment and environmental regulation: an environmental management perspective. *Journal of World Trade* 27 (4), 101–148.

Buckley, R.C. (1994) Strategic environmental assessment. *Environmental Planning and Law Journal* 11 (2), 166–168.

Buckley, R.C. (1995) Environmental auditing in F. Vanclay and D.A. Bronstein (eds) *Environmental and Social Impact Assessment.* Wiley, Chichester, pp. 283–301.

Bunce, R.G.H., Ryszkowski, L. and Paoletti, M.G. (1993) *Landscape Ecology and Agroecosystems.* Lewis Publishers, Boca Raton, FL.

Burch, W.R. Jnr, Cheek, N.H. and Taylor, L. (eds) (1972) *Social Behavior, Natural Resources and the Environment.* Harper & Row, New York.

Burde, M., Jackel, T., Dieckmann, R. and Hemker, H. (1994) Environmental impact assessment for regional planning with SAFRAN. *IFIP Transactions (B-applications in technology)* 16, 245–256.

Burdge, R.J. (1994) *A Conceptual Approach to Social Impact Assessment: collection of writings by Rabel J. Burdge and colleagues.* Social Ecology Press (PO Box 620863), Middleton, NJ.

Burdge, R.J. and Vanclay, F. (1996) Social impact assessment: a contribution to the state-of-the-art series. *Impact Assessment* 14 (1), 59–86.

Burgman, M. (2005) *Risks and Decisions for Conservation and Environmental Management.* Cambridge University Press, Cambridge.

Burk, T. and Hill, J. (1990) *Ethics, Environment and the Company.* The Institute of Business Ethics, London.

Burton, I., Kates, R. and White, G. (1977) *The Environment as Hazard.* Oxford University Press, Oxford.

Burton, M.L., Schoepfle, G.M. and Miller, M.L. (1986) Natural resource anthropology. *Human Organization* 45 (3), 261–269.

Butler, R.W. (1991) Tourism, environment, and sustainable development. *Environmental Conservation* 18 (3), 201–209.

Buttel, F.H. (1978) Environmental sociology: a new paradigm? *The American Sociologist* 13 (3), 252–256.

Button, K.J. and Pearce, D.W. (1989) *Improving the Urban Environment: how to adjust national and local government policy for sustainable urban growth.* Pergamon, Oxford.

Cahill, L.B. (ed.) (1989) *Environmental Audits* (6th edn). Government Institutes Inc, Rockville.

Cairncross, F. (1991) *Costing the Earth.* The Economist Books, London.

Cairncross, F. (1995) *Green Inc. Guide to Business and the Environment.* Earthscan, London.

Cairncross, F. and Feachem, R.G. (1983) *Environmental Health Engineering in the Tropics: an introductory text* (2nd edn). Wiley, Chichester.

Cairns, J. Jnr and Crawford, T.V. (eds) (1991) *Integrated Environmental Management.* Lewis Publishers, Chelsea, MI.

Caldwell, L.K. (1989) A constitutional law for the environment: 20 years with NEPA indicates the need. *Environment* 31 (10), 6–11, 25–28.

Caldwell, L.K. (1990) *International Environmental Policy: emergence and dimensions* (2nd edn). Duke University, Durham, NC.

Caldwell, M. (1977) *The Wealth of Some Nations.* Zed Books, London.

Cameron, J. and Fijalkowski, A. (eds) (1998) *Trade and the Environment: bridging the gap* (vols 1 and 2). Cameron May, London.

Cann, J. and Walker, C. (1993) Breaking new ground on the ocean floor. *New Scientist* 140 (1897), 24–29.

Canter, L.W. (1996) *Environmental Impact Assessment* (2nd edn). McGraw-Hill, New York.

Capra, F. (1982) *The Turning Point: science, society and the rising culture.* Wildwood House, London.

Capra, F. (1997) *The Web of Life: a new synthesis of mind and matter.* Flamingo (HarperCollins), London.

Carley, M. and Christie, I. (1992) *Managing Sustainable Development.* Earthscan, London.

Carley, M., Smith, M. and Varadarajan, S. (1991) A network approach to environmental management. *Project Appraisal* 6 (2), 66–75.

Carney, D. (1998) *Sustainable Rural Livelihoods: what contribution can we make?* DFID, London.

Carney, D. (2002) Approaches to sustainable livelihoods. ODI Poverty Briefing 02/01/02, Overseas Development Institute, London (6 pp.). Download from http://www.odi.org.uk/briefing/pov2.html.

Carney, D. and Farrington, J. (1998) *Natural Resource Management and Institutional Change.* Routledge, London.

Carroll, J.E. (1988) *International Environmental Diplomacy: the management and resolution of transfrontier environmental problems.* Cambridge University Press, Cambridge.

Carruthers, I. and Chambers, R. (1981) Rapid appraisal for rural development. *Agricultural Administration* 8 (5), 407–422.

Carson, R. (1962) *Silent Spring.* Houghton Mifflin, Boston, MA.

Carter, R.W.G. (1988) *Coastal Environments.* Academic Press, London.

Cartwright, J. (1989) Conserving nature, decreasing debt. *Third World Quarterly* 11 (2), 114–127.

Cartwright, T.J. (1991) Planning and chaos theory. *Journal of the American Planning Association* 57 (1), 44–56.

Castro, F. (1993) *Tomorrow is too Late: development and the environmental crisis in the Third World.* Ocean Press, Melbourne.

Cater, E. (1993) Ecotourism in the Third World: problems for sustainable tourism development. *Tourism Management* 14 (2), 85–90.

Cater, E. (1995) Environmental contradictions in sustainable tourism. *The Geographical Journal* 161 (1), 21–28.

Cater, E. (2000) Ecotourism, in N.J. Smelser and P.B. Bates (eds) *International Encyclopedia of the Social and Behavioral Sciences.* Elsevier, New York, pp. 4165–4168.

Cater, E. and Lowman, G. (eds) (1994) *Ecotourism: a sustainable option.* Wiley, New York.

Catton, W.R. (1994) Foundations of human ecology. *Sociological Perspectives* 37 (1), 75–95.

CEC (1992) *Towards Sustainability: the Fifth Environmental Action Programme for Europe.* Commission for the European Communities, Brussels.

Cernea, M.M. (1988) *Involuntary Resettlement in Development Projects: policy guidelines in World Bank-Financed projects* (World Bank Technical Paper No. 180). World Bank, Washington, DC.

Chadwick, M.J. and Goodman, G.T. (eds) (1975) *The Ecology of Resource Degradation and Renewal.* Blackwell Scientific, Oxford.

Chambers, N., Simmons, C. and Wackernagel, M. (2000) *Sharing Nature's Interest: ecological footprints as an indicator of sustainable sustainability.* Earthscan, London.

Chambers, R. (1983) *Rural Development: putting the last first.* Longman, Harlow.

Chambers, R. (1992) *Rural Appraisal: rapid, relaxed and participatory* (Discussion Paper No. 311). University of Sussex, Institute of Development Studies, Brighton.

Chambers, R. (1994a) The origins and practice of participatory rural appraisal. *World Development* 22 (7), 953–969.

Chambers, R. (1994b) Participatory rural appraisal (PRA): analysis of experience. *World Development* 22 (9), 1253–1268.

Chambers, R. (1994c) Participatory rural appraisal (PRA): challenges, potentials and paradigm. *World Development* 22 (10), 1437–1454.

Chambers, R. and Conway, G. (1992) *Sustainable Rural Livelihoods: practical concepts for the 21st century* (IDS Discussion Paper No. 296). Institute of Development Studies, University of Sussex, Brighton.

Chandler, R. (1988) *Understanding the New Age.* Ward, London.

Chapman, C.R. and Morrison, D. (1994) Impacts on Earth by asteroids and comets: assessing the hazard. *Nature* 367 (6458), 33–39.

Chapman, M.D. (1989) The political ecology of fisheries depletion in Amazonia. *Environmental Conservation* 16 (4), 331–337.

Chappell, J.E. (1993) Social Darwinism, environmentalism, and ideology. *Annals of the Association of American Geographers* 83 (1), 160–163.

Charnovitz, S. (1993) Environmentalism confronts GATT rules: recent developments and new opportunities. *Journal of World Trade* 27 (2), 37–53.

Charter, M. (ed.) (1992) *Greener Marketing.* Greenleaf, Sheffield.

Chatterjee, N. (1995) Social forestry in environmentally degraded regions of India: case-study of the Mayurakshi basin. *Environmental Conservation* 22 (1), 20–30.

Chen, R.S. (2005) *Natural Disaster Hot Spots: a global risk analysis.* UNU University Press, Tokyo.

Cheney, J. (1987) Eco-feminism and deep ecology. *Environmental Ethics* 9 (4), 115–145.

Cheney, J. (1989) Postmodern environmental ethics: ethics as bioregional narrative. *Environmental Ethics* 11 (2), 117–134.

Cheremisinoff, P.N. and Morresi, A.O. (1977) *Environmental Assessment and Impact Statement Handbook.* Wiley, Chichester.

Cheru, F. and Bradford, C. (eds) (2005) *The Millennium Development Goals: raising the reources to tackle world poverty.* Zed Press, London.

Chiapponi, M. (1992) Environmental management and planning: the role of spatial and temporal scales. *Ekistics* 59 (356–357), 306–310.

Child, B. (ed.) (2004) *Parks in Transition: biodiversity, rural development and the bottom line.* Earthscan, London.

Cigna, A.A. (1993) Environmental management of tourist caves: the example of Grotta di Castellana and Grotta Grande del Vento, Italy. *Environmental Geology* 21 (3), 173–180.

Ciracy-Wantrup, S.V. (1952) *Resource Conservation: economics and policies.* University of California Press, Berkeley, CA.

Clark, B.D., Gilad, A., Bisset, R. and Tomlinson, P. (eds) (1984) *Perspectives on Environmental Impact Assessment.* D. Reidel, Dordrecht.

Clark, J.R. (1996) *Coastal Zone Management Handbook.* Lewis Publishers, Boca Raton, FL.

Clark, M., Smith, D. and Blowers, A. (eds) (1992) *Waste Location: spatial aspects of waste management, hazards and disposal.* Routledge, London.

Clark, R.B. (1989) *Marine Pollution* (2nd edn). Oxford Science Publishers, Oxford.

Clark, R.B. (1991) *Water – The International Crisis.* Earthscan, London.

Clark, R.B. (1999) *Countryside Management.* Routledge, London.

Clark, W.C. (1989) Managing Planet Earth. *Scientific American* 261 (3), 19–26.

Cleary, D. (1990) *Anatomy of the Amazon Gold Rush.* Macmillan, London.

Clements, F.E. (1916) *Plant Succession. An analysis of the development of vegetation* (Publication No. 242). Carnegie Institute, Washington, DC.

Cleveland, H. (1990) *The Global Commons: policy for the planet.* The Aspen Institute and University Press of America Inc., Lanham, CO.

Clunies-Ross, T. (1993) Taxing nitrogen fertilizer. *The Ecologist* 23 (1), 13–17.

Coates, J.F. and Jarrett, J. (1989) *What Futurists Believe.* Lomond Publications Inc., Mt. Airy, MI, and The World Future Society, Bethesda, MD.

Coccossis, H. and Nijkamp, P. (eds) (1995) *Sustainable Tourism Development.* Avebury, Aldershot.

Coddington, W. (1993) *Environmental Marketing.* McGraw-Hill, New York.

Coenen, R. (1993) NEPA's impact on environmental impact assessment in European Community member countries, in S.G. Hildebrand and J.B. Cannon (eds) *Environmental Analysis: the NEPA experience.* Lewis Publishers, Boca Raton, FL, pp. 703–715.

Cohen, J.E. (1996) *How Many People Can the Earth Support?* W.W. Norton, New York.

Colby, M.E. (1990) *Environmental Management in Development: the evolution of paradigms* (World Bank Discussion Paper No. 80). World Bank, Washington, DC.

Colchester, M. (1993) Pirates, squatters and poachers – the political ecology of dispossession of the native peoples of Sarawak. *Global Ecology and Biogeography Letters* 3 (4–6), 158–179.

Coleman, G. (1987) Logical framework approach to the monitoring and evaluation of agricultural and rural development projects. *Project Appraisal* 2 (4), 251–259.

Collins, J.I. (1986) Smallholder settlement of tropical South America: the social causes of ecological destruction. *Human Organization* 45 (1), 1–10.

Commins, S.K., Lofchie, M.F. and Payne, R. (eds) (1986) *Africa's Agrarian Crisis: the roots of famine.* Lynne Rienner, Boulder, CO.

Commission of the European Community (1992) *Towards Sustainability: a European Community programme of policy and action in relation to the environment and sustainable development* COM (92) 23 (final – vol. II). CEC, Brussels.

Common, M.S. (1996) *Environmental and Resource Economics* (2nd edn). Addison Wesley Longman, Harlow.

Common, M.S. and Norton, T.W. (1994) Biodiversity, natural resource accounting and ecological monitoring. *Environmental & Resource Economics* 4 (1), 29–54.

Commoner, B. (1972) *The Closing Circle.* Knopf, New York.

Connor, R. and Dovers, S. (2004) *Institutional Change for Sustainable Development.* Edward Elgar, Cheltenham.

Conroy, C. and Litvinoff, M. (1988) *The Greening of Aid: sustainable livelihoods in practice.* Earthscan, London.

Conway, G.R. (1985a) Agroecosystem analysis. *Agricultural Administration* 20 (1), 31–55.

Conway, G.R. (1985b) The properties of agroecosystems. *Agricultural Systems* 24 (2), 95–117.

Conway, G.R. (1997) *The Doubly Green Revolution: food for all in the 21st century.* Penguin, London.

Conway, G.R. and Barbier, E.B. (1990) *After the Green Revolution: sustainable agriculture for development.* Earthscan, London.

Conway, G.R. and McCracken, J.A. (1990) Rapid rural appraisal and agroecosystem analysis, in M.A. Altieri and S.B. Hecht (eds) *Agroecology and Small Farm Development.* Westview/CRC, Boulder, CO, pp. 221–236.

Conway, G.R. and Pretty, J.N. (1991) *Unwelcome Harvest: agriculture and pollution.* Earthscan, London.

Cooper, C. (1981) *Economic Evaluation and the Environment: a methodological discussion with particular reference to developing countries.* Hodder & Stoughton, London.

Cooper, C. (1990) *Green Christianity: caring for the whole creation.* Spire (Hodder & Stoughton), London.

Cooper, D.E. and Palmer, J.A. (eds) (1990) *Spirit of the Environment: religion, value and environmental concern.* Routledge, London.

Cooper, D.E. and Palmer, J.A. (eds) (1992) *The Environment in Question: ethics and global issues.* Routledge, London.

Cooper, P.J. (1995) Toward a hybrid state: the case of environmental management in a deregulated re-engineered state. *International Journal of Administrative Sciences* 61 (2), 185–200.

Copithorne, M.D. (1991) *Circumpolar Environmental Management and Regulation: from coexistence to cooperation – international law and organizations in the post-Cold War era.* Martinus Nijhoff, Dordrecht.

Cornwall, A. (1997) *Environmental Taxes and Economic Welfare: reducing carbon dioxide emissions.* Edward Elgar, London.

Corwin, R., Heffernan, P.H. and Johnston, R.A. (1975) *Environmental Impact Assessment.* W.H. Freeman, San Francisco, CA.

Cosgrove, D. (1990) Environmental thought and action: pre-modern and post-modern. *Transactions of the Institute of British Geographers* (new series) 15 (3), 344–358.

Costanza, R. (ed.) (1991) *Ecological Economics: the science and management of sustainability.* Columbia University Press, Ithaca, NY.

Costanza, R. and Cornwell, L. (1992) The approach to dealing with scientific uncertainty. *Environment* 34 (9), 12–20, 42.

Cotgrove, S. (1982) *Catastrophe or Cornucopia?: the environment, politics and the future.* Wiley, Chichester.

Council of Economic Priorities of the United States (1989) *Shopping for a Better World: a quick and easy guide to socially responsible supermarket shopping.* Council of Economic Priorities of the United States, New York.

Council on Environmental Quality and Department of State (1982) *The Global 2000 Report to the President: entering the twenty-first century.* Penguin Books, New York.

Covello, V.T., Mumpower, J.L., Stallen, P.J.M. and Uppuluri, V.R.R. (eds) (1985) *Environmental Impact Assessment, Technology Assessment, and Risk Analysis: contributions from the psychological and decision sciences.* Springer-Verlag, Berlin.

Cramer, J. and Zegveld, W.C.L. (1991) The future role of technology in environmental management. *Futures* 23 (5), 451–468.

Croner Publications Ltd (1997) *Croner's Environmental Policy and Procedures.* Croner Publications, Kingston upon Thames (also available as CD-ROM).

Cumberland, J.H. (1990) Public choice and the improvement of policy instruments for environmental management. *Ecological Economics* 2 (2), 149–162.

Dalal-Clayton, B. (1992) *Modified EIA and Indicators of Sustainability: first steps towards sustainability analysis.* Earthscan, London.

Dalal-Clayton, B. and Sadler, B. (2005) *Strategic Environmental Assessment: a sourcebook and reference guide to international experience.* Earthscan, London.

Dale, A.P. (1992) Aboriginal councils and natural resource use planning: participation by bargaining and negotiation. *Australian Geographical Studies* 30 (1), 9–26.

Dale, T. and Carter, V.G. (1954) *Topsoil and Civilisation.* University of Oklahoma Press, Norman, OK.

Dalton, R.J. (1994) *The Green Rainbow: environmental groups in western Europe.* Yale University Press, New Haven, CN.

Dankelman, I. and Davidson, J. (1988) *Women and Environment in the Third World*. Earthscan, London.

Darling. F.F. and Dasmann, R.F. (1969) The ecosystem view of human society. *Impact of Science on Society* 19 (2), 109–121.

Darwin, C. (1859) *The Origin of Species by Means of Natural Selection: or the preservation of favoured races in the struggle of life*. John Murray, London.

Dasmann, R.F., Milton, J.P. and Freeman, P.H. (1973) *Ecological Principles for Economic Development*. Wiley, London.

Davidson, J., Myers, D. and Chakraborty, M. (1992) *No Time to Wait: poverty and the environment*. Oxfam, Oxford.

Davis, J. (1991a) *Earth First! Reader*. Peregrine Smith, London.

Davis, J. (1991b) *Greening Business: managing for sustainable development*. Blackwell, Oxford.

Davis, M. (2001) *Late Victorian Holocausts: El Niño famines and the making of the Third World*. Verso, London.

Davos, C.A. (1986) Global environmental management. *The Science of the Total Environment* 56 (6), 309–316.

D'Ayala, P., Hein, P. and Beller, W. (1990*) Sustainable Development and Environmental Management of Small Islands* (MAB Series No. 5). Parthenan Publishing Group (for UNESCO), Paris.

De Bardeleben, J. (1986) *The Environment and Marxism-Leninism: the Soviet and East German experience*. Westview Press, Boulder, CO.

De Boo, H.L. and Wiersum, K.F. (2002) *Adaptive Management of Forest Resources: principles and practices*. Conservation Policy Group, Agricultural Sciences Discussion Paper (Conference 2002), Wageningen Agricultural University, Wageningen.

De Groot, W.T. (1992) *Environmental Science Theory: concepts and methods in a one-world, problem-oriented paradigm*. Elsevier, Amsterdam.

De Miraman, J. and Stevens, C. (1992) The trade/environmental policy balance. *The OECD Observer* 176 (January/July), 25–27.

Department of Environment (1994) *UK Strategy on Sustainable Development*. HMSO, London.

DeRoy, O.C. (1997) The African challenge: Internet, networking and connectivity activities in a developing environment. *Third World Quarterly* 18 (5), 883–898.

Dery, D. (1997) Coping with 'latent time bombs' in public policy. *Environmental Impact Assessment Review* 17 (5), 413–425.

Desai, M. (1995) Greening of the HHDI? in A. McGillivray (ed.) *Accounting for Change*. The New Economics Foundation, London, pp. 21–36.

Dessler, A.E. and Parson, E.A. (2005) *The Science and Politics of Global Climate Change: a guide to the debate*. Cambridge University Press, Cambridge.

Devall, B. (1991) Deep ecology and radical environmentalism. *Society and Natural Resources* 4 (3), 247–258.

Devall, B. and Sessions, G. (1985) *Deep Ecology: living as if nature mattered*. Peregrine Smith Books, Salt Lake City, UT.

Devall, W. (1988) *Simple in Means, Rich in Ends: practicing deep ecology*. Peregrine Smith Books, Salt Lake City, UT.

Devas, N. and Rakodi, C. (1993) *Managing Fast Growing Cities: new approaches to urban planning and management*. Longman, Harlow.

Diamond, I. and Drenstein, G.F. (eds) (1990) *Reweaving the World: the emergence of ecofeminism*. Sierra Club Books, San Francisco, CA.

Diamond, J. (1998) *Guns, Germs and Steel: a short history of everybody for the last 13,000 years*. Vintage/Random House, London (1997 edn published by Jonathan Cape, London).

Diamond, J. (2005) *Collapse: how societies choose to fail or survive*. Penguin/Allen Lane, London.

Diaz, H.F. and Markgraf, V. (1992) *El Niño: historical and palaeoclimatic aspects of the Southern Oscillation*. Cambridge University Press, Cambridge.

Di Castri, F. and Hadley, M. (1985) Enhancing the credibility of ecology: can research be made more comparable and predictive? *GeoJournal* 11 (4), 321–338.

Di Castri, F. and Robertson, J. (1982) The biosphere reserve concept: 10 years after. *Parks* 6 (4), 1–6.

Dickinson, G. and Murphy, K. (1998) *Ecosystems: a functional approach*. Routledge, London.

Dinham, B. (1991) FAO and pesticides: promotion 450 or proscription? *The Ecologist* 21 (2), 61–65.

Ditton, R.B. and Goodale, T.C. (eds) (1974) *Environmental Impact Analysis: philosophy and methods*. University of Wisconsin, Sea Grants Program, Madison, WI.

Dixon, J.A., James, D.E. and Sherman, P.B. (eds) (1989) *The Economics of Dryland Management*. Earthscan, London.

Dobson, A. (1990) *Green Political Thought: an introduction*. Routledge, London.

Dobson, A. (1994) Ecologism and the relegitimation of socialism. *Radical Philosophy* 67 (1), 13–19.

Dobson, A. (1996) *Green Political Thought: an introduction* (2nd edn). Routledge, London.

Dodds, F. (ed.) (1997) *The Way Forward: beyond 'Agenda 21'*. Earthscan, London.

Dodds, F. (ed.) (2001) *Earth Summit 2002*. Earthscan, London.

Dodds, S. (1997) Towards a 'science of sustainability': improving the way ecological economics understands human well-being. *Ecological Economics* 23 (2), 95–111.

Döös, B.R. (1994) Environmental degradation, global food production, and risk for large-scale migrations. *Ambio* XXIII (2), 124–130.

Döös, B.R. (1997) Can large-scale environmental migrations be predicted? *Global Environmental Change* 7 (1), 41–61.

Dorney, R.S. (1989) *The Professional Practice of Environmental Management* (posthumously edited by L.S. Dorney). Springer, New York.

Dorney, R.S. and McLellan, P.W. (1984) The urban ecosystem: its spatial structure, its scale relationships, and its subsystem attributes. *Environments* 16 (1), 9–20.

Douglas, I. (1983) *The Urban Environment*. Edward Arnold, London.

Douglas, M. and Wildavsky, A. (1990) *Risk and Culture*. Westview, Boulder, CO.

Dovers, S.R. and Handmer, J.W. (1995) Ignorance, the precautionary principle, and sustainability. *Ambio* xxiv (2), 92–97.

Dower, N. (ed.) (1989) *Ethics and Environmental Responsibility*. Avebury (Gower Books), Aldershot.

Dowling, R. (1993) An environmentally based planning model for regional tourism development. *Journal of Sustainable Tourism* 1 (1): 1–37

Doxiadis, C. (1968) *Ekistics: an introduction to the science of human settlements*. Hutchinson, London.

Doxiadis, C. (1977) *Ecology and Ekistics*. Elek Books, London.

Dresner, S. (2002) *The Principles of Sustainable Development*. Earthscan, London.

Dudley, N., Jeanrenaud, J.-P. and Sullivan, F. (1995) *Bad Harvest? The timber trade and the degradation of the world's forests*. Earthscan, London.

Duffy, R. (2002) *A Trip too Far: ecotourism, politics and exploitation*. Earthscan, London.

Duinker, P.N. (1989) Ecological monitoring in environmental impact assessment: what can it accomplish? *Environmental Management* 13 (6), 797–805.

Dunlap, R. and Mertig, A. (eds) (1992) *American Environmentalism: the US environmental movement 1970–1990*. Taylor & Francis, Philadelphia, PA.

Dunn, B.C. and Steinemann, A. (1998) Industrial ecology and sustainable communities. *Journal of Environmental Planning and Management* 48 (6), 661–673.

Eagles, P.F.J. (2002) *Sustainable Tourism in Protected Areas: guidelines for planning and management*. UNEP, World Tourist Organisation, and IUCN, IUCN, Gland.

Earll, R.C. (1992) Common sense and the precautionary principle: an environmentalist perspective. *Marine Pollution Bulletin* 24 (2), 182–186.

Easter, K.A., Dixon, J.A. and Hufschmidt, M.M. (eds) (1986) *Watershed Resource Management: an integrated framework with studies from Asia and the Pacific*. Westview, Boulder, CO.

Eckersley, R. (1988) The road to ecotopia? Socialism versus environmentalism. *The Ecologist* 18 (4–5), 142–148.

Eckholm, E.P. (1976) *Losing Ground: environmental stress and world food prospects.* Pergamon Press, Oxford.

Ecologist (1993) *Whose Common Future? Reclaiming the commons.* Earthscan, London.

Eden, M.-J. and Parry, J.T. (eds) (1996) *Land Degradation in the Tropics: environmental and policy issues.* Pinter, London.

Edington, J.M. and Edington, M.A. (1986) *Ecology, Recreation and Tourism.* Cambridge University Press, Cambridge.

Edwards, F.N. (ed.) (1992) *Environmental Auditing: the challenge of the 1990s.* University of Calgary Press, Calgary.

EEC (1993) EMAS No.1836/93 (EEC Council Regulation). Council of the European Communities, Luxembourg.

Ehrlich, P.R. (1970) *The Population Bomb.* New Ballantine Books, New York.

Ehrlich, P.R., Ehrlich, A.H. and Holdren, J.P. (1970) *Ecoscience: population, resources, environment.* Freeman, San Francisco, CA.

Ekins, P. (1992a) *Wealth Beyond Measure: an atlas of new economics.* Gaia Books, Bideford.

Ekins, P. (1992b) *A New World Order: grassroots movements for global change.* Routledge, London.

El-Hinnawi, E. (1985) *Environmental Refugees.* UNEP, Nairobi.

Elkholy, O.A. (2001) Trends in environmental management in the last 40 years, in T. Munn (ed.) *Encyclopedia of Global Environmental Change* (vol. 4). Wiley, Chichester, pp. 25–20.

Elkington, J. and Burke, T. (1989) *The Green Capitalists.* Victor Gollancz, London.

Elkington, J. and Hailes, J. (1988) *The Green Consumer Guide: from shampoos to champagne.* Victor Gollancz, London.

Elkington, J. and Knight, P. (1991) *The Green Business Guide.* Victor Gollancz, London.

Elliott, H. (1997) A general statement of the tragedy of the commons. *Population and Environment* 18 (6), 515–531.

Elsworth, S. (1984) *Acid Rain.* Earthscan, London.

Elton, C.S. (1958) *The Ecology of Invasions by Animals and Plants.* Chapman & Hall, London.

Engstrom, K. (1992) The recycling of plastic bottles collected from public waste. *Endeavour* 16 (3), 117–121.

EPA (1988) *Annotated Bibliography on Environmental Auditing.* US Environmental Protection Agency, Office of Policy and Education, Washington, DC.

Erickson, S.L. and King, B.J. (1999) *Fundamentals of Environmental Management.* Wiley, New York.

Eriksson, E. (1989) Acid deposition and its effects. *Ambio* xviii (3), 184–191.

Erkman, S. (1997) Industrial ecology: an historical view. *Journal of Cleaner Production* 5 (1–2), 1–10.

Eröcal, D. (ed.) (1991) *Environmental Management in Developing Countries* (Development Centre Seminars). Development Centre, OECD, Paris.

Esty, D.C. (1994) *Greening the GATT: trade, environment, and the future.* International Institute for Economics, Washington, DC.

Evanari, M., Shanan, L. and Tadmore, N. (1982) *The Negev: the challenge of a desert.* Harvard University Press, Cambridge, MA.

Evans, B. (1995) Local environment policy: sustainability and 'Local Agenda 21'. *Area* 27 (2), 163–164.

Evernden, N. (1985) *The Natural Alien.* University of Toronto Press, Toronto.

Ewert, A.W., Backer, D.C. and Bissix, G.C. (2004) *Integrated Resource and Environmental Management: the human dimensions.* CABI, Wallingford.

Fagan, B.M. (1999) *Floods, Famines, and Emperors: EL Niño and the fate of civilizations.* Basic Books, New York.

Fagan, B.M. (2000) *The Little Ice Age: how climate made history 1300–1850.* Basic Books, New York.

Fagan, B. (2004) *The Long Hot Summer: how climate changed civilisation.* Basic Books, New York.

Fairhead, J. and Leach, M. (1996) *Misreading the African Landscape: society and ecology in a forest–savanna mosaic.* Cambridge University Press, Cambridge.

Fairlie, S. (1992) Long distance, short life: why big business favours recycling. *The Ecologist* 22 (6), 276–283.

Fairweather, P.G. (1993) Links between ecology and ecophilosophy, ethics and the requirements of environmental management. *Australian Journal of Ecology* 18 (1), 3–19.

Falkenmark, M. and Rockstrom. J. (2004) *Balancing Water for Humans and Nature.* Earthscan, London.

FAO (Food and Agriculture Organization of the UN) (1978) *Report on the Agroecosystems Zones Project, vol. 1: Methodology and Results for Africa* (World Soil Research Paper No. 48). FAO, Rome.

FAO (1986) *Strategies, Approaches and Systems in Integrated Watershed Management.* FAO, Rome.

FAO (1988) *Guidelines for Economic Appraisal of Watershed Management Projects.* FAO, Rome.

FAO (2001a) *The State of Food and Agriculture 20001.* FAO, Rome.

FAO (2001b) *The State of Food and Agriculture 2001.* FAO, Rome. Part III (pp. 199–226) *Economic Impacts of Transboundary Plant Pests and Animal Diseases.*

Farber, S., Moreau, R. and Templet, P. (1995) A tax incentive tool for environmental management: an environmental scorecard. *Environmental Economics* 12 (3), 183–189.

Farmer, A. (1997) *Managing Environmental Pollution.* Routledge, London.

Farvar, M.T. and Milton, J.P. (eds) (1972) *The Careless Technology: ecology and international development.* Natural History Press/Doubleday, Garden City, New York.

Fava, J.A. (1994) Life-cycle assessment: a new way of thinking. *Environmental Toxicology and Chemistry* 13 (6), 853–854.

Febvre, L. (1924) *A Geographical Introduction to History.* Routledge & Kegan Paul, London.

Feeney, D., Berkes, F., McCay, B.J. and Acherson, J.M. (1990) The tragedy of the commons – 22 years later. *Human Ecology* 18 (1), 1–19.

Feitelson, E. (1992) An alternative role for economic instruments: sustainable finance for environmental management. *Environmental Management* 16 (3), 299–307.

Fennell, D.A. (1999) *Ecotourism: an introduction.* Routledge, London.

Fennell, D.A., Dowling, R.K. and Kingston Bodde, A. (2003) *Ecotourism Policy.* CABI Publications, Wallingford.

Ferguson, A.R.B. (1999) The logical foundations of ecological footprints. *Environment, Development and Sustainability* 1 (2), 149–156.

Feshbach, M. and Friendly, A. (1992) *Ecocide in the USSR.* Arum, London.

Finsinger, J. and Marx, J.F. (1996) Eco-Management and Audit Scheme (EMAS): opportunities and risks for insurance companies. *International Journal of Environment and Pollution* 6 (4–6), 491–499.

Finsterbusch, K., Ingersol L.J., and Llewellyn, L.G. (eds) (1990) *Methods for Social Impact Analysis in Developing Countries.* Westview, Boulder, CO.

Fitter, R. and Scott, P. (1978) *The Penitent Butchers.* The Fauna and Flora Preservation Society, London.

Fitzgerald, C. (1993) *Environmental Management Information Systems.* McGraw-Hill, New York.

Flamm, B.R. (1973) A philosophy of EIA: toward choice among alternatives. *Journal of Soil and Water Conservation* 28 (3), 201–203.

Foder, I. and Walker, G. (eds) (1994) *Environmental Policy and Practice.* Hungarian Academy of Sciences, Pecs.

Foley, G. (1988) Deep ecology and subjectivity. *The Ecologist* 18 (4–5), 120–123.

Fong, C.O. (1985) Integrated population-development programme performance: the Malaysian FELDA experience. *Journal of Developing Areas* 19 (2), 149–169.

Forman, R.T.T. and Godron, M. (1986) *Landscape Ecology.* Wiley, New York.

Forrest, R.A. (1991) Japanese aid and the environment. *The Ecologist* 21 (1), 24–32.

Forrest, W. and Morison, A. (1991) A government role in better environmental management. *Science of the Total Environment* 108 (1–2), 51–60.

Forsund, F.R. and Strom, S. (1988) *Environmental Economics and Management: pollution and natural resources.* Croom Helm, London.

Fortlage, C.A. (1990) *Environmental Assessment: a practical guide*. Gower, Aldershot.

Foucat, V.S.A. (2002) Community-based ecotourism management moving towards sustainability, in Ventenilla, Oaxaca, Mexico. *Ocean and Coastal Management* 45 (8), 511–529.

Fowler, C. and Mooney, P. (1990) *The Threatened Gene: food, politics, and the loss of genetic diversity*. Lutterworth Press, Cambridge.

Fox, M. (1983) *Original Blessing*. Bear and Co., Santa Fe, CA.

Fox, M. (1989) A call for a spiritual renaissance. *Green Letter* 5 (1), 4, 16–17.

Fox, W. (1984) Deep ecology: a new philosophy of our time? *The Ecologist* 14 (5–6), 194–200.

Fox, W. (1995) *Towards a Transpersonal Ecology: developing new foundations for environmentalism*. Green Books (Resurgence), Totnes.

Francis, J.M. (1996) Nature conservation and the precautionary principle. *Environmental Values* 5 (3), 257–264.

Frankel, B. (1987) *The Post-Industrial Utopias*. Polity Press, Cambridge.

Frankel, O.H., Brown, A.-H.D. and Burdon, J.J. (1995) *The Conservation of Plant Biodiversity*. Cambridge University Press, Cambridge/Earthscan, London.

Frankl, P. and Rubik, F. (2000) *Life Cycle Assessment in Industry and Business: adoption patterns, applications and implications*. Springer, New York.

Franklin, W.E. (1995) Life-cycle assessment – a remarkable tool in the era of sustainable resource and environmental management. *Resources Conservation and Recycling* 14 (3–4), R5–R7.

Fraser, L.H. and Keddy, P.A. (eds) (2005) *The World's Largest Wetlands: ecology and conservation*. Cambridge University Press, Cambridge.

Fraser-Darling, F. (1963) The unity of ecology. *British Association for the Advancement of Science* 20, 297–306.

Freeman, C. and Jahoda, M. (eds) (1978) *World Futures: the great debate*. Martin Robertson, Oxford.

Freestone, D. (1994) The road from Rio: international environmental law after the Earth Summit. *Journal of Environmental Law* 6 (2), 193–218.

French, P. (1998) *Coastal and Estuarine Management*. Routledge, London.

Frenkel, S. (1994) Old theories in new places: environmental determinism and bioregionalism. *Professional Geographer* 46 (3), 289–295.

Freudenburg, W.R. (1986) Social impact assessment. *Annual Review of Sociology* 12, 451–478.

Freudenburg, W.R. (1989) Social scientists' contributions to environmental management. *Journal of Social Issues* 46 (1), 133–152.

Friedman, J. and Weaver, C. (1979) *Territory and Function: the evolution of regional planning*. Edward Arnold, London.

Friend, A.M. (1996) Sustainable development indicators: exploring the objective function. *Chemosphere* 33 (9), 1865–1887.

Frosch, R.A. (1995) The industrial ecology of the 21st century. *Scientific American* (September), 178–181.

Frosch, R.A. and Gallopoulos, N. (1989) Strategies for manufacturing. *Scientific American* 261 (3), 94–98.

Fuller, R.B. (1969) *Operating Manual for Spaceship Earth*. Southern Illinois University Press, Carbondale, IL.

Fuller, T. (2000) Will small become beautiful? *Futures* 32 (1), 79–89.

Funkhauser, S. (1995) *Valuing Climate Change: the economics of the greenhouse*. Earthscan, London.

Funtowicz, S.O. and Ravetz, J.R. (1991) A new scientific methodology for global environmental issues, in R. Costanza (ed.) *The Ecological Economics*. Columbia University Press, Ithaca, NY, pp. 137–152.

Funtowicz, S.O. and Ravetz, J.R. (1992) The good, the true and the post-modern. *Futures* 24 (11), 963–976.

Funtowicz, S.O. and Ravetz, J.R. (1993) Science for the post-normal age. *Futures* 25 (7), 739–755.

Funtowicz, S.O. and Ravetz, J.R. (1994) The worth of a songbird: ecological economics as a post-normal science. *Ecological Economics* 10 (3), 197–207.

Furguson, A.R.B. (1999) The logical foundations of ecological footprints. *Environment, Development and Sustainability* 1 (2), 149–156.

Gallagher, K.P. and Werksman, J. (eds) (2002) *Earthscan Reader in International Trade and Sustainable Development*. Earthscan, London.

Gardiner, J.E. (1989) Decision making for sustainable development: selected approaches to environmental assessment and management. *Environmental Impact Assessment Review* 9 (4), 337–366.

Gare, A.E. (1995) *Postmodernism and the Environmental Crisis*. Routledge, London.

Garlick, J.P. and Keay, R.W. (eds) (1970) *Human Ecology in the Tropics*. Pergamon Press, Oxford.

Garrod, B. and Chadwick, A. (1996) Environmental management and business strategy: towards a new strategic paradigm. *Futures* 28 (1), 37–50.

Garrod, B. and Feyall, A. (1998) Beyond the rhetoric of sustainable tourism. *Tourism Management* 19 (3), 199–212.

Gee, D., Wynn, B., Stirling, A.A. And MacGarvin, M. (2002) *The Precautionary Principle in the 20th Century: late lessons from early warnings*. Earthscan, London.

Geertz, C. (1971) *Agricultural Involution: the process of ecological change in Indonesia*. University of California Press, Berkeley, CA.

Geist, H. (2004) *The Course and Progression of Desertification*. Ashgate, Aldershot.

GEMI (1998) *Environment: value to business*. Global Environmental Management Institute, Washington, DC.

George, S. (1988) *A Fate Worse Than Debt*. Penguin, Harmondsworth.

George, S. (1992) *The Debt Boomerang: how Third World debt can harm us all*. Pluto, London.

Georgiou, S., Whittington, D., Pearce, D.W. and Moran, D. (1997) *Economic Values and the Environment in the Developing World*. Edward Elgar, London.

Geraghty, P.J. (1992) Environmental assessment and the application of an expert systems approach. *Town Planning Review* 63 (2), 123–142.

Geraghty, P.J. (1993) Environmental assessment and the application of expert systems – an overview. *Journal of Environmental Management* 39 (1), 27–38.

Gerasimov, I.P., Armand, D.L. and Yefron, K.M. (eds) (1971) *Natural Resources in the Soviet Union: their use and renewal* (translated from Russian (1963 edn) by J.I. Romanowski). Freeman, San Francisco, CA.

Ghai, D. and Vivian, J.M. (eds) (1992) *Grassroots Environmental Action: people's participation in sustainable development*. Routledge, London.

Ghatak, S. and Turner, R.K. (1978) Pesticide use in less developed countries: economic and environmental considerations. *Food Policy* 5 (2), 134–146.

Gibson, R.B. (2005) *Sustainability Assessment: criteria, processes and applications*. Earthscan, London.

Giddens, A. (1991) *Modernity and Self Identity in the Late Modern Age*. Cambridge University Press, Cambridge.

Gilmour, A.J. and Walkerden, G. (1994) A structured approach to conflict resolution in EIA: the use of adaptive environmental assessment and management (AEAM), in A.J. Gilmour and B. Page (eds) *Computer Support for Environmental Assessment* (vol. 16) (Proceedings of the IFIP Working Conference on Computer Support for Environmental Assessment 6–8 October 1993, Como, Italy). TC5/WG5.11 Transactions: B-Applications in Technology. North-Holland, Amsterdam, pp. 199–210.

Gilpin, A. (1995) *Environmental Impact Assessment: cutting edge for the twenty-first century*. Cambridge University Press, Cambridge.

Gleckman, H. and Krut, R. (1996) Neither international nor standard: the limits of ISO14001 as an instrument of global corporate environmental management. *Greener Management International* (14), 111–124.

Gliessman, S.R. (1990) *Agroecology: researching the ecological basis for sustainable agriculture*. Springer, New York.

Gliessman, S.R. (1992) Agroecology in the tropics: achieving a balance between land use and preservation. *Environmental Management* 16 (6), 681–689.

Goeden, G.B. (1979) Biogeographic theory as a management tool. *Environmental Conservation* 6 (1), 27–32.

Goldman, L.R. (ed.) (2000) *Social Impact Analysis; an applied anthropology manual*. Berg, Oxford.

Goldsmith, E. (1990) Evolution, neo-Darwinism and the paradigm of science. *The Ecologist* 20 (2), 67–73.

Goldsmith, E., Allan, R., Allaby, M., Davol, J. and Lawrence, S. (1972) *Blueprint for Survival*. Penguin, Harmondsworth (also published in *The Ecologist* 2 (1), 1–43).

Goldsmith, E., Khor, M., Norberg-Hodge, H. and Shiva, V. (eds) (1992) *The Future of Progress: reflections on environment and development* (revised edn). Green Books (Resurgence), Totnes, Devon.

Golley, F.B. (1991) The ecosystem concept: a search for order. *Ecological Research* 6 (2), 129–138.

Golley, F.B. (1993) *A History of the Ecosystem Concept in Ecology: more than the sum of the parts*. Yale University Press, New Haven, CN.

Gonzalez, O.J. (1996) Formulating an ecosystem approach to environmental protection. *Environmental Management* 20 (5), 597–605.

Goodall, B. (1995) Environmental auditing: a tool for assessing the performance of tourism firms. *The Geographical Journal* 161 (1), 29–37.

Goodland, R. and Edmundson, V. (eds) (1994) *Environmental Assessment and Development*. The World Bank, Washington, DC.

Goodman, D. and Redclift, M. (1991) *Refashioning Nature: food, ecology and culture*. Routledge, London.

Gordon, T.J. and Helmer, O. (1964) *Report on a Long Range Forecasting Study* (RAND Corporation Paper No. P-2982). RAND Corporation, Santa Monica, CA.

Gorman, M. (1979) *Island Ecology*. Chapman and Hall, London.

Gösling, S. (2000) Sustainable tourism development in developing countries: some aspects of energy use. *Journal of Sustainable Tourism* 8 (5), 410–425.

Gösling, S., Borgström Hansson, C., Hörstmeier, O. and Sagel, S. (2002) Ecological footprint analysis as a tool to assess tourism sustainability. *Ecological Economics* 43 (3), 199–211.

Gottlieb, R. (ed.) (1998) *This Sacred Earth: religion, nature and the environment*. Routledge, New York.

Gould, S.J. (1984) Toward the vindication of punctuational change, in W.A. Berggren and J.A. Van Couvering (eds) *Catastrophes and Earth History: the new uniformitarianism*. Princeton University Press, Princeton, NJ.

Gourlay, K.A. (1992) *World of Waste: dilemmas of industrial development*. Zed Press, London.

Graedel, T.E. and Allenby, B.R. (1995) *Industrial Ecology*. Prentice Hall, New York.

Graham Smith, L.G. (1993) *Impact Assessment and Sustainable Resource Management*. Longman, Harlow.

Grainger, A. (1993) *Controlling Tropical Deforestation*. Earthscan, London.

Gramling, R. and Freudenburg, W.R. (1992) Opportunity – threat, development, and adaptation: toward a comprehensive framework for social impact assessment. *Rural Sociology* 57 (2), 216–234.

Graves, J. and Reavey, D. (1996) *Global Environmental Change: plants, animals, and communities*. Longman, Harlow.

Gray, A. and Stokoe, P. (1988) *Knowledge-Based or Expert Systems and Decision Support Tools for Environmental Assessment and Management: their potential and limitations* (Final Report for the Federal Environmental Assessment and Review Office). Dalhousie University, School for Resource and Environmental Studies, Halifax (Nova Scotia, Canada).

Gray, R.H. (1990) *The Greening of Accounting: the profession after Pearce*. Chartered Association of Certified Accountants, London.

Gray, W. (1993) *Coral Reefs and Islands: the natural history of a threatened paradise*. David and Charles, Newton Abbot.

Grayson, L. (ed.) (1992) *Environmental Auditing: a guide to best practice in the UK and Europe*. Technical Communications and British Library Science and Information Service, Letchworth.

Grayson, R.B., Doolan, J.M. and Blake, T. (1994) Applications of AEAM (adaptive environmental assessment and management) to water quality in the Latrobe River catchment. *Journal of Environmental Management* 41 (3), 245–258.

Green, K. and Randles, S. (eds) (2006) *Industrial Ecology and Spaces of Innovation*. Edward Elgar, Cheltenham.

Greeno, J.L. and Robinson, S.N. (1992) Rethinking corporate environmental management. *Columbia Journal of World Business* 27 (3), 223–232.

Greer, J. and Bruno, K. (1997) *Greenwash: the reality behind corporate environmentalism*. Third World Network, Penang/Appex Press, New York.

Grey, W. (1986) A critique of deep ecology. *Journal of Applied Philosophy* 3 (4), 211–216.

Gribbin, J. and Gribbin, M. (1996) *Fire on Earth*. Simon & Schuster, London.

Groombridge, B. and Jenkins, M. (2002) *World Atlas of Biodiversity*. University of California Press, Berkeley, CA.

Grove, R.H. (1990) The origins of environmentalism. *Nature* 345 (6270), 11–14.

Grove, R.H. (1992) The origins of Western environmentalism. *Scientific American* 267 (1), 42–47.

Grove, R.H. (1995) *Green Imperialism: colonial expansion, tropical island Edens and the origins of environmentalism, 1600–1860*. Cambridge University Press, Cambridge.

Grubb, M., Koch, M., Munson, A., Sullivan, F. and Thomson, K. (1993) *The Earth Summit Agreements: an analysis of the Rio '92 UN Conference on Environment and Development*. Earthscan, London.

Grumbine, R.E. (1994) What is ecosystem management? *Conservation Biology* 8 (1), 27–38.

Grumbine, R.E. (1997) Reflections on 'What is ecosystem management?' *Conservation Biology* 11 (1), 41–47.

Guariso, G. and Page, B. (eds) (1994) *Computer Support for Environmental Impact Assessment* (vol. 16) (Proceedings of the IFIP Working Conference on Computer Support for Environmental Assessment 6–8 October 1993, Como, Italy). TC5/WG5.11 Transactions: B-Applications in Technology. North-Holland, Amsterdam.

Guariso, G. and Werthner, H. (1989) *Environmental Decision Support Systems*. Ellis Horwood-Wiley, New York.

Gumbricht, T. (1996) Application of GIS in training for environmental management. *Journal of Environmental Management* 46 (1), 17–30.

Gunderson, L., Holling, C.S. and Light, S.S. (eds) (1995) *Barriers and Bridges to the Renewal of Regional Ecosystems*. Columbia University Press, New York.

Gunn, D.L. and Stevens, J.G.R. (1976) *Pesticides and Human Welfare*. Oxford University Press, Oxford.

Gupta, A. and Asher, M.-G. (1998) *Environment and the Developing World: principles, policies and management*. Wiley, Chichester.

Gutman, P.S. (1994) Involuntary resettlement in hydropower projects. *Annual Review of Energy and the Environment* 19 (2), 189–210.

Haab, T.C. and McConnel, K.E. (2002) *Valuing Environmental and Natural Resources: the economics of non-market valuation*. Edward Elgar, Cheltenham.

Haas, P.M. (ed.) (2003) *Environment in the New Global Economy*. Edward Elgar, Cheltenham.

Haefele, E. (1973) *Representative Government and Environmental Management*. Johns Hopkins University Press, Baltimore, MD.

Haeuber, R. (1996) Setting the environmental policy agenda: the case of ecosystem management. *Natural Resources Journal* 36 (1), 1–28.

Haigh, N. and Irwin, F. (eds) (1987) *Integrated Pollution Control*. The Conservation Foundation, Washington, DC.

Haines-Young, R., Green, D.R. and Cousins, S. (eds) (1993) *Landscape Ecology and Geographic Information Systems (GIS)*. Taylor & Francis, London.

Hall, A.L. (1978) *Drought and Irrigation in North East Brazil*. Cambridge University Press, Cambridge.

Hall, A.L. (1994) Grassroots action for resettlement planning: Brazil and beyond. *World Development* 22 (12), 1793–1810.

Hall, C.M. and Lew, A.A. (eds) (1999) *Sustainable Tourism: a geographical perspective*. Addison Wesley Longman, Harlow.

Hallman, D.G. (1994) *Ecotheology: voices from South and North* (World Council of Churches Publications). Orbis Books, New York.

Hamilton, K., Pearce, D.W., Atkinson, G., Gomez-Lobo, A. and Young, C. (1994) *The Policy Implications of Natural Resource and Environmental Accounting* (CSERGE Working Paper GEC 94–18). University College London, London.

Hamlyn, M. (1992) El Niño fathers distant drought. *The Times*, 29 May 1992, p. 12.

Hannigan, J.A. (1995) *Environmental Sociology: a social constructionist perspective.* Routledge, London.

Hanley, N. and Spash, C.L. (1994) *Cost–Benefit Analysis and the Environment.* Edward Elgar, London.

Hanley, N., Moffat, I., Faichney, R., and Wilsson, M. (1999) Measuring sustainability: a time series of alternative indicators for Scotland. *Ecological Economics* 28 (1), 55–73.

Hanson, J.A. (1977) Towards an ecologically-based economic philosophy. *Environmental Conservation* 4 (1), 3–10.

Harcourt, W. (ed.) (1994) *Feminist Perspectives on Sustainable Development.* Zed Press, London.

Hardin, G. (1968) The tragedy of the commons. *Science* 162 (3859), 1243–1248.

Hardin, G. (1974a) *The Ethics of a Lifeboat.* American Association for the Advancement of Science, Washington, DC.

Hardin, G. (1974b) Lifeboat ethics: the case against helping the poor. *Psychology Today* 8 (1), 38–43, 123–126.

Hardin, G. (1985) Human ecology: the subversive, conservative science. *American Zoologist* 25 (2), 469–476.

Hardoy, J.E., Mitlin, D. and Satterthwaite, D. (1992) *Environmental Problems in Third World Cities.* Earthscan, London.

Hare, B. (1991) Environmental impact assessment: broadening the framework. *Science of the Total Environment* 108 (1–2), 17–32.

Harremoës, P., Gee, D., MacGarvin, M., Stirling, A., Keys, J., Wynne, B. and Guedes Vaz, S. (2002) *The Precautionary Principle in the 20th Century: late lessons from early warnings.* Earthscan, London.

Harris, C.M. and Meadows, J. (1992) Environmental management in Antarctica: instruments and institutions. *Marine Pollution Bulletin* 25 (9–12), 239–249.

Harris, N. (1990) *Environmental Issues in the Cities of the Developing World* (DPU Working Papers No. 20). Development Planning Unit, University College London, London.

Harris, R. and Leiper, N. (eds) (1995) *Sustainable Tourism: an Australian perspective.* Butterworth-Heinemann, Oxford.

Harrison, L.L. (ed.) (1984) *The McGraw-Hill Environmental Auditing Handbook: a guide to corporate environmental risk management.* McGraw-Hill, New York.

Harrison, P. (1987) *The Greening of Africa: breaking through in the battle for land and food.* Paladin, London.

Harrison, P. (1990) Too much life on Earth? *New Scientist* 126 (1717), 28–29.

Harrison, P. (1992) *The Third Revolution: population and a sustainable world.* Penguin, Harmondsworth.

Harte, D. (1992) Environmental law, in M. Newson (ed.) *Managing the Human Impact on the Natural Environment.* Belhaven, London, pp. 56–79.

Hartshorn, G.S. (1991) Key environmental issues for developing countries. *Journal of International Affairs* 44 (2), 393–402.

Hartwick, J.M. (1990) Natural resources, national accounting and economic depreciation. *Journal of Public Economics* 43 (3), 291–304.

Harvey, B. and Hallett, J.D. (1977) *Environment and Society: an introductory analysis.* Macmillan, London.

Harvey, D. (1989) *The Condition of Postmodernity: an enquiry into the origins of cultural change.* Blackwell, Oxford.

Harvey, T., Mahaffey, K.R., Velazquez, S. and Dourson, M. (1995) Holistic risk assessment: an emerging process for environmental decisions. *Regulatory Toxicology and Pharmacology* 22 (2), 110–117.

Hasan, S. and Khan, M.A. (1999) Community-based environmental management in a megacity: considering Calcutta. *Cities* 16 (2), 103–110.

Hasler, R. (1999) *An Overview of the Social, Ecological and Economic Achievements of Zimbabwe's CAMPFIRE Programme* (IIED 'Evaluating Eden' Discussion Papers No. 3). International Institute for Environment and Development, London.

Hassan, L. and Anglestam, P. (1991) Landscape ecology as a theoretical basis for nature conservation. *Landscape Ecology* 5 (4), 191–201.

Hawken, P. (1993) *The Ecology of Commerce: how businessmen can save the planet.* Weidenfeld & Nicolson, London.

Hayter, T. (1989) *Exploited Earth: Britain's aid and the environment.* Earthscan, London.

Hecht, J. (1991) Asteroid 'airburst' may have devastated New Zealand. *New Scientist* 132 (1789), 19.

Hector, T. (1996) Biotechnology and environmental management. *Biochemical Education* 24 (2), 112–113.

Heer, J.E. and Hagerty, D.J. (1977) *Environmental Assessments and Impact Statements.* Van Nostrand Reinhold, New York.

Heinberg, R. (2004) *Powerdown: options and actions for a post-carbon world.* Clairview Books, Sussex.

Heintzenberg, J. (1989) Arctic haze: air pollution in polar regions. *Ambio* 18 (1), 50–55.

Heiskanen, E. (2002) The institutional logic of life cycle assessment. *Journal of Cleaner Production* 10, 427–437.

Hemmelskamp, J. and Brockmann, K.L. (1997) Environmental labels: the German 'Blue Angel'. *Futures* 29 (1), 67–76.

Henderson, H. (1981a) Thinking globally, acting locally: ethics for the dawning solar age, in H. Henderson (ed.) *The Politics of the Solar Age: alternatives to economics.* Anchor-Doubleday Books, New York, pp. 354–405.

Henderson, H. (1981b) *The Politics of the Solar Age: alternatives to economics.* Anchor-Doubleday Books, New York.

Henderson, H. (1994) Paths to sustainable development: the role of social indicators. *Futures* 26 (2), 125–137.

Henderson, H. (1996) *Building a Win–Win World.* Kumarian Press, Bloomfield, CT.

Henning, D.H. and Mangun, W.R. (1989) *Managing the Environmental Crisis: incorporating competing values in natural resource administration.* Duke University Press, Durham, NC.

Henry, R. (1996) Adapting United Nations agencies for Agenda 21: programme coordination and organisational reform. *Environmental Politics* 5 (1), 1–24.

Herberlein, T.A. (1989) Attitudes to environmental management. *Journal of Social Issues* 45 (1), 37–57.

Hershkovitz, L. (1993) Political ecology and environmental management in the Loess Plateau, China. *Human Ecology* 21 (4), 327–353.

Hester, G.L. (1992) Electric and magnetic fields: managing an uncertain risk. *Environment* 34 (1), 6–11, 25–32.

Hewett, J. (ed.) (1995) *European Environmental Almanac.* Earthscan, London.

Heyward, V. (1995) *Global Biodiversity Assessment.* Cambridge University Press, Cambridge.

Higgs, A.J. (1981) Island biogeographic theory and nature reserve design. *Journal of Biogeography* 8 (2), 117–124.

Hildebrand, S.G. and Cannon, J.B. (eds) (1993) *Environmental Analysis: the NEPA experience.* Lewis Publishers, Boca Raton, FL.

Hildyard, N. (1991) An open letter to Edouard Saouma, Director General of FAO. *The Ecologist* 21 (2), 43–46.

Hildyard, N. (1992) Liberation ecology, in E. Goldsmith, M. Khor, H. Norberg-Hodge and V. Shiva (eds) *The Future of Progress: reflections on environment and development.* Green Books (Resurgence), Totnes, pp. 154–160.

Hill, A.R. (1987) Ecosystem stability: some recent perspectives. *Progress in Physical Geography* 11 (3), 313–333.

Hill, C. (1972) *The World Turned Upside Down: radical ideas during the English Revolution.* Temple Smith, London (Penguin edn 1975).

Hill, D., Smith, N. and Kenneth, I. (1994) *Small Companies, Partnership and Shared Experiences in Improving Environmental Performance* (Proceedings of the 1994 Business, Society and Environment Conference). University of Nottingham, Nottingham.

Hill, J.C. (1964) Puritanism, capitalism and the scientific revolution. *Past and Present* 29 (1), 88–97.

Hill, M.K. (1998) *Understanding Environmental Pollution*. Cambridge University Press, Cambridge.

Hillary, R. (ed.) (1995) *The Role of the Environmental Manager*. Stanley Thornes, Cheltenham.

Hindmarsh, R. (1990) The need for effective assessment: sustainable development and the social impacts of biotechnology in the Third World. *Environmental Impact Assessment Review* 10 (1–2), 195–208.

Hirsch, A. (1993) Improving conservation of biodiversity in NEPA assessments. *The Environmental Professional* 15 (1), 103–115.

HMSO (1956a) *Volta River Project, I: Report of the Preparatory Commission* (Government of UK and Gold Coast). HMSO, London.

HMSO (1956b) *Volta River Project, II: Appendices to reports of the Preparatory Commission* (Government of UK and Gold Coast). HMSO, London.

HMSO (1971) *Report of the Roskill Commission on the Third London Airport*. HMSO, London.

Hodge, I. (1995) *Environmental Economics: individual incentives and public choices*. Macmillan, Basingstoke.

Hofer, T. and Messerlie, B. (2006) *Floods in Bangladesh: history, dynamics and the role of the Himalayas*. United Nations University Press, Tokyo.

Holden, C. (1991) World Bank Environment Fund. *Science* 251 (4996), 870.

Holdgate, M.W. (1990) Antarctica: ice under pressure. *Environment* 32 (6), 4–9, 30–33.

Holdridge, L.R. (1964) *Life Zone Ecology*. Tropical Science Center, San Jose, CA.

Holdridge, L.R. (1971) *Forest Environments in Tropical Life Zones: a pilot study*. Pergamon Press, Oxford.

Holland, A.-C.S. (2005) *The Water Business: corporations versus people*. Zed Press, London.

Holling, C.C. (1973) Resilience and stability of ecological systems. *Annual Review of Ecology and Systematics* 4 (1), 1–23.

Holling, C.S. (ed.) (1978) *Adaptive Environmental Assessment and Management* (revised edn 1980). Wiley, New York.

Holling, C.S. (1987) Simplifying the complex: the paradigm of ecological function and structure. *European Journal of Operational Research* 30 (2), 139–146.

Holling, C.S. (2005) *Adaptive Environmental Assessment and Management* (updated edn). Blackburn Press, Miami, FL.

Holmberg, J. (ed.) (1992) *Policies for a Small Planet*. Earthscan, London.

Holmberg, J., Thomson, K. and Timberlake, L. (1993) *Facing the Future: beyond the Earth Summit*. Earthscan, London.

Holmes, T. and Scoones, I. (2000) *Participatory Environmental Policy Process: experience from North and South* (IDS Working Papers No. 113). Institute of Development Studies, University of Sussex, Brighton. Downloadable http://www.eldis.org/static/Doc7863.htm – accessed April 2005.

Homer-Dixon, T.F. (1991) On the threshold: environmental changes as causes of acute conflict. *International Security* 16 (2), 76–116.

Honey, M. (1999) *Ecotourism and Sustainable Development: who owns paradise?* Island Press, Washington, DC.

Honey, M. (2002) *Ecotourism and Certification: setting standards in practice*. Island Press, Washington, DC.

Honghai Deng (1992) Urban agriculture as urban food supply and environmental protection subsystems in China. Paper to the 1992 Conference on Planning for Sustainable Urban Development, University of Wales Cardiff, Cardiff.

Hopgood, S. (1998) *American Foreign Policy and the Power of the State*. Oxford University Press, Oxford.

Hornsby, M. (1989) Farming's nitrate nightmare. *The Times*, 11 December 1989, p. 11.

Horowitz, M.M. (1991) Victims upstream and down. *Journal of Refugee Studies* 2 (2), 164–181.

Horton, S. and Memon, A. (1997) SEA: the uneven development of the environment. *Environmental Impact Assessment Review* 17 (3), 163–175.

Htun, N. (1990) EIA and sustainable development. *Impact Assessment Bulletin* 8 (1–2), 16–23.

Hudson, N. (1981) *Soil Conservation*. Batsford Books, London.

Hudson, N. (1992) *Land Husbandry*. Batsford Books, London.

Hufschmidt, M.M. (1986) A conceptual framework for watershed management, in K.A. Easter, J.A. Dixon and M.M. Hufschmidt (eds) *Watershed Resource Management: an integrated framework with studies from Asia and the Pacific*. Westview Press, Boulder, CO, pp. 17–31.

Huggett, R. (1990) The bombarded Earth. *Geography* 75 (2), 114–127.

Hughes, B.B. (1980) *World Modeling: the Mesarovic–Pestel world model in the context of its contemporaries*. Lexington Books, Lexington, MA.

Hughes, D. (1992) *Environmental Law* (2nd edn). Butterworths, London.

Hülsberg, W. (1988) *The German Greens: a social and political profile* (English trans. by G. Fagan). Verso, London.

Hunt, D. and Johnson, C. (1995) *Environmental Management Systems: principles and practice*. McGraw-Hill, London.

Hunt, E. (2004) *Thirsty Planet: strategies for sustainable water management*. Zed Press, London.

Hunt, L. (1998) Send in the clouds. *New Scientist* 158 (2136), 28–33.

Hunter, S. and Leyden, K.M. (1995) Beyond NIMBY: explaining opposition to hazardous-waste facilities. *Policy Studies Journal* 23 (4), 601–619.

Huntington, E. (1915) *Civilisation and Climate*. Yale University Press, New Haven, CN.

Hurley, A. (1995) *Environmental Inequalities: class, race, and industrial pollution in Gary, Indiana, 1945–1980*. University of North Carolina Press, Chapel Hill, NC.

Hutchinson, A. and Hutchinson, F. (1997) *Environmental Business Management: sustainable development in the new millennium*. McGraw-Hill, New York.

Hutton, J. and Dickson, B. (eds) (2000) *Endangered Species, Threatened Convention: the past, present and future of CITES*. Earthscan, London.

Independent Commission on International Development Issues (1980) *North–South: a programme for survival*. Pan, London.

International Chamber of Commerce (1989) *Environmental Auditing* (ICC Publication No. 468). International Chamber of Commerce, Paris.

International Chamber of Commerce (1991) *ICC Guide to Effective Environmental Auditing* (ICC Publication no. 483). International Chamber of Commerce, Paris.

International Chamber of Commerce (1993) Business Charter for Sustainable Development: principles for environmental management. *Environmental Conservation* 20 (1), 82–83.

Irvine, S. (1989) Consuming fashions: the limits of green consumerism. *The Ecologist* 19 (3), 88–93.

Irwin, A. (2001) *Sociology and the Environment: a critical introduction to society, nature and knowledge*. Polity Press, Cambridge.

Isard, W. (1972) *Ecologic-Economic Analysis for Regional Development*. Free Press, New York.

IUCN Inter-Commission Task Force on Indigenous Peoples (1997) *Indigenous Peoples and Sustainability: cases and actions*. International Books, Utrecht.

IUCN, UNEP and WWF (1980) *World Conservation Strategy: living resource conservation for sustainable development*. International Union for Conservation of Nature and Natural Resources, Gland.

IUCN, UNEP and WWF (1991) *Caring for the Earth: a strategy for sustainable living*. Earthscan, London.

Ives, J.D. (1987) The theory of Himalayan environmental degradation – its validity and application challenged by recent research. *Mountain Research and Development* 7 (2), 185–199.

Ives, J.D. and Messerli, B. (1989a) *The Himalayan Dilemma: reconciling development and conservation*. Routledge, London.

Ives, J.D. and Messerli, B. (1989b) *The Theory of Himalayan Environmental Degradation. What is the nature of the perceived crisis?* Routledge, New York.

Ives, J.H. (1985) *The Export of Hazards*. Routledge & Kegan Paul, London.

Jackson, A.R.W. and Jackson, J.M. (1996) *Environmental Science: the natural environment and human impact*. Longman Scientific and Technical, Harlow.

Jackson, C. (1993) Environmentalism and gender interests in the Third World. *Development and Change* 24 (4), 649–678.

Jackson, S.L. (1997) *ISO14000 Implementation Guide: creating an integrated management system*. Wiley, Chichester.

Jacob, M. (1994) Sustainable development and deep ecology: an analysis of competing traditions. *Environmental Management* 18 (4), 477–488.

Jacobs, P. and Sadler, B. (eds) (1989) *Sustainable Development and Environmental Assessment: perspectives on planning for a common future*. Canadian Environmental Assessment Research Council, Hull, Quebec, Canada.

Jakeman, A.J., Beck, M.B. and McAleer, M.J. (eds) (1993) *Modelling Change in Environmental Systems*. Wiley, Chichester.

James, D.E. (1994) *The Application of Economic Techniques in Environmental Impact Assessment*. Kluwer Academic, London.

Jansen, K. and Vellema, S. (eds) (2004) *Agribusiness and Society: corporate responses to environmentalism, market opportunities and public regulation*. Zed Press, London.

Janssen, R. (1995) *Multiobjective Decision Support for Environmental Management*. Kluwer Academic, Dordrecht.

Jansson, B.-O. (1972) *Ecosystem Approach to the Baltic Problem* (Ecological Research Committee Bulletin No. 16). Ekologicommitten, Statens Naturvetenskapliga Forskningsrad, Stockholm.

Jasch, C. (2000) Environmental performance evaluation and indicators. *Journal of Cleaner Production* 8 (1), 79–88.

Jeffery, R. and Vira, B. (eds) (2001) *Conflict and Cooperation in Participatory Natural Resource Management*. Palgrave, New York.

Jeffrey, D.W. and Madden, B. (eds) (1991) *Bioindicators and Environmental Management*. Academic Press, London.

Jenkins, B.R. (1991) Changing Australian monitoring and policy practice to achieve sustainable development. *Science of the Total Environment* 108 (1), 33–50.

Jensen, M.E., Bourgeron, P., Everett, R. and Goodman, I. (1996) Ecosystem management: a landscape ecology perspective. *Water Resources Bulletin* 32 (2), 203–216.

Jiggins, J. and Rowling, N. (2000) Adaptive management: potential and limitations for ecological governance. *Journal of Agricultural Resources Governance and Ecology* 1 (1), 28–42.

Jodha, N.S. (1991) *Rural Common Property Resources: a growing crisis* (Gatekeeper Series No. S24). International Institute for Environment and Development, London.

Johnson, D.L. and Lewis, L.A. (1995) *Land Degradation: creation or destruction*. Blackwell, Oxford.

Johnson, J. (1990) Waste that no one wants. *New Scientist* 127 (1733), 50–55.

Johnson, S.P. (1993) *The Earth Summit – the United Nations Conference on Environment and Development (UNCED)*. Graham & Trotman, London.

Johnson, W.H. and Steere, W.C. (eds) (1974) *The Environmental Challenge*. Holt, Rinehart & Winston, New York.

Jolly, V. (1978) The concept of environmental management. *Development Forum* 8 (2), 13–26.

Jones, M.I. and Greig, L.A. (1985) Adaptive environmental assessment and management: a new approach to environmental impact assessment, in V.W. Maclaren and J.B. Whitney (eds) *New Directions in Environmental Impact Assessment in Canada*. Methuen, Toronto, pp. 21–42.

Jorgensen, S.E. and Goda, T. (1986) Scope and limit in the application of biological models to environmental management. *Ecological Modelling* 32 (1–3), 237–240.

Jorgensen, S.E., Halling-Sorensen, B. and Nielsen, S.N. (1996) *Handbook of Environmental and Ecological Modeling*. CRC Press, Boulder, CO, and Springer, New York.

Jorissen, J. and Coenen, R. (1992) The EEC Directive on EIA and its implementation in the EC member states, in A.G. Colombo (ed.) *Environmental Impact Assessment* (vol. 1). Kluwer, Dordrecht, pp. 1–14.

Jull, P. (ed.) (1994) *Surviving Columbus: indigenous peoples, political reform, and environmental management in North Australia.* North Australia Research Unit, Australian National University, Darwin.

Kabeer, N. (1994) *Reversed Realities: gender hierarchies in development thought.* Verso, London.

Kahn, H., Brown, W. and Martel, L. (eds) (1976) *The Next 200 Years.* Abacus, London.

Kalapakian, J. (2004) *Identity, Conflict and Cooperation in International River Systems.* Ashgate, Basingstoke.

Kalof, L. (ed.) (2005) *The Earthscan Reader in Environmental Values.* Earthscan, London.

Karagozoglu, N. and Lindell, M. (2000) Environmental management: testing the win–win model. *Journal of Environmental Planning and Management* 43 (6), 817–830.

Karl, H. (1994) Better environmental future in Europe through environmental auditing. *Environmental Management* 18 (4), 617–621.

Kasperson, J.X. and Kasperson, R.E. (eds) (2001) *Global Environmental Risk.* Earthscan, London.

Kasperson, J.X., Kasperson, R.E. and Turner II, B.L. (eds) (2003) *Regions at Risk: comparisons of threatened environments.* UN University Press, Tokyo.

Kassas, M. (1999) Rescuing drylands: a project for the world. *Futures* 31, 945–958.

Kates, R.W. (1978) *Risk Assessment of Environmental Hazards.* Wiley, Chichester.

Kates, R.W. and Haarman, V. (1992) Where the poor live: are the assumptions correct? *Environment* 34 (4), 4–11, 25–28.

Kates, R.W. and Hohenemser, C. (eds) (1982) *Technological Hazard Assessment.* Oelgeschlager, Gunn and Hain, Cambridge, MA.

Kearns, L. (1996) Saving the creation: Christian environmentalism in the United States. *Sociology of Religion* 57 (1), 55–70.

Keating, M. (1993) *The Earth Summit's Agenda for Change: a plain language version of Agenda 21 and the other Rio Agreements.* Centre for Our Common Future, Geneva.

Keen, M., Brown, V.A. and Dyball, R. (eds) (2005) *Social Learning in Environmental Management: building a sustainable future.* Earthscan, London.

Kennedy, P. (1993) *Preparing for the Twenty-First Century.* Harper Collins, London.

Keeney, R.L. and von Winterfeldt, D. (2001) Appraising the precautionary principle – a decision analysis perspective. *Journal of Risk Research* 4 (2), 191–202.

Kennedy, I. (1992) *Acid Soil and Acid Rain* (2nd edn). Wiley, Chichester.

Keohane, R.O. and Levy, M.A. (1996) *Institutions for Environmental Aid: pitfalls and promise.* MIT Press, Cambridge, MA.

Kerr, R.A. (1972) Huge impact led to mass extinction. *Science* 257 (5072), 878–880.

Kershaw, K.A. (1973) *Quantitative and Dynamic Plant Ecology* (2nd edn). Edward Arnold, London.

Khagram, S. (2005) *Dams for Development: transnational struggles for water and power.* Cornell University Press, Ithaca, NY.

Khan, H., Brown, W. and Martel, L. (1976) *The Next 200 Years: a scenario for America and the world.* William Morrow, New York.

Khoo, C.H. (1991) Environmental management in Singapore. *Environmental Monitoring and Assessment* 19 (1–3), 127–130.

Khor, M. (2004) *Intellectual Property, Biodiversity and Sustainable Development: resolving the difficulties.* Zed Press, London.

Kiely, R. and Marfleet, P. (eds) (1998) *Globalisation and the Third World.* Routledge, London.

Kimmins, J.P. (1993) Ecology, environmentalism and green religion. *Forestry Chronicle* 69 (3), 285–289.

Kirkman, R. (1997) Why ecology cannot be all things to all people: the 'adaptive radiation' of scientific concepts. *Environmental Ethics* 19 (4), 375–390.

Kirkpatrick, D. (1990) Environmentalism: the new crusade. *Fortune*, 12 February 1990, pp. 24–30.

Kivell, P., Roberts, P. and Walker, G. (eds) (1988) *Environment, Planning and Land Use.* Ashgate, Aldershot.

Kiy, R. and Wirth, J.D. (eds) (1998) *Environmental Management on North America's Borders.* Texas A & M University Press, Austin, TX.

Klare, M.T. (2001) *Resource Wars: the new landscape of global conflict*. Henry Holt, New York.

Klassen, R.D. and McLaughlin, C.P. (1996) The impact of environmental management on firm performance. *Management Science* 42 (8), 1199–1214.

Klee, G.A. (ed.) (1980) *World Systems of Traditional Resource Management*. Edward Arnold, London.

Klijn, F., De Waal, R.W. and Oude Voshaar, J.H. (1995) Ecoregions and ecodistricts: ecological regionalizations for The Netherlands' environmental policy. *Environmental Management* 19 (6), 797–813.

Kneese, A.V. (1977) *Economics and the Environment*. Penguin, Harmondsworth.

Knight, P. (1997) Rally round the standard. *Green Futures*, December 1997, pp. 38–41.

Kobori, I. and Glantz, M.H. (eds) (1998) *Central Eurasian Water Crisis: Caspian, Aral, and Dead Seas*. UN University Press, Tokyo.

Kok, E., O'Laoire, D. and Welford, R. (1993) Environmental management at the regional level. A case study: the Avoca–Avonmore Catchment Conversion Project. *Journal of Environmental Planning and Management* 36 (3), 377–394.

Komarov, B. (1981) *The Destruction of Nature in the Soviet Union*. Pluto Press, London.

Koninklijk Instituut voor de Tropen (1990*) Environmental Management in the Tropics: an annotated bibliography 1985–89*. Royal Tropical Institute, Amsterdam.

Kopitsky, J.J. and Betzenberger, E.T. (1987) Bankers debate . . . should banks lend to companies with environmental problems? *The Journal of Commercial Bank Lending* 68 (11), 3–13.

Kotlyakov, V.M. (1991) The Aral Sea basin: a critical environmental zone. *Environment* 33 (1), 4–9, 36–38.

Koutstaal, P. (1997) *Economic Policy and Climate Change: tradeable permits for reducing carbon emissions*. Edward Elgar, Cheltenham.

Koziel, I. and Saunders, J. (eds) (2001) *Living off Biodiversity: exploring livelihoods and biodiversity issues in natural resource management*. IIED, London.

Kozlowski, J.M. (1986) *Threshold Approach in Urban, Regional and Environmental Planning*. University of Queensland Press, St Lucia, Queensland, Australia.

Kozlowski, J.M. (1990) Sustainable development in professional planning: a potential contribution of the EIA and UET concepts. *Landscape and Urban Planning* 9 (3), 307–332.

Kriebel, D., Tickner, J., Epstein, P., Lemons, J., Levins, R., Loechler, E., Quinn, M., Rudel, R., Schettler, T. and Stoto, M. (2001) The precautionary principle in environmental science. *Environmental Health Perspectives* 109 (9), 871–876.

Krimsky, S. and Goulding, D. (eds) (1992) *Social Theories of Risk*. Praeger, Westport, CT.

Kropotkin, P. (1974) *Fields, Factories and Workshops*, ed. C. Ward (original 1899 in Russian–English edition 1899 published by Freedom Press, London). Unwin, London – with addition of the word 'tomorrow' to the end of the title.

Krutilla, J.V. and Fisher, A. (1975). *The Economics of Natural Environments*. Johns Hopkins University Press, Baltimore, MD.

Kuik, O. and Verbruggen, H. (1991) *In Search of Indicators of Sustainable Development*. Kluwer, Dordrecht.

Kundstier, J.H. (2005) *The Long Emergency: surviving the end of the oil age, climate change, and other converging catastrophes of the twenty-first century*. Atlantic Monthly Press, New York.

Kunstadter, P., Bird, E.C.F. and Sabhasri, S. (eds) (1989) *Man in the Mangroves: the socio-economic situation of human settlements in mangrove forests*. United Nations University Press, Tokyo.

Kurth, W. (1992) The mounting pile of waste paper. *The OECD Observer* 174, 27–30.

Kuzmiak, D.T. (1991) A history of the American environmental movement. *The Geographical Journal* 157 (3), 265–278.

Lafferty, W.M. (ed.) (2004) *Governance for Sustainable Development: the challenge of adapting form to function*. Edward Elgar, Cheltenham.

Lafferty, W.M. and Meadowcroft, J. (eds) (1996) *Democracy and the Environment: problems and prospects*. Edward Elgar, Cheltenham.

Lagadic, L., Caquet, T., Amard, J.-C. and Ramade, F. (eds) (2000) *Use of Biomarkers for Environmental Quality Assessment.* Science Publishers, Enfield, NH.

Lajeunesse, D., Domon, G., Drapeau, P., Cogliastro, A. and Bouchard, A. (1995) Development and application of an ecosystem management approach for protected natural areas. *Environmental Management* 19 (4), 481–495.

Lal, R. (1995) *Sustainable Management of Soil Resources in the Humid Tropics.* United Nations University Press, Tokyo.

Lane, P. and Peto, M. (1995) *Blackstone's Guide to the Environment Act 1995.* Blackstone Press, London.

Lang, R. (ed.) (1986) *Integrated Approaches to Resource Planning and Management.* University of Calgary Press, Calgary.

Lang, T. and Heasman, M. (2004) *Food Wars: the global battle for our mouths, minds and markets.* Earthscan, London.

Lang, T. and Hines, C. (1993) *The New Protectionism: protecting the future against free trade.* Earthscan, London.

Lang, T. and Yu, D. (1992) Free trade versus the environment: a debate. *Our Planet* (UNEP) 4 (2), 12–13.

Laurence, W. (2005) Razing America. *New Scientist* 188 (2521), 34–39.

Lawrence, D.P. (1997) Integrating sustainability and environmental impact assessment. *Environmental Management* 21 (1), 23–42.

Leach, G. and Mearns, R. (1989) *Beyond the Woodfuel Crisis: people, land and trees in Africa.* Earthscan, London.

Leach, M. and Mearns, R. (eds) (1996) *The Lie of the Land: challenging received wisdom on the African environment* (International African Institute, London). James Currey and Indiana University Press, London.

Leach, M., Joekes, S. and Green, C. (1995) Editorial: gender relations and environmental change. *IDS Bulletin* 26 (1), 1–8.

Leach, M., Mearns, R. and Scoones, I. (eds) (1997) Community-based sustainable development: consensus or conflict? *IDS Bulletin* 28 (4), 1–96.

Ledgerwood, G., Street, E. and Therevel, R. (1992) *Environmental Audit and Business Strategy.* Pitman, London.

Lee, N. (1978) Environmental impact assessment of projects in EEC countries. *Journal of Environmental Management* 6 (1), 57–71.

Lee, N. (1982) The future development of environmental impact assessment. *Journal of Environmental Management* 14 (1), 71–90.

Lee, N. and Walsh, F. (1992) Strategic environmental assessment: an overview. *Project Appraisal* 7 (3), 126–137.

Lee, R.G., Field, D.R. and Burch, W.R. Jnr (1990) *Community and Forestry: continuities in the sociology of natural resources.* Westview Press, Boulder, CO.

Leistritz, L.F., Coon, R.C. and Hamm, R.R. (1995) A microcomputer model for assessing socio-economic impacts of development projects. *Impact Assessment* 12 (4), 373–384.

Leitmann, J. (1993) Rapid urban environmental assessment: toward environmental management in cities of the developing world. *Impact Assessment Review* 11 (3), 225–260.

Leonard, H.J. (with Yudelman, M., Stayker, J.D., Browder, J.O., De Boer, A.J., Campbell, T. and Jolley, A.) (1989) *Environment and the Poor: development strategies for a common agenda* (US-Third World Policy Perspectives No. 11) (Overseas Development Corporation). Transaction Books, New Brunswick, NJ.

Leopold, A. (1949) *A Sand County Almanac.* Oxford University Press, London.

Leu, W.S., Williams, W.P. and Bark, A.W. (1995) An environmental evaluation of the implementation of environmental assessment by UK local authorities. *Project Appraisal* 10 (2), 91–102.

Levett, R. (1993) *A Guide to the Eco-Management and Audit Scheme for UK Local Government.* HMSO, London.

Levett, R. (1996) From eco-management and audit (EMAS) to sustainability management and audit (SMAS). *Local Environment* 1 (3), 329–334.

Levine, L. (1993) Gaia – goddess and idea. *Biosystems* 31 (2–3), 85–92.

Levy, D.L. and Newell, P.J. (2004) *The Business of Global Environmental Governance.* MIT Press, Baltimore, MD.

Lewin, R. (1992) How to destroy the doomsday asteroid. *New Scientist* 134 (1824), 12–13.

Lewin, R. (1993) *Complexity: life at the edge of chaos.* Dent, London.

Lewis, C. (ed.) (1996) *Managing Conflicts in Protected Areas.* IUCN, Gland.

Lewis, M. (1992) *Green Delusions: an environmentalist critique of radical environmentalism.* Duke University Press, Durham, NC.

Light, A. (ed.) (1998) *Social Ecology After Bookchin.* Guilford Press, New York.

Likens, G.E. (1992) *The Ecosystems Approach: its use and abuse.* Ecology Institute, Oldendorf (Germany).

Lindberg, K. and Hawkins, D. (eds) (1993) *Ecotourism: a guide for planners and managers.* The Ecotourism Society, Bennington, VT (1st edn 1988).

Linear, M. (1982) Gift of poison: the unacceptable face of development aid. *Ambio* xi (1), 2–8.

Linear, M. (1985) *Zapping the Third World: the disaster of development aid.* Zed Press, London.

Linnerooth-Bayer, J., Loftstedt, R.E. and Jostedt, G. (eds) (2005) *Transboundary Risk Management.* Earthscan, London.

Lintott, J. (1996) Environmental accounting: useful to whom and for what? *Ecological Economics* 16 (3), 179–190.

Lipscombe, N. and Thwaites, R. (2003) Contemporary challenges for ecotourism in Vietnam. *Tourism Review International* 7 (1), 23–35.

Lise, W. (1995) Participatory development, a path to ecological preservation: an approach with case-studies and modelling. *International Journal of Environment and Pollution* 5 (2–3), 243–259.

Lister, C. (1996) *European Union Environmental Law: a guide for industry.* Wiley, Chichester.

Litke, S. and Day, J.C. (1998) Building local capacity for stewardship and sustainability. *Environment* 25 (2–3), 91–109.

Little, P.D. and Horowitz, M.M. (1987) *Lands at Risk in the Third World: local level perspectives.* Westview Press, Boulder, CO.

Llewellyn, L.G. and Freudenburg, W.R. (1989) Legal requirements for social impact assessment – assessing the social-science fallout from Three Mile Island. *Society and Natural Resources* 2 (3), 193–208.

Local Government Management Board (1991) *Environmental Auditing in Local Government: a guide and discussion paper.* Local Government Management Board, Luton.

Local Government Management Board (1992) *Local Agenda 21 – Agenda 21: a guide for local authorities in the UK.* Local Government Management Board, Luton.

Local Government Management Board (1994) *Agenda 21, a Guide for Local Authorities in the UK.* Local Government Management Board, Luton.

Loehle, C. and Osteen, R. (1990) IMPACT – an expert system for environmental impact assessment. *AI Applications in Natural Resource Management* 4 (1), 35–43.

Lomborg, B. (2001) *The Sceptical Environmentalist: measuring the real state of the world.* Cambridge University Press, Cambridge (published in Danish 1998).

Lomborg, B. (ed.) (2004) *Global Crises, Global Solutions: priorities for a world of scarcity.* Cambridge University Press, Cambridge.

Lovelock, J.E. (1979) *Gaia: a new look at life on Earth.* Oxford University Press, Oxford.

Lovelock, J.E. (1988) *The Ages of Gaia: a biography of our living Earth.* Oxford University Press, Oxford.

Lovelock, J.E. (1992) *Gaia: the practical science of planetary medicine.* Gaia Books, London.

Lovelock, J.E. and Margulis, L. (1973) Atmospheric homoeostasis by and for the biosphere: the Gaia hypothesis. *Tellus* 26 (1), 2.

Low, N. and Gleeson, B. (1998) *Justice, Society and Nature: an explanation of political ecology.* Routledge, London.

Lowe, E.A., Warren, J.L. and Moran, S.R. (2000) *Discovering Industrial Ecology: an executive briefing and sourcebook.* Battelle Press, Columbus, OH.

Lowe, J. and Lewis, D. (1980) *The Economics of Environmental Management*. Phillip Alan, Oxford.

Lucas, C. and Woodin, M. (2004) *Green Alternatives to Globalization: a manifesto*. Pluto Press, London.

Lück, M. (2002) Looking into the future of ecotourism and sustainable tourism. *Current Issues in Tourism* 5 (3–4), 371–374.

Ludwig, H.F., Ludwig, R.G., Anderson, D.R. and Garber, W.F. (1992) Appropriate environmental standards in developing nations. *Water Science and Technology* 25 (9), 17–30.

Lyell, C. (1830) *Principles of Geology, Being an Attempt to Explain the Former Changes of the Earth's Surface by Reference to Changes Now in Operation* (3 vols). Published between 1830 and 1834. John Murray, London.

Lynas, M. (2004) *High Tide: news from a warming world*. Flamingo, London.

Lynch, K. (2004) *Rural–Urban Intervention in the Developing World*. Routledge, London.

Lyon, R.M. (1989) Transferable discharge permit systems and environmental management in developing countries. *World Development* 17 (8), 1299–1312.

MacArthur, R.H. and Wilson, E.O. (1967) *The Theory of Island Biogeography*. Princeton University Press, Princton NJ.

McAuslan, P. (1991) The role of courts and other judicial bodies in environmental management. *Journal of Environmental Law* 3 (2), 195–205.

McBurney, S. (1990) *Ecology into Economics Won't Go*. Green Books, Bideford.

MacCarthy, F. (1994) *William Morris: a life for our times*. Faber and Faber, London.

McConnel, F. (1996) *The Biodiversity Convention: a negotiating history*. Kluwer Law International, London.

McCool, S.F. (2002) *Tourism, Recreation and Sustainability: linking culture and the environment*. CABI Publications, Wallingford.

McCormick, J.F. (1989) *Reclaiming Paradise: the global movement*. Indiana University Press, Bloomington, IN.

McCormick, J.F. (1992) *The Global Environmental Movement*. Belhaven, London.

McCormick, J.F. (1993) Implementation of NEPA and environmental impact assessment in developing nations, in S.G. Hildebrand and J.B. Cannon (eds) *Environmental Analysis: the NEPA experience*. Lewis Publishers, Boca Raton, FL, pp. 716–727.

McCormick, J.F. (1997) *Acid Earth: the politics of acid pollution* (3rd edn). Earthscan, London.

McCulloch, J. (ed.) (1990) *Cities and Global Climate Change*. Climate Institute, Washington, DC.

McDavid, L. (1995) Safety net: the Sacred Earth Network uses the Internet to protect the environment in the former Soviet Union. *Internet World* 6 (8), 70.

McDonagh, P. and Prothero, A. (1997) *Green Management: a reader*. Dryden Press, London.

McDonald, G.T. and Brown, A.L. (1990) Planning and management processes and environmental assessment. *Impact Assessment Bulletin* 8 (1–2), 261–274.

McEldowney, J.F. (1996) *Environment and the Law*. Addison Wesley Longman, Harlow.

McEvoy III, J. (1971) A comment: conservation an upper-middle class social movement. *Journal of Leisure Research* 3 (2), 127–128.

McGill, S. (1999) *Environmental Risk Management*. Routledge, London.

McGillivray, J.W. (1994) *Environmental Measures*. Environmental Challenge Group, London.

McGillivray, J.W. (1996) *Natural Resource and Environmental Economics*. Addison Wesley Longman, Harlow.

McGranaham, G. (1991) *Environmental Problems and the Urban Household in Third World Countries*. The Stockholm Environmental Institute, Stockholm.

McGregor, D., Simon, D. and Thompson, D. (2005) *The Peri-urban Interface: approaches to sustainable natural and human resource use*. Earthscan, London.

McGregor, J. (1993) Refugees and the environment, in R. Black and V. Robinson (eds) *Geography and Refugees: patterns and processes of change*. Belhaven, London, pp. 157–170.

McGregor, J. (1994) Climate-change and involuntary migration: implications for food security. *Food Policy* 19 (2), 120–132.

McHarg, I. (1969) *Design with Nature.* Doubleday (Natural History Press), Garden City, NY.

McKenna & Co. (1993) *Environmental Auditing – a Management Guide* (2 vols). Intelex Press, London.

MacKenzie, D. (1998) Waste not. *New Scientist* 159 (2149), 26–30.

McLain, R.J. and Lee, R.G. (1996) Adaptive management: promises and pitfalls. *Environmental Management* 20 (4), 437–448.

McMichael, A.J., Haines, A., Sloof, R. and Kovats, S. (eds) (1996) *Climate Change and Human Health.* WHO, Geneva.

McNeill, J. (2000) *Something New Under the Sun: an environmental history of the twentieth century.* Penguin, London.

McNeely, J.A. (1989) How to pay for conserving biological diversity. *Ambio* xviii (6), 303–318.

McNeely, J.A., Miller, K.R., Reid, W.V., Mittermeier, R.A. and Werner, T.B. (1990) Strategies for conserving biodiversity. *Environment* 32 (3), 16–20, 36–40.

Maddox, J. (1972) *The Doomsday Syndrome.* Maddox Educational, London.

Mahony, R. (1992) Debt-for-nature swaps: who really benefits? *The Ecologist* 21 (3), 97–103.

Mainguet, M. (1994) *Desertification: natural background and human mismanagement* (2nd edn). Springer-Verlag, Berlin.

Malmqvist, B. and Rundle, S. (2002) Threats to the running water ecosystems of the world. *Environmental Conservation* 29(2), 134–153.

Maltby, E. (1986) *Waterlogged Wealth: why waste the world's wet places?* Earthscan, London.

Malthus, T.R. (1798) *An Essay on the Principle of Population as it Affects the Future Improvement of Mankind.* J. Johnson, London.

Manheim, B.S. Jnr (1994) NEPA's overseas application. *Environment* 36 (3), 43–45.

Mannion, A.M. (1991) *Global Environmental Change: a natural and cultural environmental history.* Longman, Harlow.

Mannion, A. (1996) The new environmental determinism. *Environmental Conservation* 21 (1), 7–8.

Mara, D. and Cairncross, S. (1990) *Guidelines for the Safe Use of Wastewater and Excreta in Agriculture and Aquaculture.* World Health Organisation, Geneva.

Margaris, N.S. (1987) Desertification in the Aegean Islands. *Ekistics* 54 (323/324), 132–137.

Margerum, R.D. and Born, S.M. (1995) Integrated environmental planning and management: moving from theory to practice. *Journal of Environmental Planning and Management* 38 (3), 371–391.

Marion, M. (1996) *Environmental Issues and Sustainable Futures: a critical guide to recent books, reports, and periodicals.* World Futures Society, Bethesda, MD, and Norton, New York.

Markandya, A. and Halsnaes, K. (eds) (1992) *Climate Change and Sustainable Development: prospects for developing countries.* Earthscan, London.

Marsh, G.P. (1864) *Man and Nature: or physical geography as modified by human action.* Charles Scribner, New York.

Marsh, J.S. (1994) GATT and the environment. *Global Environmental Change* 4 (2), 91–95.

Marsh, W.M. (1991) *Landscape Planning: environmental applications* (2nd edn). Wiley, New York.

Marshall, G. (2005) *Economics for Collaborative Environmental Management: renegotiating the commons.* Earthscan, London.

Marten, G. (2001) *Human Ecology: basic concepts for sustainable development.* Earthscan, London.

Martens, W.J.M. (1998) *Health and Climate Change: monitoring the impacts of global warming and ozone depletion.* Earthscan, London.

Marti, J. and Ernst, G. (eds) (2005) *Volcanoes and the Environment.* Cambridge University Press, Cambridge.

Martin, P.S. (1967) Prehistoric overkill, in P.S. Martin and H.E. Wright (eds) *Pleistocene Extinctions.* Yale University Press, Newhaven, CT, pp. 75–12.

Martin, P.S. and Klein, R.G. (eds) (1984) *Quaternary Extinctions.* University of Arizona Press, Tucson, AZ.

Martin, S. (2002) Professionals and sustainability. *Journal of the Institution of Environmental Sciences* 11 (3), 6–7.

Martinez-Alier, J. (2003) *The Environmentalism of the Poor: a study of ecological conflicts and valuation.* Edward Elgar, Cheltenham.

Maskrey, A. (1989) *Disaster Mitigation: a community based approach* (Oxford Development Guidelines No. 3). Oxfam, Oxford.

Maslin, M. (2004) *Global Warming: a very short introduction.* Oxford University Press, Oxford.

Mason, P. (1993) Spare tyres drop into uncharted waters. *The Times,* 5 Feburary 1993.

Matthews, W.H., Perkowski, J.C., Curtis, F.A. and Martin, W.F. (1976) *Resource Management for Environmental Management and Education.* MIT Press, Cambridge, MA.

Max-Neef, M. (1982) *From the Outside Looking In.* Dag Hammarskjöld Foundation, Copenhagen.

Max-Neef, M.A. (1986) *Economía descalza: señales desde el mundo invisible.* Nordan, Stockholm.

Max-Neef, M.A. (1992a) *From the Outside Looking In: experiences in 'barefoot economics'* (originally published in 1982 by Dag Hammarskjöld Foundation, Copenhagen). Zed Press, London.

Max-Neef, M.A. (ed.) (1992b) *Real-Life Economics: understanding wealth creation.* Routledge, London.

Maxwell, J.W. and Reuveny, R. (eds) (2005) *Trade and Environment: theory and policy in the context of EU employment and economic transition.* Edward Elgar, Cheltenham.

Maxwell, S. (1986) Farming systems research: hitting a moving target. *World Development* 14 (1), 65–77.

Mayda, J. (1993) Historical roots of EIA? *Impact Assessment Bulletin* 11 (4), 411–415.

Meadows, D.H., Meadows, D.L. and Randers, J. (1992) *Beyond the Limits: global collapse or sustainable future?* Earthscan, London.

Meadows, D.H., Randers, J. and Meadows, D. (2004) *The Limits to Growth: the 30-year update.* Earthscan, London.

Meadows, D.H., Meadows, D.L., Randers, J. and Behrens, W.W. III (1972) *The Limits to Growth (a report to the Club of Rome's project on the predicament of mankind).* Universal Books, New York.

Mellanby, K. (1970) *Pesticides and Pollution.* Fontana, London.

Mercer, K.G. (1995) An expert system utility for environmental impact assessment in engineering. *Journal of Environmental Management* 45 (1), 1–23.

Merchant, C. (1980) *The Death of Nature: women, ecology and the scientific revolution.* Harper & Row, New York.

Merchant, C. (1992) *Radical Ecology: the search for a livable world.* Routledge, London.

Merchant, C. (1996) *Earthcare: women and the environment.* Routledge, New York.

Mesarovic, M. and Pestel, E. (1975) *Mankind at the Turning Point* (the Second Report to the Club of Rome). Hutchinson, London (published 1974 in USA).

Messerschmidt, D.A. (1986) 'Go to the people': local planning for natural resource development in Nepal. *Practising Anthropology* 7 (4), 1–15.

Middleton, N. (1995) *The Global Casino: an introduction to environmental issues.* Arnold, London.

Middleton, N. and O'Keefe, P. (2003) *Rio Plus Ten: politics, poverty and the environment.* Pluto Press, London.

Mies, M. (1986) *Patriarchy and Accumulation on a World Scale.* Zed Press, London.

Mies, M. and Shiva, V. (1993) *Ecofeminism.* Zed Press, London.

Mikesell, R.F. (1995) *Economic Development and the Environment.* Cassell, London.

Milani, B. (2000) *Designing the Green Economy: the post-industrial alternative to corporate globalization.* Rowman & Littlefield, Lanham, MD.

Miller, G.T., Jnr (1990) *Living in the Environment: an introduction to environmental science* (6th edn). Wadsworth, Belmont, CA.

Miller, G.T. Jnr (1991) *Environmental Science: sustaining the Earth* (3rd edn). Wadsworth, Belmont, CA.

Miller, R.I. (1978) Applying island biogeographic theory to an East African reserve. *Environmental Conservation* 5 (3), 191–195.

Milne, R. (1988) Conservation clouds choke Britain's forests. *New Scientist* 117 (1604), 27.

Milton, K. (ed.) (1993) *Environmentalism: the view from anthropology.* Routledge, London.

Milton, K. (1996) *Environmentalism and Cultural Theory: exploring the role of anthropology in environmental discourse.* Routledge, London.

Ministerie VROM (1989) *National Environmental Policy Plan: to choose or lose?* SDU, The Hague.

Mintel (1990) *The Green Consumer.* Mintel, London.

Mitchell, B. (1979) *Geography and Resource Analysis.* Longman, Harlow.

Mitchell, B. (1997) *Resource and Environmental Management.* Addison Wesley Longman, Harlow.

Mitchell, J.K. (ed.) (2005) *Crucibles of Hazard: mega-cities and disasters in transition.* United Nations University Press, Tokyo.

Mitsch, W.J. and Gosselink, J.G. (1993) *Wetlands* (2nd edn). Van Nostrand Reinhold, New York.

Mnaksakanian, R.A. (1992) *Environmental Legacy of the Former Soviet Union.* Centre for Human Ecology, Edinburgh.

Moffatt, I. (1990) The potentialities and problems associated with applying information technology to environmental management. *Journal of Environmental Management* 30 (3), 209–220.

Moghissi, A.A. (1995) Eco-environmentalism vs human environmentalism. *Environment International* 21 (3), 253–254.

Mollison, B. and Holmgren, D. (1978) *Permaculture: a designer's manual.* Tagari Publications, Hobart, Tasmania.

Mollinson, B. and Holmgren, D. (1979) *Permaculture 2: practical design for town and country in permanent agriculture.* Tagari Publications, Hobart, Tasmania.

Momsen, J.H. (1991) *Women and Development in the Third World.* Routledge, London.

Montgomery, D.R. (1995) Input- and output-oriented approaches to implementing ecosystem management. *Environmental Management* 19 (2), 183–188.

Montgomery, J.D. (1990a) Environmental management as a Third World problem. *Policy Sciences* 23 (2), 163–176.

Montgomery, J.D. (1990b) Environmental management as behavioral policy. *Canadian Public Administration – Administration Publique du Canada* 33 (1), 1–16.

Moran, E.F. (1981) *Developing the Amazon.* Indiana University Press, Bloomington, IN.

Moran, E.F. (ed.) (1990) *The Ecosystem Approach in Anthropology: from concept to practice.* University of Michigan Press, Ann Arbor, MI.

Morris, D. (1990) Free trade the great destroyer. *The Ecologist* 20 (5), 190–195.

Morris, W. (1891) *News from Nowhere.* Reeves and Turner, London (more recent edn published by Routledge & Kegan Paul, London).

Morrison, D.E. (1986) How and why environmental consciousness has trickled down, in A. Schnaiberg, N. Watts and K. Zimmerman (eds) *Distributional Conflicts in Environmental Resource Policy.* Gower, Aldershot, pp. 187–220.

Moser, W. and Peterson, J. (1981) Limits to Obergurgl's growth. *Ambio* x (2–3), 68–72.

Moss, T.H. and Stills, D.L. (eds) (1981) *Three Mile Island Nuclear Accident: lessons and implications* (Annals of the New York Academy of Sciences vol. 365). New York Academy of Sciences, New York.

Mott, J.J. and Tothill, J.C. (1985) *Ecology and Management of the World's Savannas.* Commonwealth Agricultural Bureaux, Slough.

Mougeot, L.J.A. (2005) *Agropolis: the social, political and environmental dimensions of urban agriculture.* Earthscan, London.

Mowforth, M. and Munt, I. (1998) *Tourism and Sustainability: new tourism in the Third World.* Routledge, London.

Moxen, J. and Strachan, P. (1995) The formulation of standards for environmental management systems: structural and cultural issues. *Greener Management International* 12 (October 1995), 32–48.

Müeller, F.G. and Ahmad, Y.J. (eds) (1982) *Integrated Physical, Socio-economic and Environmental Planning.* Tycooly International, Dublin.

Mueller-Dombois, D. (1975) Some aspects of island ecosystem analysis, in F.B. Golley and E. Medina (eds) *Tropical Ecological Systems: trends in terrestrial and aquatic research.* Springer-Verlag, Berlin, pp. 353–366.

Mueller-Dombois, D., Bridges, K.W. and Carson, H.L. (eds) (1981) *Island Ecosystems. Ecological organization in selected Hawaiian communities.* Hutchinson Ross, Stroudsburg, VI.

Mumme, S.P. (1992) New directions in United States–Mexican transboundary environmental management – a critique of current proposals. *Natural Resources Journal* 32 (2), 539–562.

Munasinghe, M. and McNeely, J.A. (eds) (1994) *Protected Area Economics and Policy: linking conservation and sustainable development.* World Bank, Washington, DC.

Munasinghe, M. and Swart, R. (2005) *Primer on Climate Change and Sustainable Development: facts, policy analysis, and applications.* Cambridge University Press, Cambridge.

Munda, G., Nijkamp, P. and Rietveld, P. (1994) Qualitative multicriteria evaluation for environmental management. *Environmental Economics* 10 (2), 97–112.

Musgrave, R.K. (1993) The GATT tuna–dolphin dispute: an update. *Natural Resources Journal* 33 (4), 957–976.

Myers, N. (1985) Economics and ecology in the international arena; the phenomenon of 'linked linkages'. *Ambio* xv (5), 296–300.

Myers, N. (1992) Population/environment linkages: discontinuities ahead. *Ambio* xxi (1), 116–118.

Myers, N. (1993) Environmental refugees in a globally warmed world. *BioScience* 43 (11), 752–761.

Myers, W.L. and Shelton, R.L. (1980) *Survey Methods for Ecosystem Management.* Wiley, New York.

Naess, A. (1973) The shallow and the deep, long range ecology movement, a summary. *Inquiry* 16 (1), 95–100.

Naess, A. (1988) Deep ecology and ultimate premises. *The Ecologist* 18 (4–5), 128–132.

Naess, A. (1989) *Ecology, Community and Lifestyles: outline of an ecosophy* (translated and edited by D. Rothenberg). Cambridge University Press, Cambridge.

Naiman, R.J. (1992) *Watershed Management: balancing sustainability and environmental change.* Springer-Verlag, London.

Napier, C. (ed.) (1998) *Environmental Conflict Resolution.* Cameron May, London.

Narveson, J. (1995) The case for free-market environmentalism. *Journal of Agricultural and Environmental Ethics* 8 (2), 145–156 (pp. 126–144 present a case against).

Nash, J. and Ehrenfeld, J. (1997) Codes of environmental management practice: assessing their potential as tools for change. *Annual Review of Energy and the Environment* 22 (5), 487–535.

Nellor, D.C.L. (1987) Sovereignty and natural resource taxation in developing countries. *Economic Development and Cultural Change* 35 (2), 367–392.

Nelson, N. and Wright, S. (1995) *Power and Participatory Development: theory and practice.* Intermediate Technology Publications, London.

Neumayer, E. (2003) *Weak Versus Strong Sustainability: exploring the limits of two opposing paradigms* (2nd edn). Edward Elgar, Cheltenham.

Newby, H. (1990) Environmental change and the social sciences. Paper to 1990 Annual Meeting of the British Association for the Advanacement of Science (mimeo).

Newson, M. (ed.) (1992) *Managing the Human Impact on the Natural Environment: problems and processes.* Belhaven, London.

Ngwa, N.E. (1995) The role of women in environmental management: an overview of the rural Cameroonian situation. *GeoJournal* 35 (4), 515–520.

Nicholson, E.M. (1970) *The Environmental Revolution.* Hodder & Stoughton, London.

Nijkamp, P. (1980) *Environmental Policy Analysis.* Wiley, New York.

Nijkamp, P. (1986) Multiple criteria analysis and integrated impact analysis. *Impact Assessment Bulletin* 4 (3–4), 226–261.

Nijkamp, P. and Soeteman, F. (1988) Ecologically sustainable economic development: key issues for strategic environmental management. *International Journal of Social Economics* 15 (3–4), 88–102.

Nisbet, R. (1982) *Prejudices: a philosophical dictionary.* Harvard University Press, Cambridge, MA.

Nixon, A. and Harrop, O. (1998) *Environmental Impact Assessment.* Routledge, London.

Nolch, G. (1994) Australia could become target for environmental refugees. *Search* 25 (9), 269.

Nollman, J. (1990) *Spiritual Ecology.* Bantam Books, London.

Norris, S. (1990) The return of impact assessment: assessing the impact of regional shopping centre proposals in the UK. *Papers in Regional Science* 69, 101–119.

North, K. (1992) *Environmental Business Management: an introduction.* International Labour Office, Geneva.

North, R.D. (1995) *Life on a Modern Planet: a manifesto for progress.* Manchester University Press, Manchester.

Norton, B.G. (1991) *Towards Unity among Environmentalists.* Oxford University Press, Oxford.

Oates, J.F. (1995) The dangers of conservation by rural development: a case study from the forests of Nigeria. *Oryx* 229 (2), 115–122.

Oates, J.F. (1999) *Myth and Reality in the Rain Forest: how conservation strategies are failing West Africa.* University of California Press, Berkeley, CA.

O'Callaghan, P.W. (1996) *Integrated Environmental Management Handbook.* Wiley, Chichester.

O'Connor, M. (1997) The internalization of environmental costs: implementing the polluter pays principle in the European Union. *International Journal of Environment and Pollution* 7 (4), 450–482.

ODA (1984) Checklist for screening environmental aspects in aid activities (mimeo, amended 1984). Overseas Development Administration, London.

ODA (1989a) *The Environment and the British Aid Programme.* Overseas Development Administration, London.

ODA (1989b) *Manual of Environmental Appraisal* (revised edn 1992). Overseas Development Administration, London.

Odum, E.P. (1975) *Ecology: the link between the natural and the social sciences* (2nd edn). Holt, Rinehart & Winston, London.

Odum, E.P. (1983) *Systems Ecology: an introduction.* Wiley, New York.

OECD (1975) *The Polluter Pays Principle: definition, analysis and implementation.* Organization for Economic Cooperation and Development, Paris.

OECD (1991) *Environmental Indicators.* Organization for Economic Cooperation and Development, Paris.

OECD (1993) *Coastal Zone Management: integrated policies.* Organization for Economic Cooperation and Development, Paris.

OECD (2005) *Environmental Management in Eastern Europe, Caucasus and Central Asia.* Organization for Economic Cooperation and Development, Paris.

Ohlsson, L. (ed.) (1995) *Hydropolitics: conflicts over water as a development constraint.* Zed Press, London.

Oldfield, S. (1988) *Buffer Zone Management in Tropical Moist Forest.* IUCN, Gland.

Oliver, S. (1994) The political ecology of deforestation – the case of the Basilicata region of Italy. *Environmental Conservation* 21 (1), 72–75.

Omara-Ojungu, P.H. (1992) *Resource Management in Developing Countries.* Longman, Harlow.

Omernik, J.M. (1987) Ecoregions of the coterminous United States. *Annals of the Association of American Geographers* 77 (1), 118–125.

Ooi, J.B. (1983) *Natural Resources in Tropical Countries.* Singapore University Press, Singapore.

Oppenheimer, S. (2003) *Out of Eden: the peopling of the world.* Constable & Robinson, London.

Orams, M.B. (1995) Towards a more desirable form of ecotourism. *Tourism Management* 16 (1), 3–8.

O'Riordan, T. (1976) *Environmentalism.* Pion, London.

O'Riordan, T. (1979) The scope of environmental risk management. *Ambio* viii (6), 260–264.

O'Riordan, T. (1991) The new environmentalism and sustainable development. *Science of the Total Environment* 108 (1–2), 5–15.

O'Riordan, T. (ed.) (1995) *Environmental Science for Environmental Management.* Longman, Harlow.

O'Riordan, T. and Cameron, J. (1994) *The History and Contemporary Significance of the Precautionary Principle.* Earthscan, London.

O'Riordan, T. and Cameron, J. (eds) (1995) *Interpreting the Precautionary Principle.* Cameron & May, London.

O'Riordan, T. and Rayner, S. (1991) Risk management for global environmental change. *Global Environmental Change* 1 (2), 91–108.

O'Riordan, T. and Stoll-Kleemann, S. (eds) (2001) *Biodiversity, Sustainability and Human Communities.* Cambridge University Press, Cambrdige.

O'Riordan, T. and Turner, R.K. (eds) (1983) *An Annotated Reader in Environmental Planning and Management.* Pergamon Press, Oxford.

Orstrom, E. (1990) *Governing the Commons: the evolution of institutions for collective action.* Cambridge University Press, New York.

Ortolano, L. and Shepherd, A. (1995) Environmental impact assessment, in F. Vanclay and D.A. Bronstein (eds) *Environmental and Social Impact Assessment.* Wiley, Chichester, pp. 3–30.

Osborn, F. (1948) *Our Plundered Planet.* Little, Brown, Boston, MA.

Osborn, F. (1953) *Limits of the Earth.* Little, Brown, Boston, MA.

Osborn, F. and Bigg, T. (1998) *Earth Summit II: outcomes and analysis.* Earthscan, London.

Osuagwu, L. (2002) TQM strategies in a developing economy: empirical evidence from Nigerian companies. *Business Process Management Journal* 8 (2), 160–181.

Owen, L. and Unwin, T. (eds) (1997) *Environmental Management: readings and case studies.* Blackwell, Oxford.

Page, S. and Dowling, R.K. (2001) *Ecotourism.* Addison Wesley Longman, Harlow.

Palmer, A.R. (1999) Ecological footprints: evaluating sustainability. *Environmental Geosciences* 6 (4), 200–204.

Parasuraman, S. (1994) Economic and social marginalization of irrigation project displaced people. *Journal of Rural Development* 13 (2), 227–242.

Park, A. (1997) Trees, people, food and soil: a case study of participatory development in Malawi. *Forestry Chronicle* 73 (2), 221–227.

Park, C. (1980) *Ecology and Environmental Management.* Dawson & Sons, Folkstone.

Park, C. (1987) *Acid Rain: rhetoric and reality.* Methuen, London.

Park, C. (1997) *The Environment: principles and applications.* Routledge, London.

Parkin, S. (1989) *Green Parties: an international guide.* Heretic Books, London.

Parson, E.A. (1995) Integrated assessment and environmental policy making. In pursuit of usefulness. *Energy Policy* 23 (4–5), 463–476.

Parsons, J.J. (1985) On 'bioregionalism' and 'watershed consciousness'. *Professional Geographer* 37 (1), 1–6.

Partidário, M. do R. (1996) Strategic environmental assessment: key issues emerging from recent practice. *Environmental Impact Assessment Review* 16 (1), 31–55.

Partidário, M. do R. and Clark, R. (eds) (2000) *Perspectives on Strategic Environmental Assessment.* Earthscan, London.

Paruccini, M. (ed.) (1995) *Applying Multiple Criteria Aid for Decision to Environmental Management* (Euro Course Series: Environmental Management vol. 3). Kluwer Academic, Dordrecht.

Pasoff, M. (1991) Ecotourism re-examined. *Earth Island Journal* 6 (2), 28–29.

Pasqualetti, M.J. (ed.) (1990) *Nuclear Decommissioning and Society: public links to a new technology.* Routledge, London.

Paterson, D. (1993) Did Tibet cool the world? *New Scientist* 139 (1880), 29–33.

Patlis, J.M. (1992) The Multilateral Fund of the Montreal Protocol: a prototype for financial mechanisms in protecting the global environment. *Cornell International Law Journal* 25 (1), 181–183.

Patricos, N.N. (1986) *International Handbook on Land Use Planning.* Greenwood Press, Westport, CT.

Patterson, A. (1990) Debt-for-nature swaps and the need for alternatives. *Environment* 32 (10), 4–13, 31–34.

Patterson, A. and Theobald, K.S. (1995) Sustainable development, Agenda 21 and the new local governance in Britain. *Regional Studies* 29 (8), 773–778.

Payoyo, P.B. (ed.) (1994) *Ocean Governance: sustainable development of the seas.* United Nations University Press, Tokyo.

Pearce, D.W. (1992) Sustainable development and environmental impact appraisal, in CISP (ed.) *Forum Valutazione: semestrale a cura del Comitato Internazionale per lo Sveluppo dei Popoli* 4. CISP, Rome, pp. 13–22.

Pearce, D.W. (1994) Commentary. *Environment and Planning-A* 26 (9), 1329–1338.

Pearce, D.W. (1995) *Blueprint 4: capturing global environmental value.* Earthscan, London.

Pearce, D.W. and Barbier, E. (2002) *Blueprint for a Sustainable Economy.* Earthscan, London.

Pearce, D.W. and Brisson, I. (1993) BATNEEC: the economics of technology-based environmental standards, with a UK case illustration. *Oxford Review of Economic Policy* 9 (4), 24–40.

Pearce, D.W. and Turner, R.K. (1990) *Economics of Natural Resources and the Environment.* Harvester Wheatsheaf, Hemel Hempstead.

Pearce, D.W., Barbier, D. and Markandya, A. (1990) *Sustainable Development: economics and environment in the Third World.* Edward Elgar, London.

Pearce, D.W., Markandya, A. and Barbier, E.B. (1989) *Blueprint for a Green Economy.* Earthscan, London.

Pearce, D.W., Barbier, E., Markandya, A., Barrett, S., Turner, R.K. and Swanson, T. (1991) *Blueprint 2: greening the world economy.* Earthscan, London.

Pearce, F. (1989) Kill or cure? Remedies for the rainforest. *New Scientist* 123 (1682), 40–43.

Pearce, F. (1998) Burn me. *New Scientist* 156 (2109), 31–34.

Pearce, F. (2005) *Keepers of the Spring.* Island Press, Washington, DC.

Peat, D. (1988) *Superstrings and the Search for the Theory of Everything.* Cardinal-Sphere Books, London.

Peattie, K. (1995) *Environmental Marketing Management: meeting the green challenge.* Pitman, London.

Penning-Rowsell, E.C. and Fordham, M. (eds) (1994) *Floods Across Europe: hazard assessment, modelling and management.* Middlesex University Press, London.

Pentreath, R.J. (2000) Strategic environmental management: time for a new approach. *The Science of the Total Environment* 249 (1), 3–11.

Pepper, D. (1984) *The Roots of Modern Environmentalism.* Croom Helm, London.

Pepper, D. (1993) *Eco-Socialism: from deep ecology to social justice.* Routledge, London.

Pepper, D. (1996) *Modern Environmentalism: an introduction.* Routledge, London.

Pereira, H.C. (1989) *Policy and Practice in the Management of Tropical Watersheds.* Westview Press, Boulder, CO.

Perera, L. and Amin, A. (1996) Accommodating the informal sector: a strategy for urban environmental management. *Journal of Environmental Management* 46 (1), 3–15.

Perez-Trejo, F., Clark, N. and Allen, P. (1993) An explanation of dynamic systems modelling as a decision tool for environmental policy. *Journal of Environmental Management* 39 (4), 305–319.

Perkins, P.E. (1995) An overview of international institutional mechanisms for environmental management with reference to Arctic pollution. *Science of the Total Environment* 161 (9), 849–857.

Perman, R., Yue, M. and McGilvray, J. (1996) *Natural Resource and Environmental Economics.* Longman, Harlow.

Pernetta, J. (1993) *Monitoring Coral Reefs for Global Change.* IUCN, Cambridge.

Perrings, C., Mahler, K.G., Folke, C., Holling, C.S. and Jansson, B.O. (eds) (1995) *Biodiversity Loss: economic and ecological issues.* Cambridge University Press, Cambridge.

Pest, C. and Grabber, J. (2001) Historical analysis, a valuable tool in community-based environmental protection. *Marine Pollution Bulletin* 42 (5), 339–349.

Peterson, D.J. (1995) *The Legacy of Soviet Environmental Destruction.* Westview Press, Boulder, CO.

Pforr, C. (2001) Concepts of sustainable development, sustainable tourism, and ecotourism: definitions, principles and linkages. *Scandinavian Journal of Hospitality and Tourism* 1 (1), 68–71.

Phillimore, J. and Davidson, A. (2002) A cautionary tale: Y2K and the politics of foresight. *Futures* 34 (1), 147–157.

Pichon, F.J. (1992) Agricultural settlement and ecological crisis in the Ecuadorian Amazon frontier: a discussion of the policy environment. *Policy Studies Journal* 20 (4), 662–678.

Pichon, L. (1993) *Environmental Management for Hotels: the industry guide to best practice.* Butterworth-Heinemann, Oxford.

Pickering, K.T., Owen I. and Lewis, A. (1997) *An Introduction to Global Environmental Issues.* Routledge, London.

Pidgeon, S. and Brown, D. (1994) The role of lifecycle analysis in environmental management: general panacea or one of several useful paradigms? *Greener Management International* 7 (July), 36–44.

Pierce, J.T. (1990) *The Food Resource.* Longman, Harlow.

Pigou, A.C. (1920) *The Economics of Welfare* (1st edn). Macmillan, London.

Pimbert, M.P. (1991) *Designing Integrated Pest Management for Sustainable and Productive Futures* (IIED Gatekeeper Series No. 29). IIED, London.

Pimbert, M.P. and Pretty, J. (1995) *Parks, People and Professionals: putting 'participation' into protected area management.* UN Research Institute for Social Development (Discussion Paper No. 57). UNRISD, Geneva.

Pimm, S.L. (1984) The complexity and stability of ecosystems. *Nature* 307 (5949), 321–326.

Pinchot, G. (1910) *The Fight for Conservation.* University of Washington Press, Seattle, WA.

Pinstrup-Andersen, P. (1982) *Agricultural Research and Technology in Economic Development.* Longman, London.

Pirages, D. (1978) *The New Context for International Relations: global ecopolitics.* Duxbury Press, North Scituate, MS.

Pirages, D. (1994) Sustainability as an evolving process. *Futures* 26 (1), 197–205.

Pollard, S.J., Harrop, D.O., Crowcroft, P., Mallett, S.H., Jeffries, S.R. and Young, P.J. (1995) Risk assessment for environmental management: approaches and applications. *Journal of the Chartered Institution of Water and Environmental Management* 9 (6), 621–628.

Porritt, J. (1984) *Seeing Green: the politics of ecology explained.* Blackwell, Oxford.

Porter, R.N. and Brownlie, S.F. (1990) Integrated environmental management: a planning strategy for nature conservation developments. *South African Journal of Wildlife Research* 20 (2), 81–86.

Posey, D.A. (1990) Intellectual property rights and just compensation for indigenous knowledge: challenges to science and international law. Paper presented to the Association for Applied Anthropology, New York.

Posey, D.A. (1999) *Cultural and Spiritual Values of Biodiversity* (published for the UNEP). Intermediate Technology Publications, London.

Posner, R. (2005) *Catastrophe: risk and response.* Oxford University Press, Oxford.

Postel, S. (1992) *Last Oasis, Facing Water Scarcity.* W.W. Norton, New York.

Postel, S. (1994) Carrying capacity: Earth's bottom line, in *The State of the World 1994.* Earthscan, London.

Prasad, K. (1993) Environmental impact assessment: an analysis of the Canadian experience. *International Studies* 30 (3), 299–318.

Pratt, V. (with Howarth, J. and Brady, E.) (1999) *Environment and Philosophy.* Routledge, London.

Prestcott, R. (1996) *Coastal Zone Management: towards best practice* (Department of the Environment). HMSO, London.

Pretty, J.N. (1999) Can sustainable agriculture feed Africa? New evidence on progress, processes and impacts. *Environment, Development and Sustainability* 1 (1), 253–274.

Pretty, J.N. and Ward, H. (2001) Social capital and the environment. *World Development* 29 (2), 209–227.

Price, M.F. (ed.) (1996) *People and Tourism in Fragile Environments.* Wiley, Chichester.

Price, M.F., Jensky, L. and Iatsenia, A.A. (eds) (2004) *Key Issues for Mountain Areas.* United Nations University Press, Tokyo.

Princen, T. and Finger, M. (1994) *Environmental NGOs in World Politics: linking the local and the global.* Routledge, London.

Pritchard, D. (1993) Towards sustainability: first steps towards sustainability analysis. *Ecos* 14 (3–4), 10–15.

Pritchard, D. (1994) Is carrying capacity real, and can we use it in planning? *RSPB Conservation Review* 8, 72–76.

Pritchard, P. (1993) *Managing Environmental Risks and Liability*. Technical Communications, Letchworth.

Proost, S. and Braden, J.B. (eds) (1998) *Climate Change, Transport and Environmental Policy: empirical applications in a federal system*. Edward Elgar, London.

Prothero, R.M. (1994) Forced movements of populations and health hazards in tropical Africa. *International Journal of Epidemiology* 23 (4), 657–664.

Pryde, P.R. (1991) *Environmental Management in the Soviet Union*. Cambridge University Press, Cambridge.

Purvis, M. and Grainger, A. (eds) (2004) *Exploring Sustainable Development: geographical perspectives*. Earthscan, London.

Quammen, D. (1997) *The Song of the Dodo: island biogeography in an age of extinctions*. Pimlico (Random House), London.

Rabe, B.G. (1996) An empirical examination of innovations in integrated environmental management – the case of the Great Lakes Basin. *Public Administration Review* 56 (4), 372–381.

Rabinovitch, J. (1992) Curitiba: towards sustainable urban development. *Environment and Urbanization* 4 (2).

Rabinovitch, J. (1996) Innovative land-use and public transport policy: the case of Curitiba, Brazil. *Land Use Policy* 13 (1), 51–67.

Raghavan, C. (1990) *Recolonization: GATT, the Uruguay Round and a new global economy*. Zed Press, London.

Ramlogan, R. (1996) Environmental refugees: a review. *Environmental Conservation* 23 (1), 81–88.

Rao, B. (1991) Dominant constructions of women and nature in social science literature. Capitalism, Nature, Socialism CES/CNS Pamphlet No. 2. CES/CNES, PO Box 8467, Santa Cruz, CA.

Rapoport, R.N. (1993) Environmental values and the search for a global ethic. *Journal of Environmental Psychology* 13 (2), 173–182.

Rathgeber, E.M. (1990) WID, WAD, GAD: trends in research and practice. *The Journal of Developing Areas* 24, 489–502.

Rau, B. (1991) *From Feast to Famine: official cures and grass roots remedies to Africa's food crisis*. Zed Press, London.

Raup, D.M. (1988) Diversity crisis in the geological past, in E.O. Wilson (ed.) *Biodiversity*. National Academy Press, Washington, DC, pp. 51–57.

Raup, D.M. (1993) *Extinction: bad genes or bad luck?* (first published in 1992 by Norton, London). Oxford University Paperbacks, Oxford.

Ravenhill, J. (1986) *Africa in Economic Crisis*. Macmillan, London.

Ravetz, J. (2000a) How I got Y2K wrong. *Futures* 32 (6), 937–939.

Ravetz, J. (2000b) Integrated assessment for sustainability appraisal in cities and regions. *Environmental Impact Assessment Review* 20, 31–64.

Ravilious, K. (2005) Four days that shook the world. *New Scientist* 182 (2446), 32–35.

Raymond, K. (1996) The long-term future of the Great Barrier Reef. *Futures* 28 (10), 947–970.

Reardon, T. and Vosti, S. (1995) Links between rural poverty and the environment in developing countries: asset categories and investment poverty. *World Development* 23 (9), 1495–1506.

Redclift, M.E. (1984) *Development and the Environmental Crisis: red or green alternatives?* Methuen, London.

Redclift, M.E. (1985) *The Struggle for Resources: limits of environmental 'managerialism'* (IIUG ks 85–5). Internationale Institut für Umwelt und Gesellschaft, Berlin.

Redclift, M.E. (1987) *Sustainable Development: exploring the contradictions*. Methuen, London.

Redclift, M.E. (1992) A framework for improving environmental management: beyond the market mechanism. *World Development* 20 (2), 255–259.

Redclift, M.E. (1995) The environment and structural adjustment: lessons for policy interventions in the 1990s. *Journal of Environmental Management* 44 (1), 55–68.

Redclift, M.E. (ed.) (2005) *Sustainability: critical concepts in the social sciences – vol. III: Sustainability Indicators.* Routledge, London.

Redclift, M.E. and Benton, T. (eds) (1994) *Social Theory and the Global Environment.* Routledge, London.

Reed, D. (ed.) (1992a) *Structural Readjustment and the Environment.* Earthscan, London.

Reed, D. (1992b) *Structural Adjustment and the Environment.* Westview Press, Boulder, CO.

Reed, D. (ed.) (1996) *Structural Adjustment, the Environment, and Sustainable Development.* Earthscan, London.

Rees, J. (1985) *Natural Resources: allocation, economics and policy.* Methuen, London.

Rees, J. (1995) *Natural Resources: allocation, economics and policy* (2nd edn). Routledge, London.

Rees, W.E. (1992) Ecological footprints and appropriate carrying capacity: what urban economics leaves out. *Environment and Urbanization* 41 (2), 121–130.

Reich, C. (1970) *The Greening of America.* Random House, New York.

Reid, D.G. (ed.) (1999) *Ecotourism Development in Eastern and Southern Africa.* Weaver Press, Guelph, ON.

Reij, C., Scoones, I. and Toulmin, C. (eds) (1996) *Sustaining the Soil: indigenous soil and water conservation in Africa.* Earthscan, London.

Reisenweber, R.L. (1995) Making environmental standards more reasonable. *Environment* 37 (2), 15–32.

Rensvik, H. (1994) Who sets the environmental standards for tomorrow's industry: industry, consumers, government regulations, the green lobby? *Marine Pollution Bulletin* 29 (6–12), 277–278.

Renwick, W.H. (1988) The eclipse of NEPA as environmental policy. *Environmental Management* 12 (3), 267–272.

Repetto, R., Dower, R.C., Jenkins, R. and Geohagen, J. (1992) *Green Fees: how a tax shift can work for the environment and the economy.* World Resources Institute, Washington, DC.

Ricci, R.F. (ed.) (1981) *Technological Risk Assessment.* Martinus Nijhof, Boston, MA.

Rich, B. (1986) Environmental management and multilateral development banks. *Cultural Survival* 10 (1), 4–13.

Richards, L. and Biddick, I. (1994) Sustainable economic development and environmental auditing: a local authority perspective. *Journal of Environmental Planning and Management* 37 (4), 487–494.

Richerson, P.J. and McEvoy III, J. (1976) *Human Ecology: an environmental approach.* Duxbury Press, North Scituate, MS.

Richey, J.S., Horner, R.R. and Mar, B.W. (1985) The Delphi technique in environmental assessment 2: consensus of critical issues in environmental monitoring program design. *Journal of Environmental Management* 21 (2), 147–160.

Riddell, R. (1981) *Ecodevelopment: economics, ecology and development.* Gower, Farnborough.

Rietbergen, S. (ed.) (1993) *The Earthscan Reader in Tropical Deforestation.* Earthscan, London.

Rigby, D., Woodhouse, P., Young, T. and Burton, M. (2001) Constructing a farm level indicator of sustainable agricultural practice. *Ecological Economics* 39 (4), 463–478.

Risser, P.G. (1985) Toward a holistic management perspective. *BioScience* 35 (7), 414–418.

Ritchie, M. (1990) GATT, agriculture and environment: the US double zero plan. *The Ecologist* 20 (6), 214–220.

Ritchie, M. (1992) Free trade versus sustainable agriculture: the implications of NAFTA. *The Ecologist* 21 (5), 221–227.

Robbins, P. (2004) *Political Ecology: a critical introduction.* Blackwell, Oxford.

Robért, K.-H. (2000) Tools and concepts for sustainable development, how do they relate to a general framework for sustainable development, and to each other? *Journal of Cleaner Production* 8, 243–254.

Robinson, K.S. (1992) *Green Mars.* HarperCollins, London (there are a further two related science fiction books by the same author).

Robinson, N. (1992) International trends in environmental impact assessment. *Boston College Environmental Affairs Law Review* 19 (3), 591–621.

Rodda, A. (1991) *Women and Environment.* Zed Press, London.

Rodhe, H. and Herrera, R. (eds) (1988) *Acidification in Tropical Countries* (SCOPE No. 36). Wiley, Chichester.

Rodhe, H., Callaway, J. and Dianwu, Z. (1992) Acidification in Southeast Asia – prospects for coming decades. *Ambio* xxi (2), 148–150.

Roe, E. (1996) Why ecosystem management can't work without social science: an example from the California Northern Spotted Owl controversy. *Environmental Management* 20 (5), 667–674.

Rogers, M.F., Sinden, J.A. and De Lacey, T. (1997) The precautionary principle for environmental management: a defensive-expenditure application. *Journal of Environmental Management* 51 (4), 343–360.

Roggeri, H. (1995) *Tropical Freshwater Wetlands: a guide to current knowledge and sustainable development.* Kluwer Academic, Dordrecht.

Rondinelli, D. and Vastag, G. (2000) Panacea, common sense, or just a label? The value of ISO 14001 environmental management systems. *European Management Journal* 18 (5), 499–510.

Roszak, T. (1972) *Where the Wasteland Ends: politics and transcendence in post-industrial society.* Faber, London.

Roszak, T. (1979) *Person/Planet: the creative disintegration of industrial society.* Victor Gollancz, London.

Roth, E., Rosenthal, H. and Burbridge, P.R. (2000) A discussion of the use of the sustainability index: 'ecological footprint' for aquaculture production. *Aquatic and Living Resources* 13, 461–469.

Rothery, B. (1993) *BS7750: implementing the environment management standard and the EC eco-management scheme.* Gower, Aldershot.

Royston, M.G. (1978) The modern manager in the human environment. Paper presented to Symposium on the Malaysian Environment, RECSAM Complex, Penang, 16–20 September (mimeo).

Rudderian, W.F. and Kutzback, J.E. (1991) Plateau uplift and climatic change. *Scientific American* 264 (3), 42–50.

Ruddle, K. and Manshard, W. (1981) *Renewable Natural Resources and the Environment: pressing problems in the developing world.* Tycooly International, Dublin.

Rugman, A.M. and Kirton, J.J. (1998) *Trade and the Environment: economic legal and policy perspectives.* Edward Elgar, London.

Ruivenkamp, G. (1987) Social impacts of biotechnology in agriculture and food processing. *Development* 4 (1), 58–59.

Russo, M.V. (1999) *Environmental Management Readings and Cases.* Houghton Mifflin, Boston, MA.

Ryecroft, R.W., Regens, J.L. and Dietz, T. (1988) Incorporating risk assessment and benefit cost analysis in environmental management. *Risk Analysis* 8 (3), 415–420.

Ryle, M. (1988) *Ecology and Socialism.* Radius, London.

Sachs, C.E. (ed.) (1997) *Women Working in the Environment.* Taylor & Francis, Washington, DC.

Sachs, J.D. (ed.) (2005) *Understanding and Achieving the Millennium Development Goals: investing in development.* Earthscan, London.

Sadler, B. (1994) Environmental assessment and sustainability at the project and programme level, in R.G. Goodland and V. Edmundson (eds) *Environmental Assessment and Development.* World Bank, Washington, DC, pp. 3–19.

Sagan, C. (1997) *The Demon-Haunted World: science as a candle in the dark.* Hodder & Stoughton, London.

Sagar, A.M. and Najam, A. (1998) The human development index: a critical review. *Ecological Economics* 25 (3), 249–264.

Salafsky, N. and Wallenberg, E. (2000) Linking livelihoods and conservation: a conceptual framework and scale for assessing the integration of human needs and biodiversity. *World Development* 28 (9), 1421–1438.

Salafsky, N., Margoluis, R. and Redford, K. (2001) *Adaptive Management: a tool for conservation practitioners.* World Wildlife Fund (Biodiversity Support Programme), Geneva.

Sale, K. (1985) *Dwellers in the Land: the bioregional vision.* Sierra Club, San Francisco, CA.

Sale, K. (1991) *Dwellers in the Land: the bioregional vision* (republication). New Society Publishers, Gabriola Is., Canada, and Philadelphia, PA.

Sampson, G.P. (2005) *The WTO and Sustainable Development.* United Nations University Press, Tokyo.

Sampson, G.P. and Whalley, J. (eds) (2005) *The WTO, Trade and the Environment.* Edward Elgar, Cheltenham.

Samson, F.B. and Knopf, F.L. (eds) (1996) *Ecosystem Management: selected readings.* Springer, New York.

Sandbach, F. (1980) *Environment, Ideology and Policy.* Blackwell, Oxford.

Sandgrove, K. (1992) *The Green Manager's Handbook.* Gower, London.

Sands, P. (ed.) (1993) *Greening International Law.* Earthscan, London.

Sargeant, F. II (ed.) (1974) *Human Ecology.* North-Holland Publishers, Amsterdam.

Sarkar, A.U. and Ebbs, K.L. (1992) A possible solution to tropical troubles? Debt-for-nature swaps. *Futures* 24 (6), 653–668.

Satterthwaite, D. (ed.) (1999) *The Earthscan Reader in Sustainable Cities.* Earthscan, London.

Saull, M. (1990) Nitrates in soil and water (Inside Science Supplement No. 37). *New Scientist* 127 (1734), 4pp.

Save the Children (1995) *Toolkits: a practical guide to assessment, monitoring, review and evaluation* (compiled by L. Gosling and M. Edwards). Save the Children, London (revised edn 2004).

Savory, A. (1988) *Holistic Resource Management.* Westview (Island Press), Boulder, CO.

Schaltegger, S., Burritt, R. and Petersen, H. (2003) *An Introduction to Corporate Environmental Management: striving for sustainability.* Greenleaf Publishing, Sheffield.

Schell. L.M., Smith, M.T. and Bilsborough, A. (eds) (1993) *Urban Ecology and Health in the Third World.* Cambridge University Press, New York.

Scher, C.A. (1991) Proactive environmental management. *Environmental Progress* 10 (1), F2–F3.

Scherr, S.J. (1997) People and environment: what is the relationship between exploitation of natural resources and population growth in the South? *Forum for Development Studies* 1, 33–58.

Schibuola, S. and Byer, P.H. (1991) Use of knowledge-based systems for the review of environmental impact assessments. *Environmental Impact Assessment Review* 11 (1), 11–27.

Schmidheiny, S. (1992) *Changing Courses: a global business perspective on development and the environment.* MIT Press, Cambridge, MA.

Schneider, S.H. (1990) Debating Gaia. *Environment* 32 (4), 5–9.

Schoffman, A. and Tordini, A.M. (2000) *ISO14001: a practical approach.* Oxford University Press, Oxford.

Scholz, U. (1992) Transmigrasi – ein Desaster? Probleme und Chancen des Indonesischen Umsiedlungsprogramms. *Geographische Rundschau* 44 (1), 33.

Schramm, G. and Warford, J.J. (eds) (1989) *Environmental Management and Economic Development.* Johns Hopkins University Press, Baltimore, MD.

Schultz, J. (1995) *The Ecozones of the World: the ecological divisions of the geosphere.* Springer-Verlag, Berlin.

Schumacher, E.F. (1973) *Small Is Beautiful: a study of economics as if people mattered.* Harper and Row, New York.

Schuman, R.W. (1996) *Eco-data: using your PC to obtain free environmental information* (2nd edn). Government Institutes Inc, Rockville, MD (Maryland 20850).

Scitovsky, T. (1976) The *Joyless Economy.* Oxford University Press, Oxford.

Scolimowski, H. (1988) Eco-philosophy and deep ecology. *The Ecologist* 18 (4–5), 124–127.

Scoones, I. (1998) *Sustainable Livelihoods: a framework for analysis* (IDS Working Paper No. 72). Institute of Development Studies, University of Sussex, Brighton.

Scoones, I. and Thompson, J. (eds) (1994) *Beyond Farmer First: rural people's knowledge, agricultural research and extension practice.* Intermediate Technology, London.

Scudder, T. (2005) *The Future of Large Dams: dealing with social, environmental and political costs.* Earthscan, London.

Seager, J. (1993) *Earth Follies: feminism, politics and environment.* Earthscan, London.

Seda, M. (ed.) (1993) *Environmental Management in ASEAN: perspectives on critical regional issues.* Institute of Southeast Asian Studies, Singapore.

Seldner, B.J. and Cottrel, J.P. (1994) *Environmental Decision Making for Engineering and Business Managers.* McGraw-Hill, New York.

Selin, H. (ed.) (1995) *Nature Across Cultures: views of nature and the environment in non-Western cultures.* Kluwer Academic, Dordrecht.

Selin, S. and Chavez, D. (1995) Developing a collaborative model for environmental planning and management. *Environmental Management* 19 (2), 189–195.

Semple, E.C. (1911) *Influences of Geographic Environment.* Henry Holt, New York.

Sessions, G. (ed.) (1995) *Deep Ecology for the 21st Century: readings on the philosophy and practice of the new environmentalism.* Shambhala Press, Boston, MA.

Sewell, G.H. (1975) *Environmental Quality Management.* Prentice-Hall, Englewood Cliffs, NJ.

Seymour, S., Cox, G. and Lowe, P. (1992) Nitrates in water: the politics of the polluter pays principle. *Sociologia Ruralis* 32 (1), 82–103.

Shaner, W.W., Philipp, P.F. and Schmehl, W.R. (1982) *Farming Systems Research and Development: guidelines for developing countries.* Westview Press, Boulder, CO.

Shankar, U. (1986) Psychological dimensions of environmental management. *Current Science* 55 (6), 297.

Sharratt, P. (ed.) (1995) *Environmental Management Systems.* Institution of Chemical Engineers, Rugby.

Shaw, R.P. (1993) Warfare, national sovereignty, and the environment. *Environmental Conservation* 20 (2), 113–121.

Sheldon, C. (ed.) (1997) *ISO14001 and Beyond: environmental management systems in the real world.* Greenleaf Publishing, Sheffield.

Shepherd, A. (1995) Participatory environmental management: contradictions of process, project and bureaucracy in the Himalayan foothills. *Public Administration and Development* 15 (5), 465–479.

Shillito, D. (ed.) (1994) *Implementing Environmental Impact Management.* Institution of Chemical Engineers, Rugby.

Shiva, V. (1988) *Staying Alive: women, ecology and environment in India.* Zed Press, London.

Shiva, V. (1993) *Monocultures of the Mind: perspectives on biodiversity and biotechnology.* Zed Press, London.

Shiva, V. (2002) *Water Wars: pollution, profits and privatization.* Pluto Press, London.

Shiva, V. (2004) *Green Alternatives to Globalization: a manifesto.* Pluto Press, London.

Shiva, V. and Mies, M. (1994) *Ecofeminism..* Zed Books, London.

Shiva, V., Anderson, P., Schücking, H., Gray, A., Lohmann, L. and Cooper, D. (1991) *Biodiversity: social and ecological perspectives.* Zed Press, London.

Shrader-Frechette, K.S. and McCoy, E.D. (1994) What ecology can do for environmental management. *Journal of Environmental Management* 41 (4), 293–307.

Shrybman, S. (1990) Free trade vs. the environment: the implications of GATT. *The Ecologist* 20 (1), 30–34.

Shutkin, W.A. (1991) International human rights law and the Earth: the protection of indigenous peoples and the environment. *Virginia Journal of International Law* 31 (3), 479–511.

Sidaway, R. (2005) *Resolving Environmental Disputes: from conflict to consensus.* Earthscan, London.

Simmons, I.G. (1989) *Changing the Face of the Earth: culture, environment, history.* Blackwell, Oxford.

Simon, J.L. (1981) *The Ultimate Resource.* Princeton University Press, Princeton, NJ.

Simonis, U.E. (1990) *Beyond Growth: elements of sustainable development.* Edition Sigma (WZB), Berlin.

Simons, P. (1988) Costa Rica's forests are reborn. *New Scientist* 120 (1635), 43–47.

Sinclair, J. (1990) Rising sea-levels could affect 300 million. *New Scientist* 125 (1700).

Singh, G.S. (1997) Sacred groves in western Himalaya: an eco-cultural imperative. *Man in India* 77 (2–3), 247–257.

Singh, R.B. (ed.) (2001) *Urban Sustainability in the Context of Global Change: towards promoting healthy and green cities.* Science Publishers, Enfield, NH.

Sinner, J. (1994) Trade and the environment – efficiency, equity and sovereignty considerations. *Australian Journal of Agricultural Economics* 38 (2), 171–187.

Slocombe, D.S. (1993) Environmental planning, ecosystem science, and ecosystem approaches for integrating environment and development. *Environmental Management* 17 (3), 289–303.

Smil, V. (1983) *The Bad Earth: environmental degradation in China.* Zed Press, London.

Smit, B. and Spaling, H. (1995) Methods for cumulative effects assessment. *Environmental Impact Assessment Review* 15 (1), 81–106.

Smit, T. (2001) *Eden.* Corgi Books, London.

Smith, C.P., Feyerabend, G.B. and Sandbrook, R. (1994) *The Wealth of Communities: stories of success in local environmental management.* Earthscan, London.

Smith, D. (ed.) (1992) *Business and the Environment: implications for the new environmentalism.* Paul Chapman, London.

Smith, D. and Dawson, A. (1990) Tsunami waves in the North Sea. *New Scientist* 127 (1728), 46–49.

Smith, F. (1996) Biological diversity, ecosystem stability and economic development. *Ecological Economics* 16 (3), 191–204.

Smith, J.F. (1998) Does decentralization matter in environmental management? *Environmental Management* 22 (2), 263–276.

Smith, M. (1993) Cheney and the myth of postmodernism. *Environmental Ethics* 15 (1), 3–18.

Smith, M. (1998) *Ecologism.* University of Minnesota Press, Minneapolis, MI.

Smith, S. and Reeves, E. (eds) (1989) *Human Systems Ecology.* Westview Press, Boulder, CO.

Smuts, J.C. (1926) *Holism and Evolution* (3rd edn 1936). Macmillan, London.

Snipp, C.M. (1986) American Indians and natural resource development: indigenous peoples' land, now sought after, has promoted new Indian–White problems. *American Journal of Economics and Sociology* 45 (4), 457–474.

Socolow. R., Andrews, C., Berkhout, F. and Thomas, V. (eds) (1994) *Industrial Ecology and Global Change.* Cambridge University Press, Cambridge.

Soderstrom, E.J. (1981) *Social Impact Assessment: experimental methods and approaches.* Praeger, New York.

Solomon, B.D. (1985) Regional econometric models for environmental impact assessment. *Progress in Human Geography* 9 (3), 378–399.

Sontheimer, S. (ed.) (1991) *Women and the Environment: a reader.* Earthscan, London.

Soroos, M.S. (1993) The odyssey of Arctic haze: toward a global atmospheric regime. *Environment* 34 (10), 6–11, 25–27.

Sorsa, P. (1992) GATT and the environment. *World Economy* 15 (1), 115–134.

Soule, M.E. (ed.) (1987) *Viable Populations for Conservation.* Cambridge University Press, Cambridge.

Sowman, M., Fuggle, R. and Preston, G. (1995) A review of the evolution of environmental evaluation procedures in South Africa. *Environmental Impact Assessment Review* 15 (1), 45–68.

Spaling, H. (1994) Cumulative effects assessment: concepts and principles. *Impact Assessment* 12 (3), 231–251.

Spaling, H. and Smit, B. (1993) Cumulative environmental change: conceptual frameworks, evaluation approaches, and institutional perspectives. *Environmental Management* 17 (5), 587–600.

Spash, C.L. (1996) Reconciling different approaches to environmental management. *International Journal of Environment and Pollution* 7 (4), 497–511.

Spedding, L.S., Jones, D.M. and Dering, C.J. (eds) (1993) *Eco-management and Eco-auditing: environmental issues in business.* Chancery Law Publishers, London; Wiley, Chichester.

Spellerberg, I.F. (1991) *Monitoring Ecological Change* (2nd edn 2005) Cambridge University Press, Cambridge.

Spencer-Cooke, A. (1996) *From EMAS to SMAS.* SustainAbility Ltd, London.

Spretnak, C. (1986) *The Spiritual Dimensions of Green Politics.* Bear & Co., Boston, MA.

Spretnak, C. (1990) *Reweaving the World: the emergence of ecofeminism.* Science Book Club, San Francisco, CA.

Spretnak, C. and Capra, F. (1985) *Green Politics: the global promise* (US edn 1984). Paladin, London.

Stabler, M.J. (ed.) (1998) *Tourism and Sustainability.* CABI Publishing, Wallingford.

Stebbing, E.P. (1938) The man-made desert in Africa: erosion and drought. *Journal of the Royal African Society* Supplement (January), 3–40.

Steel, D. (1995) *Rogue Asteroids and Doomsday Comets.* Wiley, Chichester.

Steinbeck, J. (1939) *The Grapes of Wrath.* Heinemann, New York.

Steiner, D. and Nauser, M. (eds) (1993) *Human Ecology: fragments of anti-fragmentary views of the world.* Routledge, London.

Stigliani, W.M., Doelman, P., Salomons, W., Schulin, R., Schmidt, G.R.B. and Van der Zee, S.E.A.T.M. (1991) Chemical time bombs: predicting the unpredictable. *Environment* 33 (4), 4–9, 26–33.

Stiles, D. (ed.) (1995) *Social Aspects of Sustainable Dryland Management.* Wiley, Chichester.

Stocking, M.A. and Murnaghan, N. (2001) *Handbook for the Field Assessment of Land Degradation.* Earthscan, London.

Stolton, S. and Dudley, N. (eds) (1999) *Partnerships for Protection: new strategies for planning and management for protected areas.* Earthscan, London.

Stone, L., Gabric, A. and Berman, T. (1996) Ecosystem resilience, stability, and productivity: seeking a relationship. *American Naturalist* 148 (5), 892–903.

Stonehouse, D.P., Giraldez, C. and VanVuuren, W. (1997) Holistic policy approaches to natural resource management and environmental care. *Journal of Soil and Water Conservation* 52 (1), 22–25.

Stonich, S.C. and Browder, J.O. (1996) I am destroying the land – the political ecology of poverty and environmental destruction in Honduras. *Latin American Research Review* 30 (3), 123–137.

Stott, P.J. (1994) Global environmental security, energy resources and planning. *Futures* 26 (7), 741–758.

Stott, P.J. (1995) *Atoms, Whales and Rivers: global environmental security and international organization.* Nova Science Publishers, New York.

Stout, B.B. (1992) Environmental determinism. *Journal of Forestry* 90 (7), 4–5.

Stouth, R., Sowman, M. and Grindley, S. (1993) The panel evaluation method: an approach to evaluating controversial resource allocation proposals. *Environmental Impact Assessment Review* 13 (1), 13–35.

Stren, R.E. and White, R. (eds) (1992) *Sustainable Cities: urbanization and the environment in international perspective.* Westview Press, Boulder, CO.

Stuart, R. (2000) Environmental management systems in the 21st century. *Chemical Health and Safety* November–December, 23–25.

Sukopp, H., Numata, M. and Huner, A. (1995) *Urban Ecology as the Basis of Urban Planning.* Academic Publishing, The Hague.

Sullivan, C. (2002) Calculating a Water Poverty index. *World Development* 30 (7), 1195–1210.

Sunderland, T.J. (1996) Environmental management standards and certification: do they add value? *Greener Management International* 14 (April), 28–36.

Susskind, L.E. (1992) New corporate role global environmental treaty-making. *Columbia Journal of World Business* 27 (3), 62–73.

Suter, G.W. II (ed.) (1993) *Ecological Risk Assessment.* Lewis, Boca Raton, FL.

Sutton, K. (1989) Malaysia: a land settlement model in time and space. *Geoforum* 20 (3), 339–354.

Sutton, P.W. (2004) *Nature, Environment, and Society.* Palgrave Macmillan, Basingstoke.

Swain, A. (2001) Water wars: fact or fiction? *Futures* 33 (8), 769–781.

Swingland, I. (2003) *Capturing Carbon and Conserving Biodiversity: the market approach.* Earthscan, London.

Sylvan, R. and Bennett, D. (1988) Taoism and deep ecology. *The Ecologist* 18 (4–5), 148–158.

Syms, P. (1997) *Contaminated Land: the practice and economics of redevelopment.* Blackwell Science, Oxford.

Székely, A. (1990a) The International Law of Natural Resources and the Environment: a selected bibliography. *Natural Resources Journal* 30 (4), 765–917.

Székely, A. (1990b) The International Law of Natural Resources and the Environment: a selected bibliography. Part II. *Natural Resources Journal* 31 (2), 265–412.

Tamagawa, H. (ed.) (2006) *Sustainable Cities: Japanese perspectives on physical and social structures.* United Nations University Press, Tokyo.

Tansley, A.F. (1935) The use and abuse of vegetational concepts and terms. *Ecology* 16 (2), 284–307.

Taylor, B. (1991) The religion and politics of Earth First! *The Ecologist* 21 (6), 258–266.

Taylor, S.R. (1992) Green management: the next competitive weapon. *Futures* 24 (7), 669–680.

Teilhard de Chardin, P. (1959) *The Phenomenon of Man.* Harper & Row, New York.

Teilhard de Chardin, P. (1964) *The Future of Man.* Collins, London.

Thana, N.-C. and Biswas, A.K. (eds) (1990) *Environmentally Sound Water Management.* Oxford University Press, Delhi.

The Ecologist (1993) *Whose Common Future: reclaiming the commons.* Earthscan, London.

Therevel, R. (1993) Systems of strategic environmental assessment. *Environmental Impact Assessment Review* 13 (3), 145–168.

Therevel, R. (2004) *Strategic Environmental Assessment in Action.* Earthscan, London.

Therevel, R. and Partidário, M.R. (1996) *Practice of Strategic Environmental Assessment.* Earthscan, London.

Therevel, R., Wilson, E., Thompson, S., Heaney, D. and Pritchard, D. (1992) *Strategic Environmental Assessment.* Earthscan, London.

Theutenberg, B.J. (1984) *The Evolution of the Law of the Sea: a study of resources and strategy with special regard to the polar regions.* Tycooly International, Dublin.

Thomas, C. (ed.) (1994) *Rio: unravelling the consequences.* Frank Cass, London.

Thomas, D.S.G. and Middleton, N.J. (1994) *Desertification: exploding the myth.* Wiley, Chichester.

Thomas, W.L. (ed.) (1956) *Man's Role in Changing the Face of the Earth.* University of Chicago Press, Chicago, IL.

Thompson, M., Hatley, T. and Warburton, M. (1986) *Uncertainty on a Himalayan Scale.* Ethnographica, London.

Thompson, S. and Therevel, R. (eds) (1991) *Environmental Auditing* (Oxford Polytechnic, School of Planning, Working Paper No. 130). Oxford Polytechnic, Oxford.

Thomson, D.R. (1993) *The Growth of Environmental Auditing.* Institute of Environmental Assessment, East Kirkby.

Thomson, D.R. and Wilson, M.J. (1994) Environmental auditing: theory and applications. *Environmental Management* 18 (4), 605–615.

Thoreau, H.D. (ed.) (1854) *Walden, or Life in the Woods* (1854 edn, Houghton Mifflin, Boston, MA); 1960 edn – New American Library, New York.

Thukral, E.G. (1992) *Big Dams, Displaced People: rivers of sorrow, rivers of change.* Sage, New Delhi.

Tietenberg, T. (ed.) (1997) *The Economics of Global Warming.* Edward Elgar, London.

Tietenberg, T. and Folmer, H. (1998) *The International Yearbook of Environmental and Resource Economics 1998/1999: a survey of current issues.* Edward Elgar, London.

Tiffen, M. (1993) Productivity and environmental conservation under a rapid population growth: a case-study of Machakos District, Kenya. *Journal of International Development* 5 (2), 207–224.

Tiffen, M. (1995) Population density, economic growth and societies in transition: Boserüp reconsidered in a Kenyan case-study. *Development & Change* 26 (1), 31–65.

Tiffen, M., Mortimore, M. and Gichuki, F. (1994) *More People, Less Erosion: environmental recovery in Kenya*. John Wiley, Chichester.

Tilman, D. (1996) Biodiversity: population versus ecosystem stability. *Ecology* 77 (2), 350–363.

Tisdell, C. (1993) *Environmental Economics: policies for environmental management and sustainable development*. Edward Elgar, London.

Tisdell, C. (1999) *Biodiversity, Conservation and Sustainable Development: principles and practices with Asian examples*. Edward Elgar, Cheltenham.

Tisdell, C. and Sen, R.K. (eds) (2004) *Economic Globalisation: social conflicts, labour and environmental issues*. Edward Elgar, Cheltenham.

Tiwari, K.M. (1983) *Social Forestry for Rural Development*. International Book Distributors (India), Dehra Dun.

Tobey, J.A. and Smets, H. (1996) The polluter pays principle in the context of agriculture and the environment. *World Economy* 19 (1), 63–87.

Todaro, M.P. (1994) *Economic Development* (5th edn). Longman, Harlow.

Tokar, B. (1988) Social ecology, deep ecology and the future of green political thought. *The Ecologist* 18 (4–5), 132–142.

Tolba, M.K. (1982) *Development Without Destruction: evolving environmental perceptions*. Tycooly International, Dublin.

Tornell, A. and Velasco, A. (1992) The tragedy of the commons and economic growth: why does capital flow from poor to rich countries? *Journal of Political Economy* 100 (6), 1208–1231.

Toteng, E.N. (2001) Urban environmental management in Botswana: toward a theoretical explanation of public policy failure. *Environmental Management* 28 (1), 19–30.

Treshow, M. (1976) *The Human Environment*. McGraw-Hill, New York.

Treweek, J. (1995a) Ecological impact assessment. *Impact Assessment* 13 (3), 289–316.

Treweek, J. (1995b) Ecological impact assessment, in F. Vanklay and D.A. Bronstein (eds) *Environmental and Social Impact Assessment*. Wiley, Chichester, pp. 171–191.

Tribe, J. and Font, X. (2000) *Forest Tourism and Recreation: case studies in environmental management*. CABI Publishing, Wallingford.

Tribe, J., Font, X., Griffiths, N., Vickery, R. and Yale, K. (2000) *Environmental Management for Rural Tourism and Recreation*. Cassell, London.

Triggs, G.D. (ed.) (1988) *The Antarctic Treaty Regime: law, environment and resources*. Cambridge University Press, Cambridge.

Tromans, S. (1992) International law and UNCED: effects on international business. *Journal of Environmental Law* 4 (2), 189–202.

Trottier, J. and Slack, P. (eds) (2005) *Managing Water Resources*. Oxford University Press, Oxford.

Troumbis, A.Y. (1987) Disturbance in Mediterranean islands: a demographic approach to changes in insular ecosystems. *Ekistics* 54 (323–324), 127–131.

Troumbis, A.Y. (1992) Environmental management: theoretical inputs of ecology – the guest editor's introductory statement. *Ekistics* 59 (356–357), 250–259.

Trudgill, S. (1990) *Barriers to a Better Environment: what stops us solving environmental problems?* Belhaven, London.

Trzyna, T.C. (1995) *A Sustainable World: defining and measuring sustainable development*. Earthscan, London.

Tucker, K.C. and Richardson, D.M. (1995) An expert system for screeening potentially invasive alien plants in the South African fynbos. *Journal of Environmental Management* 44 (4), 309–338.

Tudge, C. (1995) *The Day Before Yesterday: five million years of human history*. Pimlico (Random House), London.

Turner, B.L. and Ali, A.M.S. (1996) Induced intensification: agricultural change in Bangladesh with implications for Malthus and Boserüp. *Proceedings of the National Academy of Sciences of the USA* 93 (25), 14984–14991.

Turner, K. and Jones, T. (eds) (1991) *Wetlands: market and intervention failures*. Earthscan, London.

Turner, M.G. and Gardiner, R.H. (eds) (1991) *Quantitative Methods in Landscape Ecology: the analysis and interpretation of landscape hererogeneity.* Springer, New York.

Turner, R.K. (1995) *Sustainable Development and Climate Change.* CSERGE, University of East Anglia, Norwich.

Turner, R.K., Pearce, D.W. and Bateman, I. (1994) *Environmental Economics: an elementary introduction.* Harvester Wheatsheaf, New York.

Turner, R.K., van den Bergh, J.C.L.M., and Bruwer, R. (eds) (2003) *Managing Wetlands: an ecological economics approach.* Edward Elgar, Cheltenham.

Turnham, D. (1991) Multilateral development banks and environmental management. *Public Administration and Development* 11 (4), 363–380.

Turton, C. and Farrington, J. (1998) Enhancing rural livelihoods through participatory watershed development in India. *ODI Natural Resource Perspectives* 34, 4pp.

Tussie, D. (1987) *The Less Developed Countries and the World Trading System: a challenge to the GATT.* Frances Pinter, London.

Tylecote, A. and Van der Straaten, J. (eds) (1997) *Environment, Technology and Economic Growth: the challenge to sustainable development.* Edward Elgar, London.

Uglow, J. (2002) *The Lunar Men: the friends who made the future.* Faber and Faber, London.

Underwood, A.J. (1995) Ecological research and (research into) environmental management. *Ecological Applications* 5 (1), 232–247.

UN (1992) *Agenda 21: programme of action for sustainable development.* United Nations, New York.

UN (1993) *Systems of National Accounts.* United Nations, New York.

UN (2002) *Living with Risk: a global review of disaster reduction.* United Nations, New York.

UNCHS (1988) *Refuse Collection Vehicles for Developing Countries.* UNCHS (Habitat), Nairobi, Kenya.

UNCTAD (1993) *Environmental Management in Transnational Corporations: report on the Benchmark Corporate Environmental Survey.* United Nations Conference on Trade and Development, United Nations, New York.

UNDP (1991) *Human Development Report* (for UN Development Programme). Oxford University Press, Oxford.

UNDP (1992) *Handbook and Guidelines for Environmental Management and Sustainable Development.* UN Development Program (UNDP), New York.

UNEP (1981) *Global Environmental Issues.* Tycooly Publishing, Dublin.

UNEP (2002) *Global Environmental Outlook 3: past, present and future perspectives.* Earthscan, London.

UNEP, NASA, USGS and University of Maryland (2005) *One Planet. Many People: atlas of our changing environment.* UNEP, Nairobi, Kenya.

UNHCR (1992) *Refugees and the Environment: special report* (special issue on Refugees No. 89). UN High Commission for Refugees, Geneva.

UN Millenium Development Library (2005) *A Home in the City.* Earthscan, London.

Usher, M.B. and Thompson, D.B. (eds) (1988) *Ecological Change in the Uplands* (British Ecological Society Special Publication No. 7). Blackwell Scientific, Oxford.

Vanclay, F. and Bronstein, D.A. (eds) (1995) *Environmental and Social Impact Assessment.* Wiley, Chichester.

Van den Bergh, J. (1996) *Ecological Economics and Sustainable Development.* Edward Elgar, Cheltenham.

Vandermeer, J. (1996) Tragedy of the commons – the meaning of the metaphor. *Science and Society* 60 (3), 290–306.

Van der Sluijs, J.P. (2002) Integrated assessment, in R.E. Munn (ed.) *Encyclopedia of Global Environmental Change*, vol. 4. Wiley, Chichester, pp. 250–253.

Van Dieren, W. (ed.) (1995) *Taking Nature into Account: towards a sustainable natural income.* Springer Verlag, New York.

Van Dyne, G.M. (ed.) (1969) *The Ecosystem Concept in Natural Resource Management.* Academic Press, New York.

Vannucci, M. (ed.) (2005) *Mangrove Management and Conservation: present and future.* United Nations University Press, Tokyo.

Van Pelt, M.J.F. (1993) *Ecological Sustainability and Project Appraisal.* Avebury, Aldershot.

Vastag, G., Kerekes, S. and Rondinelli, D.A. (1996) Evaluation of corporate environmental management approaches: a framework and application. *International Journal of Production Economics* 43 (2–3), 193–211.

Vaughn, D. (ed.) (1991) *Current EC Developments: EC environmental and planning law.* Butterworths, Oxford.

Vaughn, D. and Michle, C. (1993) *Environmental Principles for European Business.* Earthscan, London.

Velva, V., Hart, M., Greiner, T. and Crumbley, C. (2001) Indicators of sustainable production. *Journal of Cleaner Production* 9 (6), 447–452,

Victor, P.A. (1991) Indicators of sustainable development: some lessons from capital theory. *Ecological Economics* 4, 191–213.

Viles, H.A. and Spencer, T. (1995) *Coastal Problems.* Edward Arnold, London.

Vink, A.P.B. (1983) *Landscape Ecology and Land Use* (English trans. from German ed. D.A. Davidson). Longman, London.

Virtanen, Y. and Nilsson, S. (1993) *Environmental Impacts of Waste Paper Recycling.* Earthscan, London.

Vivian, J. (1994) NGOs and sustainable development in Zimbabwe: no magic bullets. *Development & Change* 25 (1), 167–193.

Vlachos, E. (1985) Assessing long range cumulative impacts, in V.T. Covello, J.L. Mumpower, P.J.M. Stallen and V.R.R. Uppuluri (eds) *Environmental Impact Assessment, Technology Assessment, and Risk Analysis: contributions from the psychological and decision sciences* (NATO ASI series G, Ecological Science, vol. 4). Springer-Verlag, Berlin, pp. 49–80.

Vogler, J. and Imber, M. (eds) (1995) *The Environment and International Relations: theories and processes.* Routledge, London.

Vogt, K.A., Gordon, J.C., Wargo, J.P., Vogt, D.J., Asbornsen, H., Palmiotto, P.A., Clark, H.J., O'Hara, J.L., Keeton, W.S., Patel-Wayland, T. and Witten, E. (1997) *Ecosystems: balancing science with management.* Springer-Verlag, New York.

Vogt, W. (1948) *Road to Survival.* William Sloane, New York.

Voisey, H., Beuermann, C., Sverdrup, L.A. and O'Riordan, T. (1996) The political significance of Local Agenda 21: the early stages of some European experience. *Local Environment* 1 (1), 33–50.

Von Moltke, K. (1992) The United Nations development system and environmental management. *World Development* 20 (4), 619–626.

Von Weizsäcker, E.U. and Jesinghaus, J. (1992) *Ecological Tax Reform: a policy proposal for sustainable development.* Zed Press, London.

Vos, C.C. and Opdam, P. (eds) (1993) *Landscape Ecology of a Stressed Environment.* Chapman & Hall, London.

Vun, L.W. and Latiff, A. (1999) Preliminary ecological impact assessment and environmental impact assessment for coastal resort development in Malaysia. *Impact Assessment and Project Appraisal* 17 (2), 140–148.

Wackernagel, M. and Rees, W. (2003) *Our Ecological Footprint: reducing human impact on the Earth.* New Society Publishers, Gabriola Island, BC, Canada.

Wackernagel, M. and Yount, J.D. (2000) Footprints for sustainable development: the next steps. *Environment, Development and Sustainability* 2 (1), 21–42.

Walker, K.J. (1989) The state in environmental management: the ecological dimension. *Political Studies* 37 (1), 25–38.

Wallace, T. and March, C. (eds) (1991) *Changing Perceptions: writings on gender and development.* Oxford University Press, Oxford.

Wallington, T.J., Schneider, W.F., Worsnop, D.R., Neilsen, O.J., Sehested, J., Debruyn, W.J. and Shorter, J.A. (1994) The environmental impact of CFC replacements – HFCs and HCFCs. *Environmental Science and Technology* 28 (7), 320A–326A.

Walters, C.J. (1986) *Adaptive Environmental Management of Renewable Natural Resources.* McGraw-Hill, New York.

Ward, B. and Dubos, R.E. (1972) *Only One Earth: the care and maintenance of a small planet.* Penguin, Harmondsworth.

Ward, P.M. (1990) *Mexico City: the production and reproduction of an urban environment.* Belhaven, London.

Ward, R.C. (1978) *Floods: a geographical perspective.* Macmillan, London.

Warford, J. and Partow, Z. (1989) Evolution of the World Bank's environmental policy. *Finance and Development* (December), 5–9.

Warren, K.J. (ed.) (1997) *Ecofeminism: women, culture, nature.* Indiana University Press, Bloomington, IN.

Warwick, C.J., Mumford, J.D. and Norton, G.A. (1993) Environmental management expert systems. *Journal of Environmental Management* 39 (4), 251–270.

Wathern, P. (ed.) (1988) *Environmental Impact Assessment: theory and practice.* Unwin Hyman, London.

Watson, A. (1991) Inside science No. 48: 'Gaia'. *New Scientist* 131 (1776), 4pp.

Watt, K.E.F. (ed.) (1966) *Systems Analysis in Ecology.* Academic Press, New York.

Watt, K.E.F. (1969) *Ecology and Resource Management: quantitative resource.* McGraw-Hill, New York.

Watts, D. (1971) *Principles of Biogeography.* McGraw-Hill, New York.

Watts, M.J. (1989) The agrarian crisis in Africa: debating the crisis. *Progress in Human Geography* 13 (1), 1–41.

Weale, A. (1993) *The New Politics of Pollution.* Manchester University Press, Manchester.

Wearing, S. and Neil, J. (1999) *Ecotourism: impacts, potentials and possibilities.* Butterworth-Heinemann, London.

Weaver, D.B. (1998) *Ecotourism in the Less Developed World.* CABI Publishing, Wallingford.

Weaver, D.B. (ed.) (2001) *The Encyclopaedia of Ecotourism.* CABI Publishing, New York.

Weaver, D.B. (2002) Asian ecotourism: patterns and themes. *Tourism Geography* 4 (2), 153–172.

Weber, M. (1958) *The Protestant Ethic and the Spirit of Capitalism* (English translation by T. Parsons – originally published in 1904 and 1905 in 2 vols). Charles Scribner, New York.

Weeks, W.W. (1997) *Beyond the Ark: tools for an ecosystem approach to conservation.* Island Press, Washington, DC.

Wehrmeyer, W. (ed.) (1996) *Greening People: human resources and environmental management.* Greenleaf, Sheffield.

Weinberg, A., Bellows, S. and Esker, D. (2002) Sustaining ecotourism: insights and implications from two successful studies. *Society and Natural Resources* 15 (4), 371–380.

Weir, D. and Shapiro, M. (1981) *Circle of Poison: pesticides and people in a hungry world.* Institute for Food and Development Policy, San Francisco, CA.

Welford, R. (1992) *Environmental Auditing: the EC Eco-Audit Scheme and the British Standard on Environmental Management Systems.* University of Bradford, Bradford.

Welford, R. (1993) Local economic development and environmental management – an integrated approach. *Local Economy* 8 (2), 130–142.

Welford, R. (1996) *Corporate Environmental Management: systems and strategies.* Earthscan, London.

Welford, R. (1997) *Hijacking Environmentalism: corporate responses to sustainable development.* Earthscan, London.

Welford, R. (2000) *Corporate Environmental Management: towards sustainable development* (Vol. 3). Earthscan, London.

Welford, R. and Gouldson, A. (1993) *Environmental Management and Business Strategy.* Pitman, London.

Wellburn, A. (1988) *Air Pollution and Acid Rain: the global threat of acid pollution.* Earthscan, London.

Wellburn, A. (1994) *Air Pollution and Climate Change* (2nd edn). Longman, Harlow.

Wells, S. (1992) *The Greenpeace Book of Coral Reefs.* Blandford, London.

Wells-Howe, B. and Warren, K.J. (1994) *Ecological Feminism.* Routledge, London.

Werner, P. (1991) *Savanna Ecology and Management: Australian perspectives and interconti-nental comparisons*. Blackwell Scientific, Oxford.

Wesche, R. (1996) Developed country environmentalism and indigenous community controlled ecotourism in the Ecuadorian Amazon. *Geographische Zeitschrift* 84 (3–4), 157–168.

West, K. (1995) Ecolabels: the industrialization of environmental standards. *The Ecologist* 25 (1), 16–20.

Westcoat, J.L. and White, G.F. (2004) *Water for Life*. Cambridge University Press, Cambridge.

Westing, A.H. (1992) Environmental refugees: growing category of displaced persons. *Environmental Conservation* 19 (3), 201–207.

Westlund, S. (1994) *Is Free Trade Compatible with Sustainable Development?* Stockholm Environmental Institute (December 1994 No. 4/94). SEI, Stockholm.

Westman, W.E. (1985) *Ecology, Impact Assessment and Environmental Planning*. Wiley, Chichester.

Weston, J. (ed.) (1986) *Red and Green: the new politics of the environments*. Pluto Press, London.

White, G.F. (1968) Organising scientific investigations to deal with environmental impacts. Paper delivered to the Careless Technology Conference, Washington. (Later published in M.T. Farvar and J.P. Milton (eds) (1972) *The Careless Technology: ecology and international develop-ment*. Natural History Press (Doubleday), New York, pp. 914–926.)

White, L. Jnr (1967) The historical roots of our environmental crisis. *Science* 15 (3767), 1203–1207.

White, R.R. (1993) *North, South, and the Environmental Crisis*. University of Toronto Press, Toronto.

White, R.R. (1994) *Urban Environmental Management: environmental change and urban design*. Wiley, New York.

White, R.R. and Whitney, J. (eds) (1992) *Sustainable Cities: urbanisation and the environment in international perspective*. Westview Press, Boulder, CO.

Whitten, J.M. and Bennett, J. (eds) (2005) *The Private and Social Values of Wetlands*. Edward Elgar, Cheltenham.

Wilkinson, C.R. and Buddemeier, R.W. (1994) *Global Climate Change and Coral Reefs: impli-cations for people and reefs* (Report of UNEP-IOC-ASPEI-IUCN Global Task Team on the Implications of Climate Change on Coral Reefs). IUCN, Gland.

Wilkinson, P. (1979) Public participation in environmental management: a case study. *Natural Resources Journal* 16 (1), 117–135.

Williams, C. (2005) A land turned to dust. *New Scientist* 186 (2502), 38–41.

Williams, P.J. (1979) *Pipelines and Permafrost: physical geography and development in the circumpolar north*. Longman, London.

Williams, P.W. and Todd, S.E. (1997) Towards an environmental management system for ski areas. *Mountain Research and Development* 17 (1), 75–90.

Willig, J.T. (ed.) (1994) *Environmental TQM* (1st edn 1992). McGraw-Hill, New York.

Wilson, E.O. (1992) *The Diversity of Life*. Penguin, Harmondsworth.

Wilson, G.A. and Bryant, R.L. (1997) *Environmental Management: new directions for the twenty-first century*. University College London Press, London.

Wilson, J. (1994) Environmentalism and political theory: toward an ecocentric approach. *Political Science* 46 (1), 129–311.

Winter, G. (1988) *Business and Environment: a handbook of industrial ecology with 22 check-lists for procedural use*. McGraw-Hill, New York.

Winter, G. (1994) *Blueprint for Green Management: creating your company's own environmental action plan*. McGraw-Hill, London.

Winter, G. (1996) *European Environmental Law: a comparative perspective*. Dartmouth Publishing, Aldershot.

Wirth, D.A. (1986) Law: the World Bank and the environment. *Environment* 28 (10), 33–44.

Wisemann, G. (1994) Europe braces for eco-audit – standardizing the acronyms. *Chemical Weekly* 154 (13), 54–56.

Wisner, B. (1990) Harvest of sustainability: recent books on environmental management (review article). *Journal of Development Studies* 26 (2), 335–341.

Woehlcke, M. (1992) Environmental refugees. *Aussenpolitik* 43 (3), 287–296.

Wohl, E. (2005) *Disconnected Rivers.* Yale University Press, New Haven, CT.

Wolf, A.T. (2003) *Hydropolitics along the Jordan River: scarce water and its impact on the Arab –Israeli conflict.* United Nations University Press, Tokyo.

Wolf, E.C. (1986) *Beyond the Green Revolution: new approaches to Third World agriculture* (Worldwatch Paper No. 73). Worldwatch Institute, Washington, DC.

Wolters, G.J.R. (1994) Integrated environmental management: the Dutch governmental view. *Marine Pollution Bulletin* 29 (6–12), 272–274.

Wood, A., Steadman-Edwards, P. and Mang, J. (eds) (2000) *The Root Causes of Biodiversity Loss.* Earthscan, London.

Wood, C.M. (1988) EIA in plan making, in P. Wathern (ed.) *Environmental Impact Assessment: theory and practice.* Unwin Hyman, London, pp. 98–114.

Wood, C.M. (1992) Strategic environmental assessment in Australia and New Zealand. *Project Appraisal* 7 (3), 137–143.

Wood, C.M. (1995) *Environmental Impact Assessment: a comparative review.* Longman, Harlow.

Wood, C.M. (ed.) (2005) *Strategic Environmental Assessment and Land Use Planning.* Earthscan, London.

Woodbridge, R. (2004) *The Next World War: tribes, cities, nations and ecological decline.* University of Toronto Press, Toronto.

Woodcock, C.E., Sham, C.H. and Shaw, B. (1990) Comments on selecting a geographic information system for environmental management. *Environmental Management* 14 (3), 307–315.

Woolston, H. (ed.) (1993) *Environmental Auditing: an introduction and practical guide.* British Library, Science Reference and Information Service, London.

Worhurst, A. (ed.) (1999) *Mining and the Environment: case studies from the Americas.* International Development Research Centre, PO Box 8500, Ottawa, Canada.

World Bank (1992) *The World Development Report: development and the environment.* Oxford University Press, Oxford.

World Bank (1994) *Economy-Wide Policies and the Environment: emerging lessons from experience.* World Bank, Washington, DC.

World Bank (2000) *The Little Green Data Book.* The World Bank, Washington, DC.

World Bank (2003) *World Development Report 2003. Sustainable Development in a Dynamic World: transforming institutions, growth, and quality of life.* Oxford University Press, Oxford.

World Bank (2005) *Miniatlas of Millennium Development Goals.* World Bank, Washington, DC.

World Commission on Environment and Development (1987) *Our Common Future* (the 'Brundtland Report'). Oxford University Press, Oxford.

World Commission on Large Dams (2000) *Dams and Development: a new framework for decision-making* (Report of the World Commission on Dams). Earthscan, London.

World Commision on Large Dams (2004) *Dams and Development: a new framework for decision* (http://www.dams.org).

Worldwatch Institute (2005) *Vital Signs 2005–2006: the trends that are shaping our future.* Earthscan, London.

Worldwatch Institute (2006) *State of the World 2006.* Earthscan, London.

WRI, IIED and UNEP (1988) *World Resources 1988–9.* Basic Books, New York.

Wright, J.R., Wiggins, L.L., Jain, R.K. and Kim, T.J. (eds) (1993) *Expert Systems in Environmental Planning.* Springer-Verlag, Berlin.

Wrisberg, N. and Udo de Haes, H.A. (eds) (2002) *Analytical Tools for Sustainable Design and Management in a Systems Perspective.* Kluwer Academic, Dordrecht.

WSSD (2002) *World Summit on Sustainable Development: plan of implementation.* WSSD, Johannesburg.

WWF (1986) *The Assisi Declarations: messages on man and nature from Buddhism, Christianity, Hinduism, Islam and Judaism.* World Wide Fund for Nature, Gland.

WWF (1995) Ecotourism: conservation tool or threat? *Conservation Issues* 2 (3), 1–10.

WWF (2000) Tourism certification still leaves much to be desired. *WWF-UK News* 29 August 2000. Worldwide Fund for Nature (UK) (http://www.wwf-uk.org/news).

Yaffee, S.L. (1996) Ecosystem management in practice: the importance of human institutions. *Ecological Applications* 6 (3), 727–730.

Yang, H. and Zehnder, A.J.B. (2002) Water scarcity and food import: a case study for southern Mediterranean countries. *World Development* 30 (8), 1413–1430.

Yar, N.T. (1990) Round the peg or square the hole? Populists, technocrats and environmental assessment in Third World countries. *Impact Assessment Bulletin* 8 (1–2), 69–84.

Yasui, H. and Kobayashi, E. (1991) Environmental management of the Seto Inland Sea. *Marine Pollution Bulletin* 23 (5), 485–488.

Yearley, S. (1991) *The Green Case: a sociology of environmental issues, arguments and politics.* Harper Collins, London.

Yost, N.C. (1990) NEPA's promise – partially fulfilled. *Environmental Law* 20 (3), 681–702.

Young, A.L. (1994) An overview of ISO-9000 application to drug, medical device, and environmental management issues. *Food and Drug Law Journal* 49 (3), 469–483.

Young, J. (1990) *Post Environmentalism.* Belhaven, London.

Young, J.E. (1990) *Discarding the Throwaway Society* (Worldwatch Paper No. 101). Worldwatch Institute, Washington, DC.

Young, L. (1985) A general assessment of the environmental impact of refugees in Somalia with attention to the Refugee Agricultural Programme. *Disasters* 9 (2), 122–133.

Young, M.D. and Solbrig, O.T. (1992) *Savanna Management for Ecological Sustainability, Economic Profit and Social Equity.* UNESCO, Paris.

Young, O. and Osherenko, G. (eds) (1993) *Polar Politics: creating international environmental regimes.* University of Columbia Press, New York.

Young, S. (1994) An Agenda 21 strategy for the UK? *Political Science* 3 (2), 325–342.

Young, Z. (2002) *A New Green Order? The World Bank and the Global Environmental Facility.* Pluto Press, London.

Zarsky, L. (1994) *Borders and the Biosphere: environment, development and world trade rules.* Earthscan, London.

Zeba, S. (1996) *The Role of NGOs in Reforming Natural Resource Management: policies in Burkina Faso* (Issues Paper, Drylands Programme, IIED No. 68). International Institute for Environment and Development, London.

Ze'ev, N. (1994) *Landscape Ecology: theory and application* (2nd edn). Springer Verlag, London.

Zetter, R. and Watson, G.B. (eds) (2002) *Sustainable Urban Design in the Developing World.* Ashgate, Aldershot.

Zhu, X., Healey, R.G. and Aspinall, R.J. (1998) A knowledge-based systems approach to design of spatial decision support systems for environmental management. *Environmental Management* 22 (1), 35–48.

Zimmerer, K.S. and Bassett, T.J. (eds) (2003) *Political Ecology: an integrative approach to geography and environment-development studies.* Guilford Press, Andover.

Zimmermann, E.W. (1993) *World Resources and Industries* (revised edn). Harper & Brothers, New York.

Zimmerman, M. (1987) Feminism, deep ecology, and environmental ethics. *Environmental Ethics* 9 (1), 21–44.

Zwarteveen, M. (1996) *A Plot of One's Own: gender relations and irrigated land allocation policies in Burkina Faso.* IIMI Abstract (Research Report No. 10). International Irrigation Management Institute, Colombo, 2pp.

Index

Note: page numbers in **bold** denote references to figures/tables/boxes.

Lightning Source UK Ltd.
Milton Keynes UK
UKOW05f2259071017
310514UK00010B/340/P